Network Automation with Nautobot

Adopt a network source of truth and a data-driven approach to networking

Jason Edelman | Glenn Matthews | Josh VanDeraa
Ken Celenza | Christian Adell | Brad Haas
Bryan Culver | John Anderson | Gary Snider

Network Automation with Nautobot

Group Product Manager: Pavan Ramchandani
Publishing Product Manager: Khushboo Samkaria
Book Project Manager: Ashwin Kharwa
Senior Editor: Athikho Sapuni Rishana
Technical Editor: Nithik Cheruvakodan
Copy Editor: Safis Editing
Proofreader: Athikho Sapuni Rishana
Indexer: Hemangini Bari
Production Designer: Aparna Bhagat
DevRel Marketing Coordinator: Marylou De Mello

First published: May 2024

Production reference: 1300424

Published by Packt Publishing Ltd.
Grosvenor House
11 St Paul's Square
Birmingham
B3 1RB, UK

ISBN 978-1-83763-786-7

www.packtpub.com

To the Nautobot and wider network automation community, for showing us that change was needed and that if you focus time, energy, and development on the right areas, good things will happen. Thank you to the Nautobot community for the continued support and the fostering of an environment that is welcoming and makes it okay to challenge the status quo.

– The authors

Foreword

George Bernard Shaw wrote in his 1905 play *Man and Superman* the age-old quip "*Those who can, do; those who can't, teach.*" It's no doubt a catchy line, but I think it misses the mark a bit. Theory can be understood without being practiced. But practice cannot be mastered alone by high-level engagement with theory.

I have been working on network automation since 2007. And I can tell you with absolute confidence that tools and methods have existed for literally decades at this point, but the vast majority of network operations are still painfully manual. Now, this fact hasn't been missed by networking vendors and would-be technology entrepreneurs. But if the pain is so acute, the ambition so strong, and the solutions so plentiful, why is this still such a struggle?

In my not terribly humble opinion, it's because the gap between theory and practice has never been wider. And even worse, with every new technology cycle, the gap gets bigger as technology after technology leads to promise after promise. All without an on-the-ground understanding of the actual networks and the people who manage them.

I have known Jason and members of the *Network to Code* team for years. As I have dabbled in strategy (yet more theory), they have doubled down on practice. And out of that practice has emerged a set of core principles accompanied by real-life experiences. It's these that bridge the gap between theory and practice. And frankly, it's what's allowed Jason and the team to develop a solution to a problem that has thus far proved difficult to tame.

This book represents the very best of their collective experience. They have captured details – including specific steps and the thought process required to succeed – that are unknowable by those who watch from the outside and merely opine on what ought to be done. They have transformed a product into a solution.

Most of us have heard the Man in the Arena quote made famous by Theodore Roosevelt. Jason and his team are active participants in the arena. And this book will help convert those spectators with the will to succeed into automation gladiators.

- Mike Bushong

Vice President, Data Center at Nokia

Contributors

About the authors

Jason Edelman is the founder and CTO at Network to Code. Observing how DevOps was radically changing IT operational models for systems administrators and developers, Jason saw an opportunity to combine existing technologies from the worlds of DevOps and software development within the networking infrastructure domain to create holistic network automation solutions. Prior to Network to Code, Jason spent a career in technical sales developing and architecting network solutions, with his last role leading efforts around SDN and programmability. Jason is also a coauthor of O'Reilly's *Network Programmability & Automation* book. He is a former CCIE and has a B.E. in Computer Engineering from Stevens Institute of Technology. He can be found on X as @jedelman8.

Glenn Matthews is a principal engineer at Network to Code and is the technical lead of the Nautobot project. Prior to Network to Code, he worked at Cisco Systems for more than a decade in software testing and software development roles with technical focuses including routing protocols, virtualization, and network automation, including the YANG Suite project. Glenn is committed to designing and developing quality software to help make the world a better place. His academic background includes a master's degree in computer science from the University of Georgia. He lives in Durham, North Carolina, with his daughter and a very persistent cat.

Josh VanDeraa is a 20-year networking veteran who has been doing network automation for the past 8 years. He has worked in large enterprise retail, travel, managed services, and most recently, professional services industries. He has worked on networks of all sizes, bringing multiple network automation solutions to the table to drive real value with Python, Ansible, and Python web framework solutions. Josh is the author of *Open Source Network Management* and maintains a blog site to provide additional content to those on the web.

Ken Celenza is VP of Network Automation Architecture at Network to Code. Ken is an experienced network and automation engineer with over 20 years of experience working in military, consulting, and enterprise environments. Ken leads client engagements at Network to Code as both a developer and an architect and serves as a mentor to network engineers.

Christian Adell is a network software engineer who has played multiple roles related to networking and IT automation. Currently, as principal architect at Network to Code, he is focused on building network automation solutions for diverse use cases, with great emphasis on open source software. He is passionate about learning and helping others to be happier, but also has more hobbies than hours in the day, so working remotely from Barcelona gives him the time and the space to achieve his dreams. Christian is co-author of O'Reilly's *Network Programmability & Automation* book.

Brad Haas is a seasoned professional serving as the Vice President of Professional Services at Network to Code. With a career spanning more than two decades, Brad has been instrumental in delivering innovative technology solutions, particularly in network automation and the integration of software-defined infrastructure. Brad is known for his advocacy of a data-informed approach to automation, ensuring technology aligns with business goals. Brad's career is distinguished by his achievement of numerous technical certifications, encompassing multiple CCIEs as well as a range of cloud certifications. His philosophy centers on using technology not just as a tool, but as a driving force for organizational transformation and growth.

Bryan Culver is an engineering manager at Network to Code, where he is currently enjoying building the team and platform behind Nautobot Cloud. He has served many roles in his career directly and indirectly related to network automation, from templating configs while racking data centers to deploying automation solutions in enterprise environments. He has a strong software engineering background, having worked in software development with start-ups and Fortune-sized companies. Outside of work, he enjoys time with his amazingly supportive wife and children, wielding power tools on any number of home renovation projects, traveling to beaches, and watching Formula 1 races.

John Anderson is a principal consultant at Network to Code and the Nautobot product owner, responsible for the direction of the project. John has 10 years of experience in network engineering and software development in higher education and global enterprise environments. He has been a maintainer and contributor to a number of network automation projects over the years. John lives in Charleston, SC and is working on a Ph.D. in computer science with a focus on zero trust network security, at Clemson University.

Gary Snider is a software engineer with 10 years of experience in network automation for global corporate networks and 10 years of experience in routing, switching, and network security. He has designed and maintained data center, branch office, and large campus networks for state and federal government. Gary is a core developer for the Nautobot project at Network to Code.

About the reviewers

Eric Chou is a seasoned technologist with over 20 years of experience. He has worked on some of the largest networks in the industry while working at Amazon, Azure, and other Fortune 500 companies. Eric is passionate about network automation, Python, DevOps, and helping companies build better security postures. Eric is the primary inventor or co-inventor of three U.S. patents in IP telephony and networking. He shares his deep interest in technology through his books, classes, and blog, and contributes to some of the popular Python open source projects.

I would like to thank my wife, Joanna, and my kids, Mikaelyn and Esmie, for inspiring me to be the best version of myself.

Tim Fiola, in automation since 2009, advocates for network engineers to embrace automation. Starting as a network engineer, he delved into Junos automation in 2009, crafting solutions and authoring *Day 1 – Navigating the Junos XML Hierarchy* for Juniper Networks. Starting out with Python in 2012, he went on to automate network planning for cloud providers and automated device upgrade workflows using SaltStack. Coauthoring *This Week – Deploying MPLS* showcased his expertise in RSVP and MPLS services. His open source project, pyNTM, simulates traffic failover in wide area networks. Joining Network to Code in 2021, he continues championing Python for network engineers, emphasizing the value of automating to free up time for high-value tasks.

I'd first like to thank my family, who supports me and tolerates my nerdy tendencies.

Professionally, I want to give a large shout-out to those technical experts who tolerated and still tolerate my persistent questions when I am having trouble understanding a complex topic. Without your patience and kindness, it'd have been a much tougher road.

Finally, thank you to the NTC team, where the work continues to challenge me and teach me every day.

Cristian Sirbu is a consultant, trainer, and community builder, with a particular interest in infrastructure design, automation, and solving business problems with technology. He's been in the industry for a while (getting his CCIE #43453 in the process), building, breaking, and fixing networks of various sorts and sizes. He currently lives in Ireland, helping businesses around the world understand network automation and learn about the technologies that drive it. Ever since being introduced to Linux back in high school, he has loved free and open source software. So, today Cristian's focus is on building the Nautobot ecosystem together with the talented folks at Network To Code and its worldwide community of practice.

Table of Contents

Part 1: Introduction to Source of Truth and Nautobot

1

Introduction to Nautobot   3

2

Nautobot Data Models   29

Part 2: Getting Started with Nautobot

3

Installing and Deploying Nautobot   67

4

Understanding the User Interface and Bootstrapping Nautobot   107

5

Configuring Nautobot Core Data Models   147

6

Using Nautobot's Extensibility Features   187

7

Managing and Administering Nautobot 253

Part 3: Network Automation with Nautobot

8

Learning about Nautobot APIs – REST, GraphQL, and Webhooks   311

9

Understanding Nautobot Integrations for NetDevOps Pipelines   371

10

Embracing Infrastructure as Code with Nautobot, Git, and Ansible   423

11

Automating Networks with Nautobot Jobs   453

12

Data-Driven Network Automation Architecture   515

Part 4: Nautobot Apps

13

Learning about the Nautobot App Ecosystem   551

14

Intro to Nautobot App Development   583

15

Building Nautobot Data Models   633

Appendix 2

Appendix 3

Preface

In an ever-changing world that is multi-vendor, multi-domain, and multi-cloud, there needs to be a consistent and holistic approach to network automation. Having a data-first approach provides consistency from day one. Consistent and uniform data powers pervasive network automation. Moreover, the process of data curation and data management is one of the most, if not the most, time-consuming tasks and problems of network automation. Consider these questions. What data should be used in a network change? Where does that data come from? The answer is, the Network Source of Truth!

A source of truth or data-first approach changes what is possible for network automation. It attacks the problem head-on and provides the path for long-term success. Network data is the foundation of defining intent and allows users to finally answer the question, what is the intended configuration (rather than what is the current configuration)?

Data is the foundation of network automation. This is made possible by adopting a Network Source of Truth strategy that defines the intended state of the network. Having clean and quality data inside the Source of Truth results in trusted data being deployed by the automation platform and onto the network.

Nautobot is an open source Network Source of Truth for enterprises looking to adopt a data-driven approach to network automation and a platform that complements any network automation journey. Nautobot is open source and has a growing open source ecosystem of Nautobot Apps that help users all over the world take back control of their network.

Come along for the ride and learn how Nautobot can be deployed as a Network Source of Truth and network automation platform to power your network automation journey.

Who this book is for

This book is for network engineers who manage and deploy networks, network automation engineers who automate networks and support network engineers, and network developers and software engineers who create software that supports network and automation teams.

What this book covers

Chapter 1, Introduction to Nautobot, is a comprehensive overview of network automation, data, and sources of truth. It introduces Nautobot and its key use cases and lays the foundation for the rest of the book.

Chapter 2, Nautobot Data Models, dives into the built-in core data models of Nautobot, highlighting the breadth and depth of Nautobot as a Network Source of Truth. It provides an understanding of the relationships between the components that comprise a network modeled in Nautobot.

Chapter 3, Installing and Deploying Nautobot, explores the architecture of Nautobot and then takes you through your first Nautobot deployment. You'll learn how to install each core component (Nautobot itself, workers, scheduler, database, etc.) and start to configure and load data into Nautobot.

Chapter 4, Understanding the User Interface and Bootstrapping Nautobot, explains how to add devices to your fresh Nautobot installation, including learning about many other attributes and models and how they relate to your inventory.

Chapter 5, Configuring Nautobot Core Data Models, dives deep into adding and configuring Nautobot with IP addresses, circuits, cabling and power management, secrets, and modeling high-availability, and covers notes, tags, the changelog, and filter forms.

Chapter 6, Using Nautobot's Extensibility Features, demonstrates how flexible Nautobot is by leveraging its extensibility feature set, which allows users to customize Nautobot to their specific network or design. You'll learn about using Git as a data source, Config Contexts and JSON schemas, relationships, and much more.

Chapter 7, Managing and Administering Nautobot, focuses on Nautobot platform administration. It enables a platform admin to best administer Nautobot using the `nautobot-server` command and manage permissions, along with tips for upgrading and troubleshooting Nautobot.

Chapter 8, Learning about Nautobot APIs – REST, GraphQL, and Webhooks, explains how Nautobot is integrated with other tools by examining its APIs. This chapter first covers its RESTful and GraphQL APIs, then goes into webhooks, setting the stage to learn about Jobs and JobHooks in *Chapter 11*.

Chapter 9, Understanding Nautobot Integrations for NetDevOps Pipelines, explores Nautobot integrations with a focus on `pynautobot` and its Ansible collection, while providing an overview of its Docker, Kubernetes, Terraform, and Go projects.

Chapter 10, Embracing Infrastructure as Code with Nautobot, Git, and Ansible, focuses on enabling users who use both Ansible and Nautobot together. It provides a deeper look at the Ansible collection, explains how to set up dynamic inventory, and then builds a playbook using various Ansible modules to perform network automation.

Chapter 11, Automating Networks with Nautobot Jobs, begins with an overview and an introduction to the Django ORM, then walks through how to create Jobs, migrate scripts to Nautobot, and create self-service forms that allow anyone to execute Jobs. Beyond setup and configuration, Job permissions, logging, and scheduling, approvals are also covered.

Chapter 12, Data-Driven Network Automation Architecture, dives into network automation architecture and highlights why data-driven network automation is the best approach to guarantee success in a network automation journey, and explains how this is accomplished with Nautobot and its surrounding ecosystem.

Chapter 13, Learning about the Nautobot App Ecosystem, demystifies the Nautobot app ecosystem and reveals all that the ecosystem has to offer, while highlighting the best is yet to come and is in the hands of the community.

Chapter 14, Intro to Nautobot App Development, provides an overview of the developer API that is used to extend Nautobot and create Nautobot apps, ranging from lightweight Nautobot apps that are only data models to full-blown apps that cater to specific outcomes.

Chapter 15, Building Nautobot Data Models, covers real-world use cases for building custom Nautobot with a case study of an organization that needs custom data models and walks through the path to create them from start to finish.

Chapter 16, Automating with Nautobot Apps, continues building the app from the previous chapter, showcasing how Jobs can be packaged with apps to create an end-to-end network automation solution.

Appendix 1, Nautobot Architecture, dives into the internal components of Nautobot, reviewing its use of Django, Celery, Beat, and databases such as Postgres and MySQL for those who want to understand Nautobot at a deeper level.

Appendix 2, Integrating Distributed Data Sources of Truth with Nautobot, introduces the problem of managing distributed data sources and explains how Nautobot can be used as part of the solution to integrate and aggregate data by using the Nautobot Single Source of Truth framework. Solving network data problems in large enterprises is not a trivial task.

Appendix 3, Performing Config Compliance and Remediation with Nautobot, explains how Nautobot Golden Config can be used to conquer the most common use cases in networking, including backups, generating intended configurations, and ultimately performing compliance and remediation.

To get the most out of this book

You should have basic network knowledge (CCNA or greater), along with at least 6-12 months' experience of using Python and Ansible for network automation, and you should be comfortable with Netmiko, NAPALM, or Nornir. You should understand how to read and use Jinja templates, YAML, and JSON.

Software/hardware covered in the book
Ubuntu 22.04
Python 3+
Ansible 2.16+
Nautobot 2.1

Many of the demos can be followed on the public Nautobot instance hosted by Network to Code at `https://demo.nautobot.com`*. This is mentioned throughout the book.*

If you are using the digital version of this book, we advise you to type the code yourself or access the code from the book's GitHub repository (a link is available in the next section). Doing so will help you avoid any potential errors related to the copying and pasting of code.

Download the example code files

You can download the example code files for this book from GitHub at `https://github.com/PacktPublishing/Network-Automation-with-Nautobot`. If there's an update to the code, it will be updated in the GitHub repository.

We also have other code bundles from our rich catalog of books and videos available at `https://github.com/PacktPublishing/`. Check them out!

Conventions used

There are a number of text conventions used throughout this book.

`Code in text`: Indicates code words in text, database table names, folder names, filenames, file extensions, pathnames, dummy URLs, user input, and Twitter handles. Here is an example: "In the device management section, search for the `WayneEnt_FW1` firewall."

A block of code is set as follows:

```
devices_url = "https://demo.nautobot.com/api/dcim/devices/"
# adds ams01-leaf-11 to the location AMS01
r = session.post(devices_url, data=json.dumps(payload))
# the UUID of the device will be saved for the next API call
device_id = r.json()["id"]
```

When we wish to draw your attention to a particular part of a code block, the relevant lines or items are set in bold:

```
payload = {
        "name": "ams01-leaf-11",
        "device_type": "74cf95a8-4233-46b9-a740-fba4f5dc88d3",
        "status": "9f38bab4-4b47-4e77-b50c-fda62817b2db",
        "role": "869267d8-7d75-4bd3-8a9e-5e6adcf200f6",
        "tenant": "1f7fbd07-111a-4091-81d0-f34db26d961d",
        "platform": "f48fd9e2-45c5-4c2f-aa54-28964edb3e1e",
        "location": "9e39051b-e968-4016-b0cf-63a5607375de"
}
```

Bold: Indicates a new term, an important word, or words that you see onscreen. For instance, words in menus or dialog boxes appear in **bold**. Here is an example: "Click on the **Interfaces** tab; if one does not exist already, you can click **Add Components** | **Add Interface**."

> **Tips or important notes**
> Appear like this.

Get in touch

Feedback from our readers is always welcome.

General feedback: If you have questions about any aspect of this book, email us at `customercare@packtpub.com` and mention the book title in the subject of your message. You can also talk to the authors directly if you join the `#nautobot` channel in the Network to Code slack. Self sign-up is at `slack.networktocode.com`.

Errata: Although we have taken every care to ensure the accuracy of our content, mistakes do happen. If you have found a mistake in this book, we would be grateful if you would report this to us. Please visit `www.packtpub.com/support/errata` and fill in the form.

Piracy: If you come across any illegal copies of our works in any form on the internet, we would be grateful if you would provide us with the location address or website name. Please contact us at `copyright@packt.com` with a link to the material.

If you are interested in becoming an author: If there is a topic that you have expertise in and you are interested in either writing or contributing to a book, please visit `authors.packtpub.com`.

Share your thoughts

Once you've read *Network Automation with Nautobot*, we'd love to hear your thoughts! Scan the QR code below to go straight to the Amazon review page for this book and share your feedback.

https://packt.link/r/1837637865

Your review is important to us and the tech community and will help us make sure we're delivering excellent quality content.

Download a free PDF copy of this book

Thanks for purchasing this book!

Do you like to read on the go but are unable to carry your print books everywhere?

Is your eBook purchase not compatible with the device of your choice?

Don't worry, now with every Packt book you get a DRM-free PDF version of that book at no cost.

Read anywhere, any place, on any device. Search, copy, and paste code from your favorite technical books directly into your application.

The perks don't stop there, you can get exclusive access to discounts, newsletters, and great free content in your inbox daily

Follow these simple steps to get the benefits:

1. Scan the QR code or visit the link below

https://packt.link/free-ebook/978-1-83763-786-7

2. Submit your proof of purchase
3. That's it! We'll send your free PDF and other benefits to your email directly

Part 1:
Introduction to Source of Truth and Nautobot

This part covers the what and why of network automation, Source of Truth, and Nautobot. It provides you with a general overview of the problems in network automation and how understanding the relationship between data and network automation changes the way you think about and approach network automation. From there, you will learn about Nautobot and how it is used to power enterprise network automation solutions, understanding key use cases and the Nautobot core data models.

This part consists of the following chapters:

- *Chapter 1, Introduction to Nautobot*
- *Chapter 2, Nautobot Data Models*

1

Introduction to Nautobot

Data-driven network automation powered by Nautobot is gaining momentum across the industry. This chapter provides the foundation required to understand the *what* and *why* of network automation and gives an overview of Nautobot and the role it can play in the greater network automation ecosystem. This chapter will start by uncovering the relationship between data and network automation and how **Source of Truth** (**SoT**), when used with Nautobot, is an integral part of the network automation journey. You'll learn what network automation is, key use cases for network automation, and why you should consider network automation, dive into SoT, and be introduced to Nautobot and the power it can provide on the journey with Nautobot as a SoT and a network automation platform.

This chapter covers the following main topics:

- Introduction to network automation
- Understanding SoT
- Nautobot overview
- Nautobot use cases
- Nautobot ecosystem

Introduction to network automation

If you're reading this book, you've realized you need to think differently about managing your network. And you are not alone. If you ask any network engineer, there is still not a day that goes by when they are not logging into a device via SSH *and doing work manually*. Over the last few decades, the most common approach to managing networks of any size, ranging from tens to thousands of devices, was connecting to the device and using the network **command-line interface** (**CLI**). The network CLI is used to gather data, troubleshoot, and make configuration changes. This remains the most common way of managing networks. However, this is changing.

Over the last 10 years, we've seen significant growth and improvements around the operational models for networks. The **software-defined networking** (**SDN**) era brought us controllers and APIs. Controllers provide APIs and fewer points of management. Rather than manage thousands of devices, it is possible to manage tens of controllers (or fewer in some cases). Independent of the number, the point is that the number of directly managed nodes continues to decrease. The SDN era also shined a light on the programmatic interfaces, or lack thereof, of network devices. We have evolved from SSH and SNMP to APIs – REST APIs, GraphQL, gRPC, and event-driven webhooks from controllers and devices. While SSH and SNMP are still the de facto standards across the industry – even for automation, progress is being made. For that, we need to recognize the progress and celebrate, but continue to demand more.

The progress around network automation has been driven by open source. Before network automation, there wasn't much use of open source in the network industry. The industry is learning from its history – that is, if you solely purchase and use vertically integrated tools, there is less flexibility and you could lose control of your network. With current trends, the belief is that those that adopt even just some open source remain in control and can extend libraries and tools as needed to ensure maximum adoption of network automation in their environment. Don't worry – we'll cover some of the most common open source tools and technologies for network automation in the *Industry trends* section of this chapter.

We'll start by exploring what network automation is, its key use cases, and the value it can provide an organization. From there, we'll dive into SoT and Nautobot.

What is network automation?

Any advanced and hot technology *always* gets flak when there are formal definitions because there are always varying opinions, and that's okay. For this book, our approach is to keep it simple. So, what is network automation? Network automation *is* next-generation network management. Period. We can talk about Python, Ansible, Nautobot, YAML, JSON, REST APIs, NETCONF, RESTCONF, YANG – the list can go on for pages. Here is the bottom line – all of these tools and technologies are being used to improve how networks are managed and consumed daily, which is, simply put, *better* network management. Network automation involves transforming operational models that can radically transform careers and technical and business operations.

One major point you should think about on your network automation journey is that it isn't just about doing *your* tasks better and more efficiently. That is only the starting point. You need to be thinking about how to expose your automation to other engineers, teams, and even non-technical people, thus enabling all parties with the self-service they need to do their job functions.

Let's assume you are automating tasks such as **operating system** (**OS**) upgrades, which involves gracefully moving traffic from one device (and circuit) to another. This is a complex workflow. Sure, this can help you when you need to upgrade a device or perform maintenance on a device, but what about exposing that automation to individual site leads? If this workflow is made more accessible, can this expand who can perform the task using your trusted automation? Does it allow you or your team to delegate a little more? How often are upgrades happening today contrasted with how often you'd like them to happen?

What about if you had automated diagnostics? What if your **Network Operations Center** (**NOC**), **Security Operations Center** (**SOC**), or service desk could go to a portal, click a button, and diagnose their most common issues? In a manual process, one person opens a ticket, and that ticket remains open and an engineer picks it up. The engineer reviews the request and sees it is a semi-common problem. Maybe they need to check with another engineer or two along the way. After a few discussions, they know where to go, which devices to log in to, and which tools to log in to. They correlate the data gathered between the devices and tools. They ensure things look good and update the ticket. Common workflows like this should be automated.

Would your leadership be astounded to learn that the countless hours needed to gather data, let alone the hours spent formatting to make it look good, can be eliminated with automation? Compliance and reporting tasks often take a lot of engineering time and effort because they involve manually gathering and processing information. Now, imagine being able to automatically create any compliance document or report you need. Documents that include pre/post change tests. Documents required for change control. Reports you need to run monthly, quarterly, or annually for compliance. Reports that verify your devices are operating as expected.

This is network automation.

Network automation use cases

We just discussed some examples of network automation to bring it to life. Now, let's look at some of the most common use cases, including the ones that were already mentioned:

- **Common config changes**: Is your team performing the same types of changes day to day, week to week, or month to month? These are changes such as adding VIPs, turning up a port, adding a VLAN to a switch port, managing firewall policies (also discussed later in this chapter), turning up a new BGP peer, updating routing preferences, adding static routes, and updating zones and ACLs. These changes are ripe for automation because they happen so frequently.

- **Common operational tasks**: These are similar to the previous use case, but they involve performing operations tasks that do not require a configuration change. Some examples include updating SSH keys and certificates on devices, performing a config save or backing up a configuration, copying files to devices, rebooting devices, checking logs, and even performing non-network device tasks such as checking and updating tickets.

- **Mass changes**: While *common config changes* are scoped to a set of devices (this could be just a few devices), *mass changes* are meant to be site, campus, regional, or global. Mass changes include changes such as updating AAA, NTP, or SNMP but could also include changing the format and structure of all interface descriptions on every device. These types of changes don't happen as frequently, but when they do, they are impactful and usually a large project.

- **Data gathering and reporting**: How often is someone you know logging into numerous devices or tools to perform health checks, troubleshooting, or simply to execute a request that comes in for application or network performance degradation? Automated data gathering, reporting, and documentation is not only one of the best use cases for network automation – it is a great area to start with since it is less impactful in the event there is bad automation (because it'd be read-only automation). It could also be added to nearly any other use case producing reports before and after changes or generating compliance reports specific to your team or organization.

- **Configuration and operational state compliance**: Compliance comes in two major flavors and can be best understood by asking the following two questions: *Is the network configured as expected?* and *Is the network operating as expected?* Configuration is easy to understand, but it does mean you'll need to understand the intended state of the network. This is where SoT and data-driven network automation comes into play. We'll cover this in more detail later in this chapter in the *Understanding SoT* section, as well as *Chapter 11*.

- **Pre/post-change state validation**: Similar to the previous compliance use case, pre/post-change state validation is more focused on a defined scope of devices. There may be automation when performing global compliance that only runs daily, but changes are happening continuously. Pre/post change state validation ensures that the network is healthy and operating as expected before and after the change.

- **Firewall policy automation**: How many firewall rules are you adding per day, week, or month? How do you know which firewalls need a new policy? How do you know where in the list of rules the new one should go? Do you know? Could you document this for a fellow engineer? Try. This is the start of firewall policy automation. While the last mile is configuring the actual firewall, the questions prior illustrate that a company's firewall rule change workflow often involves many steps before the actual configuration change.

- **OS upgrades**: While already mentioned briefly, how often are upgrades happening today contrasted with how often you'd like them to happen? How many of your devices adhere to your software standards? How many upgrades can you currently do in a single change window? Do you find yourself watching the console of devices as you upgrade them? Do you run any automation to see if devices have the required disk space before copying the new image to the device? Do you run any automation to verify the md5 checksum of the image after it is copied to ensure it isn't corrupt? Is your network at risk due to vulnerabilities left unpatched? Upgrading devices often happens when needed, versus having a defined cadence. It is never a priority. Automation changes that.

- **Greenfield sites and devices**: If you are repeating deployments, there is room for automation. It may mean adding new top-of-rack switches in the data center, it may mean adding a closet or IDF closet in a growing campus, adding a new retail location, or even a new colocation facility or **point of presence (PoP)**. Much of the automation discussed here is around the configuration of these devices, but that is the easy part. Site planning and deployment is about data curation and management not to mention each organization's business logic required for deployments.

How do you and your team know which IP addresses, VLANs, ASNs, and overall configuration should be entered on those devices? Is it from spreadsheets or a SoT? Again, more on SoT later.

- **Vendor migrations**: Have you ever not moved forward with changing vendors due to the work effort of migrating configurations? With a properly defined SoT and data strategy, this becomes trivial. Your focus becomes storing the intended state of the network using data, decoupled from any vendor-specific syntax. Syntax for a given vendor is generated by running the data through a set of vendor-specific configuration templates. In a migration, you can generate the desired state configuration for a given vendor by running the data through a different set of templates and then deploying those new configurations. Beyond configuration management, you'll also want to ensure multi-vendor operational state compliance to ensure there are no gaps in visibility during the end-to-end migration.

- **Self-service**: It is critical to think through how a given workflow will be triggered along with who the target user is. *Self-service* does not mean that it needs to be a click-button UI. It may mean an IT tool, CLI tool, pull/merge request, ChatOps, or yes, it may mean a full self-service user-friendly form. The point is that you do not need one way to expose network automation or even one way per workflow. Using an architectural and a platform approach to network automation allows you to expose the same workflow through multiple self-service interfaces. You should cater to your culture and your users. This will drive more adoption of network automation.

It is recommended to use a holistic multi-domain network automation architecture to serve as a platform to meet today's requirements. This architecture will also serve as the foundation for tomorrow's requirements. As you embark on the journey, be cautious about using different network automation architectures for different types of networks and domains. If so, it'll create more issues and give your team even more tools to manage while making it harder to unify standards and processes. In *Chapter 10*, we will talk much more about network automation architecture to ensure a consistent approach to managing networks independent of size, domain, and location.

Why automate your network?

After covering the *what*, let's take a look at the *why*. While many use cases are *horizontal* and can be used by any organization type (or verticals), the actual *why*, impact, and justification will differ per organization. Just to clarify, by *vertical*, we're referring to companies with different business types. A few examples of different verticals include financial services, pharmaceutical, retail, telco/the cloud, manufacturing, accounting/legal professional services firms, state and federal government, K-12 education, and universities.

For some verticals, the network may be the business. It may either be a business enabler or have serious consequences if the network is down. For other verticals, other factors may be a bigger concern. For this reason, the *why* is going to vary widely, and we'll cover general reasons to automate the network. Here are some common examples:

- **Lower costs**: Every leader in every business is always asked by their leaders or directly by finance if there is a way to lower costs. In reality, automation helps lower longer-term costs. The more a company can show how automation lowers costs, the greater the chances are that the automation projects get initial buy-in and long-term support. With some of the use cases already mentioned, costs can be significantly lowered. If a company truly documents each of the tasks required and the time to do each for a workflow (such as OS upgrades or troubleshooting) and verifies the most common incidents, they are going to see drastic savings in time and effort when using automation. Time equates to money. It doesn't mean anyone is getting replaced. However, it does mean that there is more time for more projects, each of which adds more value to the business. Increasing velocity without needing to hire new people is a tremendous cost savings.

- **Enhance security and reduce risk**: In today's world, security is top of mind for everyone; it's integrated into all that we do. No company wants to be the headline in the local, national, or global news. Security-focused automation ranges from automated scans, firewall provisioning, VPN connects and disconnects, compliance and remediation, governance adherence and monitoring, and patch management just to name a few. Even if you are not directly on a security team, you should ask yourself if security can be improved in your domain. Can you rotate passwords more frequently? Maybe change those SNMP community strings? The list can easily go on.

- **Provide greater insight and control**: Data is king and that includes greater visibility into your network and automation infrastructure. Automation can be used to gather data, document data, understand patterns, and compare against known baselines. Sure, there may be tools that provide this in the **user interface (UI)**. That's a great start, but what about seamless workflows that open tickets, update tickets, send emails, and send chat messages in response to network data that is outside the expected range? With automation, you have the opportunity to get the insights you need to answer the questions you have and know that the answers are contained within the network. Think about that. If you are logging into a few portals, copying data into a spreadsheet, creating Excel formulas, or creating a new document to then turn into a PDF and email, there is a better way. There is an automated way.

- **Increase business agility**: Each business and team is always trying to go faster and also perform activities that are not possible without automation. Organizations need to work smarter and more efficiently. In some cases, it may also make sense to hire more people. However, hiring more people often slows things down because, at a certain point, people can start to get in each other's way. In contrast, automation can reduce cost, improve performance/velocity, increase reliability, and do things that humans just cannot do. One example is automation-enabled self-service, which helps business stakeholders obtain the outcome they need sooner. Automation can also improve business-to-business connectivity, allowing organizations to either recognize revenue sooner (for those that are doing business over those connections, tunnels, or circuits) or start consuming a new service. Think about deploying a new application in a lab or test environment. If it takes weeks to get a new application and its network and security configurations deployed for each environment (dev, test, UAT, and so on), it may be an aggregate delay of months. This is either delaying employee or customer satisfaction or revenue. Using automation improves this and increases business agility.

In all that you do, keep automation top of mind, and try to understand the business and organization-level benefits for various leaders in your organization.

Persona-driven network automation

While we already looked at network automation use cases and the rationale for automation, let's take a different spin on use cases. There is *usually* never one network team. There are usually teams focused on day 0 or architecture and/or engineering; day 1 or implementation; day 2 or operations. These teams may even span network domains such as LAN, WAN, WLAN, or Security, depending on the size of the network. Recognizing the work of the various teams will help structure automation projects for what's possible *within your team*.

Here is a list of example projects and tasks broken down by the three types of teams often found in network organizations:

- Day 0 or architecture and/or engineering:

  - Ensure configuration standards are documented in a structured and modeled manner that is programmatically accessible

  - Ensure hardware standards are documented in a structured and modeled manner that is programmatically accessible

  - Ensure software standards are documented in a structured and modeled manner that is programmatically accessible

  - Ensure architectural and engineering tests exist within every CI pipeline – for example is there redundancy?

  - Develop automation architecture and framework used by other teams

- Day 1 or implementation:

  - Use automation to generate configurations

  - Use automation to perform configuration changes

  - Use automation for pre- and post-deployment verification

  - Use automation for continuous verification of deployment standards

- Day 2 or operations:

  - Execute network device automation for common troubleshooting tasks

  - Continuously update automation that is used for common troubleshooting tasks

  - Execute network device automation for common changes

  - Ensure automation for dynamically reading emails from ISP/NSPs for circuit notifications

  - Execute automation for gathering and collecting information from various tools and devices to aid in troubleshooting

  - Execute automation for dynamically creating, updating, and closing change management tickets

Industry trends

As we've already discussed, the CLI still dominates the industry. However, each year, month, week, and day brings us closer to transformative and better network management through the use of network automation. In this section, we'll look at several of the trends that are collectively driving the industry forward to do more with less and allow for more efficient network operations.

This list is not meant to be exhaustive, but illustrative of the trends that are driving operational efficiencies and automation:

- **SDN**: SDN took the industry by storm in the 2010s. Most modern network architectures include controllers that simplify management and visibility and provide programmatic access with APIs. Simplified management is made possible because it allows users to manage systems versus managing devices and nodes, which allows more abstract policies to be created and applied. Because they allow for fewer points of management, SDN controllers simplify workflows and integrations using the controller (versus individual device) APIs. With SDN, you may have different controllers and solutions for campuses, WAN, data centers, and the cloud. So, if you are looking for a unified network automation strategy, there will be a bit of integration that needs to happen when it comes to data and orchestration. More on this later.

- **NetDevOps**: We've learned a lot about the DevOps industry over the last 10 years. When we talk about NetDevOps, we're referring to doing DevOps but applied to network infrastructure, engineering, and operations. Here are a few examples that highlight trends:

  - Using Git-based **version control systems** (**VCSs**) such as GitHub, GitLab, or BitBucket. Using VCS enables collaboration while providing traceability and audibility on all software or file-based artifacts (templates, data files, scripts). VCS allows users to create owners of particular projects or sections of a project providing accountability to the respective teams.

  - Using **continuous integration** (**CI**). Organizations that use VCS will require basic CI. CI allows users to create tests that must pass before accepting or approving any changes. These tests focus on ensuring nothing is going to break in the automation or the application. CI can also be applied more directly to the network, enabling **network CI**.

- Implementing **network CI**. If the initial CI tests pass on code and static files, users can do tests such as pre-change analysis based on models of the network (mock devices or real equipment, if you have a larger budget), running active tests on the network (does the network need to be a certain state before making the change?), perform the actual change, and then finally ensure the network is operating as expected after the change.

While DevOps and NetDevOps can be talked about for days, the actual industry facts show that nearly every network automation project in the world includes version control, automated tests, and some level of CI. If your organization is one of the few that aren't using these three key items, be sure to explore them as soon as you can.

- **Open source**: Many open source tools are used in the DevOps ecosystem. The same holds for NetDevOps. We'll mention some of the most common tools in the *Tools and technology* point covered in this section. Regardless of the tools deployed, it is more important to understand the real *value* of open source. In the context of open source, the real value lies in its *extensibility*, *ecosystems*, and *community*. Extensibility and ecosystems can drastically change and improve what's possible on your network automation journey. Keep in mind that each of these is predicated on the fact that there is a strong community at the foundation. Extensibility is what should give you confidence that no matter what decision is made for your network, you can adapt and change to account for that decision. A *change* may be as simple as upgrading to the latest version of software, migrating from vendor A to vendor B, or migrating from a traditional network to a controller-based network. In any of these scenarios, an organization needs to be confident that its automation can be tailored, updated, or augmented for their needs. While certain commercial tools offer extensibility, it is usually limited and extensibility features tend to be in a perpetual state of *coming soon*. Ecosystems built around community also play a critical role in open source software, further enhancing what is possible with particular open source projects. Ecosystems are usually fostered around extensions, adapters, apps, or add-ons that are outside of the core open source project but are powered by it. It is these ecosystems that usually incorporate the solutions required for true multi-vendor management and automation. The point is not that everything needs to be open source, but that open source software and solutions should either lead or complement any network automation strategy. If they do not, there may be a great risk to the success of the automation journey three to five years out.

- **SoT**: Since you're reading this book, you've likely heard about SoT. In fact, the main topic of this book is Nautobot! At its core, Nautobot is a network SoT that is actively being developed specifically for network automation environments. A SoT is a growing industry trend and probably why you're reading this book, but the short overview of a SoT is that it is the location where you can define the intended state of the network. This is the truth; it is what should be. The SoT is not what is on the device or network. That is referred to as the actual or observed state. The intended state, or SoT, can be extrapolated and used to document the intended configured state and intended operational state, or even used as the place to define the intended state for monitoring thresholds and events. Overall, it allows for greater governance of network data with a focus on what should be in a manner that is often vendor-neutral. We'll spend much more time on SoT in the next section and throughout every other chapter in this book.

- **Self-service**: We covered self-service in the *Network automation use cases* section, but to restate it one more time, the notion of self-service is not one-sided. Those organizations that are successful on their network automation journey understand that it is about having the right mapping of workflows to people (consumers) and from those people to the right user interaction, or the right tool to execute and request that automation. If you get this wrong, there is a great chance to end up with network management systems that aren't used, which will take us back a few decades.

- **Streaming telemetry**: SNMP has been around for decades, and network visibility as we know it is largely based on SNMP. Streaming telemetry is what you may expect when you think about modern network visibility. In this modern era of streaming telemetry, network devices can continuously "push" or "stream" network data to a centralized location. This allows for greater visibility, querying, and trending based on data that would have normally been lost. Wouldn't it be great if the network device could send you the information you need when you need it? Wouldn't it be great if you could turn on a stream of data (collection of data points) from a series of devices on particular interfaces versus getting a response from an interface poll that may kill the device if your poll frequency is too high? Wouldn't it be great if you could build a closed-loop system that can operate in near real time? This is made possible by streaming telemetry.

- **Intent-based networking** (**IBN**): When you look at the key use cases and trends, you can start to see common components of an architecture, such as orchestration, automation, SoT, and telemetry. When these components are fully integrated, the result is an IBN. An IBN is just a comprehensive network automation architecture. It allows organizations to define intent, continuously collect network data (streaming telemetry, SNMP, show commands, and configuration data), analyze that data, ensure intent is deployed, and then react based on intent violations. The reaction to the data may be to remediate or make a change for managing capacity or minimizing the blast radius for a known issue. IBN becomes a natural progression as you start to deploy a holistic architecture for network automation.

- **Artificial intelligence** (**AI**): Our general belief is that a significant amount of automation must be implemented without AI/ML, meaning don't let flashy new tech derail projects and outcomes that are solving today's problems. That said, at the time of writing, we've seen the launch of OpenAI's ChatGPT (`https://openai.com/blog/chatgpt/`), Google's Gemini, and many more services like these. It should be obvious that AI/**machine learning** (**ML**) coupled with **natural language processing** (**NLP**) creating more digital assistants is going to have a transformative impact on where we are as an industry in 5 to 10-plus years as it gets mainstream adoption. Until then, it'll be explored and implemented by pioneers and manufacturers who can make it consumable in a turnkey and meaningful way.

- **Tools and technology**: This is always one of my favorite topics since we live in a product- and tool-centric industry, but let's look at existing tools trends for network automation. From an open source perspective, the dominant tools are Ansible, Nautobot, Batfish, and Terraform. We also see a sprinkling of Salt, but its presence is still largely seen in application and systems automation. Looking at open source from a lower-level library perspective, there is continued growth with Netmiko, NAPALM, Nornir, pyntc, ntc-templates, and scrapli. If you are using open source or building your solutions, you want to check out these projects. For example, if you need a custom Ansible module or custom Nautobot App, you're more than likely going to consume those libraries to perform your automation. From a telemetry perspective, there is also growth in various stacks that include Prometheus, Influx, Telegraf, and Grafana. Teams that have the skills or are further on their journey can use these stacks to provide greater visibility through data aggregation, data enrichment, extremely powerful queries, and a holistic view of their networks and their IT infrastructure. From a commercial tool perspective, and exclusive of SDN products, we're seeing the most adoption of Itential, IP Fabric, and Forward Networks.

Information

Interested in seeing a comprehensive list of all network automation projects, tools, and products? Check out *Awesome Network Automation* (`https://github.com/networktocode/ awesome-network-automation`).

From a trends perspective, we thought it may be worth calling out a few things that get attention at industry events and in social circles, but aren't gaining traction. The first is the direct use of YANG data models within automation tools. They are still mostly used by vendors to define their schema. Of course, there are outliers such as hyperscalers or a select few enterprises, but generally speaking, the actual use of YANG by network teams is not a trend. If you're using an API that is based on a YANG schema, we do not consider that a trend for end users, but it is a trend for certain manufacturers. We'll also call out REST APIs on network devices. While they are becoming more commonplace because the dominant majority of devices in production still don't have APIs, and instead have two or more (different APIs per vendor and OS) ways of performing automation, the majority of device-specific automation still happens via SSH.

Understanding SoT

We've already mentioned SoT a few times. It's finally time to dive in. Let's start by talking about *data*. We'll do that through the lens of making a change on the network.

Let's assume that you want to turn up a new port that's going to terminate a connection to a new building. If you look at other similar configurations on the same device, you're going to find a configuration similar to this:

```
interface vlan100
  description Routed Interface for connection to off campus house
  ip address 10.1.100.1/24

interface GigabitEthernet4/1
  description connects to och-sw-01 GigabitEthernet1/1 (off campus
house)
```

```
 switchport
 switchport access vlan 100

vlan 100
 name off_campus_house
```

Is there any other way to configure the same interface? Could we have used a routed port? Could we have configured a trunk instead? A different prefix? Sure, these are all valid possibilities. The point is that you are going to have your own standards, and they will drive your *new* configuration. When adopting a SoT approach, we need to decouple *data* from configuration syntax.

For example, the standard configuration you copy and paste becomes your template while you extract the data. That data becomes any input that changes to derive a configuration. In this example, the data is as follows:

- **SVI interface**: 100
- **SVI description**: Routed interface for connection to off-campus house
- **SVI IP address**: 10.1.100.1/24
- **Physical interface**: GigabitEthernet4/1
- **Physical interface description**: Connects to och-sw-01 GigabitEthernet1/1
- **VLAN ID**: 100

In reality, both descriptions – that is, the SVI interface and the IP address, could be removed from data inputs since they can be auto-generated from the VLAN ID. We'll see that soon. For descriptions, they can be auto-generated by having a use case or description of the project defined. Let's look at a few examples of showing this data as YAML structured data:

> **Note**
> Teaching YAML and Jinja2 is outside the scope of this book.

```
svi_interface: 100
svi_description: Routed Interface for connection to off campus house
svi_ip_address: 10.1.100.1/24
physical_interface: GigabitEthernet4/1
physical_interface_description: connects to och-sw-01
GigabitEthernet1/1 (off campus house)
vlan_id: 100
```

You may opt to nest some data, like this:

```
svi:
  interface: 100
  description: Routed Interface for connection to off campus house
  ip_address: 10.1.100.1/24
physical_interface:
  name: GigabitEthernet4/1
  description: connects to och-sw-01 GigabitEthernet1/1 (off campus
house)
vlan_id: 100
```

Going one step further, a few values could be eliminated if there is more logic in your Jinja2 template. This one also adds data for the remote peer:

```
physical_interface: GigabitEthernet4/1
vlan_id: 100
connection:
  description: Routed Interface for connection to off campus house
  remote_peer: och-sw-01
  remote_interface: GigabitEthernet1/1
```

Finally, a Jinja template that could consume this data and render a configuration snippet would look like this (focused on one of the devices):

```
interface vlan{{ vlan_id }}
 description {{ connection['description'] }}
 ip address 10.1.{{ vlan_id }}.1/24

interface {{ physical_interface }}
 description connects to {{ connection['remote_peer'] }} {{
connection['remote_interface'] }}
 switchport
 switchport access vlan {{ vlan_id }}

vlan {{ vlan_id }}
 name {{ connection['description'] }}
```

Defining SoT

After looking at a few different ways to represent data, the main point is that we have successfully decoupled data, which is shown as YAML, and syntax, which is shown as a Jinja template. The templates are built or defined by those who own the *standards*. However, data is what needs to be created or updated for any given change. Focusing on the data focuses on a change, without getting pulled into syntactical details that vary per vendor.

This data is now the SoT (*technically, the SoT would be the file that contains the data*).

With our focus on the data, now comes the real questions to ask:

- Why did we pick GigabitEthernet4/1?
- Why was VLAN 100 chosen?
- Why was 10.1.100.1 chosen?
- How did we construct the interface descriptions?

It would be fairly common if you were checking one or more spreadsheets to get this data, but it's more likely that *you just knew* because you're good at what you do and you checked the devices and connections that you most recently deployed.

The idea of a SoT is that it allows you to plan and focus on what should be. A SoT defines the desired state. With a SoT, users manage the data that's used for upcoming changes, which is then programmatically accessed by automation tools during a change. The automation tools access the data, render a network configuration, and then ensure that configuration exists on the network. On your SoT journey, you should be able to build a document that defines one tool as the authoritative source per type of data – for example, ASNs, VLANs, and so on.

Due to the breadth of network data required to manage a production network, often, one or more systems are used as an authoritative source of information to build a configuration. For example, a database might be used for inventory and IP addresses, and another that has policies used for ACLs. The authoritative source of data is the location where updates are made. This is also often referred to as a **system of record (SoR)**. *It's worth calling out that SoT and SoR are often used interchangeably:*

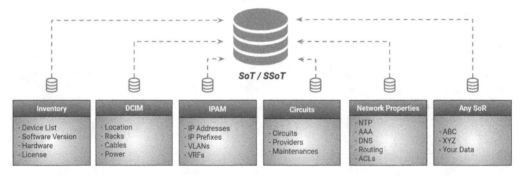

Figure 1.1 – Visualizing SoR, SoT, and SSOT

Generally speaking, the term SoT is a system that stores data from one or more SoRs. However, how often SoR and SoT are used interchangeably, the term **Single Source of Truth (SSoT)** is often used to reflect a system that is aggregating data from multiple SOR. This type of system allows relationships to be formed between these datasets and also provides one unified API that can be used to access all network data. Having this data accessed from a single API significantly lowers the amount of work required by your automation tooling. In *Appendix 2*, we review working with multiple SoTs, doing a deep dive on the Nautobot SSoT application, and discussing other designs used for managing network data.

Approaches to SoT

The previous section described the purist view and the most correct approach to understanding a SoT. It is based on the premise that *the SoT always contains the intended state*. This means that as a user, you change the data and then perform your change using that data. Of course, using automation to fetch the data is the ideal state, but even if you were using it as a documentation store, it's a step in the right direction. The gap in this approach is that the SoT does not always reflect the *actual* state of the network (maybe a user makes a manual change because they don't like automation or they are just fixing something quickly). There should be tooling built around the SoT in this approach that compares the SoT and the actual network. This provides assurance and compliance that the network is operating as expected.

> **Note**
>
> Based on the network technology deployed or your preference, another approach is also possible when implementing a SoT. The alternative is to ensure the SoT reflects what exists on the network. This approach may be used as a one-time event to turn the initial data population into a SoT. This may seem a little confusing because it goes against the purist view of SoT, but we thought it is worth calling out because it is reality.

With the growth of NetDevOps over the past few years, one common place to start with a SoT is to define data in a YAML file and version it in a Git repository. The YAML data is the intended state. That data gets rendered with one or more templates to generate the intended configuration, which is later deployed to the network. This approach provides peer review (through pull and merge requests) on the data before being merged and later deployed and also enables users to run automated tests with CI on the data providing even more assurances the data is good. This approach of defining the data first and having that drive automation is what data-driven network automation is all about.

Due to the plethora of technologies that exist today from SDN and cloud-native networking, networks are not always planned – they may be dynamic. There may be auto-scaling or dynamic policies. In these types of environments, you may prefer to see the actual state in one place. This is also possible by using a SoT. With this approach, it is more analogous to a discovery engine, but for configuration data.

It is also possible to employ a hybrid approach. This would mean certain data in the SoT is authoritative and drives the intent of the network, and other data shows what exists in certain domain managers, controllers, or clouds. The general assumption here would be that the data added via controllers or the cloud is authoritative and what is intended to be configured.

Overall, it's always worth remembering that not all purist points of view and ideals can be implemented in a network that has been evolving for 25 years. We need to take a pragmatic approach, but it is important to recognize proper definitions and terminology to ensure everyone embarking on their SoT journey is on the same page.

Keeping the purist view in mind allows us to see the relationship between network data and network automation, given the data is ultimately at the center and driving network automation. The beautiful thing about data-driven network automation is that it allows us to start thinking about abstractions and the level of intent that we want to describe the network.

Even in this book, we're talking about lower-level data, which leads to lower-level intent. However, once you've embraced data, it is possible to build abstractions around design. Consider the earlier example at a higher level of intent:

```
connection:
    source:
        device: nyc-sw-01
        interface: GigabitEthernet4/1
    destination:
        device: och-sw-01
        interface: GigabitEthernet1/1
    type: off_campus
```

In this example YAML data file, you'll notice off_campus defined as a *type*. This was not used in the prior example. With logic in your templates and automation, the right data will be generated and then populated in the SoT based on the standard off_campus designs for both required devices. You could go one step further and not even choose the devices and let the automation tell you the ports to use on particular devices that have capacity. This will take time, but it starts with repeatable standards (few to no snowflakes) and data, meaning **it starts with SoT**.

SoT tools and products

After learning more about SoT and the role of network data in network automation, we're ready to look at SoT tools and products. The fact is that there are not many tools that focus on network data specifically for network automation. Let's look at some tools that may be used in building out an overarching SoT strategy. Some are more common than others:

- **Nautobot**: It should be obvious and is likely the reason you're reading this book, but we believe Nautobot is *the* SoT for networking. With native models, extensibility, and a framework in place for aggregating data to and from other data sources, it is becoming the de facto standard for enterprises adopting a SoT for network automation. Nautobot is an open source project sponsored by Network to Code. Network to Code's mission is to continue to drive network automation around the world, one network at a time.

- **YAML files**: Usually playing a part in almost every network automation journey, they provide a solid path to getting started and understanding data-driven network automation. In *Chapter 6*, we'll look at integrating YAML files stored in a Git repository directly into Nautobot – showing that with the click of a button, those files and data can be pulled directly into Nautobot.

- **NetBox**: The motivation for Nautobot, NetBox is a solution that models and documents modern networks. NetBox is an open source project sponsored by NetBox Labs. Nautobot forked NetBox when NetBox was at v2.10 and has continued to diverge (as a hard fork (`https://producingoss.com/en/forks.html#:~:text=Hard%20forks%20(also%20sometimes%20called,line%20with%20their%20own%20vision)`) since February 2021.

- **Configuration management databases** (**CMDBs**): More often than not, CMDBs are part of a greater ITSM strategy, including ServiceNow and BMC Remedy. These tools may be used as the SoT for inventory or general asset management but are usually not used to model network configuration data due to a lack of data models, lack of skills, and how these teams are often disconnected from the network teams. These tools are often built off auto-discovery engines with a general trend toward showing what is versus the intended state.

- **Device42**: This is usually seen and adopted for **data center infrastructure management** (**DCIM**) with a focus on inventory, data center design, rack layouts, and IPAM with automated discovery. Similar to CMDBs, there is a focus on auto-discovery with a general trend toward showing what is versus the intended state, but usually not used to model actual network configurations such as routing, interfaces, and more and powering network automation solutions.

- **Infoblox and BlueCat**: Arguably the most widely deployed IPAM solutions, their focus is on IPAM. They also have discovery capabilities. They have some SoT branding and marketing, but usually, it's on discovering IPs versus defining the intent of IP schemes and having that drive automation.

These are just a select few tools that exist on the market and are being used by network teams. What we believe, and the premise for creating this book, is that Nautobot has grown immensely over the past 2 years and fills a gap in the market as an enterprise network SoT catered specifically for network automation. Through the remainder of this book, we hope you'll see what Nautobot has to offer and how it can act as the SoT and nucleus to power your data-driven network automation stack on your network automation journey.

Finally, let's dive into Nautobot.

Nautobot overview

Nautobot is an open source network SoT and automation platform that launched in February 2021. Being an open source company-sponsored project, its maintainers are from the official sponsor – Network to Code. Network to Code is a network automation solutions provider that helps clients around the world build and deploy network automation technology.

It's now been over 2 years since the launch of Nautobot and there has been significant growth, traction, and development by the Nautobot core team, as well as the community. There have been nine minor releases since inception with the second major release, 2.0, that just launched in September 2023. Nautobot 2.0 is a major milestone for the project bringing many new features and improved usability to Nautobot.

Nautobot forked NetBox in February 2021. This was due to the industry's need for a network SoT that had an immense focus on network automation with great flexibility and extensibility capabilities. Nautobot was also created to foster an ecosystem around an open source network automation platform. The details of the fork can be found at `https://blog.networktocode.com/post/why-did-network-to-code-fork-netbox/`.

Some statistics, as of March 2024, regarding the project and community are as follows:

- Over 120 releases, including two major releases, nine minor releases, and 100+ patch releases (on a defined biweekly cadence)
- Over 1,600 members in the #nautobot channel in NTC Slack (self-signup at `https://slack.networktocode.com`)
- Over 110 Nautobot blog posts on the NTC blog (blog.networktocode.com)
- Over 60 Nautobot YouTube videos in the *All Things Nautobot* playlist on the Network to Code YouTube channel

We'll highlight several key Nautobot features in this chapter but will spend a lot more time on them throughout this book.

Nautobot use cases

Before we get deep into Nautobot, let's level set on what Nautobot is as a **network SoT** and **network automation platform**. These are the **two** primary use cases for Nautobot.

These are not mutually exclusive and can be used in conjunction with other solutions. We'll review all of that and more, but let's start with the basics.

Network SoT

We already introduced the concept of a SoT and how it is the foundation for data-driven network automation. Adopting a SoT shifts the paradigm to focus on *intended* state data. At its core, Nautobot is a network SoT. What does this mean?

First off, it probably means a migration away from spreadsheets, which is a big win in itself:

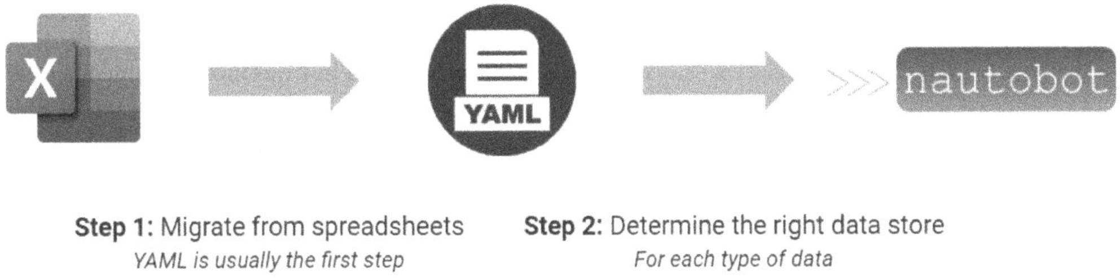

Step 1: Migrate from spreadsheets
YAML is usually the first step

Step 2: Determine the right data store
For each type of data

Figure 1.2 – Evolution of implementing a network SoT

The usual next step is YAML and then deciding which data should be in Nautobot. However, these are not mutually exclusive as Nautobot has native Git integration, which allows users to sync YAML files directly into Nautobot. Much more on that later. The following are the power of Nautobot, where you can effortlessly manage your network inventory, define locations, and organize your infrastructure according to your unique needs:

- Nautobot allows you to store network inventory-defining locations, location types, floor plans, racks, and more alongside *custom* location types. In the real world, network devices are everywhere. They are in campuses, buildings, closets, racks, ceilings, locations on a manufacturing plant floor, cars, and spaceships... the list goes on. The goal of Nautobot is to provide an opinionated way to get started but allow users to define an inventory and organization structure that makes sense to them. The Nautobot data model will be discussed in great detail in *Chapter 2*.

- Nautobot allows you to store and model your devices based on vendors (manufacturers), device models, platforms, and roles. All of these are extensible and customizable for your environment. For example, common roles are leaf and spine for the data center, but if you use different roles or naming conventions, it is as simple as adding them.

- Nautobot allows you to store your IP Addresses and prefixes with support for namespaces that allow for overlapping IP space. This is an area where there may be existing solutions in place, such as Infoblox or BlueCat, as mentioned earlier in this chapter. However, IP addresses are required for assignment to interfaces and policies in Nautobot. With the Nautobot SSoT app, it's possible to synchronize data from third-party systems into Nautobot, giving you flexibility if you need it. Having this data aggregated in Nautobot streamlines your automation initiatives.

- Nautobot allows you to store and model circuit data ranging from circuit providers to individual circuits and then allows you to attach them to specific interfaces on a device. Going one step further, it is possible to use the Nautobot Circuit Maintenance app to dynamically parse and read circuit notification emails from providers and update Nautobot accordingly attaching that notification to a circuit and a device.

- Nautobot embraces extensibility by allowing users to add *any* model to Nautobot to store the data they need and how they need it. For example, there are already open source Nautobot applications for Nautobot that allow you to store security ACLs, BGP routing protocol configuration, and device life cycle information such as End-of-Sale/End-of-Life data in Nautobot. This means that as the Nautobot core project continues to evolve, the community and users around the world can add data models they need to continue to store the intent needed to drive their network.

- Nautobot allows users to define the *relationships* that make sense for them. Nautobot has a defined data model, but *relationships* allow users to associate unrelated object types. For example, you can map a VLAN to a rack; you can map an IP address to a device (remember, IPs are assigned to interfaces); you can map a circuit to an IP address; when using Nautobot apps such as Device Lifecycle Management, you can map contracts to devices, and more. The list goes on.

- With flexibility in mind, Nautobot supports a Data Validation API that allows users to write any logic required to accept and add data to Nautobot. While many users use the Data Validation app, which allows for RegEx and ranges in the UI, the Data Validation API allows you to write any Python logic to ensure *your* standards and governance are enforced – for example, naming conventions, preventing certain data from being deleted, and more. All of your data standards can be codified and enforced so that bad data never finds its way into Nautobot.

This is just a glimpse into how Nautobot is a network SoT. The following visual also shows firsthand how Nautobot can power data-driven network automation:

Figure 1.3 – Codifying network designs through data enables network automation

As a network SoT focused on network automation, Nautobot has many features that showcase how it can seamlessly integrate into **NetDevOps environments**. Let's look at a few of those features as a precursor of what will be covered throughout this book:

- **APIs**: From REST APIs to GraphQL to webhooks, data in Nautobot is very accessible. The REST APIs provide your traditional **Create, Read, Update, and Delete** (**CRUD**) operations. GraphQL provides an extremely efficient and user-friendly way to query the exact data you want. Rather than parse through large data sets from a REST API, GraphQL allows users to query for the exact element or elements needed. We'll cover APIs in much more detail in *Chapter 8*.

- **Native Git integration**: Nautobot supports the ability to use NetDevOps workflows, allowing you to store files in a Git repository; then, in the UI, you can configure Nautobot to clone those specific repositories. You can store YAML data, Nautobot jobs, and export templates in a repository and easily clone into Nautobot all from the UI. This ensures you can run CI on your repositories, perform peer reviews, and then, once merged, *sync* those updates into Nautobot.

- **Job automation**: Nautobot Jobs are arbitrary Python code that can be used to perform any task you would script, including analysis of the data in Nautobot and simplifying data management and population, though they can be used to perform actual network automation tasks. Jobs also simplify creating self-service forms to streamline the adoption of network automation. Jobs also supports Job Hooks, which are similar to webhooks, in that when there is a change to data in Nautobot, a job can be triggered. *Chapter 10* is fully dedicated to jobs, so there's much more to come on this topic.

- **Secrets integration**: To perform network automation, there need to be integrations with secrets, credentials, SSH keys, and API tokens. There needs to be intent on which secrets are needed for a location or device. Nautobot has native secrets integration to map secrets to environment variables or files on the system, while also providing more advanced features with the Nautobot Secrets Providers app, which includes dynamic integration with HashiCorp Vault, AWS Secrets Manager, and many more Enterprise Secrets Management tools. This allows users to rotate and change secrets in secrets management or vault platforms with Nautobot fetching them as automation is performed.

- **Flexible location models and dynamic groups**: Nautobot supports flexible location models and allows you to filter on many different attributes. However, Nautobot also supports *dynamic groups*, which are based on the metadata of a given object. With automation, you likely need to automate based on predefined criteria. For example, you may need to automate all devices that are in a given region, are a given device type, and have a given status. So, the next time a device enters *that* status, it's automatically part of that group, so targeting that *dynamic group* simplifies the automation required. Rather than checking the devices, device types, and statuses, you're simply querying for devices in that logical group.

These are merely five ways Nautobot embraces network automation as a first-class citizen. All of these and many more will continue to be covered throughout this book.

Network automation platform

Nautobot is also a network automation platform, thus going beyond a SoT. Let's take a look at this in more detail to understand what this means.

Nautobot jobs

The first major feature to be aware of for Nautobot being a network automation platform is the support of Nautobot jobs.

Nautobot jobs offer users the ability to create self-service forms in a matter of minutes. Self-service is needed to drive the adoption of network automation; Nautobot jobs are the foundation of Nautobot's platform strategy. Imagine having data stored in Nautobot and you want to verify that it is on the device:

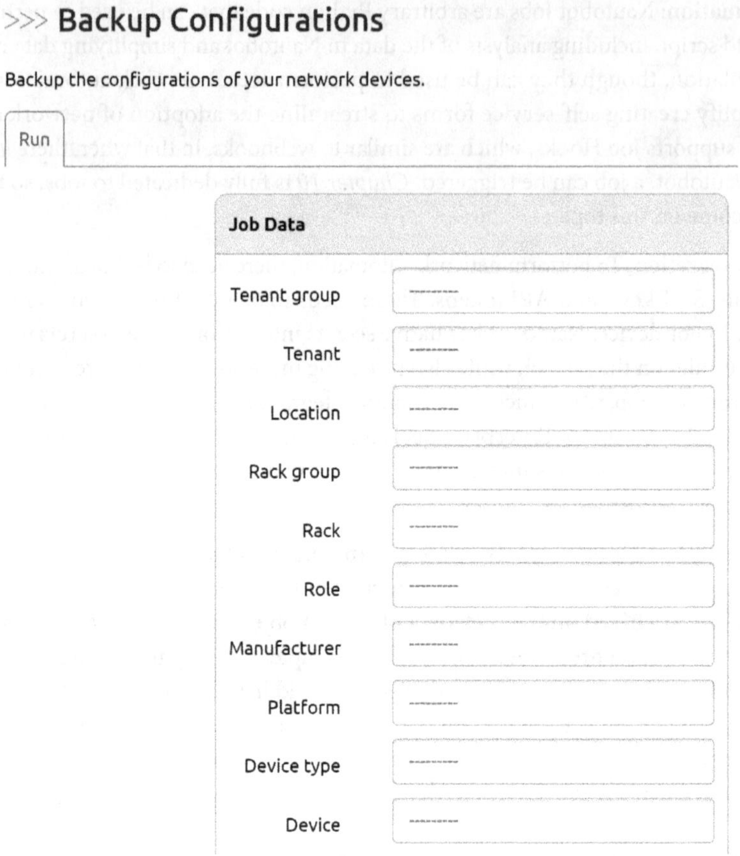

>>> **Backup Configurations**

Backup the configurations of your network devices.

Run

Job Data	
Tenant group	---------
Tenant	---------
Location	---------
Rack group	---------
Rack	---------
Role	---------
Manufacturer	---------
Platform	---------
Device type	---------
Device	---------

Figure 1.4 – Example of a self-service job form

Usually, there is a need to create some code or automation somewhere, often in another tool. Based on size or scale, that may be needed; but for many environments, tying it into Nautobot as a job makes sense because the data is already there. Keeping in mind that jobs are Python code, that *code* can be stored as a job in a Git repository and easily integrated into Nautobot, thus providing self-service to any user that needs to execute it. This is just a basic example, but any automation task that can be built as a script can be deployed as a Nautobot job. There are already Nautobot integrations to Nornir, which is one of the most common Python-based network automation frameworks in the open source community.

Nautobot apps

Beyond Nautobot jobs, Nautobot as a Platform has a powerful developer API that allows users to create Nautobot apps. Nautobot apps enable users to create APIs, create new views and pages, and create any data model required in Nautobot. Nautobot Apps are what encapsulate specific functionality and are the entities that are created for specific use cases. Thus, apps can be as lightweight as only modeling and storing new data – maybe you want to model and store SNMP data, maybe you want to model load balancers, and so on. Apps can be heavier-weight Python applications that perform actual network automation tasks:

> **Note**
>
> Nautobot Apps is the new name for Nautobot Plugins. You may see older commentary online and in the code base that says the word plugin, but that is referring to what is now called Nautobot Apps.

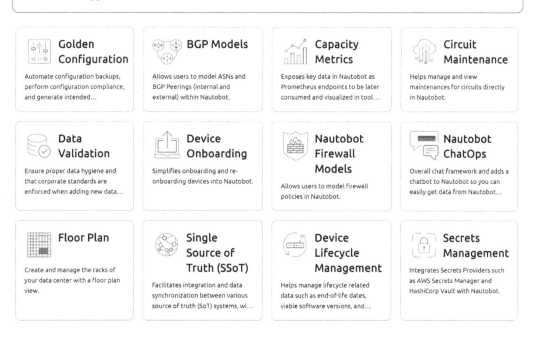

Figure 1.5 – Overview (subset) of Nautobot apps

Nautobot apps leverage the power of Nautobot as a Platform. Using Nautobot as a Platform to construct a network automation application allows users to focus on the actual development without doing the heavy lifting of creating an application from scratch. You get to take advantage of Nautobot APIs, RBAC, logging, GraphQL, relationships, Git as a data source, SSO, and the list goes on. What this means is you can add your own items in the navbar, insert menu items in existing dropdowns, insert new pages, and insert new tables and widgets on detailed object pages. This flexibility allows you to tailor Nautobot to your liking by building Nautobot apps driven by your requirements. Nautobot apps are built at a more accelerated rate than building custom stand-alone applications.

There are already numerous Nautobot apps in the open source community and this number continues to grow. Let's explore some of them.

Nautobot ecosystem

The Nautobot ecosystem is comprised of Nautobot apps that solve a variety of use cases. There are already 15+ open source Nautobot apps written by Network to Code and numerous others written by individuals in the community. Keep in mind that a Nautobot app can be as lightweight or as robust as needed to address the requirements at hand. Here are some examples of different types of applications that could be built using the Nautobot App developer API:

- Lightweight examples:

  - Create a database table, UI views, and API to manage NTP

  - Create a new page (UI view) to aggregate data from devices and VLANs the way you want to see it based on relationships

  - Create Nautobot jobs that are distributed through a Nautobot app

  - Create a command runner that fires off commands to selected devices that are already in Nautobot

- Robust examples:

  - Create an application to store, manage, and deploy firewall policies (inclusive of database tables, views, and APIs).

  - Create an application to discover and crawl the network (inclusive of database tables, views, and APIs).

  - Create an application that performs network configuration backups, generates intended configurations, and performs compliance (which, by the way, exists already in the Golden Config app!). You'll get a deep dive into Golden Config (`https://github.com/nautobot/nautobot-app-golden-config`) with Nautobot in *Appendix 3*.

> **Note**
>
> There is also a Nautobot app template in the form of a cookie-cutter GitHub repository (`https://github.com/nautobot/cookiecutter-nautobot-app`) that helps anyone create a new app.

If you can't see it already, the opportunities are endless with Nautobot apps.

As mentioned previously, the Nautobot ecosystem already consists of many Nautobot Apps. We'll take a look at a summary of a few of them while diving into a few of these in *Chapter 13*:

- **Golden configuration**: Automates configuration backups, performs configuration compliance, and generates intended configurations (`https://github.com/nautobot/nautobot-app-golden-config`).

- **Floor plan**: Allows users to create a floor plan of their data center or other locations of the racks and devices that exist within Nautobot (`https://github.com/nautobot/nautobot-app-floor-plan`).

- **Version control**: Allows users to have change (workflow) management with approvals when managing data within Nautobot powered by a Dolt database. This is in an alpha state, but watch out for the announcement of the official release (`https://github.com/nautobot/nautobot-app-version-control`).

- **Design builder**: Allows users to create data-driven designs (such as small, medium, and large sites) that then allow you to deploy a new device/site/location with that design, automatically generating the desired data for that design based on your data standards.

- **BGP models**: Allows users to model ASNs and BGP peerings (internal and external) within Nautobot (`https://github.com/nautobot/nautobot-app-bgp-models`).

- **Capacity metrics**: Exposes key data in Nautobot as Prometheus endpoints to be later consumed and visualized in tools such as Grafana (`https://github.com/nautobot/nautobot-app-capacity-metrics`).

- **Circuit maintenance**: Helps manage and view circuit maintenance directly in Nautobot (`https://github.com/nautobot/nautobot-app-circuit-maintenance`).

- **Data validation**: Ensures proper data hygiene and that corporate standards are enforced when adding new data to Nautobot (`https://github.com/nautobot/nautobot-app-validation-engine`).

- **Device life cycle management**: Helps manage life cycle-related data such as end-of-life dates, viable software versions, and maintenance contract information (`https://github.com/nautobot/nautobot-app-device-lifecycle-mgmt`).

- **Device onboarding**: Simplifies onboarding and re-onboarding devices into Nautobot (`https://github.com/nautobot/nautobot-app-device-onboarding`).

- **Firewall models**: Allows users to model firewall policies in Nautobot (`https://github.com/nautobot/nautobot-app-firewall-models`).

- **Secrets providers**: Integrates secrets providers, such as AWS Secrets Manager and HashiCorp Vault, with Nautobot (`https://github.com/nautobot/nautobot-app-secrets-providers`).

- **SSoT**: Facilitates integration and data synchronization between various SoT systems, with Nautobot acting as a central clearinghouse for data. Open source integrations exist for ServiceNow, Cisco ACI, Infoblox, IP Fabric, and Arista CloudVision, but integrations can be written for any remote system. Note that these integrations used to exist as their own dedicated GitHub projects, but were recently consolidated into the main SSoT project. SSoT will be covered in greater detail in *Appendix 2* (`https://github.com/nautobot/nautobot-app-ssot`).

- **Nautobot ChatOps**: Provides an overall chat framework and adds a chatbot to Nautobot so that you can easily get data from Nautobot directly from chat, including Slack, Microsoft Teams, Webex Teams, and Mattermost. This also has out-of-the-box chat integrations for Grafana, IP Fabric, Cisco Meraki, Cisco ACI, Ansible AWX, Arista CloudVision, and Palo Alto Panorama. Note that these integrations used to exist as their own dedicated GitHub projects but were recently consolidated into the main ChatOps project (`https://github.com/nautobot/nautobot-app-chatops`).

Summary

This chapter provided a general overview of data-driven network automation with Nautobot. It started by reviewing key use cases for network automation before highlighting the important relationship between data and network automation. It should be evident that getting an understanding of the data that drives network automation should not be understated and that having good, clean data will simplify the overall network automation journey. Finally, this chapter provided an overview of Nautobot and its two key use cases – SoT and network automation platform, and how both are further enhanced through its developer API and the Nautobot ecosystem that continues to grow with open source apps such as Firewall Models and Golden Config.

In the next chapter, we'll explore and start to understand the data models at the core of Nautobot.

2
Nautobot Data Models

At the core of Nautobot are two main use cases that were covered in *Chapter 1*: a network automation platform and a network **Source of Truth (SoT)**. While it is true that certain classes of automation can be built void of specific structured data, it is generally true that the SoT powers much of the automation that provides the most business value.

The mantra that holds in Nautobot is that the automation that you rely on is only as good as the worst data you feed into that machine. It is this fundamental truth that drives the SoT aspect of Nautobot. In this chapter, we will focus on the critical data models and relationships of the data in Nautobot that are used to power your network automation stacks.

The following are the main topics that will be covered in this chapter:

- Nautobot data models overview
- Network device inventory data models
- IPAM data models
- Circuits data models
- Data model extensibility
- Custom data models

While this book is focused primarily on enabling you to make effective use of Nautobot in your network automation journey through deeply technical hands-on topics, this chapter will first lay the foundation for the network data model that Nautobot provides. This understanding is key to how you will later use the data model to build automation capabilities through its consumption, and even by extending it to meet your specific needs

Nautobot data models overview

Before we begin, let's touch base on a few terms and concepts. First, *data modeling* refers to defining business requirements for the expression and relationships of data in Nautobot. Thus, the output of our efforts is the *network data model*, which we will dive into now. Such a comprehensive data model is naturally broken down into several high-level data domains, such as the inventory, circuits, and IP addresses. While we logically compartmentalize our network data model to make it easier to manage, it still comprises many cross-model relationships that span the boundaries of the domains. For example, a router has interfaces and those interfaces have Layer-3 IP addresses. These relationships are the key aspect of a comprehensive network SoT that positions it as a better way to manage data than numerous siloed spreadsheets or even disconnected systems.

Data model summary

We will begin our journey through the Nautobot data model with a high-level review of the landscape. Here you will note some of the data domains we spoke about earlier, and hopefully will appreciate the need to logically break the data model up in this way.

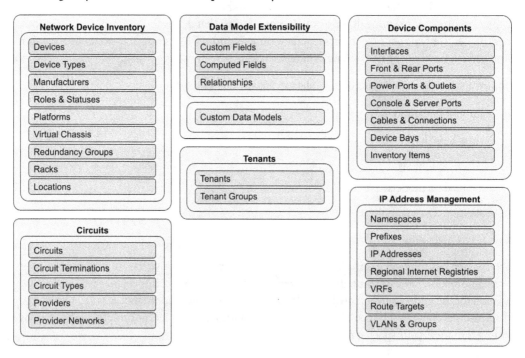

Figure 2.1 – Nautobot data model overview

As you can see, Nautobot affords the ability to track a tremendous amount of specific networking data out of the box. In this chapter, we will dig into each of these domains in more detail. Still, though, you are likely wondering about other parts of the networking world not covered in the preceding list. This is natural, and rest assured that Nautobot has you covered through its extensibility features (covered in *Chapter 6*) and the ability to build custom Nautobot data models (covered in *Chapter 15*), which allow for additional models to be provided through a variety of means. You'll see more on that at the end of the chapter, and even more in *Part 4*, where you'll get to see firsthand how to extend Nautobot.

For now, we will take a closer look at each of these models, their common and important attributes, and what you can do with them.

Network device inventory data models

The foundation of the *network data model* is rooted in the primary entities that a network organization cares about, namely devices. You will find the Device model in Nautobot to be the proverbial heart of the operation, with many ancillary models such as Interfaces associated with it, and tie-ins to other important areas of the data model. You will soon see that there are many aspects of metadata surrounding the Device model that go into constructing a logical and robust view of the world.

Devices

Tracking network devices is arguably the most popular feature of Nautobot and serves as the basis for many other data modeling and automation activities. In Nautobot, a device can represent many different types of network assets, from the obvious rack-mounted router or switch, to firewalls, and even servers. Going further, it is perfectly valid to track virtual networking appliances using the Device model. This also means that a device need not be constrained to a physical rack at all (though tracking racks affords other possibilities that we will discuss later).

Devices in Nautobot track common attributes such as the hostname, serial number, asset tag, and primary IP address (with options for both IPv4 and IPv6). Here is an example of a **Devices** table from Nautobot:

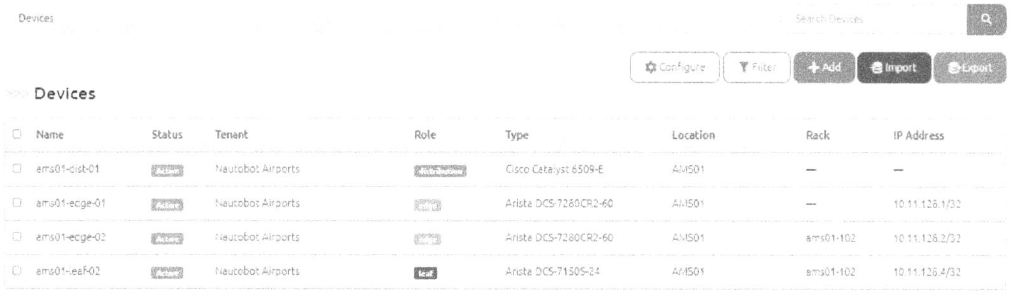

Figure 2.2 – Nautobot Devices table (list) view

Beyond the obvious, we also associate a device with a physical location (more on that in a moment). We can also attribute tenant ownership, which is its own area of the data model. As network engineers, we usually want to know the purpose of a device, and for that, the device model has a linkage to the role model, which is user-definable based on your network. Likewise, the **Status** field allows your organization to define lifecycle values based on how you operate the network. The Platform relationship is commonly employed to designate the software family the device is running, such as Cisco IOS or Juniper JunOS. As in the real world, devices can be located inside a rack, where we track which rack, which position in the rack, and on which side or face the device is installed. Finally, a device has a linkage to a Device Type, which represents the hardware model. In terms of the data model, we could say that devices are instances of a Device Type, and this works in much the same way as if we took a piece of hardware off a shelf or pallet and deployed it on a rack.

You can start to get the sense that the relationships in the data model are what makes Nautobot interesting as a network SoT. Still, though, at this point, we have described only the ability to create Device records with some specific attributes, and while that is certainly important in asset inventory and even automation contexts, we could do that with a spreadsheet! So let's explore some of the more intricate data models and their features in Nautobot.

Device components

With devices being an anchor point in the data model, device components primarily make up relationships to other parts of the model, for example, tracking interfaces and other port types. There are other general component types such as Inventory Items and additional use cases with Device Bays, such as chassis child devices.

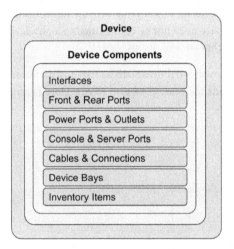

Figure 2.3 – Visual of a device model in Nautobot

We'll now take a look at these other components that can be mapped back to a given device.

Interfaces

Probably the most commonly used device component in Nautobot is the Interface model. Network interfaces play several vital roles in the real world of networking. They provide physical connectivity between devices, logical addressing, and Layer 2 management. So too, Nautobot supports all of these use cases and more. Nautobot can model both the physical and logical interfaces with support for most common form factors and configurations, including LAGs, bridges, and parent/child virtual relationships. Here is a snapshot into an **Interfaces** table inside Nautobot:

Figure 2.4 – Glimpse into the interfaces of a device

We'll dive much deeper into what's possible in later chapters in the book, but take note of the icons on the right-hand side of each row. You can perform a trace when there are cables connecting two interfaces in Nautobot, which proves to be a valuable function.

Layer-3 addressing is covered with support for primary and multiple secondary IP addresses per interface. You can also specify the 802.1Q mode with lists of tagged and untagged VLANs. You can enable or disable an interface, and make use of customizable statuses that allow for use cases such as tracking the provisioning state of an interface in some business process—it's up to you. Perhaps the most interesting usage of interfaces in Nautobot, though, is the ability to connect them to other interfaces or components, thereby creating a cable, which means you can track your entire physical cable plant if you wish, or simply indicate that two interfaces are connected "in some way." You can also flag such connections as reserved, allowing you to plan capacity (you might also do this by having a "reserved" status on the interface, depending on your use case).

Front and rear ports

The ability to model a cable plant is achieved through the usage of front and rear ports. In practice, these are combined to create patch panels and fiber cassettes in which the front port accepts the connection to an interface or other front port. A rear port is mapped to a front port and allows for multiplexing to model the bundles of cables or shrouded fiber runs that go between termination panels.

Power ports and outlets

Like the physical cable plant, Nautobot can also be used to track the power plant. This starts with power ports and outlets that model PSUs and the corresponding PDUs they plug into within a rack. Here is a view of **Power Ports** within Nautobot:

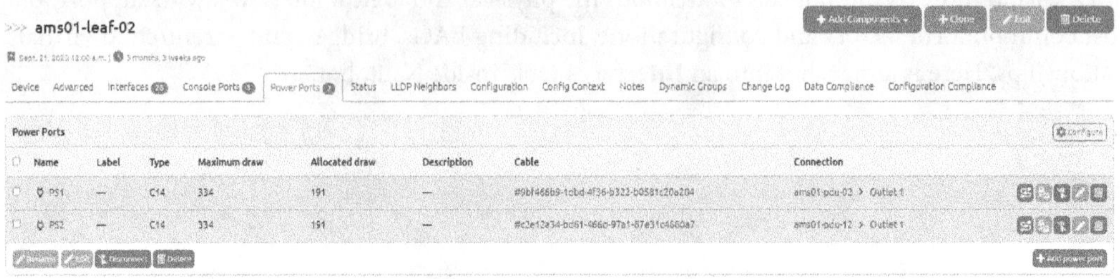

Figure 2.5 – Power Ports tab on a detailed device view

In this model, Nautobot tracks the power draw so you can budget the PDUs and connect them to power feeds and power panels to track the power type (phase, etc.) and distribution. The budget and utilization calculations are visible in a few different areas, such as viewing racks, allowing for effective capacity planning. It's worth noting that as with any of the device component feature sets, you are not required to use them if you simply want to track an inventory of devices.

Console and console server ports

Console and server ports follow the same basic principle as their power-related counterparts but allow you to model the actual console port(s) on a regular network device, but also the console server devices themselves, and the relationship between the two via a connection. You will note the ability to designate the port type, such as DB-25, RG-45, USB-A, and so on, as can be seen in the following screenshot showing the adding of a console port to an instance of a device:

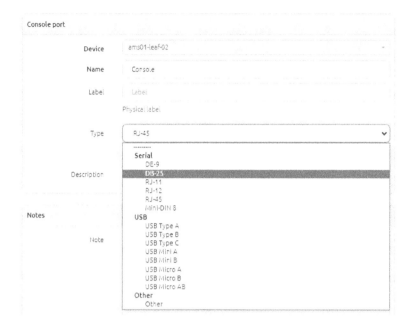

Figure 2.6 – Console port Type options

It's in many of these little details, which are usually never tracked anywhere, that Nautobot tends to shine.

Cables and connections

Having now covered many of the various port types, it is important to understand how they can be connected together. Every connection between device components is represented as a cable, embodying a direct physical link between two termination points. These points could range from console ports to patch panels, or between network interfaces. Each cable is defined by two endpoints, often referred to as A and B, but it's important to note that cables in Nautobot are inherently direction-agnostic, meaning the order of terminations doesn't impact their function. Cables can connect to a variety of objects, including instances of Circuit terminations, Console Ports, Interfaces, Pass-through Ports, Power Feeds, Outlets, and Ports. For each cable, details such as the type, label, length, and color can be assigned. Additionally, an operational *Status* is required for every cable, with default statuses including **Active**, **Planned**, and **Decommissioning**. This comprehensive approach allows for detailed tracking and management of the physical connections in a network.

While the power to track an entire physical cable plant is present, sometimes it is not warranted or necessary. It is perfectly acceptable and possible to treat cables as abstracted connections between Device Interfaces, ignoring the physical aspect, but retaining the context of the connected interfaces.

Nautobot also provides a tracing feature for cables. Users can trace a cable from either of its endpoints, either through the UI or using a REST API endpoint. Here is an example of a simple cable trace:

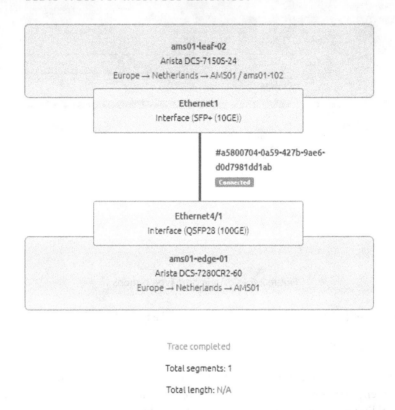

Figure 2.7 - Visual of a cable trace in Nautobot

This function follows the path of the connected cable from one termination point to another. If a cable connects to a pass-through port and there's another cable connected to the peer port, Nautobot continues tracing the path until it reaches a non-pass-through or unconnected termination point. This tracing capability is crucial for mapping out the physical path of connectivity across a network, aiding in troubleshooting and network documentation. An interesting aspect of cable tracing is its ability to trace through circuits. For instance, if a cable path includes a circuit, the tracing will show the connection from a device interface to the circuit's termination points, providing a clear view of how different network elements are interconnected through physical cabling. This feature enhances the understanding of network topology and the role of each physical connection within it.

The following diagram shows an example of modeling a cable plant that includes patch panels.

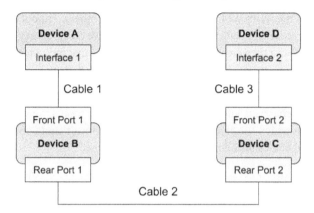

Figure 2.8 – Visualizing device connectivity with a patch panel (Device B and C) in use

Device A is connected to Device D through two patch panels, B and C. The rear ports of B and C represent the riser cable between the two panels.

This next example shows a cable path trace across a circuit, connecting two Device Interfaces.

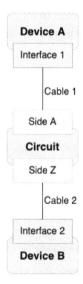

Figure 2.9 – Cable path trace across a circuit that connects two device interfaces

Device bays

Device bays diverge from the norm we have just discussed with other components. They are used as the basis to model hardware chassis-based devices. In this way, we create an instance of a device that represents the chassis itself and then create bays within that chassis that individually accept other child devices. It is very important to consider the specific set of use cases that are intended with this model.

Device bays adopt the ability to create parent/child relationships between child devices and the chassis that houses them, but the chassis is intended to be "dumb" in this model. That is, the intent is more to model blade servers in which the chassis provides housing, power, and connectivity, but has no other relation to the child blades installed in the bays, or vice versa.

This means in the networking world, device bays would not typically be the best way to model a chassis-based switch or router that contains several line cards. The litmus test for this distinction is to ask whether the chassis device has a single management IP from which you configure and control the entire device across all line cards. If you do have a case where you are managing line cards independently of one another (thereby logically managing each line card as its own device), device bays might be an acceptable means of modeling such devices. But we typically find that to be rare. Instead, it would be more appropriate to create the chassis network device as normal, but create all of the interfaces across the line cards as discrete interfaces on the chassis (named accordingly) and track the line cards as Inventory Items on the device. There are ways to bulk-rename interfaces if you need to move line cards around. Chassis devices are also not meant to model distinct network devices with a shared control plane, like a Cisco StackWise switch. The virtual chassis model is suited for that purpose and will be discussed later in this chapter.

Figure 2.10 - Usage of device bays to model a chassis-based device and its relationship to child devices, such as blade compute servers

Inventory items

Speaking of inventory items models, they are a way to associate *any* other type of component to an individual instance of a device for tracking. Normally, we would think of things such as hard drives, CPUs, PCI cards, and so on—basically, anything ancillary to the device itself that you want to track for asset inventory purposes. As you might imagine, we have the option to attribute a manufacturer, part ID, serial number, and asset tag to Inventory Items. Inventory items can also have their own parent/child relationships, which helps in tracking things such as optical transceivers in line cards. In the near future, Nautobot will allow more direct modeling on device modules, such as chassis line cards and their direct relationships to device interfaces. In doing so, the Inventory Item data model will evolve to allow hierarchical relationships and more meaningful tracking of ancillary device components.

Using device components in Nautobot

In order to add or manage device components in the Nautobot UI, you add interfaces, device bays, and any other component within or under a device as shown here:

Figure 2.11 – Add device components for a single device

However, you may be thinking that you want to apply those components across all similar devices and device types. That is possible by also adding components at the device-type level as shown here:

Figure 2.12 – Add device components for a device type

Now we'll dive deeper into device types.

Device types

Device types are closely related to devices, in that they represent the hardware model, or type, of the device. But in Nautobot this manifests in modeling several aspects of the hardware model. You have self-explanatory attributes such as the manufacturer and model number, but the real power of device types comes from the tracking of component templates. Basically, we take all of the port-based component models, plus the device bays previously discussed, and create simplified versions of them called **component templates**. You can see them at the bottom of device-type details pages as shown in the following screenshot.

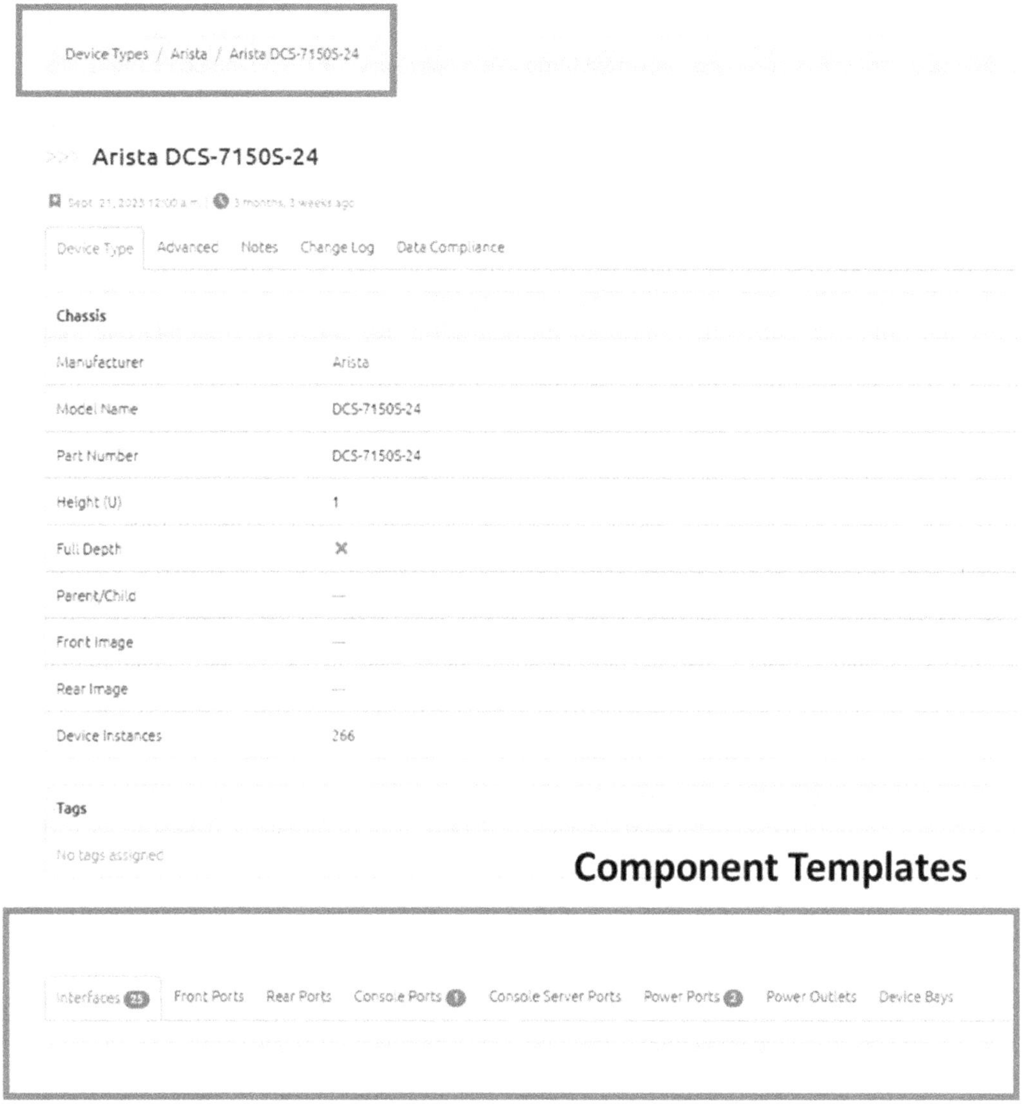

Device Types / Arista / Arista DCS-7150S-24

Arista DCS-7150S-24

Sept. 21, 2023 12:00 a.m.   3 months, 3 weeks ago

Device Type | Advanced | Notes | Change Log | Data Compliance

Chassis

Manufacturer	Arista
Model Name	DCS-7150S-24
Part Number	DCS-7150S-24
Height (U)	1
Full Depth	✕
Parent/Child	—
Front Image	—
Rear Image	—
Device Instances	266

Tags

No tags assigned

Component Templates

Interfaces 24 | Front Ports | Rear Ports | Console Ports 1 | Console Server Ports | Power Ports 2 | Power Outlets | Device Bays

Figure 2.13 - Components templates for a device type

Don't worry, you'll have plenty of time to navigate the UI in the next few chapters.

The idea is that we create a representation of a device as that particular hardware is shipped to you from the manufacturer. It is important to note that a device type is specifically void of any deployment type logic or anything that would be used to distinguish two devices of the same type. In this way, they are true templates of the devices that we instantiate, or rather deploy, in our network. For instance, if we use Juniper EX3400-48P switches in our network, we would create a device type for the model number, and then add 48 interfaces to that template. When we have several device types, which is certainly common in most networks, we end up with a library of types to choose from when creating devices; and yes, they can be imported from shareable definitions. Because we have defined templates, when you do create an actual instance of a device, you are asked which device type it uses. This causes all of the template components to be copied into the new device that gets created, effectively jump-starting the definition of that device in Nautobot.

Manufacturer

The manufacturer of a device is tracked as an attribute of the device type. The model itself is very simple, namely tracking just the name of the entity, such as Cisco or Palo Alto Networks. But having a separate entity in the data model allows for more complex use cases, as we will later discuss with elements such as custom fields and relationships.

You might ask why the *manufacturer* is only an attribute of a device type and not also, or exclusively, of the device. The answer lies in understanding data normalization, which is an advanced topic of data modeling and ultimately out of the scope of our discussion. But we point it out to say that great care has gone into the design of the core data model in Nautobot. In this case, the normalization is explained by pointing out that a device is always associated with a device type. And since a device type carries the understanding of the manufacturer for that type of hardware model, we can then infer the manufacturer for a given device without having to track it directly on the device model.

Roles and statuses

Now that we have described how to use the device model to create an inventory of devices, what can we do with it? One of the most basic questions we often need to answer in our network inventory is, *"What does this device do on the network?"* The role model is the primary mechanism to express that in the Nautobot data model. The role model itself is simple, but again, you will see the power of model extension later. Out of the box you get to specify the name of the role, a description, and a color to visually distinguish roles in the web UI. This means that the definition of roles is entirely up to you as a user. Your organization might create roles such as `switch`, `router`, or `firewall`, or more complex roles such as `dmz-edge-peer` – the point is that it is up to you. Once roles have been defined, you assign them to devices and begin consuming that added context in your SoT and automation endeavors.

Likewise, statuses reflect the administrative state of devices or interfaces. Just as with roles, you can create organizational-significant status values, or can use the ones provided out of the box, such as **Active**, **Offline**, or **Staged**. Both the role and status models in Nautobot are actually generic in nature and are used across several other use cases such as IPAM and circuit tracking. This gives platform administrators a central place to manage this type of metadata, and reduce duplication where the same values might be relevant across multiple uses. Similar to roles, statuses are covered in more detail in *Chapter 6*.

Platform

The platform model is an example in the core data model of something abstract that can be implemented to suit your needs, or simply ignored. This model is commonly used to specify the software family or even the specific version that a device is running. This is of course done by defining a platform instance and then optionally relating it to a device. The platform model also has special attributes related to Nautobot's built-in NAPALM integration feature set and these are used to tell NAPALM what driver and configuration options to use for a given device connection. More on that later.

Virtual chassis

Virtual chassis is a device deployment–specific model that allows you to track switch stacks or instances of multiple devices that share a common control plane, such as the Cisco StackWise or Juniper virtual chassis products. In Nautobot, you create a virtual chassis by grouping two or more member devices, specifying a master device from that group, and assigning membership priority values. Then, when dealing with components of the virtual chassis, we have an aggregate of all components (like interfaces) across all members. Virtual chassis are covered in more detail in *Chapter 5*.

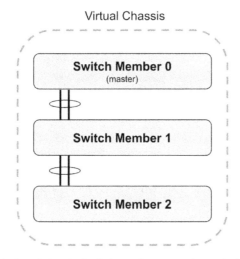

Figure 2.14 – A virtual chassis depicting a three-member switch stack (such as Cisco StackWise or Juniper VC, and their backplane connectivity)

Device redundancy groups

Related in nature but serving a different set of networking use cases is the device redundancy group model. The intent here is to model **High Availability** (**HA**) topologies involving separate control planes. Examples of this are two routers participating in an ECMP topology or perhaps a set of firewalls with some form of proprietary failover mechanism. We are tracking clusters of devices that have some form of HA, so an individual device may only be a member of one group at a time, and carries a membership priority value for that group. The group itself designates an HA strategy, either active/active or active/passive.

The following diagram shows one example use of device redundancy groups wherein pairs of routers are participating in a redundant ECMP topology.

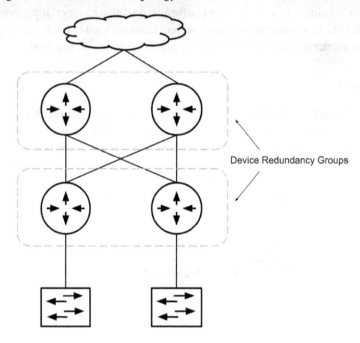

Figure 2.15 – Redundant ECMP topology using device redundancy groups

Note that device redundancy groups are covered in much more detail in *Chapter 5*.

Interface redundancy groups

Interface redundancy groups are similar in concept, but are designed to group interfaces that share a single virtual address, typically used in redundancy protocols such as HSRP or VRRP to provide fault-tolerant default gateways in networks. These groups must be created prior to assigning interfaces to them, ensuring proper setup and configuration. While their primary design is for first-hop redundancy protocols, they are versatile enough to represent any grouping of redundant interfaces, regardless of addressing or not. Adding interfaces to these groups requires setting a priority value for each interface, dependent on the redundancy protocol used, such as a range of 1 to 255 for HSRP. Optional features include associating an IP address with the group to act as the virtual address, specifying the redundancy protocol (such as HSRP, VRRP, GLBP, or CARP), using secrets groups to store sensitive information such as authentication keys, and including a protocol group ID, which can be either an integer or a text label up to 50 characters long. Like device redundancy groups, the flexibility in the model means you can use the feature in a number of ways. For example, it is not uncommon to use interface redundancy groups to express circuit redundancy at a site by terminating circuits to interfaces participating in a group.

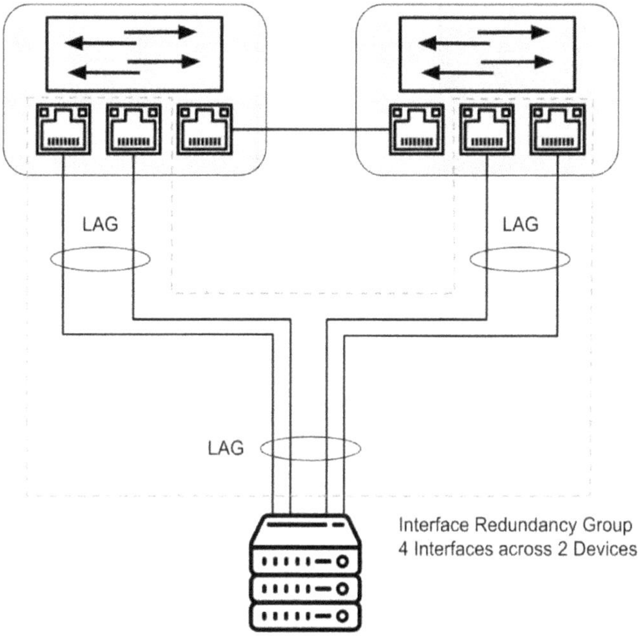

Figure 2.16 – Two-switch MC-LAG topology

The diagram shows a two-switch MC-LAG topology, where an interface redundancy group is used to track the aggregate of interface members across devices and is where config such as virtual Layer-3 addresses live.

Racks

Racks are another cornerstone of tracking a physical device plant. In Nautobot, a rack specifies its position within a location and its dimensions, in terms of width and total number of units that can be populated. It is not required to make use of racks at all in Nautobot, as evidenced by the data model relationship between devices and racks being options. Here is an example of viewing an example of a rack in Nautobot:

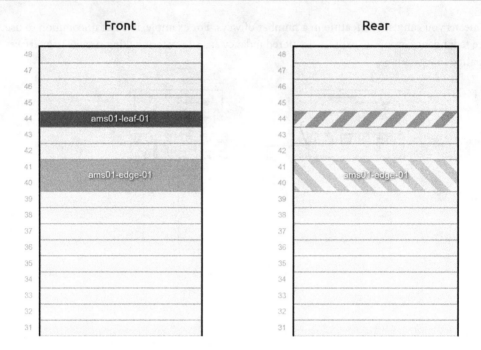

Figure 2.17 – Front and rear rack views shown on a detailed rack view

If you do wish to "rack" a device, you specify the relevant attributes on the Device instance. Those are the rack to assign the device to, the base (bottommost) used to install the device inside the rack, and the face (front or back) of the rack. The height of the device (measured in whole rack units) is specified in the device type. You have the option of reserving space in a rack, which will block devices from occupying such space at the data model level.

Going beyond just viewing racks in Nautobot, you can also populate a floor plan with racks to view an end-to-end layout of a given room, as can be seen in the following screenshot:

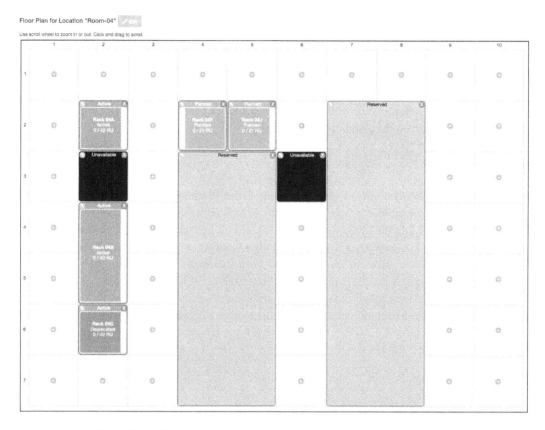

Figure 2.18 – Data center floor plan view with the Nautobot floor plan app

In order to place and view racks on a floor plan, you need to use the Nautobot floor plan app.

Rack groups

Rack groups allow you to build a nestable hierarchy of rack groups that in practice are used to represent entities such as network closets, floors in a building, rows and cages in a data center, and so on. Their use is optional, but a rack may be assigned to only one group.

Locations

Locations are an example of an organizational model that is ancillary to the network data model but still very useful in aggregating assets and other aspects of our network data. Broadly speaking, locations are any entity in which you might place network assets or associate some context, such as a prefix or VLAN. The location model is very flexible, allowing you to define a hierarchy that suits your needs. An example might be countries divided by business regions, or you might create states in the USA. The Location model is also great for modeling more fine-grained physical entities such as the buildings on a campus, floors in those buildings, or even network closets on those floors. Locations are not limited to just the physical world, though; you can also use them to designate logical areas in your network. Locations have a name, a linkage to their parent (within the user-defined hierarchy), and a set of metadata fields, such as contact name, email, phone number, address, and so on. A few models in Nautobot require assignment to a location, including devices and racks. Several others allow the optional association to a location, including VLANs, and in some cases (like VLANs) certain feature sets are augmented based on the assignment of a location.

Location type

The location type model is the underpinning of the user-definable hierarchy. You define a set of types and how they relate to one another in terms of the parent/child tiering. So, from our earlier example, a hierarchy of **Country** | **City** | **Building** would be represented as three location types with the relevant parent relationships set. Location types also designate what other types of objects may relate to a given location. For example, if you wanted to ensure that all devices are assigned to a building, and not simply a country, the location type is the place to define that constraint. The following screenshot shows an example that nests three location types for a higher-education use case in a university. It includes a **Campus** | **Building** | **IDF** hierarchy, and here it is shown in the Nautobot UI:

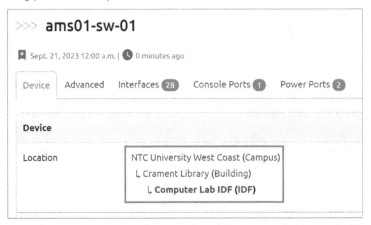

Figure 2.19 – Viewing user-defined location types in use for a single device

These location types are completely configurable to make sure you and your user base understand the data in your vocabulary.

Tenants

In Nautobot, tenants represent distinct groupings of resources, commonly used for administrative segmentation within an organization. They are typically utilized to symbolize individual customers or internal departments. A wide range of objects within Nautobot can be assigned to tenants, including locations, racks, rack reservations, devices, VRFs, prefixes, IP addresses, VLANs, circuits, clusters, and virtual machines. This assignment is crucial for indicating the ownership or association of a specific object with a particular tenant, thereby organizing the network resources efficiently.

For instance, if a rack is exclusively serving a specific customer, it would be assigned to the tenant instance representing that customer. In a service provider scenario, a container network may be split into many child subnets, each of which is assigned to a particular tenant that relates to a customer environment.

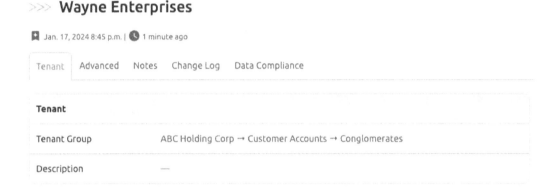

Figure 2.20 – Viewing a tenant definition inside of a user-defined tenant group hierarchy

Additionally, tenants can be grouped together for better organization. Custom groups such as "Customers" and "Departments" can be created, with the assignment of tenants to these groups being optional. This grouping allows for more structured and understandable organization within Nautobot, especially in scenarios where a clear distinction between different operational or customer segments is necessary.

Furthermore, Nautobot supports recursive nesting within tenant groups, enabling the creation of a multi-level hierarchy. For example, under a broader "Customers" group, there could be subgroups for individual tenants categorized by specific products or account teams. This hierarchical approach to organizing tenants provides a flexible and scalable way to manage and represent various entities and their relationships within an organization's network infrastructure.

IPAM data models

The next major area of the network data model is for tracking **IP Address Management** (**IPAM**) data. This includes IP addresses and subnets (called *prefixes* in Nautobot), but also other related models such as VRFs, route targets, and VLANs.

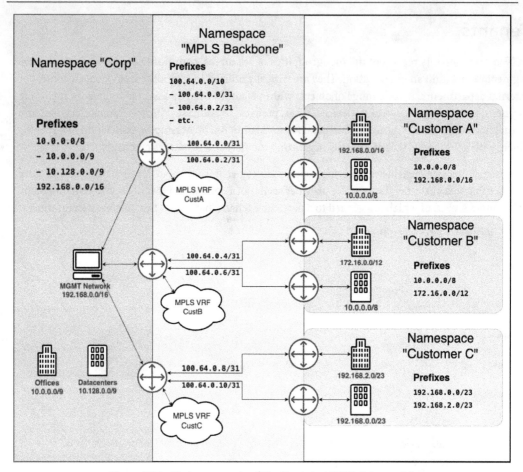

Figure 2.21 – Various aspects of the Nautobot IPAM data model

The preceding diagram describes the various aspects of the Nautobot IPAM data model. Take note of the manner in which namespaces, devices, prefixes, and VRFs all play a part. In particular, this diagram shows the uses of overlapping address space and duplicate network deployment scenarios.

Namespaces

In the IPAM domain in Nautobot, namespaces are integral for grouping and managing VRFs, prefixes, and IP addresses. They act as boundaries or constraints for ensuring uniqueness and avoiding duplication within IPAM data, where use of overlapping address space occurs. The single, default namespace might suffice for simpler networks without overlapping prefixes or duplicate IP addresses, even with thousands of IPAM records. However, in more complex scenarios, such as those of managed service providers or large enterprises, multiple namespaces become essential to accurately model the network and differentiate between records that might appear duplicate.

Each namespace in Nautobot is identified by a name and a description, and optionally, it can be linked to a specific location for informational purposes. Within any given namespace, there must be only one record for each distinct VRF, prefix, or IP address. While a single record, such as a VRF or a virtual IP address, may be utilized in various parts of the network—such as being configured on multiple devices or assigned to multiple interfaces—it is treated as a singular network entity in these instances, aligning with Nautobot's data-modeling approach.

Nautobot's implementation of namespaces particularly addresses scenarios where what appears to be the same VRF, prefix, or IP address might actually represent distinct entities within a network. This is especially relevant in situations such as corporate mergers, where overlapping network spaces from different entities might need to coexist in parallel namespaces rather than being merged into a single namespace. Another example could be an MSP that deploys network segments in an identical fashion for each of its customers. This functionality ensures accurate and distinct representations of network components in complex or evolving network environments.

Prefixes

A prefix represents an IPv4 or IPv6 network defined by a network address and mask in CIDR notation, such as `192.0.2.0/24`. It includes only the network portion of an IP address, meaning all bits outside the mask must be zero, except for `/32` IPv4 and `/128` IPv6 prefixes, which can represent specific IP addresses. Each prefix is unique within a specified namespace and can optionally be linked to a location or associated with one or more VRF instances. Those not assigned to a VRF are deemed part of the global VRF within their namespace.

Prefixes have assigned statuses to indicate their operational state. Default statuses include **Active** for in-use prefixes, **Reserved** for future use, and **Deprecated** for those no longer in use. Additionally, prefixes can have an optional role, which is customizable and represents their function, such as distinguishing between production and development environments. A prefix may also be linked to a VLAN, aiding in correlating address spaces with Layer-2 domains. VLANs can have multiple prefixes associated with them. Prefixes can be assigned to an **Regional Internet Registry** (**RIR**) for tracking authorization to use certain public IP spaces. The `date_allocated` field is available to mark the allocation date of a prefix, whether by an RIR or for internal assignment.

The prefix model includes a `type` field with three options: *Container*, *Network* (default), and *Pool*. A *Pool* type indicates that every IP within the range is assignable, while *Network* type assumes the first and last IP addresses in an IPv4 prefix are unusable. Nautobot organizes prefixes and IP addresses into a hierarchy using the `parent` field. A *Container* type prefix should only have a *Container* parent; a *Network* type should have a *Container* parent; and a *Pool* type should have a *Network* parent. Any prefix can be a root prefix, meaning it has no parent. This hierarchical structure aids in managing and understanding the relationships between different network segments.

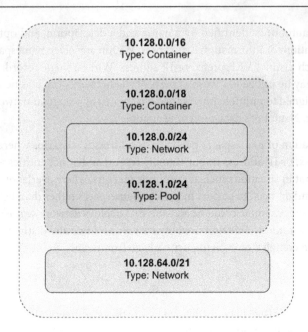

Figure 2.22 – IPAM prefix hierarchy showing how /16 could be carved
up using Container, Network, and Pool prefix types

IP addresses

An IP address is its own model and is defined as a single host address, either IPv4 or IPv6, along with its subnet mask, mirroring its real-world configuration on an interface. These IP addresses are automatically organized under parent prefixes in accordance with the IP hierarchy. They do not have direct assignments to namespaces or VRFs; instead, they inherit these attributes from their parent prefix.

Each IP address in Nautobot can be assigned an operational status and a functional role. The default statuses are the same as those provided for prefixes. Functional roles, which are conceptual and not customizable, include options such as 'Loopback', 'Secondary', 'Anycast', 'VIP', and various roles for redundancy protocols such as VRRP, HSRP, and GLBP. These roles help to indicate special attributes or uses of an IP address. IP addresses can also be classified by types indicating specific functions, such as 'Host', 'DHCP', or 'SLAAC'. The default type is "host". This classification aids in identifying the specific function or behavior of an IP address within the network.

An IP address in Nautobot can be linked to interfaces of devices or virtual machines, and an interface may have multiple IP addresses assigned to it. Furthermore, each device or virtual machine can designate one of its interface IPs as its primary IP for each address family (IPv4 and IPv6). Nautobot also supports the designation of IP addresses as inside addresses for **Network Address Translation (NAT)**. This feature is useful for denoting translations between public and private IP spaces, and the relationship is maintained bidirectionally.

With regard to data modeling, the IP address model has a direct relationship to its parent prefix, and this relationship aids in many areas, including the performance of working with deeply nested address space.

RIRs

RIRs are recognized as authorities responsible for allocating globally routable IP address space. The primary RIRs are ARIN (North America), RIPE (Europe, the Middle East, and parts of Central Asia), APNIC (Asia-Pacific region), LACNIC (Latin America and the Caribbean), and AFRINIC (Africa). In addition to these, Nautobot also treats certain RFCs, such as RFCs 1918 and 6598 that define address spaces for internal use, as equivalent to RIRs. This categorization acknowledges these RFCs as authorities owning specific address ranges. Nautobot provides flexibility in managing RIRs. Users can create custom RIRs and assign prefixes to them as needed. The RIR model in Nautobot includes a Boolean flag to indicate whether the RIR is designated for private IP space allocation only.

For practical application, consider an organization that has been allocated a specific IP range, such as `7.128.0.0/16`, by ARIN. This organization also uses internal addressing as defined by RFC 1918. In Nautobot, the organization would create two RIRs, one named "ARIN" for the publicly routable space and another named "RFC 1918" for internal addressing. Subsequently, prefixes corresponding to these address spaces would be created and assigned to their respective RIRs. This approach allows for organized management of IP spaces, both public and private, ensuring clear documentation and tracking of address allocations within the network.

VRFs

In Nautobot, a **Virtual Routing and Forwarding** (**VRF**) object is used to represent a VRF domain, which functions as an isolated routing table. VRFs are instrumental in segmenting networks, commonly used for isolating different customers or organizations within a network, or for managing overlapping IP address spaces, such as multiple instances of the `10.0.0.0/8` space.

Each VRF is given a name and a route distinguisher, which must be unique within the namespace to which the VRF is assigned. VRFs in Nautobot can be associated with specific tenants, aiding in organizing and managing the IP space according to customer or internal user groups. This association is particularly useful for service providers or large organizations managing multiple customer networks. Additionally, VRFs can have import and export route targets. These are used in Layer-3 VPNs (L3VPNs) to control the exchange of routes (or prefixes) among different VRFs, facilitating the selective sharing of routing information across different segments of the network.

In terms of IP address management, prefixes and their contained IP addresses can be assigned to one or more VRFs within their namespace. This flexibility allows for the alignment of IP address usage with the specific requirements of different parts of the network. Any prefix or IP address not assigned to a specific VRF is considered part of the implied "global" VRF within its namespace. It is important to note that this "global" VRF is distinct from any other "global" namespace, which might contain several different VRFs.

Route targets

A route target is used as an extended BGP community for controlling route redistribution among VRF tables, especially in L3VPNs. Each route target is assigned a unique name following the format prescribed by RFC 4364, similar to VRF route distinguishers. In Nautobot, route targets are linked to individual VRFs as either import or export targets to accurately model route exchange in an L3VPN. Additionally, route targets can be optionally associated with a tenant and *tagged*, providing further organizational capabilities within the network.

> **Note**
> For more advanced BGP modeling capabilities, see the Nautobot BGP app.

VLANs and VLAN groups

VLANs are used to represent isolated Layer-2 domains, as defined by IEEE 802.1Q. Each VLAN is identified by a unique name and a numeric ID that ranges from 1 to 4094. VLANs in Nautobot can be assigned to a location, a tenant, or a VLAN group. Each VLAN is required to have an assigned status and, like prefixes, can also be assigned a functional role.

VLANs have relationships to prefixes and device interfaces. Prefixes, of course, track which Layer-3 networks reside on the VLAN. Nautobot also allows proper modeling of 802.1Q by way of tracking tagged and untagged VLANs on individual interfaces. While one might expect a relationship directly between VLAN and device, the interface tagging is more meaningful to the device configuration and through understanding of the overall network data model, we are able to derive all VLANs relevant to a device, through its interfaces.

VLAN groups in Nautobot serve as a means to organize VLANs. Each VLAN group can be optionally linked to a specific location. VLAN groups are particularly useful for enforcing uniqueness among VLANs: within a group, each VLAN must have a unique ID and name. In terms of network management, this is helpful in distinguishing discrete VLANs when repeated in various network segments/topologies that are designed to look the same way, such as branch offices.

Please also note that VLANs not assigned to any group can have overlapping names and IDs, even if they belong to the same location. For instance, it's possible to have two VLANs with the ID 123, but they cannot be part of the same VLAN group.

Circuits data models

The circuit domain is the other large area of the network model. Here you can track your providers' details and certainly your inventory of circuits, but also how those circuits connect to the network. There are also abstractions that allow you to model your provider networks, which you do not control but that play a role in connectivity.

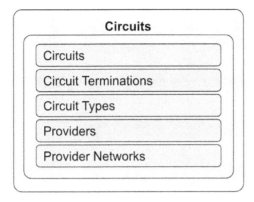

Figure 2.23 – Overview of the circuits data model provided in Nautobot

Circuits

Circuits in Nautobot play a crucial role in representing the physical connectivity within a network, much like the device types and devices do for hardware. At its core, a circuit in Nautobot is defined as a communications link that connects exactly two endpoints, known as A and Z terminations. This model allows for a flexible representation of a circuit's connection points; it's not uncommon to define only one termination, especially in cases where the details of the provider side, such as in internet access circuits, are not a primary concern. However, for more complex setups, such as private network connections linking customer locations, both terminations are typically modeled to accurately reflect the network's physical layout.

Every instance of a circuit in Nautobot is linked to a provider. This relationship is akin to how devices are linked to device types. Additionally, each circuit is categorized by a user-defined type, allowing for a detailed and customized classification of the network's various connections. For instance, a network might utilize internet access circuits from one provider and private MPLS circuits from another, each distinctly identified by their type. Circuits are primarily identified by the combination of their provider and unique circuit ID. Nautobot also introduces a robust status system for circuits, with default statuses encompassing the entire lifecycle of a circuit—from **Planned** and **Provisioning** to **Active**, **Offline**, **Deprovisioning**, and eventually **Decommissioned**. These statuses are also fully customizable, with more detail later in the chapter. Circuits in Nautobot can be enriched with several optional fields. These fields include the *installation date* and *commit rate*, adding layers of detail similar to how device types track attributes such as manufacturer and model number. Additionally, just as devices can be attributed to a location or tenant, circuits may also be assigned to Nautobot tenants, providing a clear demarcation of responsibility and ownership within the network's infrastructure.

Circuit terminations

A circuit termination in Nautobot is essentially the point where a circuit connects to a specific location or device, capturing the physical reality of network connectivity. A circuit in Nautobot can have up to two terminations, labeled as A and Z, echoing the common practice in networking of identifying the two ends of a link. The flexibility of having either one or two terminations allows for various use cases: a single-termination circuit is apt for scenarios where the far end of the circuit is unknown or irrelevant, such as an internet access circuit connecting to a transit provider. On the other hand, a dual-termination circuit is instrumental in tracking circuits that link two specific locations, mirroring the physical connection between them.

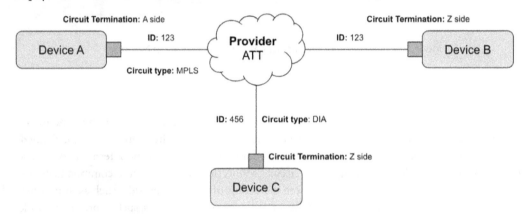

Figure 2.24 – A diagram showing various circuits terminated to devices

Each termination of a circuit is associated with either a specific location or a provider network. When linked to a location, a termination may be further detailed by connecting it via a cable to a particular device interface or port within that location, much like how devices are connected to the network. This level of granularity allows for precise tracking and management of network connections. Furthermore, each circuit termination is required to have an assigned port speed, mirroring the operational parameters of the network. Additionally, it can optionally have an upstream speed defined, especially relevant in scenarios where downstream and upstream speeds differ, such as with DOCSIS cable modems. This mirrors the attention to detail seen in other aspects of Nautobot's data model, such as the specific attributes of devices and device types.

In line with Nautobot's philosophy of closely modeling the real-world configurations, a circuit is restricted to connecting only to physical interfaces. This means circuits cannot terminate at virtual interfaces, such as LAG interfaces. In scenarios where a circuit connects to a LAG, each physical member of the LAG must have its separate circuit, and each one must be modeled discretely. This approach ensures that the Nautobot model stays true to the actual physical layout and functioning of the network, ensuring accuracy and clarity in network documentation and management.

Circuit types

Circuit types in Nautobot are entirely customizable, offering a high degree of flexibility to adapt to the specific needs and terminologies of different network environments. The primary purpose of defining circuit types is to convey the nature of the service being delivered over a particular circuit. By categorizing circuits based on their function, network administrators can easily understand and manage the various types of connectivity within their infrastructure. This is especially important in complex networks where different types of circuits serve distinct roles.

Examples of commonly defined circuit types in Nautobot include the following:

- Internet Transit
- Out-of-Band Connectivity
- Peering
- Private Backhaul

Circuit providers

A circuit provider is defined as any entity that facilitates connectivity, whether between different locations or within a single location in Nautobot. This broad definition encompasses a variety of entities, not limited to traditional carriers, but also including internet exchange points, and even organizations that are direct peering partners. The role of a circuit provider is fundamental in the configuration of circuits in Nautobot.

>>> Providers

	Name	ASN	Account number	Circuits
☐	AT&T	7018	—	0
☐	Cogent	174	—	0
☐	Deutsche Telekom	3320	—	0
☐	GTT	3257	—	0

Figure 2.25 – Circuit Providers table (list) view

Each circuit must be linked to a specific provider, ensuring that there's clear documentation of the entity responsible for the connectivity service. Providers can be detailed with additional attributes, enhancing the depth of information available for network management, such as ASNs, customer account numbers, or contact information.

Provider networks

A provider network is used to model an abstract part of a network's topology. It's a conceptual representation, similar to elements seen in a network topology diagram (often depicted as a cloud). This model is particularly useful for describing segments of a network that are managed or owned by a third-party provider, rather than the physical infrastructure within a user's direct control. For instance, a provider network can be used to represent a service provider's MPLS network. This representation is essential for understanding and documenting how external network services integrate with and support the user's internal network. Each provider network must be assigned to a specific provider, and circuits in Nautobot can terminate at a provider network, allowing for accurate modeling of how externally provided circuits integrate into the overall network.

Data model extensibility

So far we have covered the network data model that comes out of the box with Nautobot. As you can see, that data model has broad applicability to many network operators, but alas, it does not offer data points for every eventuality in tracking a modern network, nor does it actually try. Nautobot attempts to strike a balance by providing in the core data model what makes sense most of the time, and also offering features that can be used to extend the network data model to meet an organization's specific wants and needs. We collectively call this part of Nautobot the data model extensibility feature set. There are many features in this area and you will learn more about them in later chapters, but it is worth pointing out some of the most important now: custom and computed fields, relationships, config contexts, and Nautobot Apps (which offer custom data models).

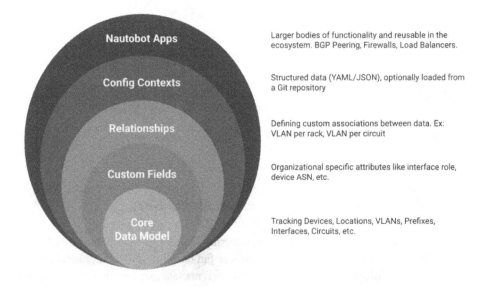

Figure 2.26 – Fundamental features of the data model extensibility capabilities in Nautobot

Custom fields

Custom fields offer a flexible way to store additional attributes for objects that are unique to specific use cases or organizational needs. Each model in Nautobot, such as devices, is represented as a table in the database with standard attributes as columns. As Nautobot evolves, new columns are added to these tables to accommodate new attributes. However, for more specialized data requirements that are not universally applicable, custom fields come into play. An example would be if an organization wished to track a purchase order number on each device, they could add a **PO Number** custom field that exists only in their instance of Nautobot, but appears as an attribute on the device model.

Custom fields in Nautobot can be created either through the **user interface** (**UI**) under **Extensibility | Miscellaneous | Custom Fields** or via the REST API. Here is one example that shows two custom fields, **Old Site ID** and **Type of Site**:

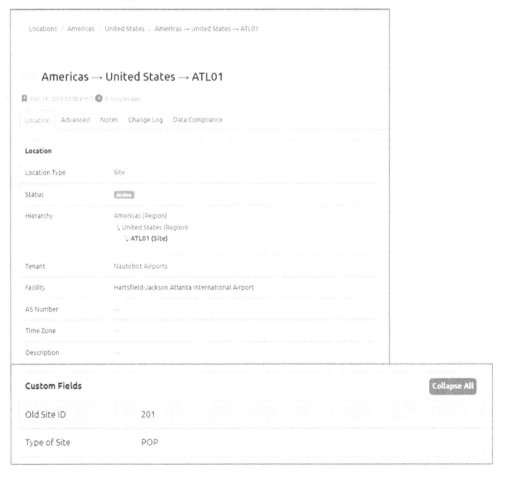

Figure 2.27 – Custom fields shown on a detailed page view

Custom fields support different object types and are also supported via the REST and GraphQL APIs. We will spend more time creating custom fields in *Chapter 6*.

Computed fields

Computed fields in Nautobot are conceptually similar to custom fields but with a significant distinction: their values are generated dynamically using existing data from the Nautobot database, combined with Jinja templating and filters. This feature is particularly useful for creating fields that require the synthesis of data from different sources within Nautobot.

For instance, consider a scenario in an automation system where there's a need for a field in the device model that concatenates the device's name with its location name, and then converts the entire string to uppercase. To achieve this, a Jinja template can be defined for the computed field. In this example, the template would look as follows:

```
{{ obj.name }}_{{ obj.location.name | upper }}
```

In this template, `{{ obj.name }}` fetches the name of the device, and `{{ obj.location.name | upper }}` takes the name of the location associated with the device and converts it to uppercase. The underscore `_` between them acts as a separator. When Nautobot processes this template for a particular instance of a device, it automatically generates the value for this computed field based on the current data of the device and its location. If a device's name was `nyc-rtr-01` and the location name was `NewYork`, it would yield the computed field of `NYC-RTR-01_NEWYORK`.

Here is another example you'll learn how to create in *Chapter 6* that concatenates a few circuit details into a single computed field:

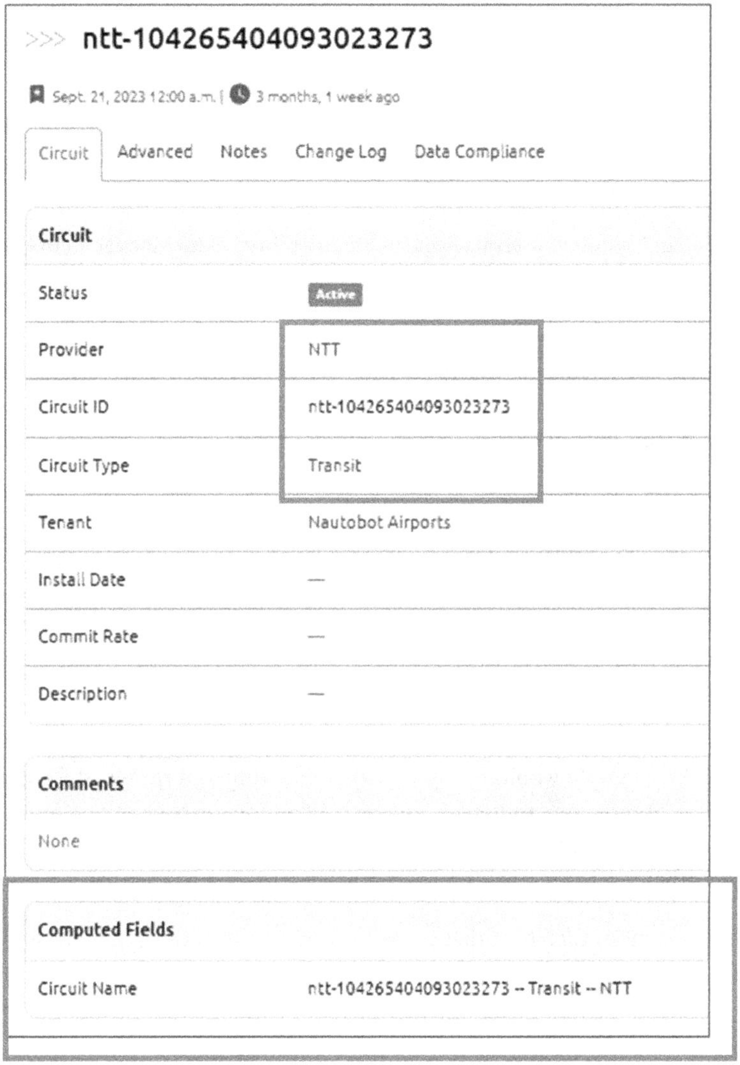

Figure 2.28 – Viewing a computed field on a detailed page view

This is made possible by the following computed field template:

```
{{ obj.cid }} -- {{ obj.circuit_type }} -- {{ obj.provider }}
```

Relationships

The relationships feature in Nautobot is designed to create custom linkages or relationships between objects in the database, allowing users to reflect unique business logic or specific relationships relevant to their network or data. This feature is akin to defining custom fields for attributes, but instead focuses on establishing specific connections between different objects.

To set up a relationship in Nautobot, users navigate to **Extensibility | Data Management | Relationships** from the left sidebar menu. There are three primary types of relationships that can be defined:

- **Many-to-many relationships**: In this type, both sides of the relationship can be connected to multiple objects. An example of this is the connection between VLANs and devices. A single VLAN can be associated with multiple devices, and similarly, a device can have multiple VLANs connected to it. This type of relationship is useful for representing complex interconnections where multiple entities share links.

- **One-to-many relationships**: This type involves one side of the connection being limited to a single object, while the other can have multiple. A typical example is the relationship between a controller (such as a parent switch) and its supplicants (such as FEXes). In most cases, a FEX is uplinked to just one parent switch, but a parent switch can have multiple FEXes connected to it. This kind of relationship is useful for hierarchical or dependency-based structures within the network.

- **One-to-one relationships**: In this scenario, there can be only one object on either side of the relationship. An example would be an IP address serving as a router-id for a device. Each device can have at most one router-id, and each IP address can serve as a router-id for only one device. This type is suitable for exclusive pairings where each entity in the pair is uniquely linked to the other.

Here is one example shown in *Chapter 6* that creates a relationship between a prefix and a circuit:

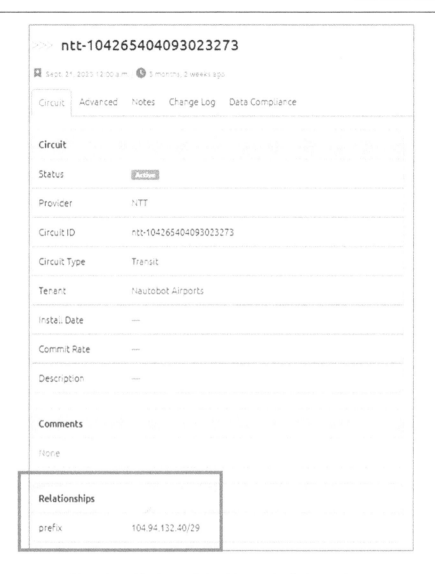

Figure 2.29 – Viewing a relationship on a detailed page view

Relationships can be used to showcase business logic or to simplify and streamline viewing the data in a simpler manner.

Config contexts

Config contexts are arbitrary YAML/JSON data that you can apply to any device. Imagine being able to create and associate any YAML data structure to a device in Nautobot, validate that data with a JSON schema definition, and access that data from the Nautobot APIs. This is what config contexts offer and allow you to greatly extend and add to the Nautobot core data model. Config contexts are great for the myriad of configuration parameters that might not otherwise logically fit within the data model. They also offer the unique ability to tie data to device attributes, such as the device's location or manufacturer. We'll cover more on config contexts in *Chapter 6*.

Custom data models

The data model extensibility features offer a great deal of flexibility in building a network SoT that fits your unique requirements, but sometimes it is necessary to go beyond even what these powerful features allow. For this, Nautobot offers a pluggable architecture that allows developers to build their own Nautobot Apps that provide custom business logic, and importantly, fully *custom data models*. You will learn about Nautobot app development in *Chapters 15* and *16*, but know that you have at your disposal the full set of tools, used in creating the core network data model that we covered in this chapter, to build your own additional models for managing any type of data you can imagine.

Summary

The data models in Nautobot are quite robust and advanced the deeper you dive in to understand the relationships between objects. This chapter spent time articulating the model to ensure you have an understanding of the data model that you'll be configuring and automating for the remainder of the book. The more you understand the data model, the smoother experience you will have with managing networks with Nautobot.

In the next chapter, we will bootstrap a Nautobot deployment, getting you up and running with Nautobot as fast as possible.

Part 2: Getting Started with Nautobot

This part begins the hands-on Nautobot journey. From installation to administration, and everything in between, you will get step-by-step instructions to get up and running Nautobot. You will start to navigate the Nautobot UI and learn about its core and extensibility feature sets, but also start to understand the key features that allow you to tailor and manage individual Nautobot deployments.

This part consists of the following chapters:

- *Chapter 3, Installing and Deploying Nautobot*
- *Chapter 4, Understanding the User Interface and Bootstrapping Nautobot*
- *Chapter 5, Configuring Nautobot Core Data Models*
- *Chapter 6, Using Nautobot's Extensibility Features*
- *Chapter 7, Managing and Administering Nautobot*

3

Installing and Deploying Nautobot

In the first two chapters, we learned about the need for network automation and a Source of Truth with Nautobot. We then provided an overview of the data model and relationships of the data inside Nautobot. Now that you understand why Nautobot is an important piece of your network automation strategy, it's time to get Nautobot up and running with hands-on exercises and start to see the data model come to life.

In this chapter, you will learn how to install Nautobot and start to populate it with some data. Both of these topics will be covered in even more detail in the next chapter of the book, including production-ready deployments and all the features that Nautobot provides for modeling your network to drive your network operations by its design.

But first things first – you need to understand the basic components of the Nautobot stack before progressing.

The following are the main topics that will be covered in this chapter:

- Nautobot architecture overview
- Installing Nautobot
- Loading data into Nautobot

Nautobot architecture overview

In this chapter, we will guide you through the installation of the major Nautobot components to get Nautobot up and running. These components are depicted in the following figure:

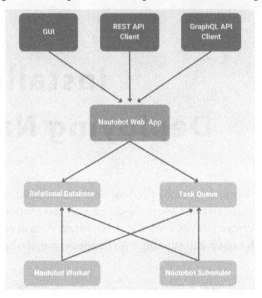

Figure 3.1 – Nautobot application stack

Here are the basic components of Nautobot with some details:

- **Nautobot web app**—This is the main application that serves as the Source of Truth. It offers external integrations (e.g., a graphical user interface, a REST API, and a GraphQL API), and also defines the data model's persistence in the database, and handles Nautobot job management (i.e., asynchronous tasks execution in Nautobot). In summary, this is what you typically think of as Nautobot when you log in to the Web UI.

- **Nautobot worker**—This is the Nautobot component in charge of executing Nautobot Jobs. Workers are also used to execute webhooks. Workers are a major component responsible for the Nautobot automation platform capabilities.

- **Nautobot scheduler**—This complements the Nautobot orchestration by handling the scheduling of predefined tasks (i.e., those stored in the database) to be taken care of by the Nautobot workers.

- **Relational database**— This is where the data is stored for persistence and connected for effective consumption.

- **Task queue**—This is used to manage background Nautobot tasks and allow the asynchronous execution of jobs in Nautobot. The task queue is where the Nautobot workers listen to get new tasks (triggered by Nautobot). In summary, the task queue is what connects the Nautobot web app with the Nautobot workers.

> **Note**
>
> For enterprise production deployments, you'll want to investigate the implementation of load balancers, proxies, and caching solutions.
>
> The Nautobot application is built on top of the Django project (`https://www.djangoproject.com/`). Django is a Python web framework that offers a pragmatic but extensible solution to build web applications. Django comes with numerous extras to implement common tasks without reinventing the wheel, and Nautobot extends it with many helpers implemented to facilitate developing network automation solutions. You'll see the power of Nautobot's extensibility using Django in *Part 4* of the book, which covers Nautobot Apps.

Installing Nautobot

Nautobot can be installed on a multitude of technologies ranging from bare-metal servers, virtual machines, and containers to cloud-native environments. Our focus in this chapter is a traditional installation on a Linux virtual machine (Ubuntu Server 20.04), installing Nautobot and its required dependencies including a local PostgreSQL database and a Redis in-memory data store used by the task queue(s).

> **Note**
>
> The Nautobot application can run on any Linux distribution that satisfies its requirements and it's regularly tested with the most recent versions of Ubuntu.

This chapter is focused on getting Nautobot up and running as fast as possible so you can experience all that it has to offer. Production considerations and overall Nautobot system administration are reviewed in more detail in *Chapter 6*.

There are other available references to spin up a Nautobot application using containers:

- Nautobot Lab is an all-in-one Docker container that allows a user to quickly get an instance of Nautobot up and running with minimal effort (`https://github.com/nautobot/nautobot-lab`).

- Nautobot Docker Compose is a docker-compose stack with one container for each one of the basic features: the Nautobot app, Nautobot worker, in-memory data store (Redis), and database (PostgreSQL). You can find an example of this in action in *Chapter 14* (`https://github.com/nautobot/nautobot-docker-compose`).

- Nautobot Helm Charts is a collection of Helm charts to deploy Nautobot into a Kubernetes cluster (`https://github.com/nautobot/helm-charts`).

This book uses a native Linux installation as also documented in the official Nautobot docs (`https://docs.nautobot.com/projects/core/en/stable/installation`), which you can check to clarify, complement, or extend the instructions in this chapter.

Getting Nautobot up and ready on Ubuntu

This section guides you through the installation of Nautobot on an Ubuntu 22.04 machine.

> **Note**
>
> Following this guide should give you a seamless Nautobot installation experience, but when you try it in a slightly different environment (e.g., a different Linux distribution), you may get some unexpected issues. To solve them, we recommend joining the public NTC Slack (`https://networktocode.herokuapp.com/`) where the community will help you to sort out any inconveniences.

First, we'll verify the Ubuntu machine being used using `uname`:

```
root@nautobot-dev:~# uname -vors
Linux 6.5.0-15-generic #15~22.04.1-Ubuntu SMP PREEMPT_DYNAMIC Fri Jan
12 18:54:30 UTC 2 GNU/Linux
```

Installing dependencies

Before getting started with the Nautobot installation, you need to make sure that all following required dependencies are installed:

- Python 3.8 or greater
- Git
- Redis
- PostgreSQL

We'll now show how to install each dependency in turn.

System packages – Python and Git

Being a Python application (Django framework), Nautobot requires a Python3 development environment and `pip` (the **Python package installer**). To install them in Ubuntu, use the `apt` package manager (after running an update of the list of packages via `update`):

```
root@nautobot-dev:~# apt update -y
... omitted output ...
```

```
180 packages can be upgraded. Run 'apt list --upgradable' to see them.

root@nautobot-dev:~# apt install -y git python3 python3-pip python3-
venv python3-dev
... omitted installation output ...
```

Redis setup

As pointed out in the *Nautobot architecture overview* section earlier in this chapter, Nautobot requires a task queue to manage the Nautobot jobs that are triggered by the Nautobot web application, and executed by Nautobot workers. This task queue is implemented by the Celery (https://docs. celeryq.dev/) task queue library, which uses a Redis (https://redis.io/) data store as its backend.

> **Note**
>
> Nautobot also uses Redis for caching, but its primary role is to support the task queue.

You can install Redis via apt:

```
root@nautobot-dev:~# apt install -y redis-server
... omitted installation output ...
```

Once installed, to validate that the Redis server service is running, you can use the redis-cli ping command (hopefully, you will get a PONG response):

```
root@nautobot-dev:~# redis-cli ping
PONG
```

By default, if no host and port are defined, Redis uses localhost and port TCP/6379. You can verify this by using lsof:

```
root@nautobot-dev:~# lsof -i:6379
COMMAND          PID      USER      FD   TYPE DEVICE   SIZE/OFF NODE
NAME
redis-ser   28766 redis    6u   IPv4
188733                            0t0     TCP localhost:6379 (LISTEN)
redis-ser   28766 redis    7u   IPv6
188734                            0t0     TCP localhost:6379 (LISTEN)
```

Database (PostgreSQL) setup

Nautobot needs a database to offer powerful relationship features and persist the data. After all, Nautobot is a source of truth and is meant to store data! There is official support for PostgreSQL and MySQL. Both are fully functional, but we are using PostgreSQL here for brevity. More information about MySQL can be found in the official Nautobot docs at `https://docs.nautobot.com/projects/ core/en/stable/installation/nautobot/#install-mysql-client-library`.

PostgreSQL is installed as an Ubuntu package:

```
root@nautobot-dev:~# apt install -y postgresql
... omitted installation output ...
```

After a successful installation, similar to what we did for Redis, we can double-check that the service is listening on the default port TCP/5432 using `lsof`:

```
root@nautobot-dev:~# lsof -i:5432
COMMAND    PID      USER    FD    TYPE DEVICE SIZE/OFF NODE NAME
postgres 31149 postgres     3u   IPv6 203134      0t0  TCP
localhost:postgresql (LISTEN)
postgres 31149 postgres     4u   IPv4 203135      0t0  TCP
localhost:postgresql (LISTEN)
```

Next, we need to create a database for Nautobot with an associated username and password.

> **Tip**
> Make sure to write these credentials down as you will use them later to configure Nautobot.

Let's log in to the `postgres` CLI using `psql`:

```
root@nautobot-dev:~# sudo -iu postgres psql
psql (14.10 (Ubuntu 14.10-0ubuntu0.22.04.1))
Type "help" for help.

postgres=#
```

After logging into the `postgres` CLI, there are a few steps to set up the database for Nautobot. Let's walk through them now:

1. Create a database with the name `nautobot`:

   ```
   postgres=# CREATE DATABASE nautobot;
   CREATE DATABASE
   ```

2. Create a user and password:

   ```
   postgres=# CREATE USER nautobot WITH PASSWORD 'nautobot123';
   CREATE ROLE
   ```

> **Note**
>
> Please note the credentials shown are only examples and should not be used for more than a lab environment.

3. Grant permissions to access the database:

```
postgres=# GRANT ALL PRIVILEGES ON DATABASE nautobot TO
nautobot;
GRANT
```

4. Connect to the nautobot database:

```
postgres=# \connect nautobot
You are now connected to database "nautobot" as user "postgres".
```

5. Grant permissions to create a schema:

```
nautobot=# GRANT CREATE ON SCHEMA public TO nautobot;
GRANT
```

6. Exit the postgres CLI by typing \q and press *Enter*:

```
postgres=# \q

root@nautobot-dev:~#
```

7. Validate access as the granted user.

    To validate access to the postgres database, use the psql command with the username and password from *step 2* and connect to the nautobot database. The resulting prompt should show the nautobot database denoted with the nautobot=> prompt.

```
root@nautobot-dev:~# psql --username nautobot --password --host
localhost nautobot
Password:
psql (12.16 (Ubuntu 12.16-0ubuntu0.20.04.1))
SSL connection (protocol: TLSv1.3, cipher: TLS_AES_256_GCM_
SHA384, bits: 256, compression: off)
Type "help" for help.

nautobot=>
```

With the dependencies up and running, it's time to focus on the Nautobot application itself.

Installing the Nautobot application

This section walks through the steps to install the main Nautobot application with a few prerequisites, such as creating a Nautobot system user and setting up a Python virtual environment for Nautobot.

Creating a Nautobot system user

Create a specific user on the Linux machine to run Nautobot; in this example, we use nautobot. While this is not a requirement, it is good practice to create a user/group per service to have more granular control of the required service. This system user will own all of the Nautobot files. Additionally, the Nautobot web services will be configured to run under this user account.

The following code snippet also creates the /opt/nautobot directory. This directory is the default NAUTOBOT_ROOT directory. More on this in *Chapter 7*.

We'll also set it as the home directory for the nautobot user:

```
root@nautobot-dev:~# useradd --system --shell /bin/bash --create-home
--home-dir /opt/nautobot nautobot
root@nautobot-dev:~#
```

Let's verify the /opt/nautobot directory has the proper owner home directory and permissions (with ls):

```
root@nautobot-dev:~# eval echo ~nautobot
/opt/nautobot
root@nautobot-dev:~#
root@nautobot-dev:~# ls -l /opt/ | grep nautobot
drwxr-x--- 2 nautobot nautobot 4096 Jan  5 10:13 nautobot
```

Setting up the virtual environment

A Python virtual environment (https://docs.python.org/3/library/venv.html), or virtualenv, functions as a container for a designated set of Python packages. By using a virtualenv, you can create isolated environments tailored to individual projects, preventing any interference with system packages or other ongoing projects (which is recommended).

We're going to create the virtualenv in our NAUTOBOT_ROOT as the nautobot user to populate the /opt/nautobot directory with a self-contained Python environment:

```
root@nautobot-dev:~# sudo -u nautobot python3 -m venv /opt/nautobot
```

Next, we'll update the nautobot user .bashrc with the NAUTOBOT_ROOT environment variable, so it will always be set without having to do it manually. This is helpful because the NAUTOBOT_ROOT environment variable is used to define the filesystem path to store Nautobot files (e.g., Jobs definitions, uploaded images, Git repositories, etc.). We'll use this path much more throughout the book:

```
root@nautobot-dev:~# echo "export NAUTOBOT_ROOT=/opt/nautobot" | sudo
tee -a ~nautobot/.bashrc
```

Now, if you change the user from root to the nautobot user (with sudo -iu), you can see how the prompt changes and verify the environmental variable is set as expected:

```
root@nautobot-dev:~# sudo -iu nautobot
nautobot@nautobot-dev:~$ echo $NAUTOBOT_ROOT
/opt/nautobot
```

Installing Nautobot

Now, we are ready to install Nautobot!

Nautobot should be installed as the nautobot user—do not install Nautobot as root. Everything from here going forward related to the Nautobot installation is done as the nautobot user. (You are already there if you have been following along.)

Before we install anything into the virtualenv, we want to make sure that pip is running the latest version.

We will also install wheel, a Python library that instructs pip to consistently attempt the installation of wheel packages when they are accessible. A wheel denotes a pre-compiled Python package, enhancing installation speed and safety by eliminating the need for development libraries or gcc (the C compiler) on your system, particularly for compiling more advanced Python libraries. This is another good practice when managing Nautobot:

```
nautobot@nautobot-dev:~$ pip3 install --upgrade pip wheel
Collecting pip
... omitted output for brevity ...
Successfully installed pip-23.2.1 wheel-0.41.2
```

And, finally, install Nautobot (it will take around 1–2 minutes) via pip. We are using a specific version (i.e., 2.1.4) for this book to avoid picking the latest version, which could introduce new features that might change the output:

```
nautobot@nautobot-dev:~$ pip3 install nautobot==2.1.4
Collecting nautobot
  Downloading nautobot-2.1.4-py3-none-any.whl.metadata (9.8 kB)
... omitted long output ...
```

```
Successfully installed Django-3.2.23 GitPython-3.1.40 Jinja2-3.1.2
nautobot-2.1.4 netaddr-0.8.0 netutils-1.6.0
... omitted output for brevity...
```

> **Note**
> It is possible to install additional extras such as MySQL, LDAP, NAPALM, remote_storage, SSO, and so on with pip3 install nautobot[napalm,mysqlclient], or, if you want all of them, nautobot[all]. This is covered in more detail in *Chapter 6*.

You should now have a pretty awesome nautobot-server command in your environment. This will be your gateway to all things Nautobot! Run it to confirm the installed version:

```
nautobot@nautobot-dev:~$ nautobot-server --version
2.1.4
```

> **Note**
> This section introduces a few nautobot-server commands, but they are covered in much greater detail in *Chapter 6*.

Since we are officially done with the installation, it is time to configure Nautobot.

Nautobot configuration settings

We now have a Nautobot application installed on a Ubuntu server, but we still need to configure several things to get started (for instance, which database to use).

Nautobot comes with the nautobot_config.py configuration file. This is a Python file where several constants and settings are defined that influence how Nautobot functions.

You can bootstrap and initialize the config file with nautobot-server init:

```
nautobot@nautobot-dev:~$ nautobot-server init
Nautobot would like to send anonymized installation metrics to the
project's maintainers.
These metrics include the installed Nautobot version, the Python
version in use, an anonymous "deployment ID", and a list of one-way-
hashed names of enabled Nautobot Apps and their versions.
Allow Nautobot to send these metrics? [y/n]: y
Installation metrics will be sent when running 'nautobot-server post_
upgrade'. Thank you!
Configuration file created at /opt/nautobot/nautobot_config.py
```

> **Note**
>
> Nautobot comes with an option to send anonymized metrics to the Nautobot development team to better enable them to make decisions based on which versions of Python installed, how popular certain apps are, and more.

At this point, you have a brand-new `nautobot_config.py` file ready to customize for your installation. Let's check it out:

```
nautobot@nautobot-dev:~$ cat /opt/nautobot/nautobot_config.py
import os
import sys

from nautobot.core.settings import *  # noqa F401,F403
from nautobot.core.settings_funcs import is_truthy, parse_redis_
connection

###########################
#                                                                         #
#            Required settings           #
#                                                                         #
###########################

... omitted output for brevity
```

Because we already defined NAUTOBOT_ROOT, the `nautobot-server init` command created the new `nautobot_config.py` file at the location using NAUTOBOT_ROOT, e.g. `/opt/nautobot/nautobot_config.py` because NAUTOBOT_ROOT is `/opt/nautobot/`.

You can observe that most of the code in the config file is commented out. These are just placeholders (with helping comments and sane defaults) to let you customize the file to your specific requirements.

Most of the configuration settings must be configured as follows:

- Commenting out the code lines (remove the # at the beginning of the line)
- Defining proper environmental variables

The `nautobot_config.py` config file is split into different sections. To get started, we'll start with the *Required settings* section, as it contains all the major settings that Nautobot needs to actually function.

Let's edit `$NAUTOBOT_ROOT/nautobot_config.py` and head over to the *Required settings* section to tweak the required settings, uncommenting *all* of them (use your preferred text editor, such as vi). These are the elements to look for in the file:

```
nautobot@nautobot-dev:~$ vi $NAUTOBOT_ROOT/nautobot_config.py
...
```

```
ALLOWED_HOSTS = os.getenv("NAUTOBOT_ALLOWED_HOSTS", "").split(" ")
...
CACHES = {
...
CONTENT_TYPE_CACHE_TIMEOUT = int(os.getenv("NAUTOBOT_CONTENT_TYPE_
CACHE_TIMEOUT", "0"))
...
CELERY_BROKER_URL = os.getenv("NAUTOBOT_CELERY_BROKER_URL", parse_
redis_connection(redis_database=0))
...
DATABASES = {
# DO NOT FORGET TO UNCOMMENT ALL REQUIRED SETTINGS
...
```

While uncommenting those lines, you may have noticed that most of the settings can be defined via environmental variables (without updating this file anymore), and others come with some defaults that can be changed in this file (for example, the type of the database being set to PostgreSQL).

Thus, let's start by defining the environmental variables we need to change (i.e., the whitelist for IPs with access to Nautobot and the database credentials):

```
nautobot@nautobot-dev:~$ echo "export NAUTOBOT_ALLOWED_HOSTS=*" | tee
-a ~nautobot/.bashrc
export NAUTOBOT_ALLOWED_HOSTS=*

nautobot@nautobot-dev:~$ echo "export NAUTOBOT_DB_USER=nautobot" | tee
-a ~nautobot/.bashrc
export NAUTOBOT_DB_USER=nautobot

nautobot@nautobot-dev:~$ echo "export NAUTOBOT_DB_
PASSWORD=nautobot123" | tee -a ~nautobot/.bashrc
export NAUTOBOT_DB_PASSWORD=nautobot123
```

After defining them, to bring them to life, you have to refresh your bash profile. Then, you can verify the environmental variables are available:

```
nautobot@nautobot-dev:~$ source ~/.bashrc
nautobot@nautobot-dev:~$ env | grep NAUTOBOT
NAUTOBOT_ALLOWED_HOSTS=*
NAUTOBOT_DB_PASSWORD=nautobot123
NAUTOBOT_ROOT=/opt/nautobot
NAUTOBOT_DB_USER=nautobot
```

> **Note**
>
> NAUTOBOT_ALLOWED_HOSTS set to ["*"] allows accessing the server from everywhere
> (that has network connectivity). That's convenient for a quick start, but this value is not
> recommended for production deployments.

Similar to having many more nautobot-server commands, there are many more options to
configure Nautobot, and here we only set the minimum ones to get you up and running as quickly as
possible. In *Chapter 7*, you will discover more options with which you can tailor Nautobot's behavior
to your needs.

Launching Nautobot

Before Nautobot can run, there are a few more steps needed.

Performing database migrations

Database migrations (creating the database schema of tables and relationships, and adding initial data)
must be performed to prepare the database for use (at this point, we have a nautobot database but
without any structure defined). Nautobot (and Nautobot apps) leverage database tables and data, and
this structure is translated to migration instructions that need to be deployed in the database. This is
what the nautobot-server migrate command does. It reads the instructions, e.g. migrations,
and prepares the database:

```
nautobot@nautobot-dev:~$ nautobot-server migrate
Operations to perform:
  Apply all migrations: admin, auth, circuits, contenttypes, database,
dcim, django_celery_beat, django_celery_results, extras, ipam,
sessions, social_django, taggit, tenancy, users, virtualization
Running migrations:
  Applying contenttypes.0001_initial... OK
  Applying contenttypes.0002_remove_content_type_name... OK
  Applying auth.0001_initial... OK

... omitted output for brevity ...

  Applying users.0007_alter_objectpermission_object_types... OK
  Applying virtualization.0026_change_virtualmachine_primary_ip_
fields... OK
10:29:55.991 INFO    nautobot.extras.utils :
  Created Job "System Jobs: Export Object List" from
<ExportObjectList>
10:29:55.998 INFO    nautobot.extras.utils :
  Created Job "System Jobs: Git Repository: Sync" from
<GitRepositorySync>
```

```
10:29:56.005 INFO    nautobot.extras.utils :
  Created Job "System Jobs: Git Repository: Dry-Run" from
<GitRepositoryDryRun>

nautobot@nautobot-dev:~$
```

> **Note**
>
> The `migrate` command is not a one-time command that only applies to the first Nautobot bootstrap. Every time a new release is deployed it may (likely) contain new migrations to apply to incorporate new data models or update existing ones.

At this point, the `nautobot` database is ready to use. Thus, the Nautobot application can connect the application logic to the database. And one of the first pieces of data we have to provide to Nautobot is a user account so that we can log in. Let's do that.

Creating a Nautobot superuser

Nautobot does not come with any predefined user accounts. You'll need to create an administrative *superuser* account to be able to log in to Nautobot for the first time and start managing it:

```
nautobot@nautobot-dev:~$ nautobot-server createsuperuser
Username: admin
Email address:
Password:
Password (again):
Superuser created successfully.
nautobot@nautobot-dev:~$
```

The user credentials (i.e., name and hashed password) are stored in the database to allow access to the Nautobot application.

Next, we have to add more data that is not being added by the migrations.

Collecting Nautobot static files

The database migrations we ran above didn't include data that is not stored in the database but in other storage places that are required by Nautobot. There are some static data that Nautobot leverages, such as the following:

- `media`: For storing uploaded images and attachments (such as device-type images)
- `static`: The home of the CSS, JavaScript, and images used to serve the web interface

As you will see later in this book, other Nautobot features such as Git repositories and jobs are also stored as static files, so they also need to be collected.

For each of the static files, there is a corresponding setting in `nautobot_config.py`, but by default, they will all be placed in NAUTOBOT_ROOT.

The `collectstatic` command takes care of establishing the file structure and importing the appropriate files:

```
nautobot@nautobot-dev:~$ nautobot-server collectstatic

957 static files copied to '/opt/nautobot/static'.

nautobot@nautobot-dev:~$
```

The same process applies to the Nautobot apps that come with static data. Also, like database migrations, you may need to run this command when some new static data is added in a new release.

And, with these three steps (i.e., database migration, superuser, and static data collection) done, you are ready to validate that Nautobot is ready.

Nautobot check

Nautobot comes with a check command to validate the configuration to detect common problems and provide hints for how to fix them.

The checks are automatically run when running a development server using `nautobot-server runserver`, but not when running in production using **WSGI** (short for **Web Server Gateway Interface**, which connects the Python code with a web server). We cover WSGI in the next section:

```
nautobot@nautobot-dev:~$ nautobot-server check
System check identified no issues (0 silenced).
nautobot@nautobot-dev:~$
```

At this point, we should be able to run Nautobot's development server for testing. We can check by starting a development instance:

```
nautobot@nautobot-dev:~$ nautobot-server runserver 0.0.0.0:8080
--insecure
10:51:19.247 INFO    django.utils.autoreload :
  Watching for file changes with StatReloader
Performing system checks...

System check identified no issues (0 silenced).
January 05, 2024 - 10:51:19
Django version 3.2.23, using settings 'nautobot_config'
Starting development server at http://0.0.0.0:8080/
Quit the server with CONTROL-C.
10:51:19.848 INFO    nautobot :
  Nautobot initialized!
```

Now you can connect to the name or IP of the server (as defined in ALLOWED_HOSTS) on port 8080. Enter the IP address of the Ubuntu server and 8080 into your web browser, e.g., https://10.x.y.x:8080. You will be greeted with the Nautobot home page.

> **Note**
>
> Do not use this server in a production setting. The development server is for development and testing purposes only. It is neither performant nor secure enough for production use.

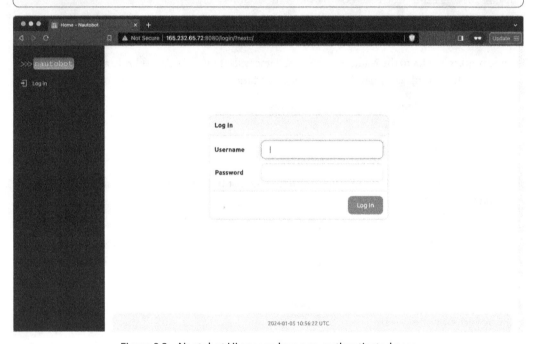

Figure 3.2 – Nautobot UI as seen by a non-authenticated user

In the terminal, you can see the logging output of the Nautobot application. This is really helpful for understanding what's going on inside the server application. For example, in the following log output we see that because no session was established, it redirects the user to the login page:

```
... from the previous terminal ...
Starting development server at http://0.0.0.0:8080/
Quit the server with CONTROL-C.
10:57:38.718 INFO     nautobot :
  Nautobot initialized!
10:57:40.680 INFO     django.server :
  "GET / HTTP/1.1" 302 0
10:57:40.783 INFO     django.server :
```

```
  "GET /login/?next=/ HTTP/1.1" 200 10605
 10:57:40.854 INFO    django.server :
  "GET /template.css HTTP/1.1" 200 3984
```

Logging in to Nautobot

Try logging in using the *superuser* account you created before. Once authenticated, you'll be able to see the UI:

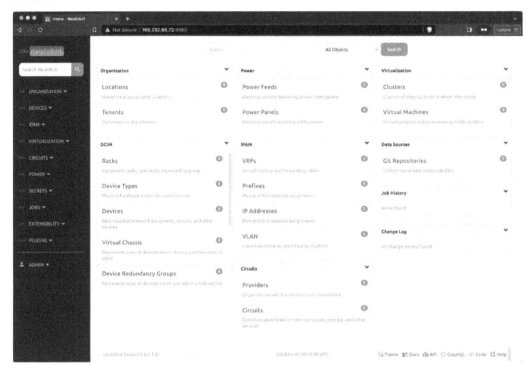

Figure 3.3 – The Nautobot UI as seen by an administrator

> **Note**
>
> You can stop the development server (started by `nautobot-server runserver`) by typing *Ctrl+C* in your terminal.

So far, we have the Nautobot web application up and running. Remember that we started the PostgreSQL database and the Redis server before. However, as introduced previously in the *Nautobot architecture overview* section, there is still one component that hasn't been started yet. That is the Nautobot worker. The Nautobot worker is not started by the `runserver` command. We'll cover how to start a Nautobot worker in the next section.

Nautobot worker

Nautobot workers are in charge of picking up the background tasks triggered by the Nautobot platform and executing them asynchronously. You can think of workers as the automation engine within Nautobot. This gives you, as a network engineer, plenty of options to implement network automation tasks.

The technology behind the Nautobot worker is based on the Python Celery (`https://docs.celeryq.dev/`) library (a distributed task queue framework). Even though we talk about the Nautobot worker in singular (as we did for the Nautobot web service), the production implementation of workers is based on scaling them out horizontally, running as many worker processes as needed.

If you rerun the Nautobot web server from the previous section and try to run one of the default jobs (for instance, the "Export Object List" for the "Circuits"), you will get the following error: **Unable to run or schedule job: Celery worker process not running**.

Figure 3.4 – Error when trying to run a job in Nautobot without a Celery worker running

The `nautobot-server celery` command is used to directly invoke Celery. This command behaves exactly as the Celery command-line utility does but launches it through Nautobot's environment to share the Redis and database connection settings transparently.

Execute the `nautobot-server celery worker` command to instantiate a worker.

You need to do this in a new terminal in parallel to the `runserver` command:

```
nautobot@nautobot-dev:~$ nautobot-server celery worker

 -------------- celery@nautobot-dev v5.3.6 (emerald-rush)
--- ***** -----
-- ******* ---- Linux-5.4.0-122-generic-x86_64-with-glibc2.29 2024-01-
05 12:44:29
- *** --- * ---
- ** ---------- [config]
- ** ---------- .> app:         nautobot:0x7f85aac77460
- ** ---------- .> transport:   redis://localhost:6379/0
- ** ---------- .> results:
```

```
- *** --- * --- .> concurrency: 2 (prefork)
-- ******* ---- .> task events: ON
--- ***** -----
 -------------- [queues]
                .> default           exchange=default(direct)
key=default
```

With the Nautobot worker, if you rerun the job, you can see how a worker is available in the task queue:

Figure 3.5 – Job task queue with one available Nautobot worker

And then, after running the job, you get a successful execution (without data, because we have not started adding it yet!):

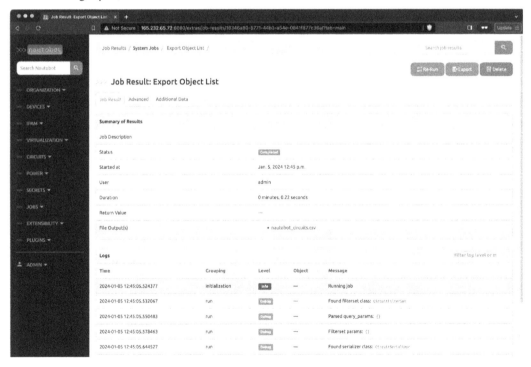

Figure 3.6 – Successful job execution in the Nautobot UI

Nautobot web service

Like other Python applications, Nautobot runs as a WSGI application behind an HTTP server.

WSGI is a standardized interface for Python web applications. It functions as the glue between the code (i.e., the logic) and the HTTP server that gets the connections.

Nautobot comes preinstalled with uWSGI as its WSGI server, however, other WSGI servers are available and should work similarly well. As you saw previously when logging into your new Nautobot server, this is not needed for development, but is highly recommended for production.

Before starting the Nautobot web service with uWSGI, we need to define the new required configuration.

uWSGI setup

You can configure uWSGI in many ways, but our goal is to focus on the major configuration parameters to get up and running and highlight the need for uWSGI.

First, we have to create the uwsgi.ini file in the NAUTOBOT_ROOT directory:

```
nautobot@nautobot-dev:~$ touch $NAUTOBOT_ROOT/uwsgi.ini
nautobot@nautobot-dev:~$ ls -l $NAUTOBOT_ROOT/uwsgi.ini
-rw-rw-r-- 1 nautobot nautobot 0 Jan  5 12:49 /opt/nautobot/uwsgi.ini
```

In this initial setup without advanced features (such as using a reverse proxy), it's important to use the http mode instead of socket, as commented in the note in the next snippet.

This is the recommended base configuration for uwsgi.ini. It listens for web requests on port 8001 (you can change this if required):

```
[uwsgi]
; The IP address (typically localhost) and port that the WSGI process
should listen on
http = 0.0.0.0:8001
; In a production environment, you would likely have a reverse-proxy
in from (e.g. NGINX), in this case,
; you should use "socket" instead of "http", and if it's running
locally, use the localhost address.

; Fail to start if any parameter in the configuration file isn't
explicitly understood by uWSGI
strict = true

; Enable master process to gracefully re-spawn and pre-fork workers
master = true
```

```
; Allow Python app-generated threads to run
enable-threads = true

;Try to remove all of the generated files/sockets during shutdown
vacuum = true

; Do not use multiple interpreters, allowing only Nautobot to run
single-interpreter = true

; Shutdown when receiving SIGTERM (default is respawn)
die-on-term = true

; Prevents uWSGI from starting if it is unable to load Nautobot
(usually due to errors)
need-app = true

; By default, uWSGI has rather verbose logging that can be noisy
disable-logging = true

; Assert that critical 4xx and 5xx errors are still logged
log-4xx = true
log-5xx = true

; Enable HTTP 1.1 keepalive support
http-keepalive = 1
```

> **Note**
>
> For extra details and options for uWSGI, refer to the official uWSGI documentation at `https://uwsgi-docs.readthedocs.io/en/latest/Configuration.html`.

With this configuration, we can use the `nautobot-server start` management command that directly invokes uWSGI using the defined configuration file (`uwsgi.ini`). Notice that we are using the full path to make it explicit and simplify later reuse when we return to it to make it a service later in the chapter:

```
nautobot@nautobot-dev:~$ /opt/nautobot/bin/nautobot-server start --ini
/opt/nautobot/uwsgi.ini

[uWSGI] getting INI configuration from /opt/nautobot/uwsgi.ini
[uwsgi-static] added mapping for /static => /opt/nautobot/static
*** Starting uWSGI 2.0.23 (64bit) on [Fri Jan  5 12:51:55 2024] ***
```

```
... ommitted output for brevity ...

uWSGI http bound on 0.0.0.0:8001 fd 4
uwsgi socket 0 bound to TCP address 127.0.0.1:44539 (port auto-
assigned) fd 3
Python version: 3.8.10 (default, Nov 22 2023, 10:22:35)  [GCC 9.4.0]
--- Python VM already initialized ---
Python main interpreter initialized at 0x285c930
python threads support enabled
your server socket listen backlog is limited to 100 connections
your mercy for graceful operations on workers is 60 seconds
mapped 145808 bytes (142 KB) for 1 cores
*** Operational MODE: single process ***
12:51:55.666 INFO    nautobot :
  Nautobot initialized!
WSGI app 0 (mountpoint='') ready in 0 seconds on interpreter 0x285c930
pid: 13538 (default app)
spawned uWSGI master process (pid: 13538)
spawned uWSGI worker 1 (pid: 13542, cores: 1)
12:51:55.683 INFO    nautobot.core.wsgi :
  Closing existing DB and cache connections on worker 1 after uWSGI
forked ...
spawned uWSGI http 1 (pid: 13543)
```

At this point, you can connect to your server (`http://10.x.y.z:8001`) and get access to Nautobot using a proper production setup (instead of the embedded Django development server).

Running Nautobot as Linux services

In this section, we explain how to run Nautobot application services as Linux services instead of simply running the debug server available in Nautobot (as done via the `nautobot-server runserver` command).

Running applications as Linux services has many benefits. Let's highlight two of them that we think are more significant than the rest:

- **Reliability**: The application is automatically started during system startup
- **Resource and dependency management**: You can define the resources (e.g., CPU and memory limits) and the dependencies between different services (for example, the database should be running before the web service starts)

There are many different Linux system management tools. In this book, we use the very popular systemd (`https://github.com/systemd/systemd`).

We already set up two Linux services, Redis and PostgreSQL, which were registered automatically during installation.

You can verify the registered Linux services with the `systemctl` status command:

```
root@nautobot-dev:~$ systemctl status redis-server postgresql
● redis-server.service - Advanced key-value store
     Loaded: loaded (/lib/systemd/system/redis-server.service;
enabled; vendor preset: enabled)
     Active: active (running) since Fri 2024-01-05 10:07:45 UTC; 2h
47min ago
       Docs: http://redis.io/documentation,
             man:redis-server(1)
   Main PID: 7579 (redis-server)
      Tasks: 4 (limit: 2339)
     Memory: 2.5M
     CGroup: /system.slice/redis-server.service
             └─7579 /usr/bin/redis-server 127.0.0.1:6379


● postgresql.service - PostgreSQL RDBMS
     Loaded: loaded (/lib/systemd/system/postgresql.service; enabled;
vendor preset: enabled)
     Active: active (exited) since Fri 2024-01-05 10:08:40 UTC; 2h
46min ago
   Main PID: 9253 (code=exited, status=0/SUCCESS)
      Tasks: 0 (limit: 2339)
     Memory: 0B
     CGroup: /system.slice/postgresql.service
```

Let's now add the Nautobot web application as another Linux service.

Running the Nautobot web service as a Linux service

Having the WSGI set up already from the last section, it's time to convert our application into a Linux service to gain the benefits mentioned above. Thus, we have to define our Nautobot web service with `root` privileges (notice the different prompt).

First, we create a `nautobot.service` file in `/etc/systemd/system/nautobot.service`:

```
root@nautobot-dev:~# touch /etc/systemd/system/nautobot.service
root@nautobot-dev:~#
```

Next, let's add the following systemd configuration to the newly created `nautobot.service` file. You can see how the exact command used before to run the WSGI server is defined in `ExecStart`:

```
[Unit]
Description=Nautobot WSGI Service
```

```
Documentation=https://docs.nautobot.com/projects/core/en/stable/
After=network-online.target
Wants=network-online.target

[Service]
Type=simple
Environment="NAUTOBOT_ROOT=/opt/nautobot"
Environment="NAUTOBOT_ALLOWED_HOSTS=*"
Environment="NAUTOBOT_DB_PASSWORD=nautobot123"
Environment="NAUTOBOT_DB_USER=nautobot"

User=nautobot
Group=nautobot
PIDFile=/var/tmp/nautobot.pid
WorkingDirectory=/opt/nautobot

ExecStart=/opt/nautobot/bin/nautobot-server start --pidfile /var/tmp/
nautobot.pid --ini /opt/nautobot/uwsgi.ini
ExecStop=/opt/nautobot/bin/nautobot-server start --stop /var/tmp/
nautobot.pid
ExecReload=/opt/nautobot/bin/nautobot-server start --reload /var/tmp/
nautobot.pid

Restart=on-failure
RestartSec=30
PrivateTmp=true

[Install]
WantedBy=multi-user.target
```

Finally, it's time to reload the systemd daemon with the new configuration and enable the service. Remember to stop the server running in any other terminal from before:

```
root@nautobot-dev:~# systemctl daemon-reload
root@nautobot-dev:~# systemctl enable --now nautobot
Created symlink /etc/systemd/system/multi-user.target.wants/nautobot.
service → /etc/systemd/system/nautobot.service.
```

Assuming everything worked, systemd is now handling the Nautobot web service (still on port 8001 and via HTTP):

```
root@nautobot-dev:~# systemctl status nautobot.service
● nautobot.service - Nautobot WSGI Service
     Loaded: loaded (/etc/systemd/system/nautobot.service; enabled;
vendor preset: enabled)
```

```
        Active: active (running) since Fri 2024-01-05 13:01:30 UTC; 8s
ago


... ommitted output for brevity ...

Jan 05 13:01:32 nautobot-dev nautobot-server[13796]: mapped 145808
bytes (142 KB) for 1 cores
Jan 05 13:01:32 nautobot-dev nautobot-server[13796]: *** Operational
MODE: single process ***
Jan 05 13:01:32 nautobot-dev nautobot-server[13796]: 13:01:32.924
INFO    nautobot :
Jan 05 13:01:32 nautobot-dev nautobot-server[13796]:   Nautobot
initialized!
... output omitted for brevity ...
```

But we can't forget the Nautobot worker services that take care of the automation tasks that run on the Nautobot platform. Let's look at those now.

Running Nautobot workers as Linux services

Let's repeat the same approach we did for the Nautobot web service, but now for the Nautobot workers:

```
root@nautobot-dev:~# touch /etc/systemd/system/nautobot-worker.service
root@nautobot-dev:~#
```

We can reuse most of the service definition from before, changing the [Unit] and [Service] sections as follows:

```
[Unit]
Description=Nautobot Celery Worker
Documentation=https://docs.nautobot.com/projects/core/en/stable/
After=network-online.target
Wants=network-online.target

[Service]
Type=simple
Environment="NAUTOBOT_ROOT=/opt/nautobot"
Environment="NAUTOBOT_ALLOWED_HOSTS=*"
Environment="NAUTOBOT_DB_PASSWORD=nautobot123"
Environment="NAUTOBOT_DB_USER=nautobot"

User=nautobot
Group=nautobot
PIDFile=/var/tmp/nautobot-worker.pid
```

```
WorkingDirectory=/opt/nautobot

ExecStart=/opt/nautobot/bin/nautobot-server celery worker --loglevel
INFO --pidfile /var/tmp/nautobot-worker.pid

Restart=on-failure
RestartSec=30
PrivateTmp=true

[Install]
WantedBy=multi-user.target
```

Running the Nautobot scheduler as Linux services

There is another Celery component used in Nautobot, Celery Beat (https://docs.nautobot.com/projects/core/en/stable/installation/services/#celery-beat-scheduler), which is used as the Nautobot scheduler and is required for scheduling Nautobot jobs (this is optional for this chapter). It's very similar to the Nautobot worker setup; you just need to repeat the previous process replacing worker with beat:

```
root@nautobot-dev:~# touch /etc/systemd/system/nautobot-scheduler.
service
root@nautobot-dev:~#
```

We can reuse most of the service definition from before, changing the [Unit] and [Service] sections as follows:

```
[Unit]
Description=Nautobot Celery Scheduler
Documentation=https://docs.nautobot.com/projects/core/en/stable/
After=network-online.target
Wants=network-online.target

[Service]
Type=simple
Environment="NAUTOBOT_ROOT=/opt/nautobot"
Environment="NAUTOBOT_ALLOWED_HOSTS=*"
Environment="NAUTOBOT_DB_PASSWORD=nautobot123"
Environment="NAUTOBOT_DB_USER=nautobot"

User=nautobot
Group=nautobot
PIDFile=/var/tmp/nautobot-beat.pid
WorkingDirectory=/opt/nautobot
```

```
ExecStart=/opt/nautobot/bin/nautobot-server celery beat --loglevel
INFO --pidfile /var/tmp/nautobot-beat.pid

Restart=on-failure
RestartSec=30
PrivateTmp=true

[Install]
WantedBy=multi-user.target
```

> **Note**
>
> Environment variables can be defined in different ways to add more security, including by creating a dynamic file with the variables and loading it with the `EnvironmentFile` option.

As we did before, it's time to load the config and enable the new service:

```
root@nautobot-dev:~# systemctl daemon-reload
root@nautobot-dev:~#
root@nautobot-dev:~# systemctl enable --now nautobot nautobot-worker
nautobot-scheduler
Created symlink /etc/systemd/system/multi-user.target.wants/nautobot-
worker.service → /etc/systemd/system/nautobot-worker.service.
root@nautobot-dev:~#
root@nautobot-dev:~# systemctl status nautobot nautobot-worker
nautobot-scheduler
● nwautobot-worker.service - Nautobot Celery Worker
     Loaded: loaded (/etc/systemd/system/nautobot-worker.service;
enabled; vendor preset: enabled)
     Active: active (running) since Fri 2024-01-05 13:03:47 UTC; 14s
ago
... output omitted for brevity ...
```

Press *Q* to quit.

Loading data into Nautobot

Once Nautobot is up and running, it's time to start using it! However, Source of Truth is useless without data. Therefore, the first step is to populate Nautobot with data.

The data defines the intended state of the network and, independent of whether the data is taken from a running network or defined in advance, there are different methods to populate Nautobot with data, as follows:

- **GUI**: Using the Nautobot **graphical user interface (GUI)** allows easy access to create objects. We will cover a few examples in this section, but more examples will come in *Chapter 5*.

- **REST API**: Nautobot exposes a fully capable HTTP REST API to access data models. Examples and details will come in *Chapter 8*.

- **SDKs**: A Python SDK (`pynautobot`) and a Go SDK (`go-nautobot`) are available to programmatically interact with the REST API. More info on these in *Chapter 10*.

- **Ansible collection**: Similar to SDKs, there is an Ansible collection to interact with Nautobot via Ansible playbooks, allowing Ansible to populate (or get data from) Nautobot. More info and examples are available in *Chapter 10*.

- **Nautobot jobs**: Nautobot jobs are Python code and allow users to create self-service forms to simplify and codify the adding of data to Nautobot. For example, a user can define the site type as **Small**, **Medium**, or **Large** and click **Submit**. The job then codifies the design and makes all the required data changes inside Nautobot. More info on creating and managing jobs will be presented in *Chapter 11*.

- **Nautobot apps**: These allow full extensibility of the data model (i.e., adding new ones) that, when combined with jobs, can run complex operations, such as the Nautobot Onboarding App that connects to network devices to import their actual states and translate them into Source of Truth objects. More about Nautobot Apps can be found in *Chapter 13*.

Finally, you also have the option to import/export the database, but you need to have a starting database to begin with. This is used for disaster recovery or environment replication. We will show you how to import/export data later in this chapter.

Using the graphical user interface

Nautobot can store and represent any network property you may need to define the intended state of your network. Some of this information will use existing data models, but sometimes you will need to extend them or create new models with Nautobot apps. However, in all the cases, the foundation of the network Source of Truth is its inventory, and that starts with devices.

In Nautobot, this object is represented by the device model as was originally covered in *Chapter 2*. So, this is going to be our focus here: getting started adding *one* device into Nautobot via its GUI. In *Chapter 5*, you will cover the other base data models in Nautobot.

The following diagram shows the required dependencies for the device model. It's necessary to understand that the dependencies have to be sorted in order. For example, to create a *Device Type*, first we need to define a *Manufacturer*.

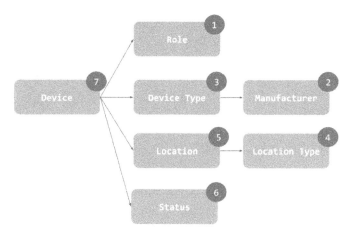

Figure 3.7 – Nautobot device object hierarchy

Now, let's embark on an exercise to explore all the steps to add a new device via the Nautobot UI. A very similar process applies to all the other object types.

Reviewing requirements for adding a device in the UI

In the sidebar navigation menu, the second menu item is **DEVICES**. When you click on it, all the related objects with devices are listed. In the first container (also named **Devices**), the first element is **Devices** and has a blue button with a plus sign (+) to add a new device. Click the + sign:

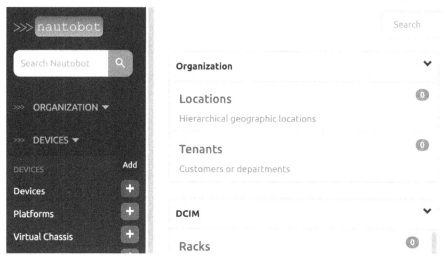

Figure 3.8 – Nautobot UI's Devices tab

In the **Add a new device** form, you can observe a few different sections: **Device, Hardware, Location, Management**, etc. Let's focus on the first one, **Device**:

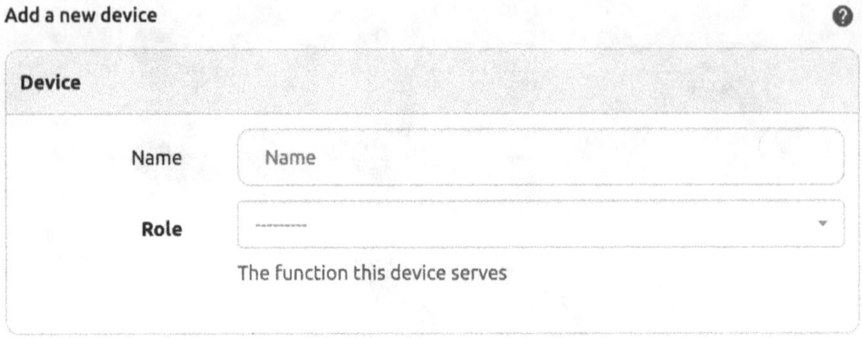

Add a new device   ?

Device
Name	Name
Role	---------- ▾
	The function this device serves

Figure 3.9 – Nautobot UI Device form

Take note of two input fields. **Name** represents the device name (a string), and **Role** is a dropdown menu that indicates that another object (of type **Role**) is expected. Notice that this attribute (**Role**) is in bold. This means that it's a required field, so you must define it when creating a new device. Because we are starting from scratch, you don't have any role created yet, so we have to create at least one role.

There are a few other required objects besides **Role** that are needed before we can actually create a device. When you're viewing the **Add a new device** form, look at all of the bold form fields. That's the shortcut to see all of the required fields that need to be filled in. If there is nothing in a given required dropdown, an object for that type must be created.

These are the required objects and form fields when creating a device:

- **Role**: This represents the logical role of the device – for instance, whether it's a leaf or a spine, or a core router versus an access switch.

- **Device Type**: This represents the model of the device, which often determines certain characteristics, such as the type of interfaces – for example, a DCS-7010T-48-F (from Arista) or an MX480 (from Juniper).

- **Location**: This represents the site where the device is installed. It can have different definitions, including a city, a building, or a room.

- **Status**: This represents the status of the device. Nautobot comes with a pre-populated list of available options: **Active**, **Decommissioning**, **Planned**, and so on. The great part is that you can add your own statuses if needed.

Let's now get started with a concrete example. We want to create a new device that acts as a border router, type A900-IAMC from Cisco, and is located at the ABC site.

Creating a (device) role in the UI

A device role is used to represent the logical grouping of the device depending on its purpose. In the example, we'll define a **Border Router** role. This role could be used to group all the network devices that are on the network edge connecting to the internet. Grouping them together will help the automation later to understand which special characteristics apply to them. For instance, external BGP configuration will only apply to these devices, and not to access campus switches.

Click the + button in the same row as **Organization | Roles** in the sidebar navigation menu.

The following values will be used in the form:

- **Name**: `Border Router`
- **Content Type(s)**: **dcim|device** is used to define which objects can use this role. Roles can apply to different Nautobot objects.
- **Color**: **Dark red**

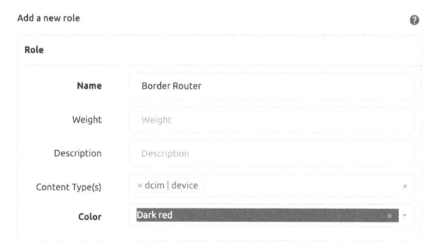

Figure 3.10 – Nautobot UI device role

As you can see, it only requires a name and a predefined color. Next, press the **Create** button at the bottom right of the page.

Adding a device type and a manufacturer in the UI

As the name suggests, a manufacturer is the vendor who built the device being added. In our example, the device type is a Cisco A900-IMA1C, so the vendor in this case is Cisco.

Click + in the same row as **Devices | Manufacturers** in the sidebar navigation menu.

The following values will be used in the form:

- **Name**: `Cisco`
- **Description**: `Cisco Systems` (not mandatory)

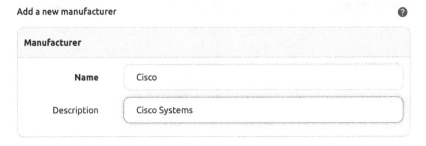

Figure 3.11 – Nautobot UI device manufacturer details

Then, press the **Create** button, and next, you can define the device type that represents the actual device's model or type. This is determined by some characteristics such as interface type and quantities. However, when you go into the device type creation form, you will notice that a **Manufacturer** value is required – fortunately, we added one in the previous step.

The following values will be used in the form:

- **Manufacturer: Cisco**
- **Model**: `A900-IAMC`
- **Height**: `1` (with **Is full depth** checked)

Click + in the same row as **Devices | Device Types** in the sidebar navigation menu, filling in the details as shown in the preceding list. Remember to scroll down and press the **Create** button when you're done!

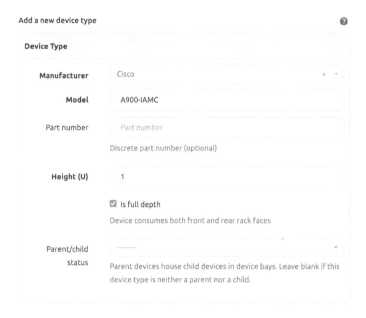

Figure 3.12 – Nautobot UI Device Type configuration

This is an extremely simple definition of a device type. Many more attributes can be defined, such as the components (e.g., interfaces) or the images of the device when rendered in a rack visualization.

Device components are any physical extension of a device – for instance, the interfaces, console ports, and power ports. These components are attached at the device level, but they can be inherited automatically from the device type. So, it makes sense to define them at the device-type level and reuse their definitions. These will be covered in *Chapter 5*.

Creating device types in bulk in the UI

As you may have noticed, defining all the attributes of a device type can take a while. Usually, the specifications are well known as they come from the vendor datasheets, so fortunately, there is an easier way to configure them in the UI. There is a community library at https://github.com/nautobot/devicetype-library that you can use to import the device type definition as a YAML file.

In the **Device Types** view, there is an **Import** feature that allows the creation of a single device type from JSON or YAML.

For instance, let's try to use the ASR-9001 definition from the previous repository with this **Import** feature. In the **Device Types** list view (**Devices | Device Types**), you can see the existing device types, and a blue **Import** button to do a bulk import:

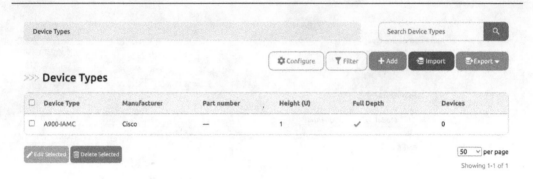

Figure 3.13 – Nautobot UI Device Types list

When you go into the **Import** view, you can provide a YAML (or JSON) dataset that defines the data necessary for this device type as shown in the following screenshot:

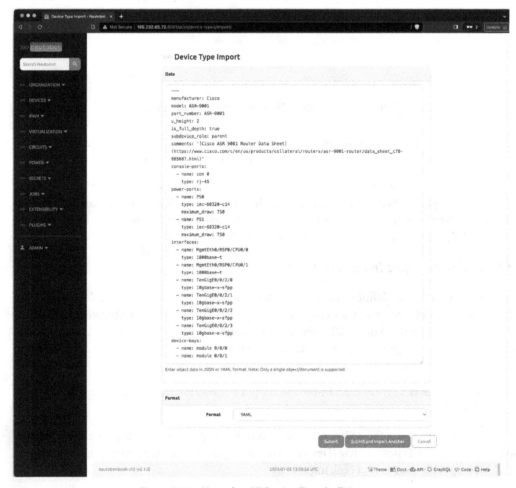

Figure 3.14 – Nautobot UI Device Type bulk import

As you can see, the manifest defines the device type attributes and the components related to this device type.

After hitting **Submit**, you will get a detailed view of the **Cisco ASR-9001** device type with all of its components. Whenever you assign this device type to a device, all these components will be replicated to each device. This means that every new device of this type will have six new interfaces.

Figure 3.15 – Nautobot UI Device Type bulk import result

Creating a location and location type in the UI

The next required reference for a device is the *location* where the device is located. As previously, you can create a location with several attributes even though only the location type, name, and status are required.

Following the same approach as before, we have to create a location type first. It only requires a name, but you can be more specific with the type of objects that can use this location or specify whether a nested relationship exists (for instance, a site that belongs to a region, or a room that belongs to a building).

Click + in the same row as **Organization | Location Types** in the sidebar navigation menu and enter the values shown in the following screenshot:

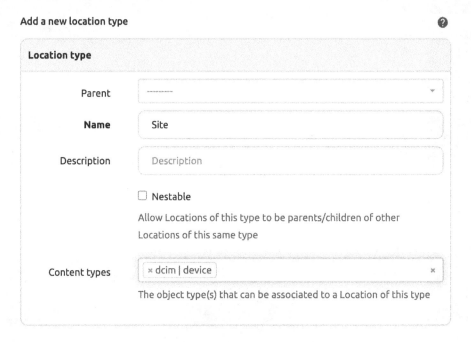

Figure 3.16 – Nautobot UI Location type form

Now, we can create our ABC location using the location type recently created.

Click + in the same row as **Organization | Locations** in the sidebar navigation menu and enter the values shown in the following screenshot:

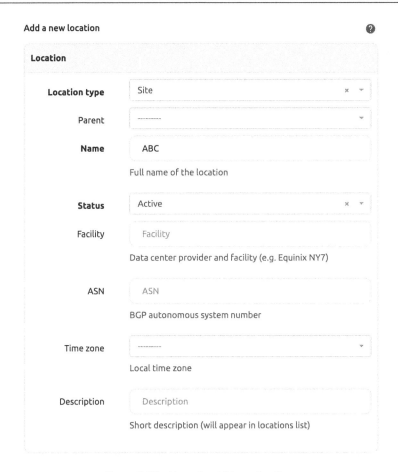

Figure 3.17 – Nautobot UI Location form

Note that status objects come pre-populated with some common values such as **Active**, **Planned**, and so on. You can extend these options with new statuses related to each type of object. When done, scroll down to the end of the form and hit **Create**.

Creating a device in the UI

Finally, you have all the necessary information available to create your very first *device*. Simply select the available options you have—at a minimum, **Role**, **Device type**, **Location**, and **Status**—and a new device will be created.

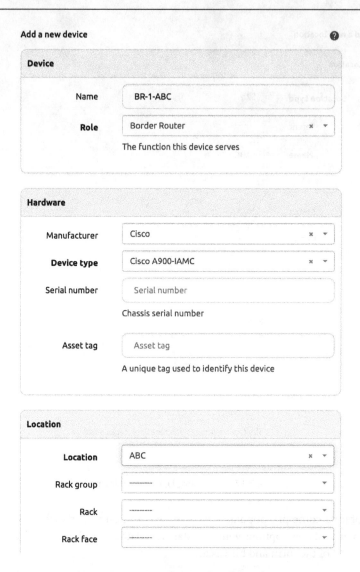

Figure 3.18 – Nautobot UI new device form

After creating a device, you will be redirected to the device's detail view (this behavior could be different depending on the version in use):

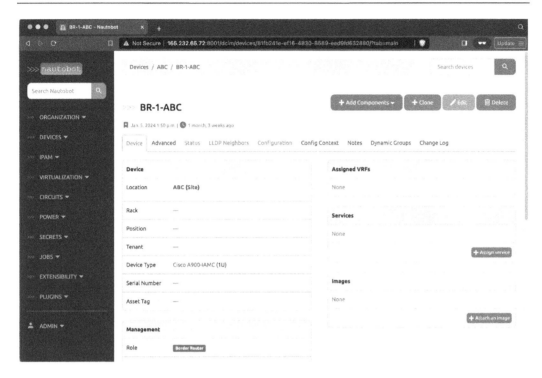

Figure 3.19 – Nautobot UI device detail view

This simple walkthrough provided a basic introduction to managing objects in the GUI. This will be extended in the next chapter.

Summary

This chapter provided a high-level view of the Nautobot architecture components and demonstrated how to get up and running with a development instance on an Ubuntu server. This is just the starting point of your Nautobot learning journey.

In the next chapter, you will better understand many of the models available in the UI and start to populate even more data using the Nautobot UI. Then, in *Chapter 6*, you'll go even further, learning about the Nautobot extensibility features.

4

Understanding the User Interface and Bootstrapping Nautobot

Building on an installation from *Chapter 3*, this chapter explores deploying and configuring Nautobot. You will get to see how to add objects to Nautobot and build out a real use case of a network while seeing firsthand the data models and relationships described in *Chapter 2*.

Managing and maintaining a network can be a challenging task, especially as the number of devices and services continues to grow. However, with the right tools and processes, you can streamline your network management and simplify the management of your inventory.

In this chapter, we will guide you through the navigation capabilities of Nautobot and its **user interface** (**UI**) and walk you through managing inventory and bootstrapping your installation. Whether you're an experienced network engineer or a newcomer to the field, this chapter will provide you with the knowledge you need to successfully deploy and use Nautobot in your network infrastructure.

The following are the main topics that will be covered in this chapter:

- Understanding the navigation and UI
- Managing inventory and bootstrapping your first installation

> **Pro tip**
> You can follow many of this chapter's examples if you have deployed your Nautobot instance from *Chapter 3*. Alternatively, you can use the always on demo.nautobot.com cluster, which resets its configuration every night.

Understanding the navigation and UI

The Nautobot UI is designed to provide a user-friendly and intuitive way to manage your network inventory. It includes a range of features and components that make it easy to navigate the application, view and edit data, and customize your experience. In this section, we will explore the main components of the Nautobot UI, which includes the following primary elements:

- Navigation menu
- Panels
- Footer navigation
- Detailed view tabs
- Customized views

By the end of this section, you'll have a thorough understanding of the UI, and be able to make the most of its features when managing your network inventory.

Navigation menu

Beginning in Nautobot 2.1, the upper navigation menu (NavBar), which was previously located at the top of the screen has moved to the left side. It provides access to the main sections of the application, including inventory, circuits, IPAM, and more. From here, you can quickly navigate to the parts of the application you need, without having to search through multiple menus or pages. The navigation menu is broken down into several major categories to help streamline your access to important and often-used features. Let's take a look at the navigation menu:

Figure 4.1 – Nautobot sidebar navigation menu

The NavBar's layout is intuitive. Its top-to-bottom sequence mirrors the natural progression you would follow during an initial setup and system configuration. This design ensures that as you advance from one category to the next, you're building upon previous information, mirroring the dependencies seen in the data models discussed earlier.

Figure 4.2 – Top-down navigation starting with ORGANIZATION

Moreover, within each drop-down menu, items are thoughtfully arranged in order of frequency of use. This prioritization contrasts with the overall NavBar organization, which, as mentioned, is more about the sequential flow of setup tasks.

> **Pro tip**
>
> If you ever find yourself lost in the navigation of Nautobot, you can click on the *Nautobot* logo at the top of the NavBar. This will take you to the home page of Nautobot.

At the bottom of the NavBar, you should see an icon with your user ID and icon.

From this menu, you will have the ability to modify your user profile and log out of the system. Additionally, depending on your role within the organization, the user icon provides access to the Nautobot administration features.

Figure 4.3 – Modify user profile and admin settings

This is where users, groups, and granular permissions (as well as API keys) can all be configured.

Next, we will explore the Nautobot home page.

Nautobot home page and panels

The Nautobot home page is your starting point for managing your infrastructure. It consists of several panels, each representing a different area of functionality within Nautobot. These panels are preconfigured for convenience and are a way for Nautobot users to see things at a glance and also have quick links available to get to configuration items quickly. The panels loosely represent the upper navigation menu and are arranged by popularity. In the following screenshot, you can see the **Organization** panel highlighted so you can see how panels are placed on the home page.

Figure 4.4 – Nautobot home page

Just above the panels, you will find two very useful features: the notification area and the universal search bar. When new versions of Nautobot are available, you will see a notification toast message appear just above the panels on the home page as you can see in the following screenshot:

Figure 4.5 – Nautobot release notifications shown in a banner

Release notifications can be enabled by going to your profile, clicking on **Admin | Config**, looking for `Release Checking`, and then adding the `https://api.github.com/repos/nautobot/nautobot/releases` URL.

Notifications can also contain links. In the example, you can see that the instructions on how to upgrade Nautobot to the latest version are included in the notification.

The universal search bar provides a quick and convenient method for finding objects within the system. This can be particularly useful if you have added thousands of items to the system. It can narrow the scope of the search as well by selecting the dropdown to the right of the search bar.

Now that we have a good understanding of the home page layout, let's take a look in further detail at what each panel provides.

Organizations

This panel displays the number of locations and tenants that have been created in Nautobot. The objects within the organizations category represent a logical grouping of assets and devices that share a common ownership or hierarchical/physical location. Each of the items provides quick links to the configuration page for the topics displayed.

Figure 4.6 – Organization shown on the Nautobot home page

Data Center Infrastructure Management (DCIM)

This panel provides a summary of the data center equipment and infrastructure that is managed by Nautobot. It includes the number of racks, device types, devices, virtual chassis, device redundancy groups, and connections that are currently being tracked. You are certain to spend a lot of time in this area, considering this is one of the primary use cases that Nautobot provides. Each of the items provides quick links to the configuration page for the topics displayed.

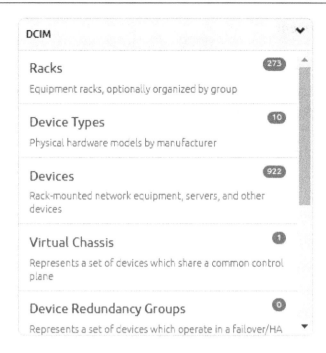

Figure 4.7 – DCIM shown on the Nautobot home page

Power

This panel provides a summary of the power infrastructure that is managed by Nautobot. It includes the number of power feeds and power panels that are currently being tracked. Power is something that's often overlooked by IT organizations, and Nautobot has a powerful way of representing these physical connections. Each of the items provides quick links to the configuration page for the topics displayed. This topic will get more coverage later on.

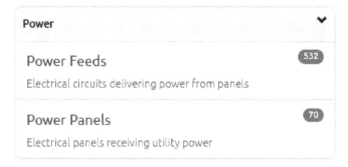

Figure 4.8 – Power shown on the Nautobot home page

IP Address Management

This panel provides a summary of the IP address space that is managed by Nautobot. It includes the number of IP addresses that are in use, as well as the number of available addresses. It also provides information on VRFs, VLANs, and Prefixes. Each of the items provides quick links to the configuration page for the topics displayed. **IP Address Management (IPAM)** is a large topic area that will also get some direct coverage later in this chapter.

Figure 4.9 – IPAM on the Nautobot home page

Circuits

This panel displays the number of circuits that have been created in Nautobot. A circuit represents a communication link and can be used to model connections between network devices. It is important to note that cables are distinct from circuits within Nautobot. We will discuss how to connect an interface to a provider as we review our first device (the internet border router), which was onboarded at the end of *Chapter 3*. Circuits will be covered in more detail later in this chapter.

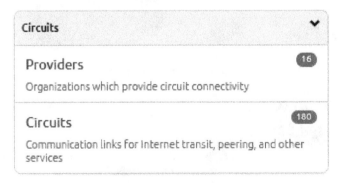

Figure 4.10 – Circuits shown on the Nautobot homepage

Virtualization

This panel displays the number of virtual compute instances and clusters of physical hosts that are being managed by Nautobot. Each of the items provides quick links to the configuration page for the topics represented in the image here. Virtualization consists of cluster definitions and virtual machines.

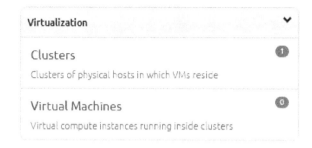

Figure 4.11 – Virtualization shown on the Nautobot home page

Data Sources

The **Data Sources** panel on the Nautobot home page provides an overview of the available Git-based data sources. A data source is a repository that provides information about devices or other providers relevant to the Nautobot environment. You can use a Git-based data source to import jobs, export templates, config contexts, and much more, all of which are covered in *Chapter 6*. When a data source is configured in Nautobot, the information is automatically imported and can be used for a variety of purposes.

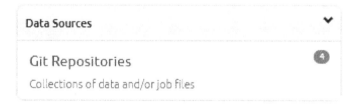

Figure 4.12 – Data Sources shown on the Nautobot home page

In the **Data Sources** panel, you can view the number of configured data sources and also view the data source configuration by clicking the quick link. It is important to note that you cannot see the synchronization status or Git repository names from this panel. The good news, however, is that these details can be found in the **Job History** section. The status of each configured data source, including the last time the data was synced with Nautobot, will show up there. Data sources will be covered in more detail in *Chapter 6*.

Job History

This panel provides a summary of the jobs that have been executed in Nautobot. It includes information on the number of successful and failed jobs, as well as the average time taken to execute a job. You can quickly get to a jobs management page by clicking the quick link on the particular job shown in the view. Jobs are covered at great length in *Chapter 11*.

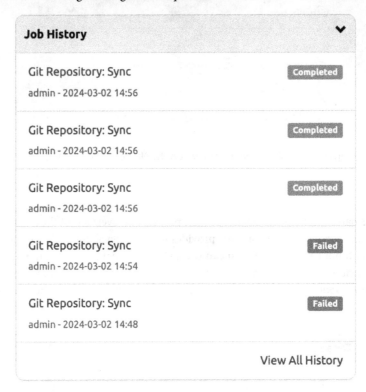

Figure 4.13 – Job History shown on the Nautobot home page

As you can see in this example, we have a Git data source Job that has failed synchronization. Something that should be reviewed and rectified! Nautobot **Job History** is not only informational, it can also be *actionable*.

Change Log

This panel displays the most recent changes that have been made to the configuration in Nautobot. This can include changes to device configurations, IP addresses, and other asset data. This can come in handy for audits and traceability.

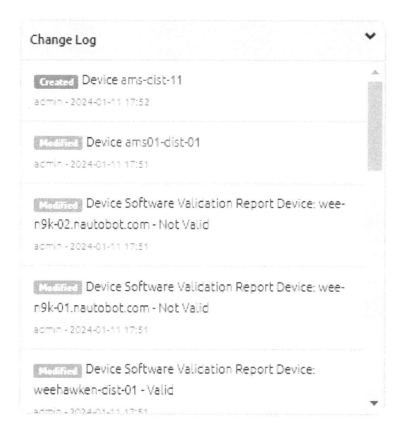

Figure 4.14 – Change Log shown on the Nautobot home page

Each of these panels provides a quick and easy way to get an overview of your infrastructure and identify any issues that need to be addressed. As you delve deeper into Nautobot's functionality, you'll find that each panel has its own set of tools and options to help you manage your environment with ease, we've only just begun scratching the surface!

> **Tip**
>
> It's important to note that Nautobot has great flexibility and allows you to use the developer API and Nautobot Apps to add your own panels and content to the home page. We cover this in *Chapters 14, 15*, and *16*.

Footer navigation

The footer navigation in Nautobot is a useful and often overlooked element of its user interface. Located at the bottom of the screen, it provides easy access to a range of important tools and functions, including the version of Nautobot you're currently running, the date and time, and quick links to other key resources.

Figure 4.15 – Footer shown at the bottom of every page

In the left corner of the footer, you'll find the Nautobot version number, which can be particularly useful when troubleshooting issues or identifying which features are available. This provides critical information on the version of Nautobot you're running, allowing you to compare it against the latest version and take advantage of any new features or enhancements.

The footer navigation also includes quick links to a range of resources that can be helpful when working within Nautobot. You can access the Nautobot documentation, API, and GraphQL interfaces, as well as the code repository and help resources, all from the comfort of the footer navigation.

In this section, we'll explore the various components of the footer navigation and discuss how they can be utilized to improve your experience working within Nautobot. From checking the Nautobot version to accessing key resources, the footer navigation provides quick and easy access to the tools and information you need to make the most of Nautobot's powerful network automation features.

Theme

The **Theme** quick link in the footer navigation was first introduced in Nautobot version 1.4.0, and has since become a popular and frequently used feature. One of the key benefits of Nautobot being an open source project is that the community can actively participate in its development, and this was particularly true with the introduction of the **Theme** toggle.

When you click the **Theme** quick link, you will see a popup allowing you to choose which option you'd prefer to choose.

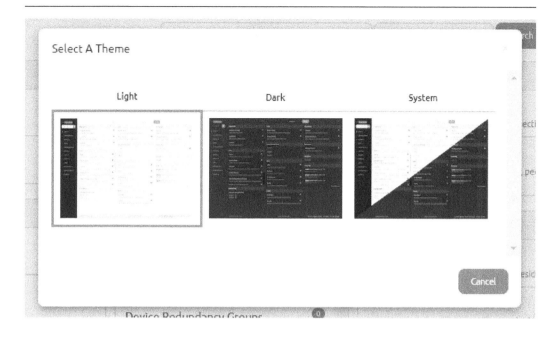

Figure 4.16 – Light mode and dark mode options

Thanks to the open source nature of Nautobot, anyone using it can view the development process behind new features such as the **Theme** toggle, including the discussion and decision-making that led up to its release. This helps to foster a sense of community and collaboration among Nautobot users and encourages active participation in the development and evolution of the project.

Docs

The Nautobot documentation page is an extensive resource that provides information on everything from getting started with Nautobot, to more advanced topics such as API documentation, customizations, and troubleshooting. The site serves as an important tool for users who want to make the most of Nautobot's many features and capabilities. Nautobot comes bundled with a local instance of the documentation that reflects the *current version* you are running. This is super convenient for scenarios where you may have highly secure and restricted network access.

The Nautobot online documentation is very well organized and has had a recent overhaul, including the apps (formerly referred to as plugin-ins) that *Network to Code* sponsors. The site is constantly updated to reflect the latest changes and updates to the platform as well as its associated apps and plugins. For the latest documentation, it is always a good idea to check out the online documentation.

> **Note**
> The link to the *latest* Nautobot online documentation for the core application and Nautobot Apps can be found at `https://docs.nautobot.com`.

API

Nautobot's API is a powerful tool that allows users to automate and streamline many of the tasks associated with managing a network.

Clicking on the **API** quick link will take you to the Nautobot API documentation page, where you can find detailed information on how to use the API to interact with Nautobot. The API documentation page provides information on the different endpoints and parameters available in the API, as well as examples of how to use the API to accomplish specific tasks.

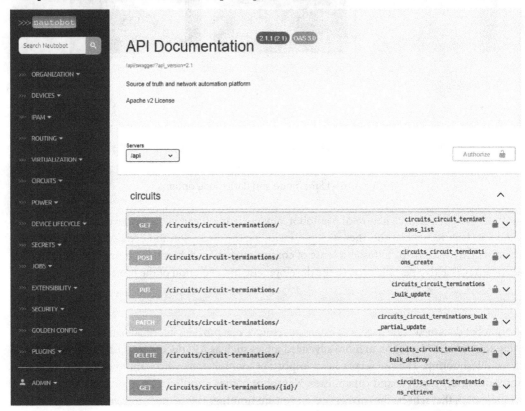

Figure 4.17 – API interactive documentation accessible within Nautobot

While the API documentation page can be a powerful tool for users who are looking to automate their network management tasks, it is also a complex resource that may take some time to fully understand. In *Chapter 8*, we'll provide an overview of the API documentation page and its various features and offer some tips for getting started with the Nautobot API. This feature is very powerful and will become a core resource for you to leverage along your network automation journey.

GraphQL

The **GraphQL** quick link in the footer navigation provides users with a convenient way to access Nautobot's GraphiQL interface. GraphiQL is a powerful tool that enables users to interactively query Nautobot's database using the GraphQL API. The GraphiQL interface features a range of capabilities, including autocompletion, error highlighting, and API documentation.

We encourage users to experiment with the GraphiQL interface by clicking the **GraphQL** quick link in the footer navigation. Once there, you'll be able to test GraphQL APIs, allowing you to query data from Nautobot.

To learn more about GraphiQL, you can visit *Chapter 8*.

Code

Clicking on the **Code** quick link will take you directly to the Nautobot repository on GitHub, where you can browse the source code, submit issues, and contribute to ongoing development efforts.

The Nautobot codebase is open source, which means that anyone can contribute to its development or examine its source code. This openness provides transparency and fosters collaboration within the Nautobot community.

If you're interested in contributing to Nautobot's development or have encountered an issue that you'd like to report, clicking the **Code** quick link is a great place to start. Additionally, reviewing the source code can be an excellent way to understand how Nautobot works and how various features and functionalities are implemented.

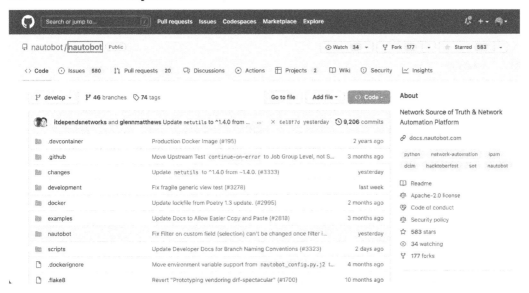

Figure 4.18 – Nautobot code repository on GitHub

> **Note**
>
> The link to the *Nautobot Open Source Code Repository* can be found at `https://github.com/nautobot/nautobot`.

Help

The **Help** quick link is a valuable resource for users looking for additional guidance and support.

When you click on the **Help** link, you will be taken to a wiki page that contains a wealth of information, including a space for living documentation and guides that change more often than the official documentation. Here you can find useful information on various topics, such as issue management, personas, definition of done, and more.

If you are interested in contributing to the Nautobot community or simply want to stay up-to-date with the latest developments, you may find the **Community** section particularly useful. This section contains an index of community-focused content such as blogs, videos, and presentations, as well as information on monthly community meetings.

The Nautobot roadmap and release schedule is also available here for those who want to keep track of new features and updates. Whether you are a new user or an experienced Nautobot contributor, we hope you will find these resources helpful.

> **Note**
>
> The link to the *Nautobot Help Wiki* can be found at `https://github.com/nautobot/nautobot/wiki`.

Table views

For almost all cases within Nautobot, when navigating through a NavBar menu the first place you'll land is on a list of items. These tables represent a quick way to see all the items within a particular category and are a great way to see objects within Nautobot and get to them quickly. Most of these list views are sortable and exportable as well, which provides a powerful way to interact with the data once it has been populated.

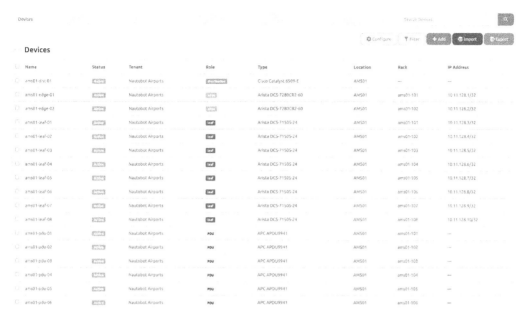

Figure 4.19 – Table view (Devices)

There are several notable capabilities to mention about the **Table** view. When a view supports it, you will see a menu above the table on the top right as shown:

Figure 4.20 – Table view (Capabilities)

Configure

This option allows you to select which column fields you would like to show within the **Table** view. It also allows you to reorder the columns to your preferences. There may be performance considerations when selecting how many columns you would like to present each time you navigate to this view, so it is important to be selective in what you want to show. When in doubt, you can always reset the view by choosing the **Reset** button in the **Table Configuration** menu.

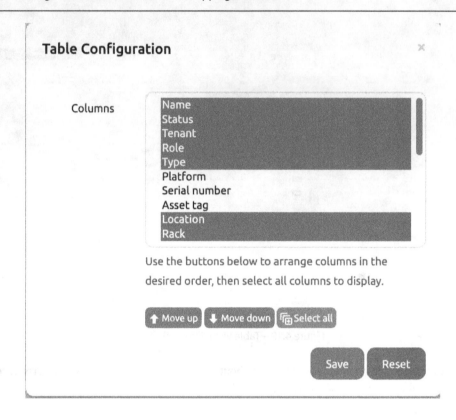

Figure 4.21 – Table view (Configure option)

Filter

The **Filter** option provides a powerful way to use data fields to selectively show what appears in the **Table** view. The following screenshot depicts an example of what you might see for valid filter options. Filter forms will be covered later in this chapter.

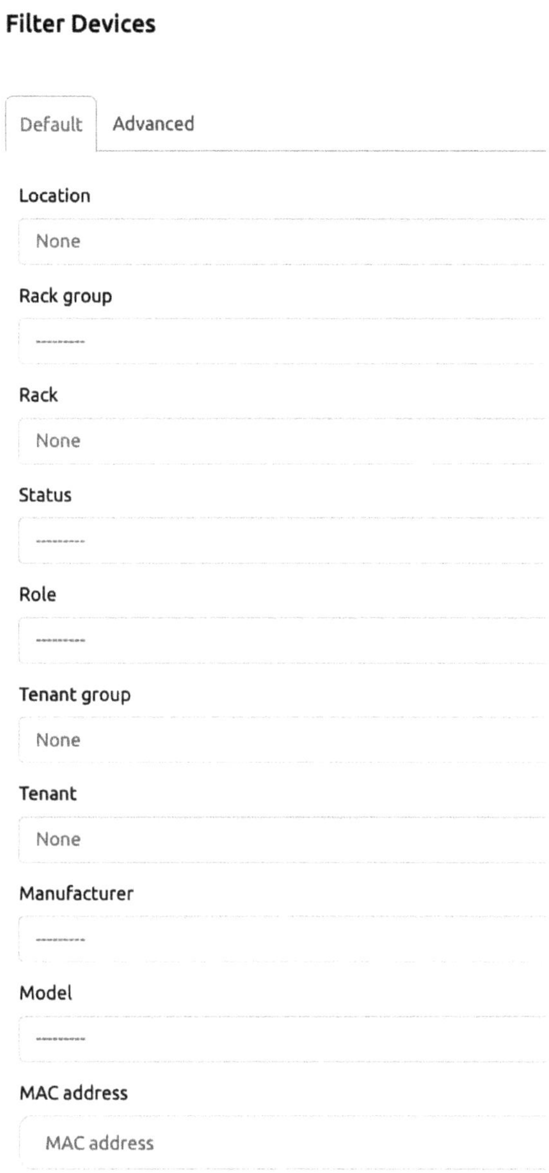

Filter Devices

Default	Advanced

Location

> None

Rack group

> ---------

Rack

> None

Status

> ---------

Role

> ---------

Tenant group

> None

Tenant

> None

Manufacturer

> ---------

Model

> ---------

MAC address

> MAC address

Figure 4.22 – Filter (Devices)

Bulk Import

This quick link will take you straight to the bulk import page for the specific object type you are on in the current **Table** view. Bulk importing is just one of many ways to populate data into Nautobot. When importing in this manner, you can see all the available fields right in the user interface and whether or not the fields are required. Further, it is important to know that the order of operations matters when importing data to Nautobot. More information on this can be found in the *Managing inventory and bootstrapping your first installation* section.

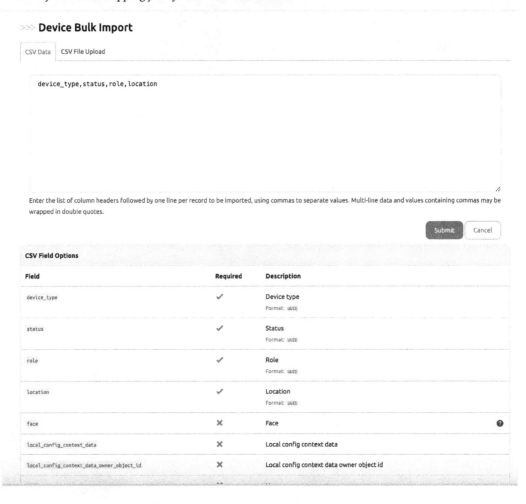

Figure 4.23 – Bulk Import page with Field Options (Devices)

Detailed views

When working with objects in Nautobot, you'll often encounter detailed views that display all the relevant information and options for a specific object, such as a device, IP address, or circuit. These detailed views are organized into tabs, which help you navigate through different categories of information and settings related to the object you are working with. In this section, we will explore the structure and functionality of the **Detailed View** tabs and discuss how they can be used to efficiently manage and configure objects in Nautobot. If you'd like to follow along, you can visit `https://demo.nautobot.com`, log in, click **DEVICES** in the NavBar, and then click the first device in the list.

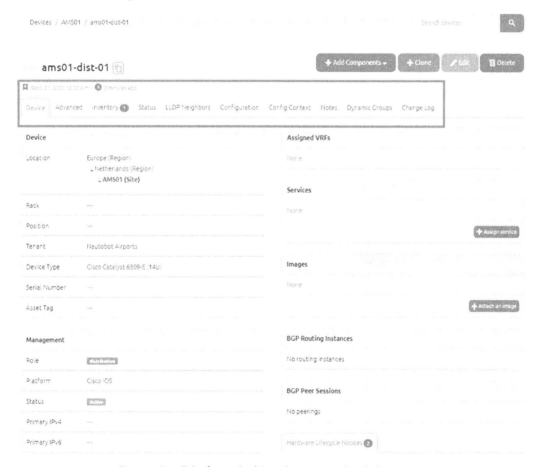

Figure 4.24 – Tabs for each object shown on a detailed page view

The tabs on the **Detailed View** page, as shown in the preceding screenshot, can vary depending on the object type, but generally, they are organized into the following categories:

- **Selected Object**: This tab provides an overview of the object's basic information, such as its name, description, and any relevant identifiers. Depending on the object type, this tab may also display related information, such as parent objects or relationships with other objects. In the preceding example, we have selected a **Device** object.

- **Advanced**: This tab displays the object's **unique identifier** (**UUID**) and the date/time it was created and last updated. These unique identifiers are leveraged extensively for things such as making API calls from automation jobs or scripts, as well as being leveraged to generate URL detail views. When viewing any object within Nautobot, the UUID is often seen in the URL of the browser itself.

- **Notes**: This tab provides a way to annotate an object within Nautobot. It can be quite useful for things such as capturing additional details about the object, as well as leaving useful verbiage for administrators to review. It can be a powerful feature when combined with automation workflows.

- **Change Log**: This tab displays a log of all changes made to the object, including when the changes were made and who made them. This can be useful for tracking the history of an object and understanding how it has evolved over time.

By leveraging the **Detailed View** tabs, you can efficiently navigate through the different aspects of an object and manage its configuration and relationships with other objects in the system. This will enable you to have a comprehensive understanding of your network infrastructure and make informed decisions about how to optimize and automate its management. Detailed tabs can help you quickly navigate to the items you need when necessary. It's important to note that detailed tabs are context sensitive and will show more tabs depending on which object you are navigating.

> **Tip**
>
> NavBar items and detailed views can all be customized using the developer API and Nautobot App development.

Managing inventory and bootstrapping your first installation

In the previous chapter, you witnessed the successful installation of Nautobot and the creation of a device. You might have observed a certain order of operations during the setup process. To help you navigate the ecosystem more efficiently, we'll delve deeper into these operations and elaborate on strategies that can streamline your Nautobot deployment. The user interface of Nautobot has been designed with user-friendly navigation in mind. Familiarizing yourself with the order of common configurations can significantly reduce the need to jump between various UI elements.

> **Note**
>
> Nautobot Location Hierarchies were introduced in version 1.4, and replaced Sites and Regions in Nautobot 2.x. This greatly simplifies the way sites, regions, locations, and organizational hierarchies are considered and also creates greater flexibility.

Identifying your data

Understanding how to structure and populate your Nautobot instance starts with categorizing the data types relevant to your network. Here's a suggested approach for kick-starting your Nautobot deployment:

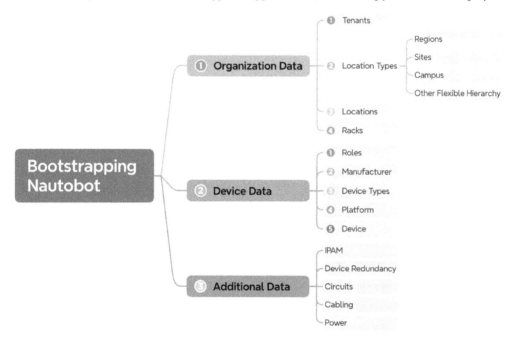

Figure 4.25 – Bootstrapping Nautobot with data

Organizational data

The following section delves into the intricacies of organizational data within Nautobot, highlighting essential components, such as tenants, locations, racks, and more. Each element plays a pivotal role in creating a cohesive and efficient network infrastructure, tailored to meet the specific needs of your organization.

For the purposes of illustrating the following examples, we will leverage an imaginary organization called *Wayne Enterprises* that has locations around the world. Bruce Wayne, the executive leader at Wayne Enterprises, needs to modernize the infrastructure at the organization and has challenged the networking team to automate "all the things." The automation team lead has recognized the benefits of a data-centric approach to automation and has recommended Nautobot to be leveraged throughout the organization.

Tenants

Tenants are useful when there's a need to segment or partition data in Nautobot based on organizational divisions, such as different departments, clients, or business units. Remember, many objects within Nautobot can reference tenants, so it's good to have them set up early on. Let's create our first tenant – *Wayne Enterprises*.

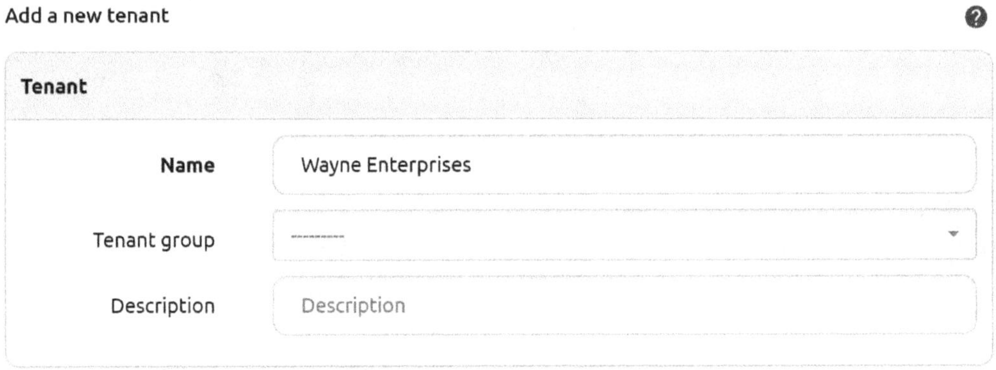

Figure 4.26 – Add new tenant form

Location types

Before you can pinpoint a device to a specific spot, you need to define the types of spots available and design the hierarchy for how you want to organize things. Location types are abstract categories that give context to the actual locations you'll define next. They can dictate what can be associated with each type. There are several optional settings within location types that allow you to control what content types of objects can be associated with them, as well as providing the ability to control whether a same location type can be nested. As an example, you may want to allow a region to have other regions within it, but maybe a site should not contain another site.

Let's create some structure for the Wayne Enterprises organization and its location hierarchy. We will create a few example location types to illustrate how you can design a flexible hierarchical structure that accurately represents your organization.

Example – regions

Regions can be configured hierarchically (nestable), so be prepared with your data and how you want to represent it. You will want to start with the high level first and work your way toward the more specific as you build out a flexible hierarchy that represents your organization. In this example, we will plan to allow regions to have other regions nested within them, and control the content types (objects) that are allowed to be associated with them.

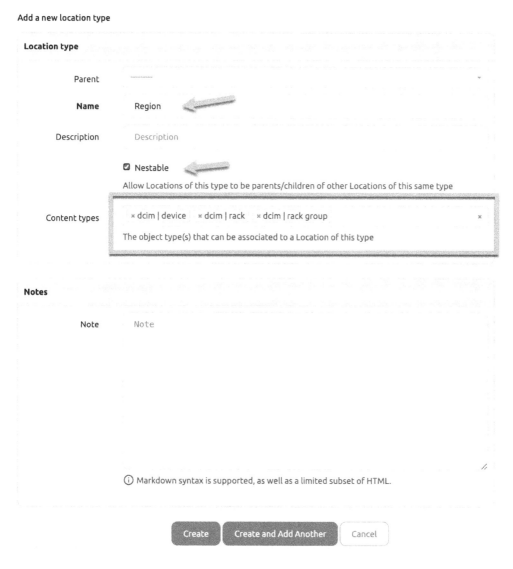

Figure 4.27 – Add new location type form for a region

Example – sites

Let's drill down one level deeper and create a location type that will represent sites. For our example, we will choose to *not* allow sites to be nestable within each other. We will also embrace the hierarchical nature of locations and be deliberate about allowing sites to be members of regions. In other words, a site must be represented with a region as its parent, as can be seen in the following example. For the site location types, we have set the content types to allow: devices, power panels, racks, and rack groups.

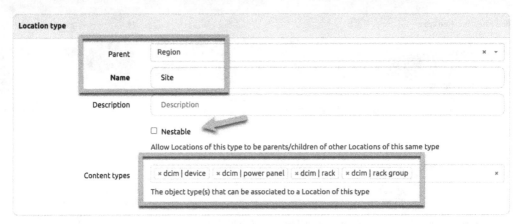

Figure 4.28 – Add a new location type (Site)

Example – campus

Lastly, let's create a location type called Campus that can represent a group of buildings. We've taken an additional step in this example to state that campuses will be members of Site.

As you can see, location types are very flexible and can get more granular depending on your needs. You can now imagine continuing to build buildings, data centers, wiring closets, and so on. This feature is very powerful and can allow for all sorts of organizations to accurately represent their structure within Nautobot.

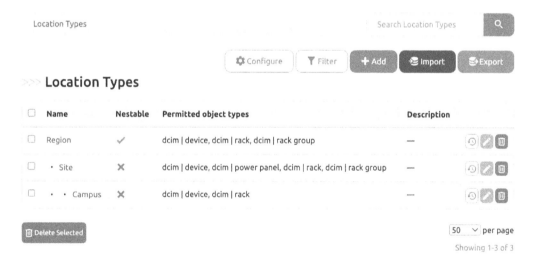

Location type

Parent	Region → Site
Name	Campus
Description	Description

☐ Nestable

Allow Locations of this type to be parents/children of other Locations of this same type

Content types   × dcim | device   × dcim | rack   ×

The object type(s) that can be associated to a Location of this type

Figure 4.29 – Location type form (campus)

OK, great! We now have the beginnings of Wayne Enterprises location structures in place. You can view your design from the list view of the **Location Types** menu, as can be seen here:

Location Types   Search Location Types   🔍

⚙ Configure   ▼ Filter   ＋ Add   🗲 Import   🗲 Export

⫸ **Location Types**

	Name	Nestable	Permitted object types	Description	
☐	Region	✓	dcim | device, dcim | rack, dcim | rack group	—	🕓 ✏ 🗑
☐	• Site	✗	dcim | device, dcim | power panel, dcim | rack, dcim | rack group	—	🕓 ✏ 🗑
☐	• • Campus	✗	dcim | device, dcim | rack	—	🕓 ✏ 🗑

🗑 Delete Selected   50 ∨ per page

Showing 1-3 of 3

Figure 4.30 – Location Type hierarchical design

Locations

After defining location types, you are now empowered to specify the actual locations. These are tangible places where your network assets are situated. Locations can be as broad as an entire country or as specific as a particular room on a floor. Let's walk through a few examples that use the location types we defined earlier.

Example – region

In this example, we will create a location called **Americas** with a location type set to **Region**. Note that the fields in bold are required fields. There are many optional fields, and you can start to understand how you might be able to leverage these fields in the future. For the purposes of this chapter, we are bootstrapping the system, so we will primarily highlight the required fields.

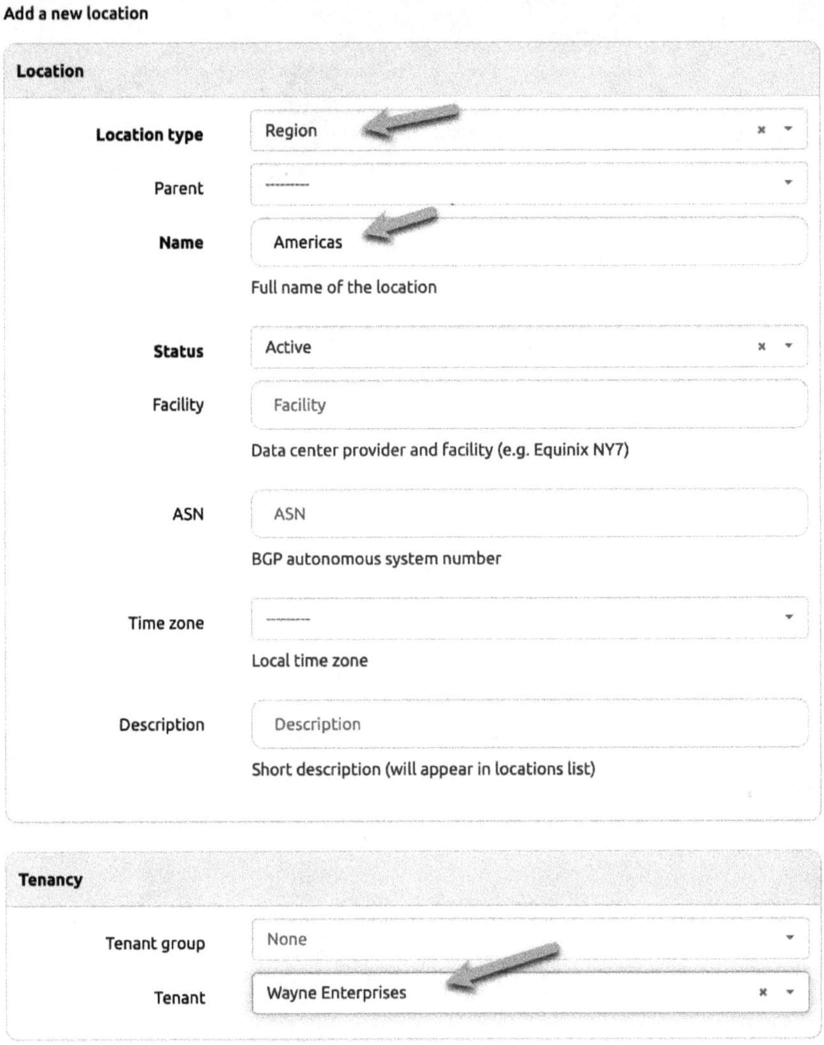

Figure 4.31 – Add new Location form (representing a Region)

Example – site

Next, we will create a site called **WayneHQ** with the **Location type** set to **Site**. This site will have the **Americas** region set as the **Parent** (*Figure 4.32*). As you create sites, think about properties such as the physical address, contact details, and any specific notes or descriptions relevant to the location. As you can see in *Figure 4.33*, **Sites** can also contain geolocation attributes, such as latitude and longitude. These attributes can be incredibly interesting when interacting with other systems, such as telemetry and observability. Imagine how powerful this can be when visualized.

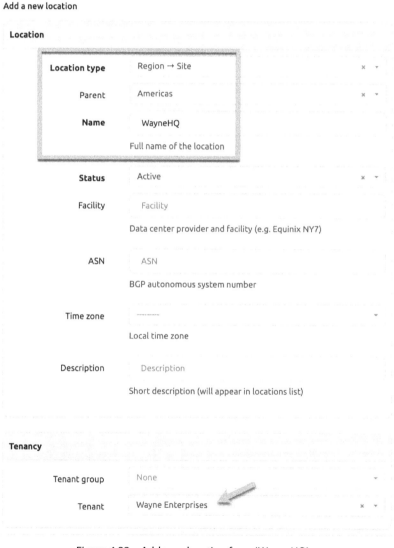

Figure 4.32 – Add new location form (WayneHQ)

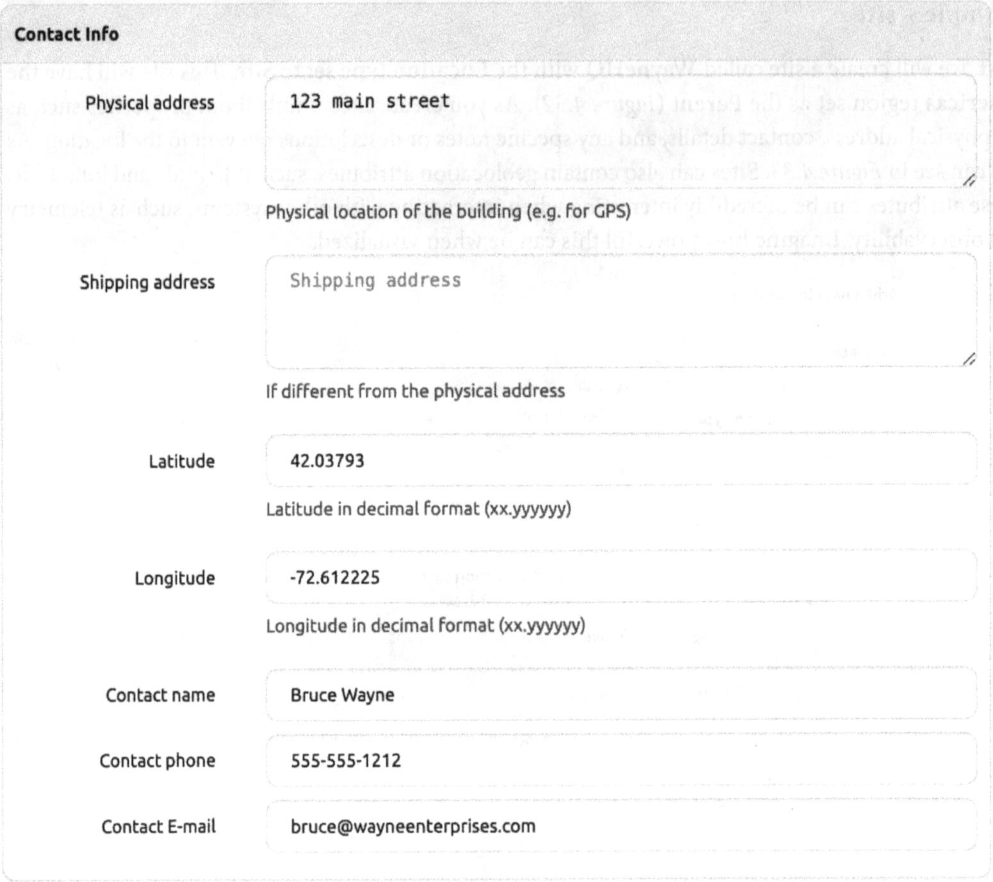

Contact Info

Physical address	123 main street

Physical location of the building (e.g. for GPS)

Shipping address	Shipping address

If different from the physical address

Latitude	42.03793

Latitude in decimal format (xx.yyyyyy)

Longitude	-72.612225

Longitude in decimal format (xx.yyyyyy)

Contact name	Bruce Wayne
Contact phone	555-555-1212
Contact E-mail	bruce@wayneenterprises.com

Figure 4.33 – Add new location form (Contact Info)

Racks (optional, but often next)

After organizing the top-level structures, such as tenants and location hierarchy, it's likely that you will want to pinpoint the precise physical structures where your devices reside. This brings us to the granularity of racks. Racks are representations of the physical structures within a data center or network room that house your devices. These can range from tall cabinets that host multiple servers to smaller wall-mounted frames for more specific equipment.

Racks have front and rear faces for device mounting and allow you to specify rail-to-rail width, ranging from 10, 19, 21, to 23 inches. Devices are then able to be located based on the rack unit specifications that are associated with the devices assigned to the rack. Coming up is a representation of a 42-unit rack located in the WayneHQ site. No devices have been loaded (yet). As you create a new rack, you can set the width and height to match your physical infrastructure.

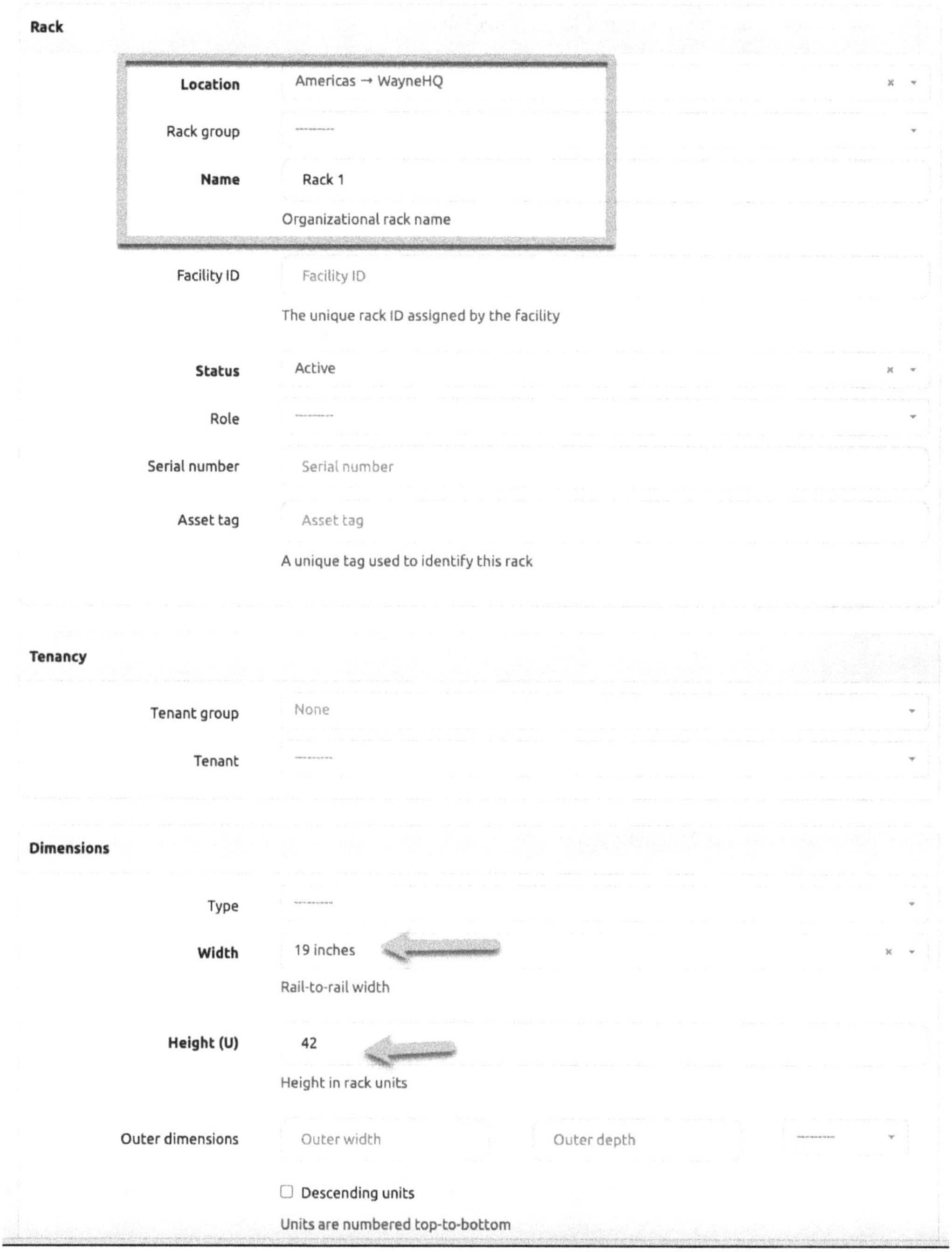

Rack

Location	Americas → WayneHQ	x ▾
Rack group	-------	▾
Name	Rack 1	
	Organizational rack name	

Facility ID	Facility ID
	The unique rack ID assigned by the facility

Status	Active	x ▾
Role	-------	▾
Serial number	Serial number	
Asset tag	Asset tag	
	A unique tag used to identify this rack	

Tenancy

Tenant group	None	▾
Tenant	-------	▾

Dimensions

Type	-------	▾
Width	19 inches	x ▾
	Rail-to-rail width	
Height (U)	42	
	Height in rack units	
Outer dimensions	Outer width Outer depth ------- ▾	
	☐ Descending units	
	Units are numbered top-to-bottom	

Figure 4.34 – Add new rack form

Once you add a rack, you'll be redirected to the rack's detailed page view:

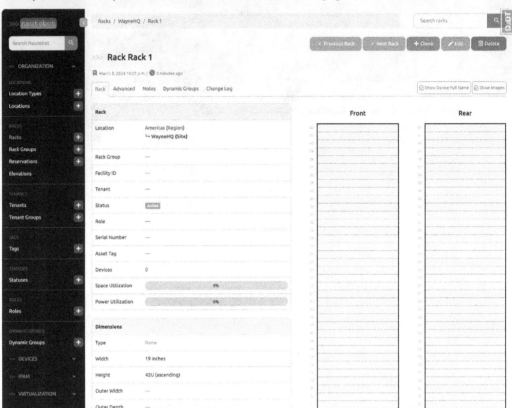

Figure 4.35 – Detailed view of a single rack

Roles

Roles have been a part of Nautobot from its inception, but the concept was previously tied to specific object types (such as device-specific roles). Nautobot 2.x introduces a powerful enhancement that decouples roles from devices and makes them available to use in other places within the system. The objects are referenced as *content types*. The following figure shows the **Table View** representation of a Nautobot instance that has been populated with various roles, which may help inspire you as to what you might leverage them for.

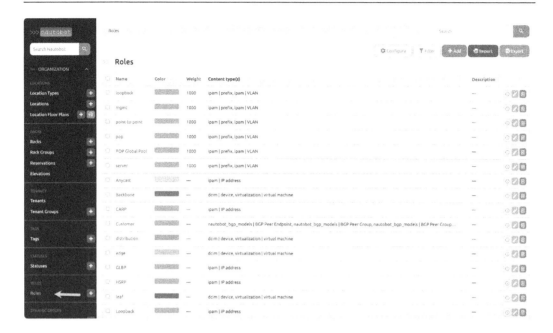

Figure 4.36 – Roles (various examples)

Device data

In this section, we'll dive into how Nautobot helps you manage device data, a key part of keeping your network organized and running smoothly. We start by looking at roles, which help you classify your devices based on their roles in the network, such as routers or servers. This step is crucial for quick and easy categorization when adding new devices. We'll also explore how to keep track of your hardware by registering manufacturers and defining device types, ensuring that each piece of equipment in your network is accurately represented. Additionally, we'll discuss the role of platforms in managing different network operating systems and their importance in network automation. Finally, we'll guide you through adding devices to Nautobot, a vital step in establishing a solid foundation for your network infrastructure. This section is all about the nuts and bolts of device data management in Nautobot, setting you up for success in your network management journey.

Role assignment

As previously mentioned in the *Organizational data* section, device roles were enhanced in Nautobot 2.x. They now leverage a more flexible data model under the hood and can be used elsewhere in the system. That said, it is common to assign roles to devices, and this section provides some details about this.

Device roles allow you to organize and classify devices based on their function on the network. Before adding devices, you will want to define *device roles* such as Router, Switch, and Server. This will facilitate faster categorization during device addition. Here is an example of a device role for **Firewall**:

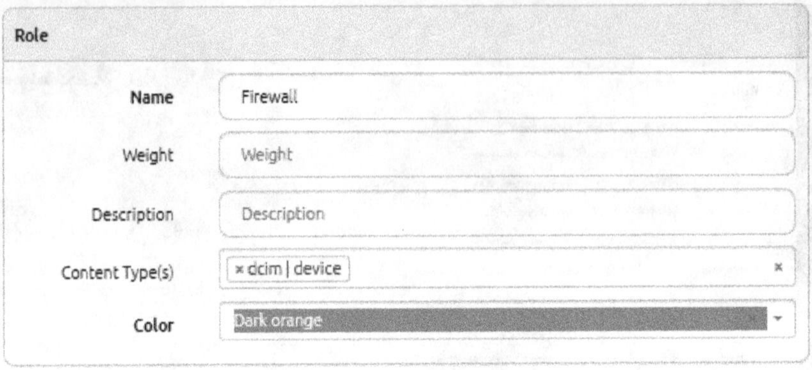

Figure 4.37 – Add new role form

When adding a new device, you can see that assigning a role is the second item in the **Add Device** form, immediately following **Name**. This is where you will take the organization-level roles and choose from those that you have created for your needs. If your **Role** was created with a content type of **dcim|device**, that means it will only show those roles in your selection option when adding the device, as can be seen in the following screenshot:

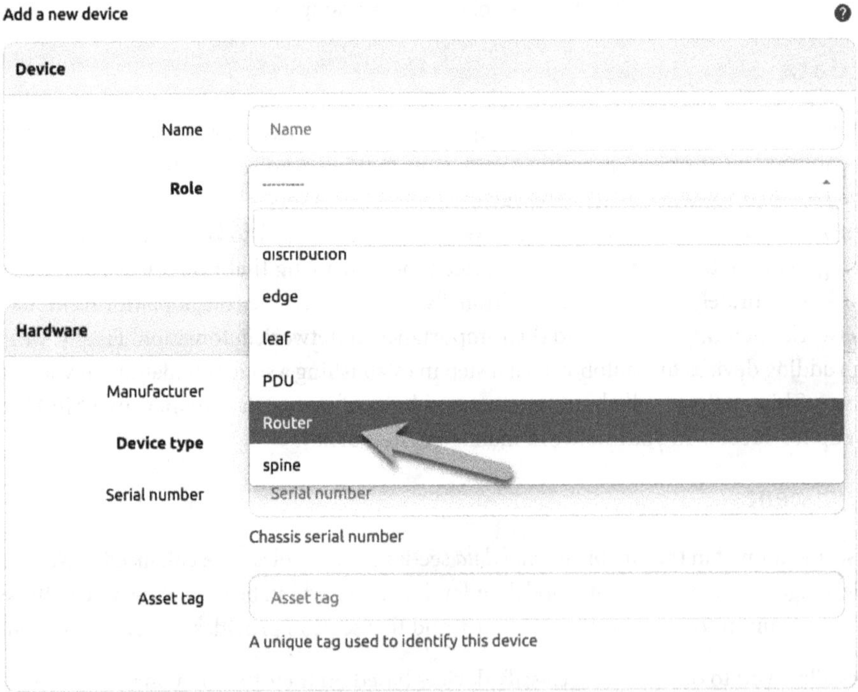

Figure 4.38 – Assigning a role to a device

Manufacturer

Identify the makers of your devices so that you can keep track of your inventory items in a more precise way. Here you can register manufacturers such as Cisco, Juniper, and Dell. This aids in linking them to device types, ensuring accurate hardware representation. Let's create a fictional vendor called *Acme* for our example.

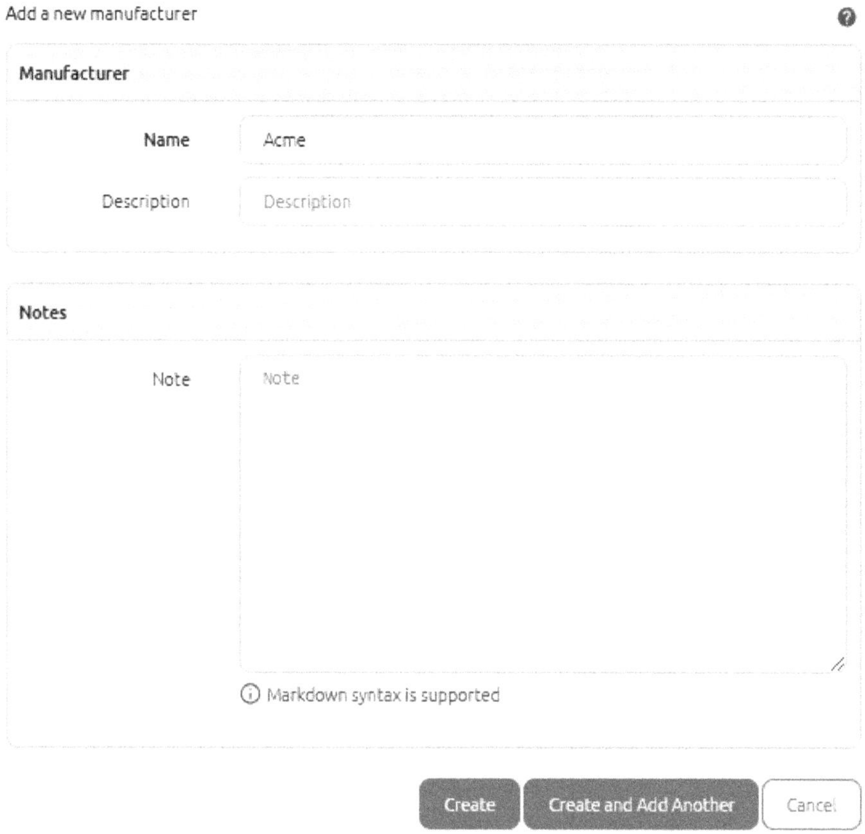

Figure 4.39 – Add new manufacturer form

Device types

Device types are meant to depict standardized hardware models. A device type might represent a specific model of a router, firewall, switch, server, etc., complete with manufacturer details and hardware specifications. Let's create an Acme Firewall model that we can use later on for Device Redundancy Groups. As you can see in the following example, we can even include pictures of the front and back of the devices:

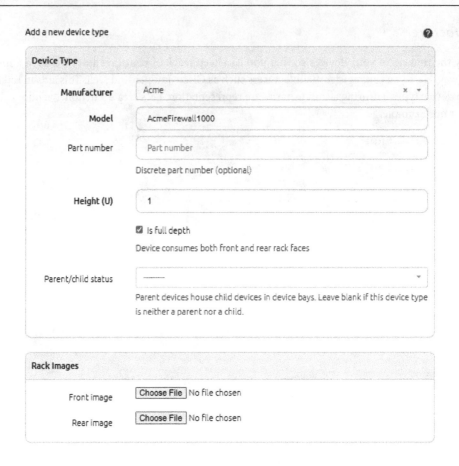

Figure 4.40 – Add new device type form

Platform

Platforms provide a way for you to organize the various network operating systems in your environment. Platforms can also be leveraged to drive automation. In fact, as you can see in the following example, we can specify a NAPALM driver. For reference, NAPALM is a Python-based network automation framework that Nautobot can leverage to launch automation jobs.

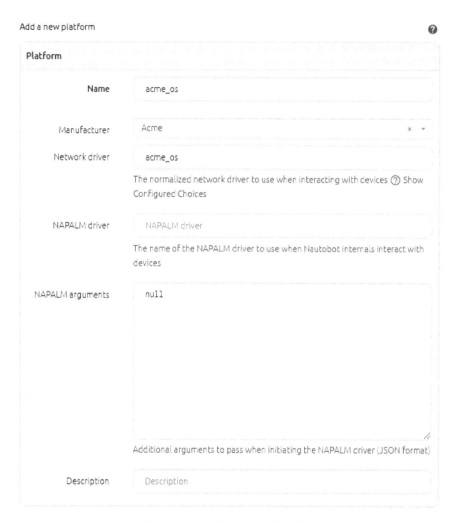

Figure 4.41 – Add new platform form

Device

Now we can create the devices themselves and start to populate Nautobot! This is a critical step in bootstrapping a fresh deployment. It will give you a great foundation to begin your journey in leveraging this data as your source of truth for many things in the future. Don't forget to associate a platform with your specific device since you've just created platforms for reference!

To add a device, navigate and expand the side navbar **Devices** menu. From here, you can simply click the *plus* button, or click **Add** from the device list view. This will bring you to the **Add a new device** form.

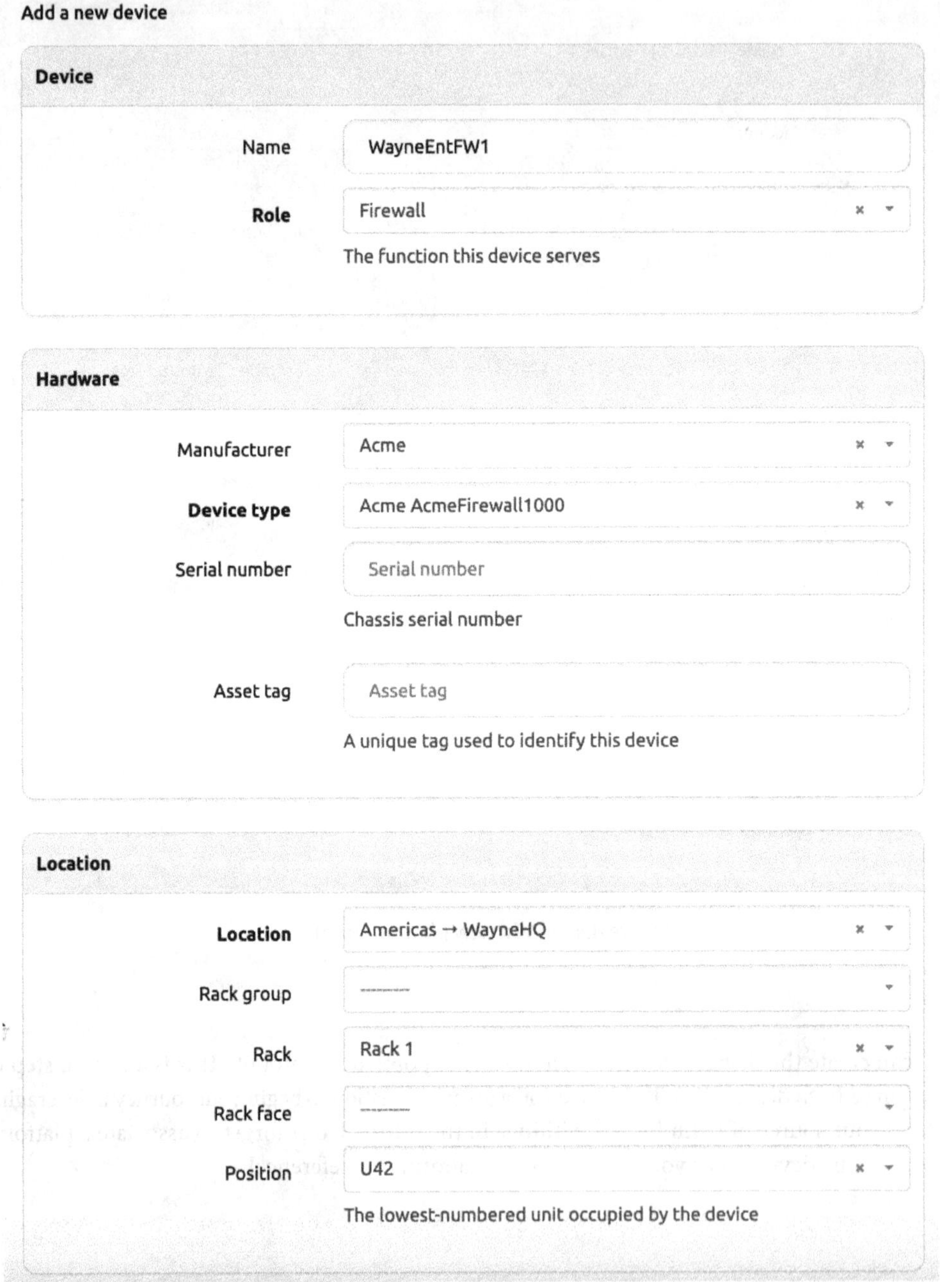

Add a new device

Device	
Name	WayneEntFW1
Role	Firewall × ▼
	The function this device serves

Hardware	
Manufacturer	Acme × ▼
Device type	Acme AcmeFirewall1000 × ▼
Serial number	Serial number
	Chassis serial number
Asset tag	Asset tag
	A unique tag used to identify this device

Location	
Location	Americas → WayneHQ × ▼
Rack group	———— ▼
Rack	Rack 1 × ▼
Rack face	———— ▼
Position	U42 × ▼
	The lowest-numbered unit occupied by the device

Figure 4.42 – Add a new device form

Congratulations! You've added a new device to your Nautobot instance! We know that Wayne Enterprises favors redundancy for their network designs, so please follow these steps and add *WayneEnt_FW2* to your system. This secondary firewall will be used later in the book to illustrate redundancy concepts.

Summary

You have now seen that we have begun to populate your Nautobot deployment with real data representing your organization's network infrastructure. This is just the beginning of your journey with a source of truth representing various elements. As you dive deeper, it is important to know that there is much more to do, depending on your use cases, such as IP address management, device redundancy, circuit management, power, and cabling. There are many possibilities to explore. Now imagine a future where all of this information is available to you for automation use cases. You could generate configurations based on location hierarchies, safely upgrade devices based on redundancy status, overlay circuit maintenance status into a telemetry solution, or even create a field installation guide for your technicians as they need to plug in devices to various ports on a system. The possibilities are boundless.

In the next chapter, we will explore how to add additional data to Nautobot and understand its key features and benefits.

5

Configuring Nautobot Core Data Models

Now that you have your system populated with data and understand how to navigate the **Nautobot user interface** (**UI**), let's explore what else is possible. Nautobot has several powerful data models that are commonly leveraged for network engineering use cases.

In this chapter, we will explore how to add additional data to Nautobot and understand its key features and benefits. Managing your inventory, modeling high-availability devices, and tracking IP addresses are a few examples of what will be covered. We will also guide you through best practices for inventory management and some of the important things to consider when adding data to your source of truth.

The following are the main topics that will be covered in this chapter:

- IP address management in Nautobot
- Modeling **high availability** (**HA**) and virtual devices
- Cabling and power management
- Incorporating power management with cabling
- Secrets management
- Using **Notes**, **Tags**, **Change log**, and **Filter forms**

IP address management in Nautobot

In the foundational days of networking, **IP Address Management** (**IPAM**) was often a manual affair, relying heavily on spreadsheets and thorough documentation. As networks expanded and became more intricate, this approach showed its limitations. Problems such as IP conflicts, misconfigurations, and occasional network downtime were not uncommon. As the stakes grew higher, the need for a more organized, automated IPAM solution came into focus.

Nautobot's IPAM capabilities respond to these challenges by offering a systematic approach to managing IP resources. Its design aims to reduce human error and ensure efficient allocation, tracking, and recording of IP addresses. Among the vast sea of IPAM tools, here are a few aspects where Nautobot distinctly shines:

- **Customization and extensibility**: Nautobot supports plugins, allowing users to customize and extend its functionalities according to specific organizational needs. This modularity ensures that Nautobot remains relevant across diverse network requirements.

- **Rich API integrations**: With a robust API, Nautobot seamlessly integrates with other tools and systems. This connectivity fosters an environment where data can flow across platforms, enhancing automation and coordination.

- **First-class integration with Source of Truth data**: Having the ability to tie IPAM-based workflows with Nautobot's source of truth data unlocks worlds of potential. We can now have deep-rooted relationships between sites, locations, devices, device types, manufacturers, interfaces, cables, switch ports, **virtual local area networks** (**VLANs**), and more. If the data is in Nautobot, you will be able to see the full picture in ways you most likely haven't in the past.

Now, let's delve into the components associated with Nautobot's IPAM capabilities.

IP addresses

In the latest update of Nautobot 2.x, IPAM has evolved, introducing new features and approaches to manage your network's IP space effectively. This chapter will guide you through these changes, focusing on the practical application of IPAM in Nautobot, including the assignment of IP addresses to devices and interfaces. You'll learn how to navigate these features, ensuring your network's IP infrastructure is organized, efficient, and up to date with the latest standards in Nautobot. Let's delve into these enhancements to understand how they can streamline your network management processes.

Figure 5.1 – Viewing IPAM in the navigation bar

Prefixes

Prefixes act as an umbrella under which a range of IP addresses reside. Represented in CIDR notation, prefixes provide a way to segment and categorize your IP space. Each prefix in Nautobot can be associated with a **regional internet registry** (**RIR**), giving insight into the origin of that address space.

Prefixes can be organized based on the following types: container, network, and pool. If a prefix's type is set to `pool`, Nautobot will treat this prefix as a range (such as a NAT pool) wherein every IP address is valid and assignable. This logic is used when identifying available IP addresses within a prefix. If the type is set to `Network`, Nautobot will assume that the first and last (network and broadcast) addresses within an IPv4 prefix are unusable.

Namespaces

Namespaces in Nautobot 2.0 represent a major advancement in managing IP addresses. They offer a more granular and organized way to handle overlapping IP spaces, which is increasingly relevant in large, multi-tenant, or cloud-based environments. By using namespaces, network administrators can segregate IP spaces according to different projects, clients, or departments, ensuring that each segment is distinct and managed effectively. This is particularly useful for organizations that have overlapping IP spaces across different network segments.

VRFs

VRFs are essential for creating isolated routing domains in a network. Each VRF can have its set of IP addresses, prefixes, and even aggregates. When associating IPs with devices, the VRF context becomes crucial. It provides clarity on which routing domain an IP and its corresponding device belong to, ensuring no overlap or conflict occurs in complex network environments.

VLANs

VLANs serve as a fundamental tool for segmenting traffic at Layer 2. Each VLAN can have associated prefixes that define the IP address space available within that VLAN. By understanding the relationship between VLANs and prefixes, network administrators can determine the IP address range usable by devices within a specific VLAN. The clear delineation of address spaces ensures effective traffic management and helps maintain network organization and efficiency.

RIRs

RIRs play a pivotal role in the distribution and administration of IP address space across the globe. Established to ensure a structured and efficient allocation of IP addresses, RIRs are regional entities that manage the IP address resources for their specific regions.

How can RIRs be leveraged in Nautobot's IPAM? The following are several reasons you might want to leverage this capability:

- **Global and local IP address management**: With the inclusion of RIRs, Nautobot is adept at handling both globally routable address space as well as private IP spaces. This offers a comprehensive view of IP address management, ensuring that no matter where your network operates, it complies with established IP allocation standards.

- **Improved organizational structure**: Linking prefixes to a specific RIR provides a view of IP allocations and their origins. It allows network teams to quickly identify where an IP space originates and how it aligns with global standards.

In Nautobot, creating an RIR is a straightforward process. While there are five top-tier RIRs managing different segments of the globe, Nautobot also has provisions to treat RFCs 1918 and 6589 as RIR-like entities due to their role in private IP space allocation. As of the time of the publication of this book, the five top-tier RIRs are as follows:

- **American Registry for Internet Numbers (ARIN)**: This RIR manages the IP address space for the United States, Canada, and a portion of the Caribbean

- **Réseaux IP Européens Network Coordination Centre (RIPE NCC)**: It oversees IP address allocation for Europe, the Middle East, and parts of Central Asia

- **Asia-Pacific Network Information Centre (APNIC)**: APNIC is the RIR for the Asia-Pacific region, encompassing a vast section of Asia and the Pacific islands

- **Latin America and Caribbean Network Information Centre (LACNIC)**: This RIR handles IP address distribution for Latin America and the Caribbean

- **African Network Information Centre (AFRINIC)**: AFRINIC manages IP addresses for the African continent

Each of these RIRs has its own policies and procedures for IP address allocation within their regions, reflecting both global standards and local requirements.

Configuring IP address management in Nautobot

Let's now take this newly acquired knowledge and put it to use with a "real-life" scenario!

Wayne Enterprises is a beacon of industry and innovation in Gotham City. With its ever-growing business units and infrastructural needs, the organization has adopted the RFC 1918 address space `10.1.0.0/16` as its designated prefix for the main headquarters. Within this prefix, a specific subnet, `10.1.2.0/24`, has been chosen for the deployment of an important project.

Our objective is to efficiently distribute and oversee this address block using Nautobot's features.

The following are the steps our team will need to accomplish this task.

IPAM configuration for Wayne Enterprises

Follow these steps to create a RIR:

1. Navigate to **IPAM | RIRs | +Add**.
2. Set the RIR name to `Private Wayne Enterprises RIR`.
3. Check the box for **Private** (internal use only).
4. Enter an optional description.
5. Click **Create**.

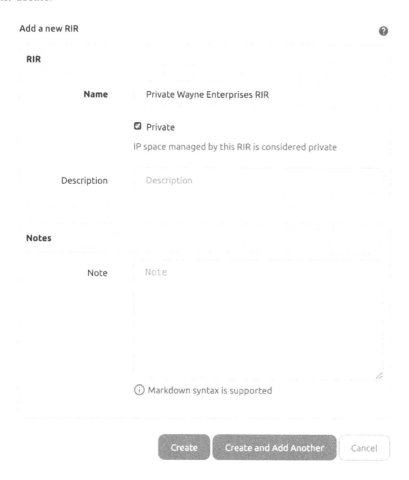

Figure 5.2 – Adding a RIR

Follow these steps to configure a VLAN with an associated prefix:

1. Navigate to **IPAM | VLANs | +Add**.
2. Set the VLAN ID (for example: 2).
3. Set the name to `Wayne HQ`.
4. Set **Status** to **Active**
5. Click **Create**.

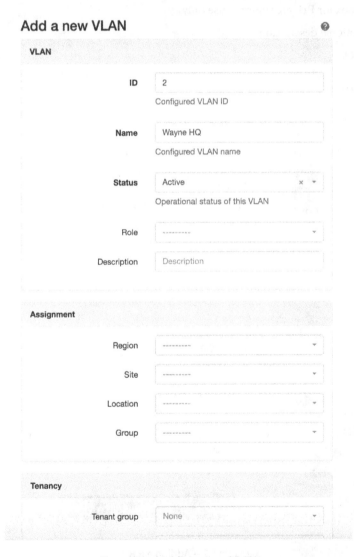

Figure 5.3 – Adding a new VLAN

Follow these steps to create a prefix:

1. Navigate to **IPAM | Prefixes | +Add**.

2. Enter **Prefix** as: `10.1.2.0/24`.

3. Set **Status** to **Active**.

4. Set **Type** to **Network**.

5. Set **VLAN** to **WayneHQ (2)**.

6. Set optional associations such as locations.

7. Click **Create**.

Figure 5.4 – Adding a new prefix

Follow these steps to create IP addresses within the prefix:

1. Navigate to **IPAM | IP Addresses | +Add**.

2. Use the **Bulk Create** tab.

3. Define the address pattern, e.g., `10.1.2.[11-240]/32` to allocate 230 individual IPs.

4. Specify a namespace.

5. Specify **Type** as **Host**.

6. Set **Status** to **Active**.

7. Click **Create**.

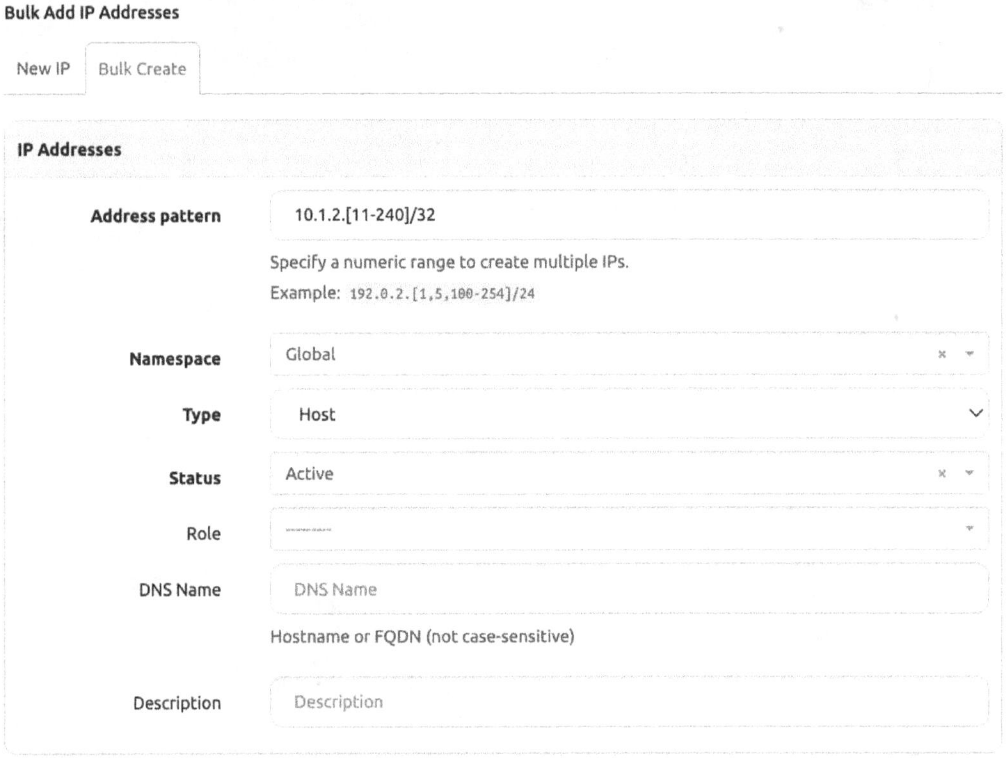

Bulk Add IP Addresses

New IP	Bulk Create

IP Addresses

Address pattern	10.1.2.[11-240]/32
	Specify a numeric range to create multiple IPs.
	Example: `192.0.2.[1,5,100-254]/24`
Namespace	Global ✕ ▾
Type	Host ⌄
Status	Active ✕ ▾
Role	--------- ▾
DNS Name	DNS Name
	Hostname or FQDN (not case-sensitive)
Description	Description

Figure 5.5 – Adding IP addresses (bulk operation)

Follow these steps to assign IP addresses to devices/interfaces:

1. Navigate to **Devices**.

2. Select the device you want to assign the IP address to.

3. Click on the **Interfaces** tab; if one does not exist already, you can click **Add Components | Add Interface**.

4. Select an IP (e.g., `10.1.2.15/32`).

5. Click **Update**.

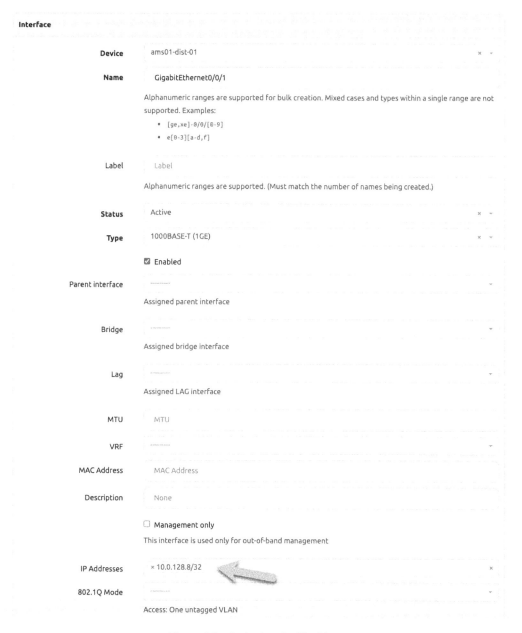

Figure 5.6 – Assigning the IP address

Congratulations!

Wayne Enterprises now is ready to go. The CEO, Bruce, has praised the networking team for their prowess in providing the new HQ with IP networking readiness. Bruce sure does love his gadgets, and now they can be attached to the network!

Nautobot's IPAM capabilities extend even further. Its integration capabilities allow it to seamlessly merge with other tools, ensuring a synchronized network management experience. Custom fields can be introduced, offering flexibility to organizations with unique requirements. Its programmability means that tasks can be automated, reducing human error and increasing efficiency.

As networks grow and become increasingly complex, there is a pressing need to ensure that the infrastructure remains online and resilient to potential disruptions. This leads us to another pivotal aspect of network management: HA. HA ensures that our systems remain operational even when unforeseen events, such as hardware failures or software glitches, come into play.

Moreover, in the era of digital transformation, where the lines between physical and digital are blurred, we can't forget the role of **virtual devices**. Virtualization offers scalability, flexibility, and cost savings, attributes which are pushing many enterprises, including Wayne Enterprises, to migrate or integrate virtual environments.

In the forthcoming sections, we'll venture into the depths of HA and explore the realm of virtual devices within Nautobot, ensuring that our networks are not just organized, but are also robust and future ready.

Modeling HA and virtual devices

In modern networks, keeping tabs on device redundancies is not just crucial—it's complex. Networks are no longer simple constructions; they're intricate labyrinths with devices often set up in redundant pairs or clusters to ensure continuous service, even if one device goes offline. This redundancy is key to ensuring HA and resilience against failures. However, effectively monitoring, tracking, and managing these redundant configurations can be challenging. Mismatched configurations between redundant devices, a lack of visibility into which devices form a redundant pair, or uncertainty about the state of a standby device can all introduce risks to a network. To help administrators tackle these challenges head on, Nautobot introduces two features: Device Redundancy Groups and virtual chassis. These tools provide a structured approach to represent and oversee the redundancy relationships and configurations within the network, reducing operational risks and simplifying management.

In addition to Device Redundancy Groups and virtual chassis, Nautobot now offers the concept of **Interface Redundancy Groups**. This feature provides a more granular and focused approach to interface redundancy, particularly for protocols such as the **Hot Standby Router Protocol** (**HSRP**) and **Virtual Router Redundancy Protocol** (**VRRP**).

The following table provides a quick guide for Nautobot users to determine which feature they should leverage for a given use case. It's important to remember that while these features provide structured methods to represent redundancy and configuration, Nautobot custom fields allow for the greatest flexibility and can be used for any use case not natively supported by built-in models.

Use Case	Device Redundancy Groups	Virtual Chassis	Interface Redundancy Groups	Consider Custom Fields
Multiple devices functioning as a single unit for redundancy (e.g., Active/Standby firewalls)	✓			
Group of devices functioning as a single logical device (e.g., stackable switches)		✓		
Interface-level redundancy for protocols such as the HSRP/VRRP			✓	
Grouping redundant power supplies across devices	✓			
Tracking physical connections between member devices in a stack		✓		
Custom metadata or specifications about devices not covered by built-in models				✓
Documenting unique configurations specific to individual devices/interfaces				✓
Identifying primary and backup devices for specific network roles	✓			
Link aggregation			✓	
Virtual Device Contexts (VDC)	✓			

Table 5.1 – Nautobot redundancy features and use cases

Device Redundancy Groups

This feature allows users to model sets of devices that work together in a redundancy setup. It could be two devices in an active/standby relationship, multiple devices in a load-sharing cluster, or even virtual devices running on shared physical hardware. It provides a structured approach to modeling, managing, and monitoring these setups effectively.

Device Redundancy Groups use cases

The following describes a few different scenarios where you can take advantage of Device Redundancy Groups.

Data center device upgrades

Scenario: Regularly updating and upgrading devices is crucial in a data center to ensure security, introduce new features, and optimize performance. However, when devices are redundant, it's vital to ensure that while one device is being upgraded, the other is fully operational to prevent any service interruptions.

The solution with Nautobot: Using Device Redundancy Groups, network engineers can meticulously plan and track upgrade cycles. It helps ensure that at least one device in a redundant pair remains operational at all times. This strategy reduces risks, ensuring smooth and uninterrupted services during upgrades.

HA for security devices

Scenario: Firewalls and **intrusion prevention systems** (**IPSs**) are commonly deployed in pairs to maintain a secure network environment without any downtime.

The solution with Nautobot: By leveraging Device Redundancy Groups, these pairs can be documented and monitored, ensuring that configuration drift doesn't occur and standby devices are always ready to take over.

Load balancer clusters

Scenario: Load balancers are often deployed in clusters to distribute incoming application traffic across multiple servers, maximizing application availability and responsiveness.

The solution with Nautobot: Such clusters can be grouped into Device Redundancy Groups, enabling an aggregated view of their operational state and ensuring uninterrupted traffic distribution.

Virtual chassis

On the other hand, the virtual chassis concept models a set of devices that share a common control plane. This is commonly seen in switch stacks that operate as a singular logical entity, even though they are multiple physical units. While each device in this setup (known as a **VC member**) can have its own position or priority, there's often a designated master device that manages the virtual chassis.

Virtual chassis use cases

The following describes a few different scenarios where you can take advantage of the virtual chassis.

Stacked switches in enterprise networks

Scenario: Switch stacking is a common practice in enterprises where multiple switches are interconnected to operate as a single logical switch.

The solution with Nautobot: The virtual chassis feature can represent this stack, allowing for holistic management and configuration. By documenting each switch's role within the stack, whether it's a master or a member, engineers can ensure operational continuity and streamlined configuration processes.

Service provider aggregation

Scenario: Service providers often deploy switches in a virtual chassis mode in aggregation layers, consolidating multiple devices into a single management plane for increased operational efficiency.

The solution with Nautobot: By leveraging the virtual chassis feature, service providers can reduce operational overhead, centralize configurations, and maintain a clear understanding of device roles and hierarchies within the aggregation layer.

Campus network design

Scenario: In large campus networks, keeping track of numerous switch stacks across various buildings can be daunting.

The solution with Nautobot: The virtual chassis feature simplifies this by offering a unified view of each stack, allowing for consistent configurations and easier troubleshooting across the entire campus.

Key differences between device redundancy and virtual chassis

After reviewing Device Redundancy Groups and virtual chassis, let's take a look at their key differences and what makes them unique.

Operational unity

While both concepts deal with multiple devices, the virtual chassis emphasizes devices that operate under a single control plane. In contrast, Device Redundancy Groups might not necessarily share a control plane but work in unison for redundancy purposes.

Autonomy of devices

Devices within a Device Redundancy Group retain their autonomy and can function independently if needed. In a virtual chassis, the devices are bound more tightly, functioning as parts of a larger logical entity.

Setting up a firewall redundancy group for Wayne Enterprises in Nautobot

Let's walk through an example setup for firewall active/passive redundancy in Nautobot to help illustrate all of the concepts we have covered thus far.

Objective: Wayne Enterprises aims to ensure HA for its data center operations. With firewalls being critical in safeguarding data, the company opts to deploy them in redundant pairs. The team intends to use Nautobot to manage and monitor these pairs effectively.

These are the prerequisites:

- Admin access to Nautobot

- **Details of the devices**: Names, IP addresses, models, and operational states

The steps to create a new Redundancy Group are as follows:

1. Navigate to **Devices | Device Redundancy Groups | +Add**.

2. Fill in the group name, for example, `WayneEnt_Firewall_Redundancy`.

3. Specify **Status** as **Active**.

4. Specify the failover strategy as **Active/Passive**.

5. Optionally configure an associated Secret Group.

6. Click the **Create** button.

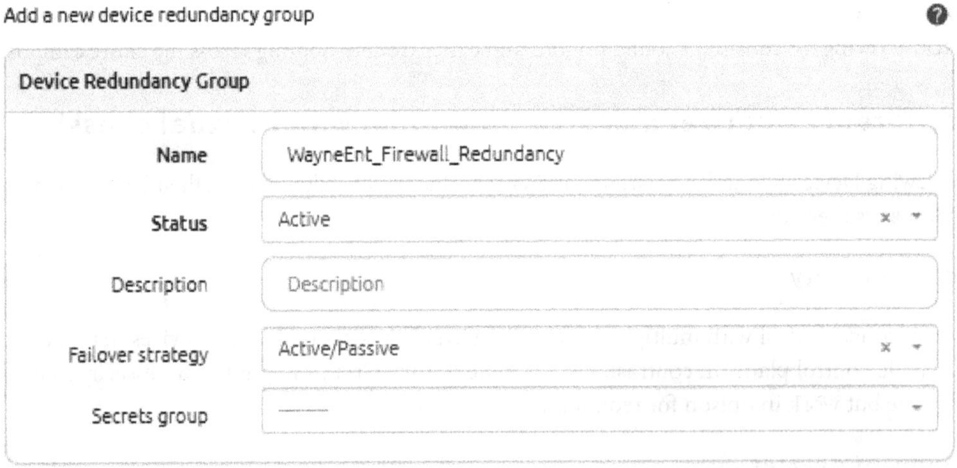

Add a new device redundancy group

Device Redundancy Group	
Name	WayneEnt_Firewall_Redundancy
Status	Active
Description	Description
Failover strategy	Active/Passive
Secrets group	———

Figure 5.7 – Add a new device redundancy group form

Next, we will add devices to the group.

7. Navigate to the active firewall device in **Devices** and click **Edit**.

8. In the device management section, search for the `WayneEnt_FW1` firewall.

9. Select the **WayneEnt_Firewall_Redundancy** Device Redundancy Group.

10. Set the status to `1`.

11. Click **Update**.

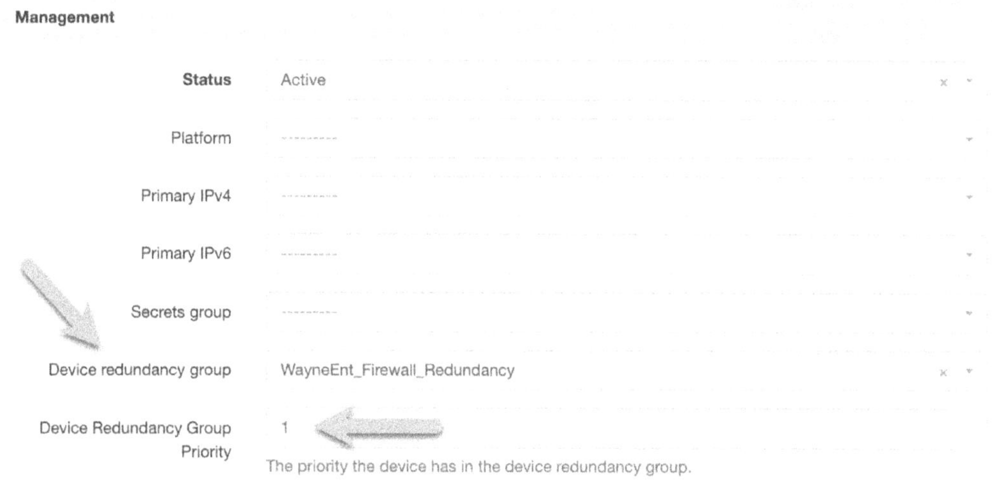

Management

Status	Active
Platform	---------
Primary IPv4	---------
Primary IPv6	---------
Secrets group	---------
Device redundancy group	WayneEnt_Firewall_Redundancy
Device Redundancy Group Priority	1

The priority the device has in the device redundancy group.

Figure 5.8 – Assign a Device Redundancy Group and priority to the primary device

12. In the device management section, search for the `WayneEnt_FW2` firewall.

13. Select the **WayneEnt_Firewall_Redundancy** Device Redundancy Group.

14. Set the status to `2`.

15. Click **Update**.

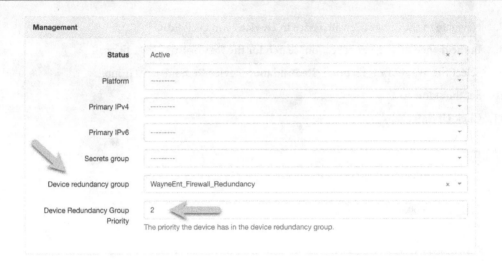

Figure 5.9 – Assign a Device Redundancy Group and priority to the secondary device

Finally, we will review and finalize the setup.

16.   Navigate back to **Devices | Device Redundancy Groups**.

17.   Click on **WayneEnt_Firewall_Redundancy**.

You should now be able to see your devices are successfully associated with the new redundancy group.

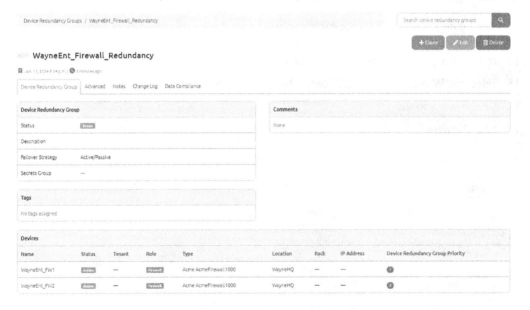

Figure 5.10 – The Device Redundancy Group detailed view

Congratulations! You've officially configured a Device Redundancy Group in Nautobot for Wayne Enterprises' redundant pair of firewalls.

Interface Redundancy Groups

Complementing the device-focused redundancy tools in Nautobot, Interface Redundancy Groups offer administrators a more detailed view of the redundancy mechanisms at the interface level. These groups cater specifically to scenarios where devices deploy protocols such as HSRP and VRRP to ensure uninterrupted network availability.

Interface Redundancy Group is a model that represents a collection of interfaces that together form a redundancy group. This can be visualized as a set of interfaces across devices that collaboratively ensure network availability, especially in scenarios where one of the interfaces (or the device hosting it) fails.

Interface Redundancy Groups use cases

The following describes a few different scenarios where you can take advantage of Interface Redundancy Groups.

VRRP in Routers for HA

Scenario: A network design requires two routers to ensure HA for a critical business application. These routers deploy the VRRP to prevent single points of failure at the interface level.

Implementation in Nautobot: Consider two routers, Router A and Router B, set up using the VRRP for interface redundancy. Both routers have an interface, GigabitEthernet 1/0/0, participating in the redundancy protocol. They share a common virtual IP address, but each has its unique physical IP and priority level for VRRP election.

As a detailed example, you may want to document how their VRRP priorities differ. Router A's vrrp 1 has a priority of 120, whereas for Router B, it's set to 100. Similarly, for vrrp 5, the priorities are 100 and 200 for Routers A and B respectively.

Value proposition: By using the Interface Redundancy Groups feature, network engineers can model and monitor the VRRP configurations across the routers. This ensures that any discrepancies or configuration drifts can be identified and rectified promptly. It offers a streamlined method to visualize and manage interface-level redundancies, making the network more resilient and easier to maintain.

Cabling and power management

Managing the myriad of cables within a data center or network environment can be challenging. Without a structured approach to documenting and tracing these cables, troubleshooting and modifications can quickly become a complicated mess. Nautobot provides a comprehensive approach to cable management, helping administrators maintain an organized, efficient, and transparent infrastructure.

Cables

In Nautobot, the essential foundation of connectivity is the cable. Each cable you define in the system stands for a tangible, physical link between two termination points. This representation is crucial for understanding how devices communicate, how power is distributed, or how data is transferred.

Key attributes of cables

Let's continue to dive in and learn more about cables:

- **Endpoints (A and B)**: While denoted as A and B for simplicity, it's important to note that cables are direction-agnostic in Nautobot. The order doesn't signify any operational difference.

- **Connection types**: Cables can connect various components such as interfaces, power ports, console ports, and circuit terminations. The diverse range of connection types ensures all physical aspects of a network or data center can be modeled.

- **Properties**: Each cable can have attributes such as type, label, length, and color. These facilitate identification and organization.

- **Status**: Nautobot allows you to assign a status to cables—**Active**, **Planned**, or **Decommissioning**. This can be useful in life cycle management and planning.

Tracing cables

One of the standout features of Nautobot is the ability to trace a cable's virtual path in the UI. This becomes invaluable when trying to understand intricate connections or troubleshoot issues. This capability is also available via the API.

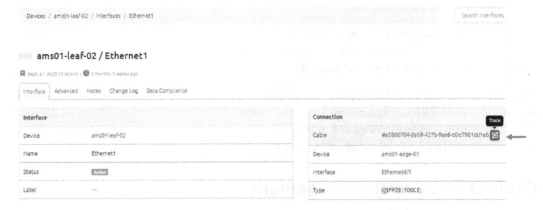

Figure 5.11 – A single interface with the ability to trace a cable

> **Note**
>
> The cable trace capability can also be accessed from the **Table** view of interfaces in addition to the individual interface view, which is shown in Figure *5.11*.

How it works

Let us see how cable tracing works in the following steps:

1. Upon selecting a cable to trace, Nautobot will depict the path from one termination point across any connected cables to the next termination.
2. If a cable connects to a pass-through port and there's a continuation from the other side, Nautobot will persist in tracing this path until it reaches a definitive endpoint or an unconnected termination.
3. This traceability can span devices or even circuits, offering a clear picture of how connectivity is structured.

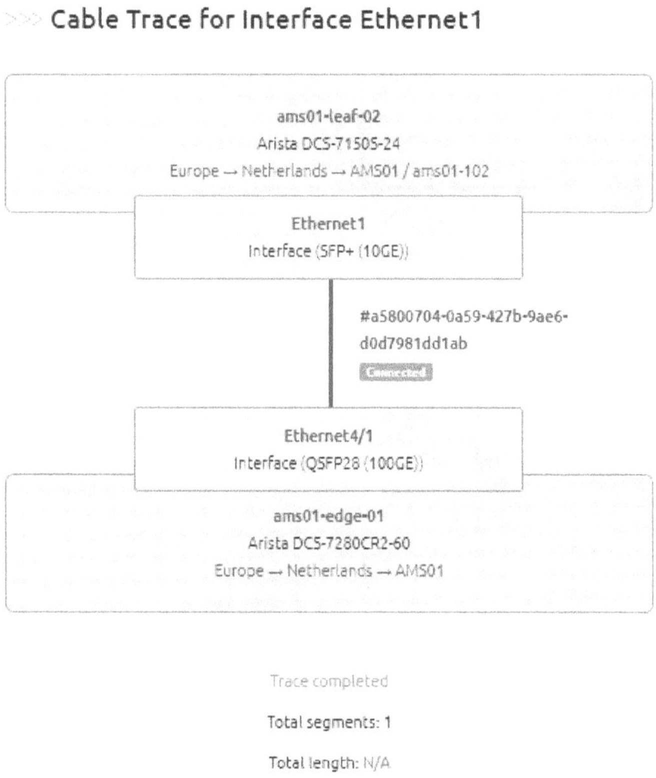

Figure 5.12 – Viewing a cable trace

Note that you can perform a trace between two devices that have multiple cables between them using a Nautobot cable trace.

Incorporating power management with cabling

Power management and cabling go hand-in-hand in any data center environment. While cabling handles the connectivity between devices, it is the power system that brings these devices to life. In Nautobot, power management is approached with the same structured methodology as cabling, ensuring that tracking and managing power across devices is both comprehensive and clear.

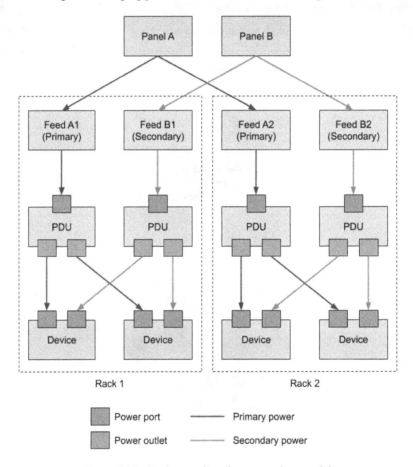

Figure 5.13 – Understanding the power data model

Power panels

The inception point of power distribution in Nautobot is the power panel. It serves as the primary distribution source that pushes electrical power through multiple power feeds.

To add a power panel, navigate to **Power** | **Power Panels** | **Add**.

Figure 5.14 – Add new power panel form

Key attributes of power panels

The following highlights the key attributes of power panels and shows you how to create one in Nautobot:

- **Location association**: Every power panel must be associated with a location. This ensures a geographic representation of where power is being sourced.

- **Rack group assignment**: Optionally, a power panel can be connected to a rack group. This helps in understanding and managing power at a micro level, especially when dealing with multiple racks.

- **Redundancy**: In mission-critical environments such as data centers, redundancy is paramount. It's common to find dual power panel setups, arranged in parallel, ensuring that there's a backup should one fail.

Power feeds

While power panels serve as the distribution hub, power feeds are the channels through which this power is routed to specific devices. This could be a server, a router, a PDU, or any device that needs electrical power.

Figure 5.15 – The Add a new power feed form

Key attributes of power feeds

The following highlights the key attributes of power feeds and how they connect to power ports. This defines whether the feed is primary or redundant. A redundant feed acts as a backup, kicking in if the primary feed fails:

- **Status**: Like cables, power feeds can have statuses such as **Offline**, **Active**, **Planned**, or **Failed**. This helps in monitoring and maintenance.

- **Electrical characteristics**: Parameters such as supply type (AC or DC), phase (single or three-phase), voltage, amperage, and maximum utilization percentage give a detailed overview of the power being supplied.

- **Connectivity**: Power feeds are linked to devices via power ports, and this connection is represented using cables. Thus, you can trace a device's power source just like you would trace its network connectivity.

Understanding the blast radius through comprehensive data

Tip

Nautobot's power management doesn't just end at tracking and managing devices' power sources. It plays a pivotal role in risk assessment and understanding the potential impact of power-related issues.

Imagine a scenario where there's a failure at a specific power panel. Without a comprehensive understanding of how power flows through the infrastructure, diagnosing the breadth and depth of such a failure could be daunting. However, Nautobot's structured approach to power management provides the following:

- **Traceability**: From a single power panel, administrators can trace which devices are impacted due to their association with specific power feeds. This immediate insight can drastically reduce downtime by focusing on troubleshooting efforts.

- **Risk mitigation**: By knowing the connections, one can simulate "what if" scenarios. For example, "What if power panel A fails? Which critical devices would be offline? Do they have redundant power feeds from another source?"

- **Proactive planning**: Understanding the complete path of power, interlinked with cabling, allows for better capacity planning and ensures that power resources are utilized efficiently without overloading any single point.

- **Informative reporting**: With the wealth of data in Nautobot, reports can be generated to provide insights on potential vulnerabilities or weaknesses in the power infrastructure. These reports can then guide infrastructure investments, ensuring HA and resilience.

By integrating power management with cabling, Nautobot offers a holistic view of the data center's physical infrastructure, allowing IT professionals to not only manage and monitor but also predict and prevent potential issues.

Secrets management

In data center operations, securing sensitive information such as access credentials or API tokens is crucial. With an increasing dependence on automation and device interactions, the need for a standardized and secure method to handle this information is more important than ever. This is where Nautobot's secrets management comes into play.

Why use secrets?

Traditionally, secrets, like device credentials or API tokens, would be stored directly in the database or application configuration. However, this is not a secure practice. To address this, Nautobot is able to provide a flexible method for managing and accessing secrets without ever directly storing the secret itself within Nautobot.

Core concepts

There are several core concepts to be familiarized with secrets management within Nautobot. Understanding the functionality will be important as you begin to leverage the platform for automation. This section will explore these key concepts:

- **Secrets**: A secret in Nautobot does not store the actual sensitive information. Instead, it stores a reference on how to retrieve that information when needed. This abstraction ensures that secrets are kept out of reach, only being retrieved when required and according to the defined method.

- **Secrets Group**: This is a collection of secrets, which can be attached to objects such as Git repositories or devices. Each secret within the group has a defined access type and secret type, making the management of multiple secrets streamlined and organized.

- **Secrets providers**: This determines how a secret value is fetched. Nautobot has built-in providers such as environment variables and text files, but also supports additional custom providers through plugins. Each provider has its own set of parameters that need to be defined to fetch a secret.

- **Templated secret parameters**: For better management and reduced data entry, Nautobot offers the option to use Jinja2 templates to modify the provider parameters based on the requesting object. This is especially handy when dealing with multiple related secrets.

Secrets versus Secrets Groups in Nautobot

This section explores the distinctions between Secrets and Secrets Groups in Nautobot. It covers their respective roles and functionalities in managing sensitive data, highlighting how they differ in configuration and application within the system.

Secrets

A "Secret" in Nautobot represents a single sensitive piece of information, such as a password, API token, or username. It doesn't store the actual secret value within Nautobot but instead stores the method to retrieve this secret from an external source when required.

Configuration: Each secret is associated with a "secrets provider." This provider determines how Nautobot retrieves the actual secret value. For instance, a secret might be retrieved from an environment variable or a specific file on the system.

Use cases: Secrets can be used in various parts of Nautobot:

- Accessing devices via protocols such as NAPALM
- Authenticating to Git repositories
- Used within custom Jobs or plugins that require sensitive data

Let's add a new secret called WAYNE_ENT_SECRET and have it source from environment variables within the Nautobot deployment. First, create the environment variable on the nautobot-server instance called export WAYNE_ENT_SECRET=capedcrusader. Next, click on **Secrets** | **Add**, and then you will be presented with the **Create new secret** form.

Figure 5.16 - Add a Secret with default providers

Security: The actual value of a secret is not readily available through the UI, REST API, or GraphQL. It's designed to be accessed internally by Nautobot processes or features that require it.

Secrets Groups

This section examines how Secrets Groups function as organizational tools, allowing for efficient management and assignment of multiple secrets within a single framework. This is essential for scenarios where multiple types of sensitive data are required for operations such as device configurations or Git integrations. In our example, we have added a Secrets Group called WayneEntSecretsGroup, which will be used for SSH access using the WAYNE_ENT_SECRET secret, which was created previously.

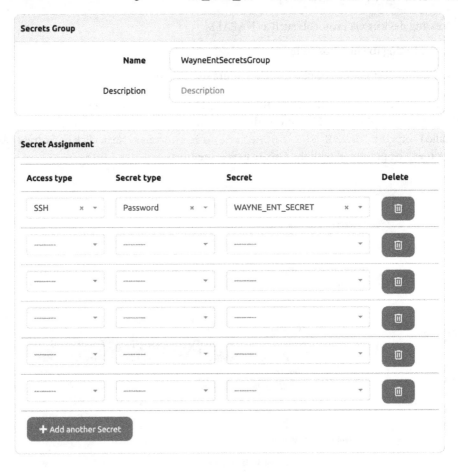

Figure 5.17 – Add new secrets group form

There are various other access types and secret types that can be selected and configured for Secrets Groups, as seen in the following images.

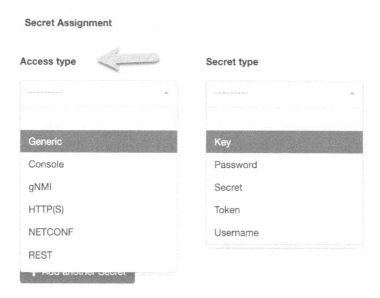

Figure 5.18 – The Access type list and the Secret type list as they appear on the Secrets Group form

A "Secrets Group" is a collection of related secrets bundled together. Think of it as a folder or a container that holds multiple secrets, each serving a distinct role or purpose.

Organization: Groups help organize and assign roles or purposes to the bundled secrets. For example, you might have a Secrets Group for "NAPALM Integration," which contains a username secret, a password secret, and an enable secret.

Attachment: Secrets Groups can be attached to various objects within Nautobot, such as Git repositories or devices. When an object references a Secrets Group, it can access all the secrets contained within that group.

Flexibility: A Secrets Group isn't limited to containing secrets of just one type or access method. It might have multiple secrets, each retrieved in a different manner (e.g., one from an environment variable and another from a text file).

Use cases: These are utilized when multiple secrets are needed for a specific operation. Here are some examples:

- A Git repository might need a token for authentication and a username for record keeping
- A device configuration might require multiple credentials, depending on the protocol or access level

Security considerations

The following are the considerations:

- **Leakage of secret values**: Nautobot's UI, REST API, and GraphQL are designed to prevent exposing actual secret values. Always retrieve the secret directly from its source, rather than through Nautobot.

- **Code and Secrets**: Any code running within Nautobot, whether it's a plugin, Job, or other integration, has potential access to secrets. It's paramount to trust the code being executed and to be aware of the implications.

- **Using object permissions with Secrets**: Nautobot's permissions model allows you to restrict which users can define, edit, and access secrets. For instance, you can limit a user to only access secrets from certain environment variables or specific file paths.

Accessing secrets in code

Retrieving a secret's value in code is straightforward using its `get_value()` method. If the secret uses templated parameters, an `obj` parameter can be provided for context. When working with a Secrets Group, the `get_secret_value()` method fetches a specific secret value from the group. The following example is done via `nautobot-server nbshell`, but could also be used in a Nautobot Job or other type of script using the built-in **object relational management** (**ORM**) within Nautobot. Accessing secrets through the REST API is not supported – for this scenario, you should plan to work with your organization's secrets management platform. HashiCorp Vault, AWS Secrets Manager, Delinea/Thycotic, and others would be popular examples.

This example leverages Secrets directly using `nbshell`:

```
from nautobot.extras.models.secrets import Secret

secret = Secret.objects.get(name="WAYNE_ENT_SECRET").get_value()
print(secret)
>>>'capedcrusader`
```

This example leveragies Secrets Group using `nbshell`:

```
from nautobot.extras.models.secrets import SecretsGroup
from nautobot.extras.choices import SecretsGroupAccessTypeChoices,
SecretsGroupSecretTypeChoices

secrets_group = SecretsGroup.objects.get(name="WayneEntSecretsGroup")
device1 = Device.objects.filter(name="WayneEntFW1")

secret = secrets_group.get_secret_value(
access_type=SecretsGroupAccessTypeChoices.TYPE_SSH,
```

```
secret_type=SecretsGroupSecretTypeChoices.TYPE_PASSWORD,
obj=device1)

print(secret)
>>>`capedcrusader`
```

Nautobot Secrets Providers app (plugin) overview

The Nautobot Secrets Providers app is an enhancement for Nautobot 1.2.0 and above that integrates Nautobot with external secrets backends. As businesses scale and diversify their infrastructure, managing secrets becomes crucial, and having them spread across various platforms can be challenging.

Why use the Nautobot Secrets Providers app?

Knowing when it makes sense to extend Nautobot's secrets management is important to understand. Let's explore some of the primary use cases and reasons that you might consider regarding the Secrets Provider app in more detail:

- **Centralized Secrets management**: This plugin allows for a consolidated approach to handle secrets from various sources such as AWS Secrets Manager, HashiCorp Vault, and Delinea/Thycotic Secret Server.

- **Flexibility**: Instead of waiting for official Nautobot updates to support new or customized secrets providers, you can extend or add new integrations more rapidly with this plugin.

- **Enhanced security**: By integrating established secrets providers, you leverage the advanced security features they offer, such as role-based access, encryption, and audit trails.

- **Operational efficiency**: Retrieving secrets directly from familiar platforms can simplify operations and reduce the need for context-switching or additional tooling.

- **Future-proofing**: As more secrets providers emerge or become popular in the industry, this plugin can be updated or extended to support them, ensuring Nautobot remains relevant and useful.

The following table compares the features and capabilities of native Nautobot Secrets versus using the Secrets Provider app.

Feature/Aspect	Native Nautobot Secrets Management	Nautobot Secrets Providers App
Source integration	Limited to Nautobot's built-in capabilities	Integrates with AWS Secrets Manager, HashiCorp Vault, Delinea/Thycotic Secret Server, and potentially more
Flexibility	Fixed feature set	Can extend/add new integrations rapidly

Feature/Aspect	Native Nautobot Secrets Management	Nautobot Secrets Providers App
Security	Standard Nautobot Security	Enhanced with third-party providers' advanced security (role-based access, encryption, and audit trails)
Operational efficiency	Requires potential multiple tool interactions	Simplifies operations by direct secret retrieval from established platforms
Future-proofing	Depends on core Nautobot updates	Can be easily updated or extended for new/emerging secrets providers
Centralized management	Natively managed within Nautobot	Centralized approach with diverse sources
Development cycle	Tied to Nautobot release cycle	More agile; not strictly tied to Nautobot's release schedule
Integration complexity	None (native to Nautobot)	Requires setup and integration with external systems

Table 5.2 – Comparing native secrets management and the Nautobot Secrets Providers app

In short, the Nautobot Secrets Providers plugin is a way to seamlessly bridge Nautobot with other well-known secrets management platforms, offering greater flexibility, security, and efficiency in managing secrets across diverse infrastructures.

> **Note**
> Nautobot Apps are constantly evolving. It is important to review the latest documentation to see what's changed, and what new capabilities might have been added in each new release.

Nautobot's secrets management provides a flexible and secure approach to handling sensitive information in the context of network automation and data center operations. By abstracting the actual secret values and offering a robust permissions model, it ensures that secrets are always kept at arm's length, only accessed when and how they should be.

Using Notes, Tags, Changelog, and Filter forms

Nautobot offers a wide array of functionalities designed to make the life of a network administrator more straightforward and the processes more intuitive. In this section, we'll delve into four particularly salient features—Notes, tags, change log, and filter forms. Each of these offers its own set of advantages to help users optimize their work and enhance their usage of the Nautobot platform.

Notes

Notes are available on many of the objects within Nautobot and can be accessed on the detailed page views. As an example, let's navigate to notes on a device in our system.

Navigate to **Devices | WayneEnt_FW1 | Notes**:

Figure 5.19 – Example note on an object in the Create/Edit form or Notes tab

What are Notes?

Notes are free-form annotations that you can add to many objects within Nautobot.

Why use Notes?

Let's explore some of the common use cases that the notes feature can be applied to. This feature can be a powerful way to represent conversation and collaboration around objects within Nautobot. Here are just a few examples of how to make the most out of this capability:

- **Documentation**: They allow for a rich description of any quirks, specifications, or unique attributes that any item in the database might have

- **Reminders**: They act as a reminder of temporary or permanent conditions that should be known to anyone interacting with the object

- **Collaboration**: They can help in communicating with team members, especially in larger teams where everyone might not be aware of specific details

> **Tip**
> A powerful way to leverage the notes feature is to collaborate with your peers around certain items within Nautobot. Notes will append to a list and stay associated with any object you add them to.

As you can see here in this example, we have a few notes about **WayneEnt_FW1**.

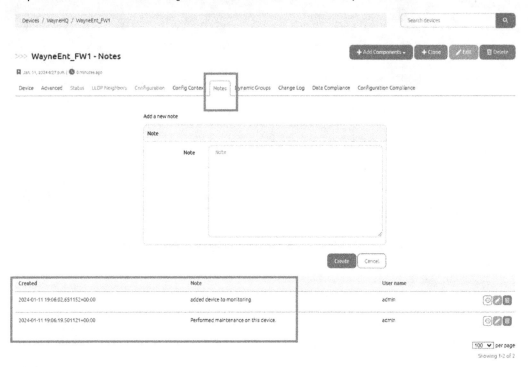

Figure 5.20 – The Notes tab

Tags

Nautobot tags offer a flexible way to add metadata to different objects, enhancing searchability and management across the system. The tag feature is extremely powerful, and to try and cover every use case would be next to impossible. As you read through this section, imagine some of the ways you could potentially leverage data when it has tags associated with it that are relevant to your organization's automation workflows. We'll explore how tags can be created, applied, and utilized effectively in Nautobot to streamline workflows and bring structure to your network data. The tag field is represented by a simply comma-separated list of values that can be associated with objects within the Nautobot platform.

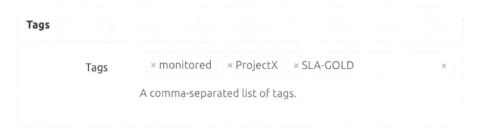

Figure 5.21 – Managing Tags

What are tags?

Tags are custom labels you can create and append to any item, helping in categorizing and marking them.

Why use tags?

As mentioned previously, the possibilities that tags introduce are somewhat boundless. The following concepts are several of the most common scenarios that are seen leveraged in the Nautobot tags feature. It is interesting to note that these concepts are also common amongst cloud service providers and cloud management platforms:

- **Quick identification**: They offer a rapid way to identify and group related items, whether they're devices, circuits, or any other object.

- **Flexible categorization**: Users can create custom tags that make sense for their organization, making the database more intuitive.

- **Bulk operations**: It makes it easier to perform operations on a group of items that share a specific tag.

Tags will be covered in greater detail in *Chapter 6*.

Change log

This section delves into the change log functionality in Nautobot, a crucial tool for tracking modifications and updates within the platform. The change log functionality offers insights into the history of changes made to various components, enhancing transparency and accountability in network management.

What is it?

The change log is a function that keeps a record of every change made within the Nautobot system, including what was modified, who did it, and when.

Why use it?

The following are the reasons for using it:

- **Accountability**: It ensures that there's a clear record of all actions, holding users accountable

- **Troubleshooting**: In case of issues, it can be a vital tool to trace back changes and identify the source of a problem

- **Security**: It can detect unauthorized or accidental changes, enhancing system security

Change logs can be viewed from various entry points within Nautobot. Change logs can be viewed under **Extensibility** | **Change Log**, which will show a list of all objects within the system. As you can see in the following figure, various object types are shown within this single view.

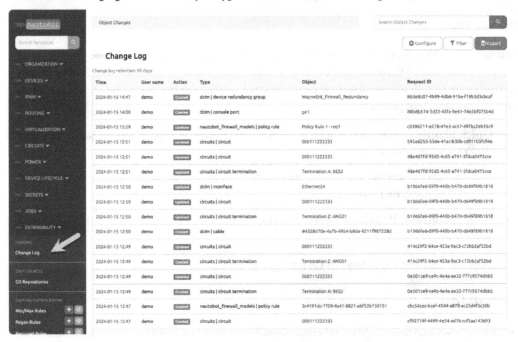

Figure 5.22 – Change Log at the system level

Change logs can also be found on the objects themselves. The following example shows how it can be navigated right from the device itself. This is a nice way to quickly see the relevancy of what has changed on a specific object while maintaining the context of its configuration.

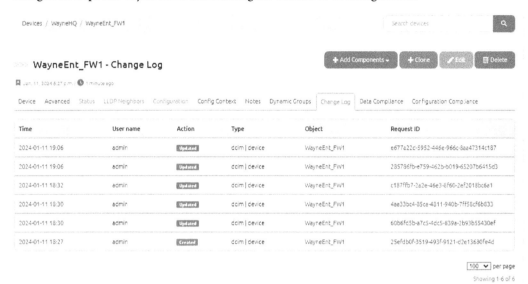

Figure 5.23 – Change log for a specific object using the Change Log tab

As you click into a change log entry, even more relevant information is shown. You can see exactly what was changed on an entity, the time it was changed, and who changed it!

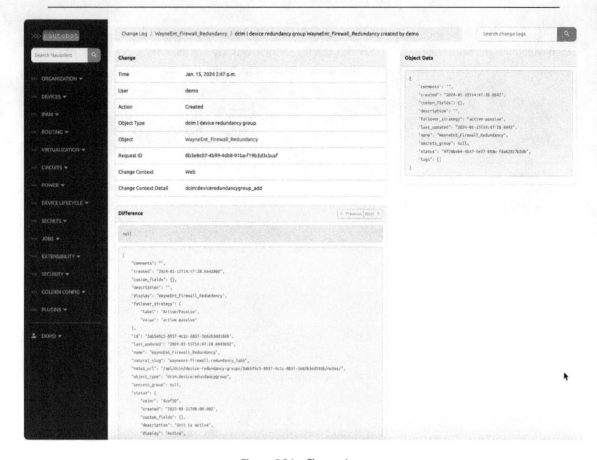

Figure 5.24 – Change Log

Filter forms

If you need some flexibility in how you find items within Nautobot, filter forms are the feature for you. This capability allows very granular control for finding objects within the system based on almost any attribute that may be associated with it. This section will explore the ways that filter forms can help you manage data more effectively within Nautobot.

What are they?

These are interfaces within Nautobot that let you filter and narrow down your view based on specific criteria.

Why use them?

There are many use cases that filter forms apply to. The following are some examples of how filter forms can build powerful navigation and search capabilities into your organization's Nautobot instance:

- **Data navigation**: They help in quickly navigating large datasets, allowing users to find exactly what they're looking for without sifting through irrelevant data

- **Efficiency**: They significantly reduce the time taken to find specific objects, especially in larger deployments

- **Custom views**: They allow for customized views based on various criteria, enabling users to create personalized data perspectives

Filter forms can be accessed on many different objects within Nautobot's UI. Let's use an example here to filter devices based on device role.

Figure 5.25 – The Filter button in the top-right corner on a detailed view page

Follow these steps:

1. Navigate to **Devices | Filter**.
2. Select **Role | Firewall**.
3. Click **Apply**.

We can now see a nice filtered view of devices that are filtered by the **Firewall** role!

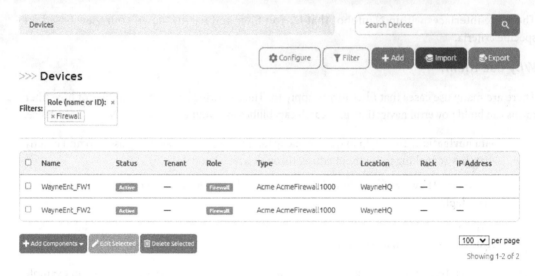

Figure 5.26 – Viewing the results of a custom filter form

Incorporating these functionalities in everyday use can make Nautobot an even more powerful tool, regardless of the specific domain of application, be it network automation or any other. They collectively enhance data interaction, promote accountability, and optimize workflow efficiency.

Best practices for inventory management

Nautobot, as a modern, extensible platform, offers a plethora of features designed to streamline and enhance the process of cataloging and managing network assets. By integrating structured data storage, comprehensive reporting, and flexible extensibility, Nautobot presents network professionals with a reliable platform for maintaining an accurate view of their infrastructure. This ensures that both operational efficiency and strategic planning are informed by up-to-date and comprehensive data.

The following are some prioritized best practices to ensure you're maximizing the benefits of Nautobot for network inventory management:

Maintain consistency with data sources

Data source syncing: Integrate Nautobot with other platforms such as monitoring or ticketing systems. This ensures that data remains consistent across all platforms. We cover this more in *Appendix 2* when we talk about Nautobot's **Single Source of Truth (SSoT)**.

Automated data entry: Use Nautobot's APIs, Jobs, and Apps (plugins) to automate the entry of data, reducing human errors. Nautobot Apps have an entire category of SSoT plugins that can synchronize data fields from various sources. Be sure to check out the Nautobot Apps documentation pages to stay up to date on what has been developed. Keep in mind that you can always create your own as well! As previously mentioned, we'll cover SSoT in *Appendix 2*.

Utilize built-in data models and custom fields

Built-in data models: Nautobot comes equipped with a diverse set of predefined data models that cater to many standard networking scenarios. Before extending or customizing, see whether your needs can be met with what's already available. These data models have been designed based on industry best practices and provide a solid foundation for most network inventory needs.

Custom fields: While the built-in data models are extensive, every organization has its specific requirements. Where the default models don't fit perfectly, use custom fields judiciously to capture additional information that's vital to your operations. We'll dive into custom fields and more ways to extend built-in data models in *Chapter 6*.

Extensible data models: Leverage Nautobot's extensibility to model complex relationships and hierarchies within your network inventory. We'll look at built-in (UI-driven) extensibility in *Chapter 6* and also cover how you can build our own native-models in *Chapter 15*.

Regularly audit with Nautobot's reporting features

Scheduled reports: Use the platform's reporting capabilities to run scheduled checks on the accuracy of your inventory data. You can build reports through the use of Jobs, covered in *Chapter 11*, and export templates, covered in *Chapter 6*.

Discrepancy identification: Use reports to identify mismatches or inconsistencies in data entries. You can even create custom Nautobot Jobs that run to handle some of this for you in an automated manner.

Access control and permissions

Role-based access: Define access roles meticulously, ensuring team members can view or modify only what's relevant to them. Make sure to put a plan together early on with your implementation so as not to create challenges later on. Remember, it's always easier to give than to take away!

Map dependencies and relationships

Device relationships: Document inter-device relationships comprehensively, ensuring clear visibility of dependencies. As you add new devices, it is good practice to keep consistent records. In fact, you can build automated workflows that take care of all of the dependency mappings for you as you add new devices into Nautobot.

Change management with Git integration

Track configurations: Monitor configuration changes over time, and keep an accessible history using Git integration. Leverage configuration contexts, which are covered in *Chapter 6*, to render the intended state of your device configurations based on hierarchy and more.

Collaboration: Encourage team members to review and comment on changes before they're finalized.

Leverage Tags and Notes

Categorization: Systematically use tags for categorizing devices by function, location, or other attributes.

Documentation: Make a habit of documenting device-specific quirks or planned changes using notes.

Engage with the Community and Apps (Plugins)

Extend functionality: Keep abreast of community developments and integrate Apps that enhance your inventory management capabilities.

Plan scalability within Nautobot

Growth management: Anticipate network growth and ensure Nautobot's data models and fields evolve correspondingly.

Train the team on Nautobot's capabilities

Regular workshops: Conduct workshops periodically to familiarize the team with new Nautobot features or optimized usage methods.

With a conscious effort to follow these best practices, Nautobot can be more than just an inventory tool—it can be a central pillar supporting effective and efficient network management and planning.

Summary

We have covered a lot of Nautobot's core functionalities in this chapter and described its commitment to efficient network inventory management. We've explored its capacity for IP addresses and prefix tracking, seen the flexibility of redundancy constructs and data models, and navigated its rich feature set that supports robust data handling. Yet, one of Nautobot's defining strengths remains its extensibility.

The next chapter will focus on this extensibility, detailing how Nautobot can be tailored to fit unique operational requirements and adapt to changing network environments. Whether it's through plugins, custom data models, or other features, Nautobot offers avenues for users to expand and customize its capabilities, ensuring it aligns closely with specific needs.

6

Using Nautobot's Extensibility Features

Building on the previous chapter, this chapter showcases how flexible Nautobot is by covering its extensibility feature set, which allows users to cater Nautobot to their specific network or design.

While there are many ways in which you can extend the Nautobot platform, this chapter will concentrate on extending the platform for improved data management.

> **Note**
> Extending the platform by means that require writing Python code is beyond the scope of this chapter, but this is covered in much more detail in *Chapters 14*, *15*, and *16*.

As you learned in *Chapter 2*, Nautobot has a predefined and opinionated data model. This data model cannot possibly account for all networks and design requirements. Take, for example, an organization that determines that a VLAN is deployed within a rack or a cluster. This is a reasonable use case for that specific organization but may not make sense for other organizations. You can take advantage of Nautobot Relationships in such use cases. Nautobot also supports the dynamic loading of YAML data from a Git repository through Config Contexts. These are just two examples of Nautobot's extensibility features.

Nautobot takes a flexible system approach with sane defaults to ensure that the maximum number of use cases can be made by network teams. There are many different approaches to this strategy.

In this chapter, we will review the following topics:

- Statuses
- Tags
- Custom fields
- Computed fields

- Custom links

- Export templates

- Config contexts (and schemas)

- Git as a data source

- Relationships

- Dynamic Groups

Most of the items that will be covered in this chapter can be found in the UI under the **Extensibility** drop-down menu. This will be seen firsthand as we cover each extensibility feature.

> **Tip**
> It is easy to conflate or mix up the names of various extensibility options with similar names. Take care not to mix up computed fields and custom fields, or custom links and custom fields, for example. Be sure to read closely as we cover each feature!

Statuses

Nautobot allows user-defined statuses to be used to facilitate business workflows around object statuses. At a high level, *Status* intends to describe some metadata on mutually exclusive states a model object can be in. As an example, a device can be in an **Active** or **Inactive** state, but not both at once.

A status is a piece of metadata that is applied to existing model object instances, often associating whether or not the object is active, decommissioned, and more. Many core models have a Status, and the Status is generally required when present. Notable exceptions are models that do not represent a physical asset, such as a Region model.

Nautobot comes with a series of statuses *out of the box*. They are applied to models in a reasonable manner, such as not having "failed" as a Status on the location model and having "provision" on the circuit model. However, the operators are free to add Statuses as they choose.

To provide a little more color, the following are the default out-of-the-box statuses for Devices and Locations:

- **Devices**: Active, Planned, Offline, Staged, Failed, Inventory, Decommissioning

- **Locations**: Active, Planned, Staging, Decommissioning, Retired

You can see the current Statuses in list view form by navigating to **Organization -> Statuses**. From there, you can find the **Add** button in the top right and use the **Edit** button or the **Delete** button on the actual Status, as you would for any other model within Nautobot:

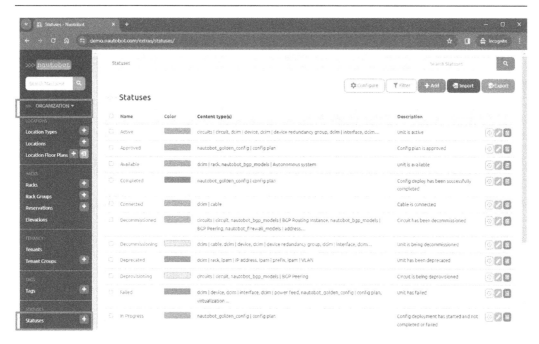

Figure 6.1 – Viewing Statuses

Managing statuses

Nautobot allows you to add, update, and delete Statuses via the UI or API.

Here are some examples of why you may want to adjust a Status:

- Change terminology to use your company's taxonomy, such as adding "**Retired**" versus **Deprecated**
- Add an additional state, such as **In-Transit**, to a Device model
- Apply an existing status to an additional model, such as adding **Staging** to the interface model
- Remove the unused state of **Provisioning**
- Add an additional state for **Awaiting-Approval** to make it easy to sort out which are pending approval
- Add a status of **Managed** to an IP address, indicating it is managed via an external DDI system
- Add a status of **Not-Managed** to indicate a Device is managed by a third party
- Add a status of **Upgrading** to help manage your ZTP workflow state machine
- Golden Config (`https://github.com/nautobot/nautobot-app-golden-config`), a popular Nautobot app, added the **Not Approved** status to indicate the Config Plan status

- Add **Available** to the VLAN model
- Add **Conflict Detected** to the VLAN model

A Status comes with the following fields:

- **Name**: The long form name of the Status
- **Description** (not required): An optional description
- **Content Types**: Content type is the Django term for a model or database table; Content Type specifies which Nautobot models the Status can apply to
- **Color**: The color, stored as a Hex value

This can be seen in the **Add Status** form:

Figure 6.2 – The Add a new status form

Applying a status

A status is required when you're adding an object to a model that has a Status – for example, Device, Location, and so on. This enables good practices for inventory management. Applying a Status via the UI is natural and mandatory, similar to other mandatory fields. You'll see the required fields in bold in the create/add forms. There are no defaults for Status, but for many models and organizations, the first available Status is **Active**. This is used as a pseudo-default.

To apply a Status to an object programmatically via the API, you must have the UUID:

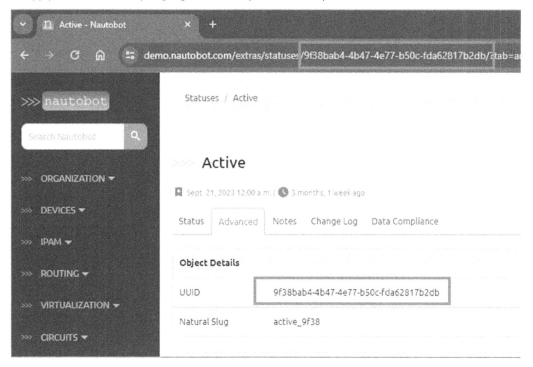

Figure 6.3 – Viewing the UUID

Using the API will be covered in more detail in *Chapter 9*, but for now, it is good to learn how to obtain the UUID.

You can apply the Status to the model object, such as a Device. You learned how to do this in *Chapters 3* and *4* when adding devices in Nautobot.

Use cases for statuses

Going beyond the examples mentioned already, let's briefly discuss more robust use cases for Status before diving deeper:

- Managing the life cycle of your circuits and comparing them to your billing
- Managing your inventory and comparing it to your support contracts
- Synchronizing IP addresses with your DDI platform based on Status
- Managing the ZTP process of your new devices

Now, let's deep dive into how the Status field can help you manage your ZTP process. Typically, a ZTP process looks like this:

- A device is powered on and reaches out based on DHCP settings for an initial configuration.
- The ZTP system applies some code or configuration.
- The device applies the configuration. If successful, it stops rebooting; if not, the system is restarted.

There are several challenges with this process. For example, there may not always be direct synchronization between the logistics of powering on the device and the configuration being ready.

Let's see how a purposefully built list of Statuses can be used to navigate the ZTP process:

- **Inventory**: When a device is extra and not currently planned to be in production
- **Planned**: When a serial number and device name are known
- **Staged**: When the device has been cabled properly
- **Provisioning**: When the device is actively being upgraded
- **Provisioned**: When the ZTP process has applied a minimal config
- **Production-Ready**: When all of the data is ready to generate a full configuration
- **Active: When the device is in production**

Let's take a closer look at what happens in this scenario:

1. A network engineer assigns a device with a serial number to a rack and adds a status of **Planned**.
2. A data center operator receives the physical device, reviews the serial number, and goes to the Nautobot detailed device view. The operator filters by devices that have a **Planned** or **Inventory** status and identifies the device. If its status is **Planned**, the operator connects the device; otherwise, it stores the device in a closet for future use.
3. Upon completion, the data center operator sets the status to **Staged**.

4. The ZTP process identifies the device by a serial number by querying Nautobot, updates the status to **Provisioning**, and applies a minimal amount of configuration, including the IP address and hostname of the device.

5. Upon completion of the ZTP process, the status is changed to **Provisioned**. A network operator reviews the data that has been applied to the device to ensure it has everything it needs for a full configuration and verifies that the generated configuration is correct.

6. The Network operator sets the status to **Production-Ready**. The orchestration system periodically determines if there are any devices in that state and deploys the full configuration.

7. Upon completion of the orchestration process, the device is set to **Active**. The **Active** status is also used to populate the monitoring system configurations.

This is an advanced use case, but you can see how no communication is required between teams to complete the cycle; it can be used to ensure the device is in monitoring as well. The use of Nautobot's Status can truly be a powerful part of your network device strategy.

Best practices for statuses

Here are a few brief best practices for statuses:

- Do not overload a status so that it implies more than one thing (use **Tags** in this case; **Tags** will be covered later in this chapter), such as using **Active-Managed** versus **Active**.

- Keep the number of statuses to a minimum to ensure that they are less likely to be applied incorrectly.

- Keep the number of statuses that are applied to a content type to a minimum.

- Be sure to programmatically update the statuses as part of the workflow, such as at the end of a ZTP process.

- In advanced cases, you can leverage data validation to ensure things such as "a serial number must be applied to have a device in the **Active** state." Data validation will be covered in *Chapter 14*.

Tags

Tags are one of the ways users can define arbitrary data without the field already being defined in the core data model. Tags differ from statuses in that they are not required and there can be many of them, accounting for different use cases. At a high level, tags are intended to describe some non-mutually exclusive metadata on a model object. These are often associated with *subscribing* to metadata, such as the interface being subscribed to the dot1x, QoS, or NAC/BYOD profile configuration. We are using the term *subscribe* here to state that the object should receive some additional metadata based on its use in the network.

As mentioned previously, only a single status can be assigned – and it should have a single purpose. However, tags can have multiple assignments. So, in the case where **Status** is set to **Active**, it may make sense to add the **Managed** tag.

Nearly all core data models in Nautobot support tags, but they are never required when creating new objects such as Devices. Nautobot comes with no predefined tags, so you must define them. You can define an arbitrary amount of tags for a given object.

Managing tags

Nautobot allows you to add, update, and delete tags. Here are some examples of why you may want to adjust a tag:

- Add a subscription to an object, such as **Service-Voice**, on a device model
- Add **Dot1x** to an interface to depict that it requires dot1x configuration
- Tag a location to a design, such as v1.1 or v2.3, for auditing purposes
- Add a tag to describe which monitoring system profile should be applied to an interface, such as managing only interfaces with the **Uplink** tag
- Add the **Unmanaged** tag to infrastructure that is not maintained by your organization, such as the service provider router in your facility
- Remove the tags that are no longer needed

A tag comes with the following fields, as can be seen in the following form for creating a new tag:

- **Name**: The long form name of the tag
- **Color**: The color, stored as a Hex value
- **Description** (not required): An optional description
- **Content Type**: Content type is the Django term for a model or database table:

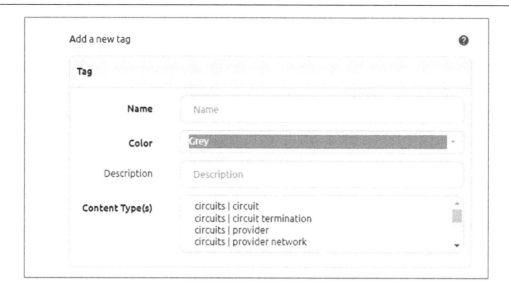

Figure 6.4 – The Add a new tag form

You can view the current tags by navigating to **Organization** > **Tags**, where you can see the list view of tags. From there, you will find the **Add** button in the top right and use the **Edit** button or the **Delete** button on the actual tag, as you would for any other model within Nautobot:

Figure 6.5 – Viewing tags

Applying a tag

You can apply a tag to a model object, such as a Device. As an example, upon going to **DCIM** > **Device** and clicking the **Add** button, you will be prompted to fill in mandatory fields and other fields, such as **Tags**. Tags reference the **Tags** model and show up as a multi-selectable field.

If a tag does not show up as an option, ensure it is both defined and that its **Content Type** is set to the current model you are looking for.

Similar to what we showed in the *Managing statuses* section, to apply a status to an object programmatically via the API, you must have the UUID.

Use cases for tags

Let's briefly discuss some common use cases for tags before deep diving into a specific use case:

- Managing the membership to services such as Voice, Data, or Video on a Device model
- Managing the application or profiled configurations, such as QoS, dot1x, or BPDU-Guard, on an interface model
- Defining how monitoring configuration is applied to a device or interface, such as **Managed** versus **Unmanaged**

Note that tags can take the conceptual form of a group of mutually exclusive fields, such as managed or unmanaged. However, in this case, no such standards are enforced. You should consider a custom field in this situation, but both are acceptable solutions.

Let's deep dive into how the **Tags** field can help you manage configuration through the use of the following tags:

- **Service-Data**
- **Service-Voice**
- **Service-Video**

The **Service-** prefix describes devices that apply the data configuration, such as subnets, IPs, and tunnels associated with their respective service. You can add these tags using the public demo instance (or your own) by navigating to **Organization** > **Tags**, as well as three new tags, by selecting `dcim | devices` as the Content Type. Then, go to the `ams01-dist-01` device. Edit the device, scroll to the bottom, and assign one or two of the new tags to that device:

Figure 6.6 – Adding a tag to ams01-dist-01

You can build a scenario with the following GraphQL query (on the public demo instance) and the associated Jinja template. This is just a brief example; we will cover GraphQL in more detail in *Chapter 8*:

```
query {
  devices(location: "AMS01") {
    name
    tags {
      name
    }
  }
}
```

This would return data in the following structure:

```
{
  "data": {
    "devices": [
      {
        "name": "ams01-dist-01",
        "tags": [
          {
            "name": "Service-Data"
          }
        ]
      },
      {
        "name": "ams01-edge-01",
        "tags": []
      },
# removed for brevity
```

By leveraging Jinja, we can conditionally generate a configuration, like this:

```
{% for device in data['devices'] %}
{% set tags = device.get('tags') | map(attribute='name') | list %}
```

```
{% if "Service-Data" in tags %}
interface Vlan 100
  description SVI for DATA VLAN
{% endif %}
{% if "Service-Voice" in tags %}
interface Vlan 300
  description SVI for VOICE VLAN
{% endif %}
{% if "Service-Video" in tags %}
interface Vlan 300
  description SVI for VIDEO VLAN
{% endif %}
{% endfor %}
```

Note that this is three separate `if` conditionals as this maps out how the service was defined – that is, they are not mutually exclusive. Conditions like this are common in the Nautobot Golden Config generation process (`https://github.com/nautobot/nautobot-app-golden-config`).

Best practices for tags

A few brief best practices for tags are as follows:

- Consider whether the tag could be applied as a custom field or a status. Custom fields will be covered in the next section.

- Tags should not be required. If the data is required, consider a custom field instead.

- Keep the number of tags to a minimum to ensure that they are less likely to be applied incorrectly.

- Keep the number of tags applied to a content type to a minimum.

- In advanced cases, you can leverage data validation to ensure things such as "a tag must be applied to firewall device types."

Custom fields

Custom fields can be thought of as "What fields would my organization want on this model if I were to build it?" You can do this without changing any code, and custom fields extend existing database tables immediately.

A custom field is data that is applied to existing model object instances, often defining configuration, inventory, or metadata. With a few exceptions, core models provide support for custom fields.

As mentioned at the beginning of this chapter, creating a data model that would apply to all organizations is impossible and wouldn't be advantageous as it would come with increased complexity – and custom fields exemplify that mantra. A custom field can define additional data such as TAC contract data on a device, something that wouldn't be reasonable to extend to all organizations. It can also be used to track Site IDs (or old IDs) and types of sites, as shown in the following screenshot:

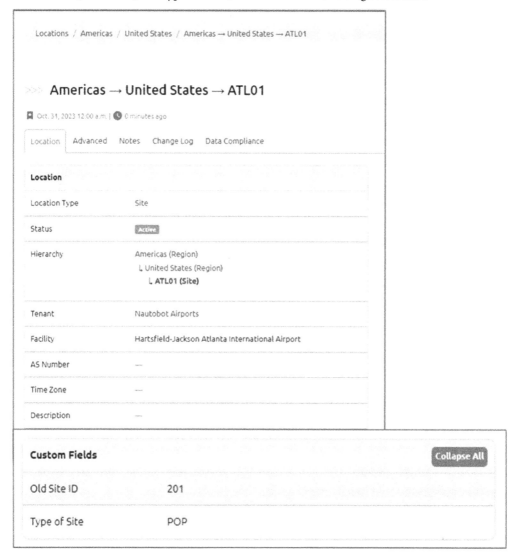

Figure 6.7 – Viewing Custom Fields on a detailed object page

In contrast to tags, which are often used for the membership of a given service or application, custom fields are predominantly used for metadata as an attribute of a given model. When there are specific enumerations of a given custom field, you have the option to choose a tag or custom field.

Managing custom fields

Nautobot allows you to add, update, and delete custom fields on a production system without the need to reboot any service via the UI. A custom field comes with the following fields or attributes:

- **Label**: Name of the field, as displayed to users.

- **Grouping**: Human-readable grouping that this custom field belongs to.

- **Key**: The internal name of this field. It is recommended to use underscores rather than dashes.

- **Type**: The type of value(s) allowed for this field, such as **Text**, **Boolean**, **Integer**, **Date**, **URL**, **Selection**, **Multiple Selection**, or **JSON**.

- **Weight**: An integer value where fields with higher weights appear lower in a form.

- **Description**: This is also used as the help text when editing models using this custom field.

- **Required**: If true, this field is required when creating new objects or editing an existing object.

- **Default**: The default value for the field (must be a JSON value).

- **Filter Logic**: Loose matches any instance of a given string; exact matches the entire field.

- **Move to Advanced Tab**: Hide this field from the object's primary information tab. It will appear in the **Advanced** tab instead.

You can view the current custom fields by navigating to **Extensibility** > **Custom Fields**. From there, you will find the **Add** button in the top right and use the **Edit** button or the **Delete** button on the actual custom field, as you would for any other model within Nautobot.

Grouping and **Move to Advanced Tab** help manage a growing list of custom fields and the user experience. **Type, Required**, and **Default** should be carefully considered when ensuring that the data behaves as required in your organization's unique requirements. When using **Selection** or **Multi Selection**, you must define the selectable fields and assign a **Weight** value to the ordering it shows. We'll look at these in the next section.

Diving into custom field attributes

Let's walk through an example of creating a new custom field for site type. The types that we'll use are **PoP, Data Center, Branch**, and **Corporate**:

- **Label**: This is what each user will see as the label when viewing the custom field in the detailed view of the object. In our case, it'll be the detailed view of a given location.

  We'll use **Site Type** as our label.

- **Grouping**: You can group custom fields so that they are within the same panel on the detailed view page. While we only have one group for now, let's create a group.

  We'll use **Site Data** as our grouping name.

- **Key**: We are going to use a key of **site_type_01**.
- **Type**: Given that we have a defined list of options – **PoP**, **Data Center**, **Branch**, and **Corporate** – we are going to use **Selection**:

Figure 6.8 – Viewing different custom fields and making selections

Now, let's scroll down and fill in the choices for our new custom field:

Figure 6.9 – Adding choices for our custom field

It's worth noting that there are many options available for **Type**, given the type of data you're looking to store in Nautobot. What you want to do is ensure that the type not only aligns with your data but also improves the user experience and data quality within Nautobot. For example, for **Site Type**, we could have chosen **Text** and used free-form data entry with guidelines for users. But I think we all know that wouldn't prove to be a reasonable choice later as it would result in inconsistency, bad user experience, and bad data that would be unusable for network automation.

- **Weight**: We'll keep the default of **100**.

- **Description**: We're going to use **The site type used by all network teams in Acme** as our description.

- **Required**: Be sure to check the **Required** check box if you want every object (in our case, location) to have **Site Type** filled in. We are going to check this.

- **Default**: We normally wouldn't have a default for this, but because this is a book (and to err on the side of showing more), we are going to make **PoP** the default choice. Note that we need to add this after we create the custom field.

- **Filter Logic**: Here, we are going to use the default value of **Loose**. This means that when you are using a filter form on a list view, it'll match any part of the string. For example, let's say that **Type of Site** is set to **Data Center**:

Figure 6.10 – Viewing the name of a custom field on the detailed object page

This will allow you to filter on `center`, `data`, or any other substring. We'll filter on `data`. Using the **Filter** button from the table list view to do so:

Figure 6.11 – Viewing the filter results with loose filter logic

There are two other options: **Exact**, in which you need an exact match, and **Disabled**, which disables filtering on a given custom field.

- **Assignment**: Since we are working with locations, let's add **dcim | location**:

Figure 6.12 – Using dcim | location in the form

Validation rules

Since our inputs aren't integers and we are using a selection field, there is no need for us to use validation rules. You should use these to ensure good data when your inputs are numbers, such as for ASNs, process IDs, or any other numerical input, or when you're using a text field and you want to apply a RegEx.

Custom field choices

You should have added this before, but you should have **PoP**, **Data Center**, **Branch**, and **Corporate** set to **70**, **80**, **90**, and **100**, respectively.

Finally, you can click **Create** to complete the process of adding a custom field. You'll be redirected to the detailed page for the newly created custom field:

Figure 6.13 – Viewing a newly created custom field

Now, click **Edit** and add **PoP** as a default in the **Default** textbox. Once you've done this, update the page:

Figure 6.14 – Setting a default selection in a custom field

You'll then see **PoP** listed as the default:

Description	The site type used by all network teams in Acme.
Required	✓
Default	['PoP']
Filter Logic	Loose

Figure 6.15 – The default selection will be set properly

If you add a new location in Nautobot, you'll see the following in the **New Location** form (you can use any arbitrary name and location type):

Figure 6.16 – Ensuring the default selection works in a New Location form

Once you add the site and navigate to the detailed page view, you'll see the following:

Custom Fields	
Old Site ID	—
Type of Site	—
⌄ **Site Data**	
Site Type	PoP

Figure 6.17 – Viewing the newly created custom field

Take note of the drop-down arrow next to **Site Data**. This was added because we used the **Grouping** attribute when adding the custom field. This will put all custom fields in the same logical grouping and allow you to toggle them up and down.

Here's another example of using the **Grouping** feature. Let's say you want to track how much a device initially costs and know how many years until it is depreciated. You could create a group called *Lifecycle* with two custom fields of **Cost** and **Depreciation (Years)**. When grouped and collapsed, it would look like this:

Figure 6.18 – Collapsed custom field grouping

Here's what it would look like when expanded:

Figure 6.19 – Expanded custom field grouping

Applying a custom field

A custom field behaves like a Nautobot Core model field in most cases, such as when adding, updating, or deleting a model instance object. Often, the only difference will be the placement on the screen, but the intention is for it to behave as close to a core model as possible, such that most users are unable to differentiate between them.

Use cases for custom fields

Let's briefly discuss some common use cases for custom fields before deep diving into a specific use case:

- Managing the subscription to services such as Voice, Data, or Video on a device model
- Adding attributes to the device, such as memory information on a network device

- Defining which group manages a device

- Defining which business unit operates a site

- Defining a separate set of contact information, such as shipping versus technical contacts on a location model

- Adding an approval status as a Boolean

- Adding upload-speed and download-speed fields to the Circuit model

- Adding an interface-profile selection to the Interface model

- Adding contact information to the Platform model

- Adding the primary Device when the Device is in a chassis

Let's review the same option use case that was described in the **Tags** field but instead using custom fields. In doing so, we can highlight the differences between the two:

- **Service-Data**

- **Service-Voice**

- **Service-Video**

See the *Use cases for tags* section to better understand the use case.

Create three custom fields that get applied to the device model. They should be of Boolean type and labeled and named as described previously. Be sure to set the default to `false`.

You can build a scenario with the following GraphQL and associated Jinja template.

Take note of how you query for custom fields using GraphQL: use `_custom_field_data` to return all custom fields for a given object. As stated earlier, GraphQL will be formally covered in *Chapter 9*:

```
query {
  devices(location: "AMS01") {
    name
    _custom_field_data
  }
}
```

```
This would return data in the following structure:
{
  "data": {
    "devices": [
      {
        "name": "ams01-dist-01",
          "_custom_field_data": {
```

```
            "service_data": true,
            "service_video": true,
            "service_voice": false
        }
    }
}
}
# removed for brevity
```

You could then construct a Jinja template, similar to how we did earlier in this chapter when using tags.

Best practices for custom fields

A few brief best practices for custom fields are as follows:

- Consider if the custom fields could better be applied as a tag or Status

- Consider if there are alternatives to using **Text**, such as **Selection** or **Multi Selection**

- Consider the user experience and use the hidden or **Move to Advanced Tab** options

- Always consider using **Default** and **Required** on the field

- Do not use custom fields on data that could be derived, such as "am-cluster" from the region; see *Computed fields* for more

- In advanced cases, you can leverage data validation to ensure things such as "a custom field must be applied to router device types"

Computed fields

Computed fields are fields that are computed at runtime. Nautobot's implementation provides access to the associated data, as well as Jinja templating, including any available Jinja filters. Computed fields intend to derive data from existing data in a consistent and centralized manner.

Imagine that you want to auto-generate an interface description from data that's already in Nautobot. One option is building this logic in some external script. Another option is using computed fields. Computed fields allow you to use Jinja templating and filters (if needed) to concatenate the value you need. In this example, you can pull data from your hostname, interfaces, and even your neighbor's data.

A computed field can be applied to existing model object instances but is never stored; instead, it's generated on the fly. A computed field can be applied to nearly all core Nautobot models and leverage Jinja syntax and any available filters to render the data you would expect.

Since computed fields are rendered on the fly, there is a computational impact that you should consider. By default, the computed fields will not be returned in an API response unless the ?include=computed_ fields query parameter is applied to it.

Managing and applying computed fields

Unlike statuses, tags, and custom fields, managing and applying are handled in the same way. Logically, it makes sense, since the fields are rendered on the fly. A computed field comes with the following fields:

- **Content Type**: Content type is the Django term for a model or database table.
- **Label**: The long form name of the computed field.
- **Key**: The unique internal key of the model.
- **Description**: An optional description.
- **Template**: The Jinja template code for the field value.
- **Fallback Value**: The value to show if the template fails to render correctly.
- **Weight**: The order in which it is shown.
- **Move to Advanced Tab**: Hide this field from the object's primary information tab. It will appear in the **Advanced** tab instead.

You can view the current computed field by navigating to **Extensibility** > **Computed Fields**. Here, you can see the list view of the computed fields. You will also see the **Add** button in the top right and use the **Edit** button or the **Delete** button on the actual computed field, as you would for any other model within Nautobot.

When filling in the template, it must be a valid Jinja template. The field will render as it would within any Jinja template engine. The following context data is available within the template when rendering a computed field:

Variable	Description
obj	The Nautobot object being displayed
debug	A Boolean indicating whether debugging is enabled
request	The current WSGI request
user	The current user (if authenticated)
perms	The permissions assigned to the user

Table 6.1 – Available list of variables

This context can be accessed via a standard Jinja variable, such as {{ request }} or {{ user }}. The links are rendered on the object detailed view (not the object list view) and are placed below the icons to clone, edit, and more.

Computed field template context

Computed field templates can utilize the context of the object the field is being rendered on. This context is available for use in templates via the `obj` keyword. As an example, for a computed field being rendered on a circuit object, you can create one field that concatenates the circuit ID, circuit type, and provider into a single string with the following computed field:

```
{{ obj.cid }} -- {{ obj.circuit_type }} -- {{ obj.provider }}
```

This would result in the following being shown in the detailed view of a given circuit:

Figure 6.20 – Creating a computed field using existing data

To try this out, you can use the following inputs to create a computed field:

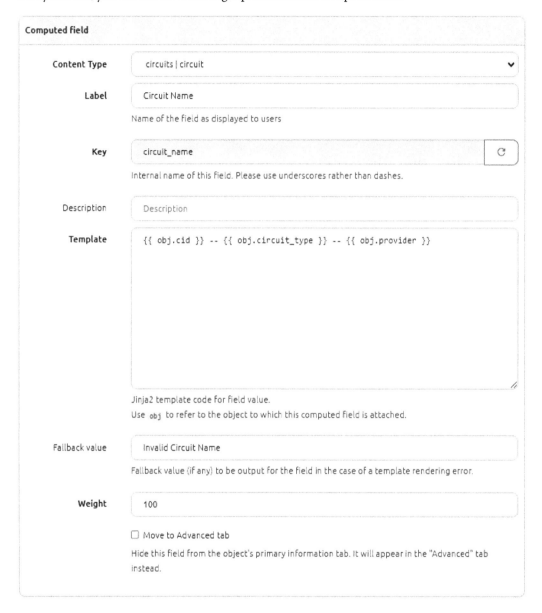

Figure 6.21 – Creating a computed field

You can also use Jinja filters in the template. A basic example is to return all capitals using the `upper` filter. See the documentation (`https://docs.nautobot.com/projects/core/en/v2.1.0/development/apps/api/platform-features/jinja2-filters/`) on built-in filters and registering custom Jinja filters in Nautobot Apps.

Again, the key to remember when starting with computed fields, and then computed fields and custom links, which we'll cover next, is that they all use `obj` as a keyword to access the object of the model type you're working on. So, if you create a computed field for `dcim | devices`, `obj` represents a device (in contrast to a circuit, as shown in the previous example).

Use cases for computed fields

Let's briefly discuss some common use cases for computed fields before deep-diving into a specific use case:

- Concatenating multiple parts of the data, such as forming **a hostname || device-type || role** as a text similar to what we showed for a circuit.

- Parsing out a name, such as getting the function of the device from the hostname – for example, getting "RT" from NYC-RT01.

- Generating PTR record from the IP address

- Generating a connection description on interfaces that are connected, resulting in data such as `Ethernet1/1.ams01-edge-01–Ethernet1/1.ams01-edge-02`

- Generating alternate views of platform data, such as for Ansible

Using the following template, where `obj` is a Device object, we can see how we can achieve the final use case outlined here:

```
{% if obj.platform.network_driver == 'arista_eos' %}
arista.eos.eos
{% elif obj.platform.network_driver == 'cisco_asa' %}
cisco.asa.asa
{% elif obj.platform.network_driver == 'cisco_ios' %}
cisco.ios.ios
{% elif obj.platform.network_driver == 'cisco_xr' %}
cisco.iosxr.iosxr
{% elif obj.platform.network_driver == 'cisco_xe' %}
cisco.ios.ios
{% elif obj.platform.network_driver == 'cisco_nxos' %}
cisco.nxos.nxos
{% elif obj.platform.network_driver == 'juniper_junos' %}
```

```
junipernetworks.junos.junos
{% else %}
{% endif %}
```

Applying that template to the Platform model and using a key of `ansible_network_os` allows you to render the expected Ansible format on the fly from a central location.

> **Note**
>
> Based on the Jinja template being used, you may need to strip or trim the whitespace in the client-side script that fetches the computed field.

Best practices for computed fields

Here are a few brief best practices for computed fields:

- Consider the computational impact of each computed field that you use
- Remove complex logic from the template and add it to a custom Jinja template filter
- Always consider using a fallback value on the field

Custom links

Custom links are user-defined URL links that show up in the detail view of the page. They provide easily creatable links that can be used by your organization to quickly navigate to other tools and websites. The following four custom links were created for easy navigation for Forward Networks, Kentik, SolarWinds, and Splunk. How cool is that? You can navigate to them for this specific device right from this page!

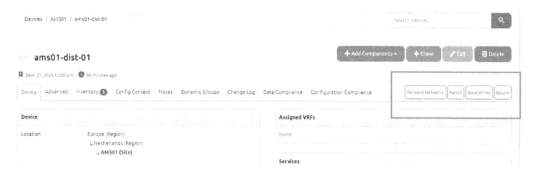

Figure 6.22 – Custom link buttons

It is also possible to change the color of a custom link button when you're creating one.

Going one step further, you can also group custom links when there are many of the same type:

Figure 6.23 – Consolidating custom links into a single "grouping" button

At a high level, custom links intend to derive the hyperlink from existing data in Nautobot in a consistent and easily accessible manner. As an example, a custom link may allow the user to quickly link to a monitoring system, as shown previously.

Similar to a computed field, a custom link can be applied to existing model object instances but is never stored; instead, it's generated on the fly. A custom link can be applied to nearly all core Nautobot models and leverage Jinja syntax and any available filters to render the data you would expect.

Since custom links are rendered on the fly, there is a computational impact that you should consider.

Managing and applying custom links

Unlike statuses, tags, and custom fields, but similar to computed fields, managing and applying custom links is handled in the same way. Logically, it makes sense, since the links are rendered on the fly, just like computed fields. A custom link comes with the following fields that you can configure or edit:

- **Content Type**: Content type is the Django term for a model or database table
- **Name**: The long form name of the custom link
- **Text**: Jinja template code for the field value
- **URL**: Jinja template code for the link URL
- **Weight**: The order in which it is shown
- **Group name**: Group one or more custom links together within one button

- **Button Class**: The class of the first link in a group will be used for the drop-down button
- **New Window**: Toggle to force the link to open in a new window

Here is a pseudo-example for navigating directly to SolarWinds from Nautobot:

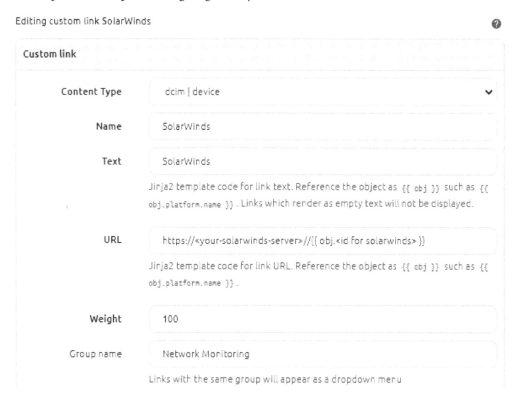

Figure 6.24 – Custom link create form

After creating the custom link, navigate to the device page (or the page of the object you chose for **Content Type**). This may require you to build a few custom Jinja filters if the tools you want to link aren't using their respective names in Nautobot. Alternatively, you can build custom fields to store the IDs of how those tools reference this object. If you do this, be sure to store these custom fields in the **Advanced** tab since they are not user-friendly attributes.

Here is another example you can test end-to-end. So far, we've been using the Nautobot public sandbox at `https://demo.nautobot.com`. There is another one available at `https://next.demo.nautobot.com`. The next demo instance is based on the next branch on GitHub, which provides the latest features coming in the next release. Both instances use the same dataset and therefore have the same UUID for objects. This means we can build a custom link from `demo.nautobot.com` that gives a quick link to `next.demo.nautobot.com`. Let's build this.

In demo.nautobot.com, add a custom link by using a **Target URL** value of https://next.demo.nautobot.com/dcim/devices/{{ obj.id }}/:

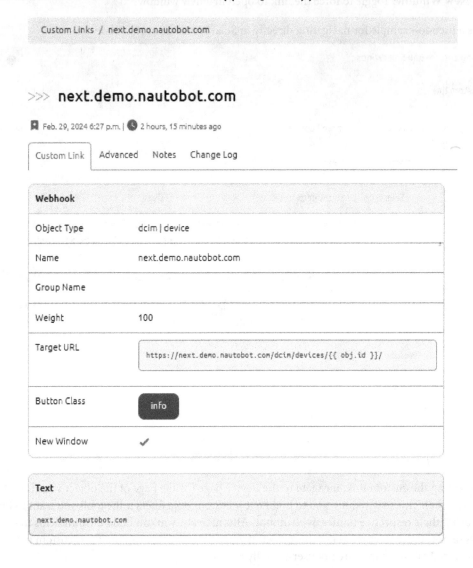

Figure 6.25 – Viewing a custom link

When you click a given device, you'll see the following custom link (see the blue button) at the top right of the page:

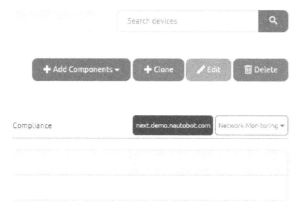

Figure 6.26 – Viewing a custom link on the device page of demo.nautobot.com

When you click on the button, you'll see a new page that'll take you to the same device in next.demo. nautobot.com! This highlights how the Jinja template was rendered, thus creating the new URL:

Figure 6.27 – Viewing the Jinja-rendered custom link

You can view all custom links by navigating to **Extensibility** > **Custom Links**:

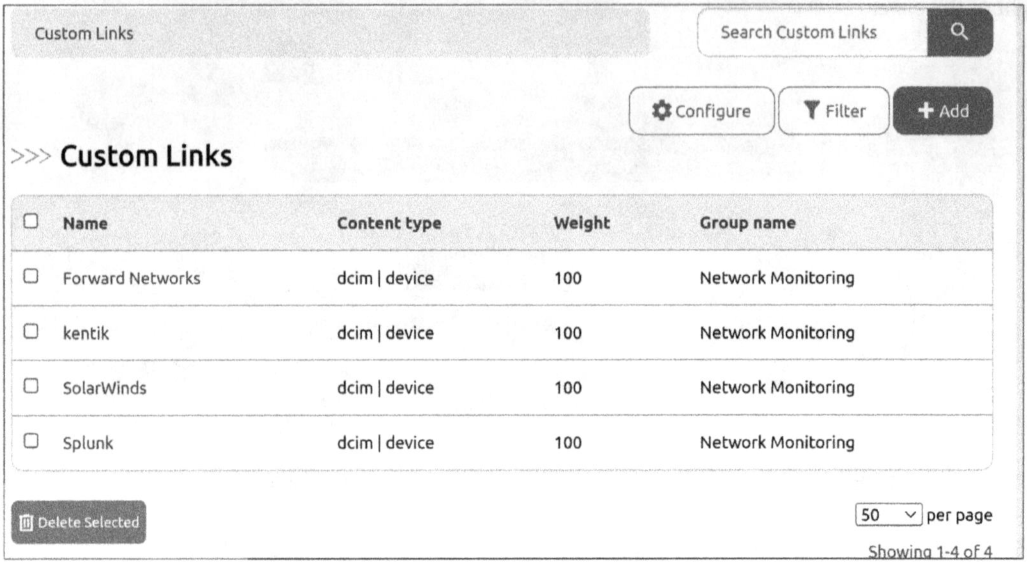
Custom Links
Search Custom Links
Configure
Filter
+ Add
>>> Custom Links
Name
Content type
Weight
Group name
Forward Networks
dcim | device
100
Network Monitoring
kentik
dcim | device
100
Network Monitoring
SolarWinds
dcim | device
100
Network Monitoring
Splunk
dcim | device
100
Network Monitoring
Delete Selected
50
per page
Showing 1-4 of 4

Figure 6.28 – Viewing custom links

From there, you will see the **Add** button at the top right and use the **Edit** button or the **Delete** button on the actual custom links, just as you would for any other model within Nautobot.

When filling in the template, it must be a valid Jinja template. The field will render as it would within any Jinja template engine, whitespace and all. The following context data is available within the template when rendering a custom link's text or URL:

Variable	Description
`obj`	The Nautobot object being displayed
`debug`	A Boolean indicating whether debugging is enabled
`request`	The current WSGI request
`user`	The current user (if authenticated)
`perms`	The permissions assigned to the user

Table 6.2 – Available variables for custom links

This context can be accessed via a standard Jinja variable, such as `{{ request }}` or `{{ user }}`. Links are rendered on the object detailed view (not the object list view) and are placed below the icons to clone, edit, and more.

Use cases for custom links

Let's briefly discuss some common use cases for custom links before deep diving into a specific use case:

- Linking to the contract details from the hardware vendor based on the serial number
- Linking to the circuit provider for circuit details
- Linking to an external monitoring system
- Building dynamic queries for a given device/object in a logging server

Using the following template on the Device model, we can see how we can achieve the last use case we described:

```
http://ntc-monitoring.networktocode.com/device/{{ obj.name }}/
```

Best practices for custom links

Here are a few brief best practices for custom links:

- Consider the computational impact of each custom link that you use
- Remove complex logic from the template to a custom Jinja template filter

Export templates

Export templates are a convenient entry point for generating and exporting reports. Export templates use Jinja templates, so if you can create a template for it, you can create a report for it!

At a high level, export templates derive reports or similar text documents based on an object, list, or table view. There is a default export template that will export the data for a given model as a CSV, but you can create custom templates as well.

Default export templates

Let's look at one default export template for Devices. You can try this out using the sandbox environment too.

Navigate to a **Devices** list, such as the table view, and click the **Export** button:

Figure 6.29 – Using the default export CSV template

After you click **Export**, you'll be redirected to the **Jobs Results** page, where you can see the status of the job that generates the CSV. Once it's been generated, you can download it:

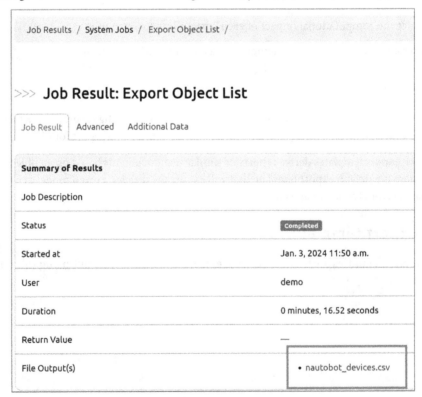

Figure 6.30 – Downloading an export template from the Job Results page

Upon clicking the `nautobot_devices.csv` link, your file will look like this:

	A	B	C	D	E	F	G
	name	display	id	object_type	natural_slug	face	local_config_context_data
	ams01-dist-01	ams01-dist-01	89b2ac3b-1853-4eeb-9ea6	dcim.device	ams01-dist-01_nautobot-air	NULL	{"cdp": false}
	ams01-edge-01	ams01-edge-01	37a938b0-bd5a-4c25-9d98	dcim.device	ams01-edge-01_nautobot-a	front	NULL
	ams01-edge-02	ams01-edge-02	9e43ed6f-0f4c-4e72-8726-	dcim.device	ams01-edge-02_nautobot-a	front	NULL
	ams01-leaf-01	ams01-leaf-01	eaeca318-23f3-4ef0-805c-	dcim.device	ams01-leaf-01_nautobot-air	front	NULL

Figure 6.31 – Viewing the downloaded file

The preceding screenshot shows seven columns of data, but the file contains 51 columns or attributes of each device! We'll show you how to build a custom export in the next section.

Use cases for export templates

The following is a list of use cases for export templates:

- Exporting the configuration required for your monitoring system
- Exporting the configuration intended for DNS or DHCP systems
- Exporting a custom CSV view with the attributes that you care about
- Exporting an Ansible inventory file
- Generating a markdown report that is versioned periodically

Now, we'll walk through how to create a new export template.

Managing and applying export templates

When creating an export template, you can configure the following fields:

- **Content Type**: Content type is the Django term for a model or database table
- **Name**: The long form name of the export template
- **Description**: An optional description
- **Template Code**: Jinja template code for the field value
- **MIME type**: Defaults to text/plain but can be set to a custom value such as text/XML
- **File extension**: Extension to append to the rendered filename

At this point, you should see a pattern – that is, you can view a list of all export templates by navigating to **Extensibility** > **Export Templates**. By default, there will be no export templates in the list:

Figure 6.32 – Viewing Export Templates

> **Note**
>
> You won't see the built-in and default export templates in this list.

When filling in the template, it must be a valid Jinja template. The field will render as it would within any Jinja template engine. The context or variable to access the data in the template is called `queryset`. It is not `obj`, as it was for custom links and computed fields, because they were single instances of an object that was used on the detailed object page. `queryset` is a list – that is, an iterable sequence of objects (for example, devices) – because it is generated on the list view page. This means that you can iterate through the queryset and obtain data for a particular (filtered or not) collection of objects of the same type.

> **Note**
>
> The term "queryset" is used because, under the hood, it is a Django queryset that is generated and used within the export template.

As usual, you can find the **Add** button in the top right of a list view. Let's create a new CSV export template that has just the device fields that we'd like – that is, **Name**, **Location**, **Status**, **Role**, **Manufacturer**, **Device Type**, and **Platform**. The columns in the default CSV file we saw earlier are the dictionary keys we are going to use within the export template, with one change: we are going to replace the double underscores with periods. This is the template we're going to use:

```
Name,Location,Status,Role,Manufacturer,Model,Platform
{%- for device in queryset %}
{{ device.name }},{{ device.location.name }},{{ device.status.name
}},{{ device.role.name }},{{ device.device_type.manufacturer.name }},
{{ device.device_type.model }},{{ device.platform.name }}
{%- endfor %}
```

After adding the template to Nautobot, you should see the following page, which provides a detailed view of the new template:

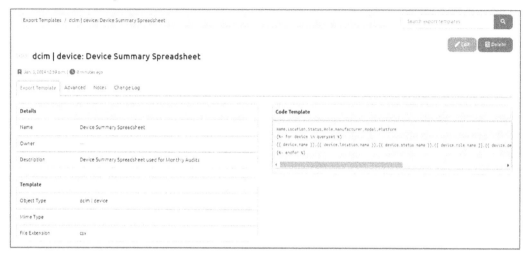

Figure 6.33 – Detailed object view of a single export template

Once the new template is created, you can navigate to the **Devices** page and click the **Export** button. You should notice that the default CSV template is called **CSV format** and the custom export templates are below it. We named our new template Device Summary Spreadsheet. Click that to generate the new report:

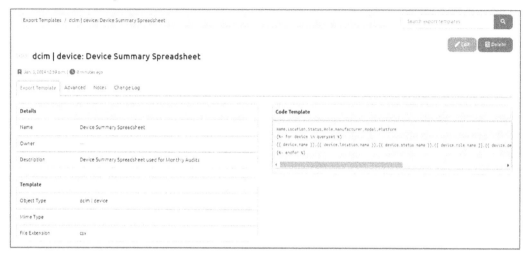

Figure 6.34 – Viewing a new export template via the Export dropdown

Once downloaded, you'll see that the resulting file that was generated:

	A	B	C	D	E	F	G
1	Name	Location	Status	Role	Manufacturer	Model	Platform
2	ams01-dist-01	AMS01	Active	distribution	Cisco	Catalyst 6509-E	Cisco IOS
3	ams01-edge-01	AMS01	Active	edge	Arista	DCS-7280CR2-60	Arista EOS
4	ams01-edge-02	AMS01	Active	edge	Arista	DCS-7280CR2-60	Arista EOS
5	ams01-leaf-01	AMS01	Active	leaf	Arista	DCS-7150S-24	Arista EOS
6	ams01-leaf-02	AMS01	Active	leaf	Arista	DCS-7150S-24	Arista EOS
7	ams01-leaf-03	AMS01	Active	leaf	Arista	DCS-7150S-24	Arista EOS
8	ams01-leaf-04	AMS01	Active	leaf	Arista	DCS-7150S-24	Arista EOS
9	ams01-leaf-05	AMS01	Active	leaf	Arista	DCS-7150S-24	Arista EOS
10	ams01-leaf-06	AMS01	Active	leaf	Arista	DCS-7150S-24	Arista EOS
11	ams01-leaf-07	AMS01	Active	leaf	Arista	DCS-7150S-24	Arista EOS
12	ams01-leaf-08	AMS01	Active	leaf	Arista	DCS-7150S-24	Arista EOS

Figure 6.35 – Viewing a newly created report created by an export template

While the examples shown generated CSV files, keep in mind that these files can be generated using any target file type. Consider markdown files stored in a Git repository and HTML files that are rendered in a web UI or a CSV, as shown here.

Best practices for export templates

Here are a few brief best practices for export templates:

- Store them in version control so that you aren't managing Jinja templates in the Nautobot UI. This will be covered in the *Git as a data source* section.

- Use the `get_config_context()` method to get data from the config contexts of devices.

- Generally, require looping on the queryset to produce a reasonable report.

- Remove complex logic from the template and add it to a custom Jinja template filter.

Config contexts

Config contexts allow you to add arbitrary YAML/JSON data to devices, in a similar fashion to how Ansible does it (through the use of host and group variables). The data can come from multiple sources and is flattened and presented as a single data structure for a given device. In Nautobot, we have config contexts (based on groups and hosts).

Imagine being able to add any YAML data you already have in a Git repository directly into Nautobot. Well, that's possible with config contexts. Here is an example of what that could look like in Nautobot (being viewed as YAML):

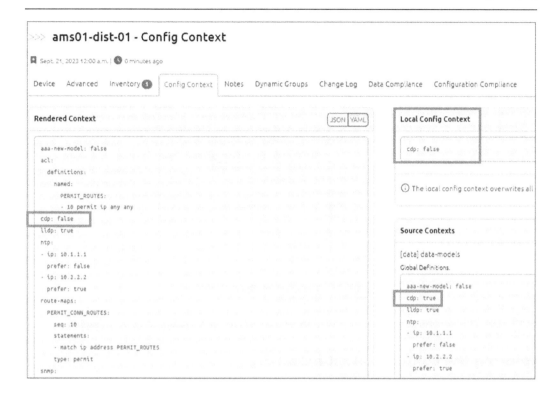

Figure 6.36 – Understanding the Config Context tab in a detailed device view

On the left-hand side, you can view a completed rendered config context (similar to a rendered device's facts and variables in Ansible). On the right-hand side, you can view your **Local Config Context** (scoped for a single device, such as a host variable) and other source contexts – think "group variables" in Ansible. If you aren't familiar with Ansible, these would be other services or groups that have data and variables applied.

Highlighted in red in the preceding screenshot, you can see that there is a config context that has `cdp` set to `True`, but in the **Local Config Context** section, it is `false`. Therefore, in the final **Rendered Context**, it is `false`.

At a high level, config contexts allow you maximum flexibility in associating any data you need to a given device. It is done via some intelligence, applying a weighted rendering to the data to optimize your data and provide an advanced JSON data structure. As an example, you may want to define a list of NTP servers that are the same per region and not have to define it on every device. **Config Context** allows you to do just that. Alternatively, think about SNMP contacts per site – config contexts make this achievable.

Note that at the time of writing, config contexts only apply to device and virtual machine models. Data is stored and applied as JSON but can be viewed as either JSON or YAML.

Be sure to review the few existing config contexts in the sandbox environment (demo.nautobot.com) by navigating to **Extensibility** > **Config Contexts**:

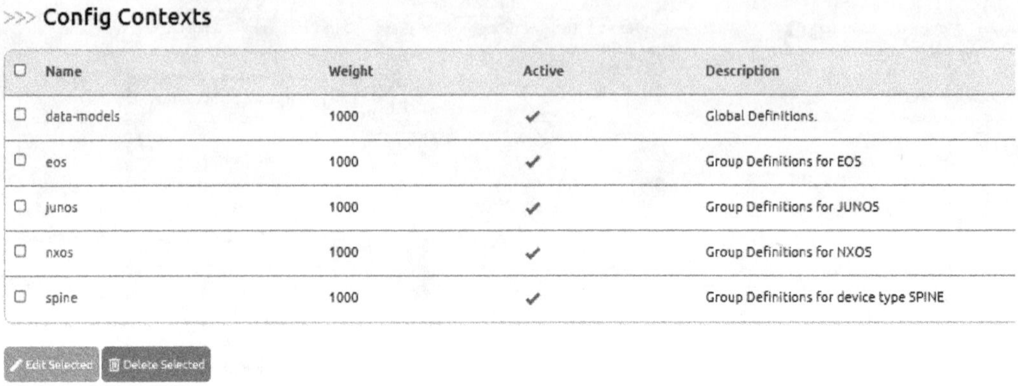

>>> **Config Contexts**

	Name	Weight	Active	Description
☐	data-models	1000	✔	Global Definitions.
☐	eos	1000	✔	Group Definitions for EOS
☐	junos	1000	✔	Group Definitions for JUNOS
☐	nxos	1000	✔	Group Definitions for NXOS
☐	spine	1000	✔	Group Definitions for device type SPINE

✏ Edit Selected   🗑 Delete Selected

Figure 6.37 – Viewing Config Contexts

Additionally, config context data can be saved in Git. This will be covered later in this chapter.

Exploring the config context hierarchy

A config context is data that can be applied to the instance or a hierarchical set of data by grouping based on the following models:

- Locations
- Role
- Device type
- Platform
- Cluster group
- Cluster
- Tenant group
- Tenant
- Device redundancy groups
- Tag

This is where the real power of config contexts comes into play. As an example, you can have NTP for the NYC location that is weighted more favorably than NTP for the Americas location/region, and it will render the more specific NYC NTP data. For further details, please refer to the documentation at `https://docs.nautobot.com/projects/core/en/stable/models/extras/configcontext/#hierarchical-rendering`.

Note that if the same key-value pair exists in multiple config contexts, the config context data is merged with a replacement on the root key – that is, the data is not merged.

Devices and virtual machines may also have a local config context defined. This local context will always take precedence over any separate config context objects that apply to the device/VM. This is useful in situations where we need to call out a specific deviation in the data for a particular object.

Managing and applying config contexts

Config Context data is managed individually but is rendered via the hierarchical model described previously on a device or virtual machine basis. **Config Context** comes with the following fields:

- **Name**: The name of the config context.
- **Weight**: The priority of the config context. The highest config context with the highest weight wins.
- **Description**: An optional description.
- **Schema**: The ability to enforce data standards (covered in more detail shortly).
- **Is active**: A Boolean to determine if this should be rendered.
- **Assignment**: The group described previously (**Region**, **Site**, **Role**, and so on).
- **Data**: The JSON-formatted data structure to apply.

As stated earlier, you can view your config contexts by navigating to **Extensibility** > **Config Contexts**, where you can see a list view of them. From there, you can find the **Add** button at the top right and use the **Edit** button or the **Delete** button on the actual config context, as you would for any other model within Nautobot.

The following is an example of an **Add a new config context** form that will create a config context for all devices that have a **Platforms** value of **Cisco NX-OS**:

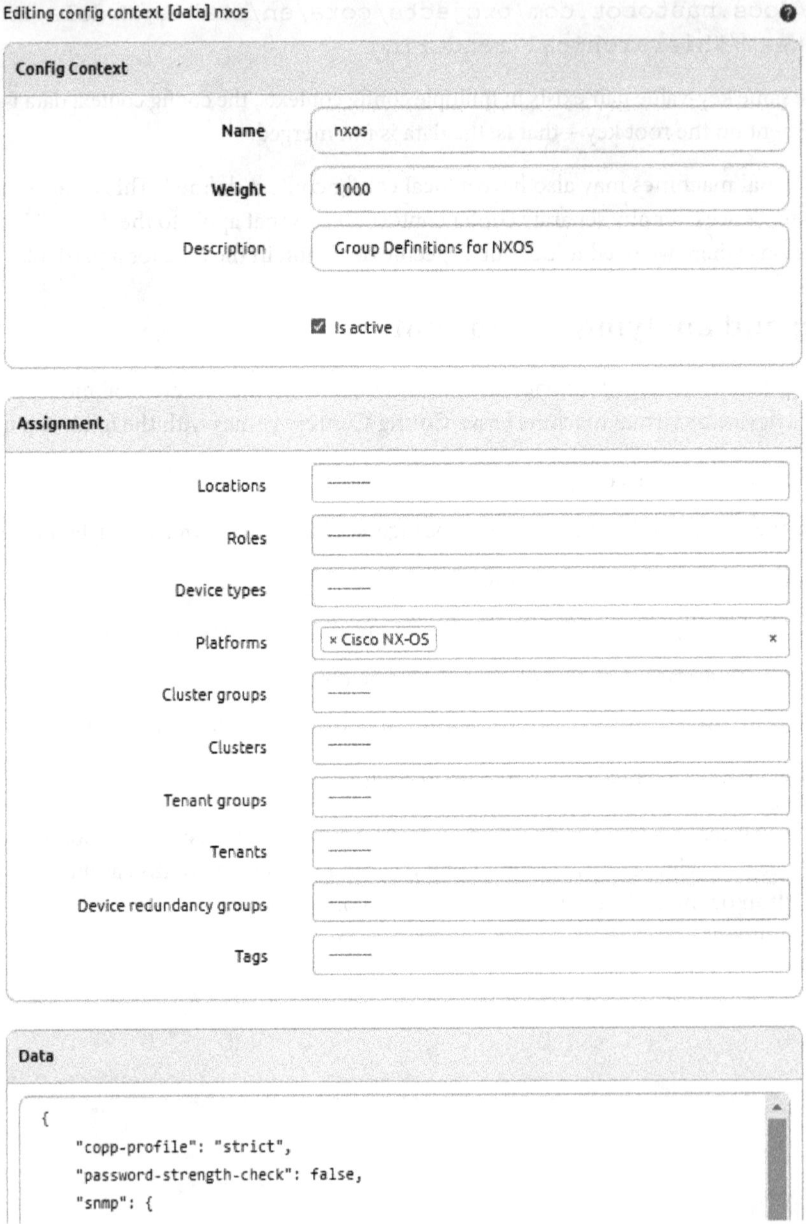

Editing config context [data] nxos

Config Context

Name	nxos
Weight	1000
Description	Group Definitions for NXOS

☑ Is active

Assignment

Locations	---------
Roles	---------
Device types	---------
Platforms	× Cisco NX-OS ×
Cluster groups	---------
Clusters	---------
Tenant groups	---------
Tenants	---------
Device redundancy groups	---------
Tags	---------

Data

```
{
    "copp-profile": "strict",
    "password-strength-check": false,
    "snmp": {
```

Figure 6.38 – Config Context form

You can update local config contexts on the device or virtual machine itself by using the standard **Add** and **Edit** buttons on the model instance, as shown here:

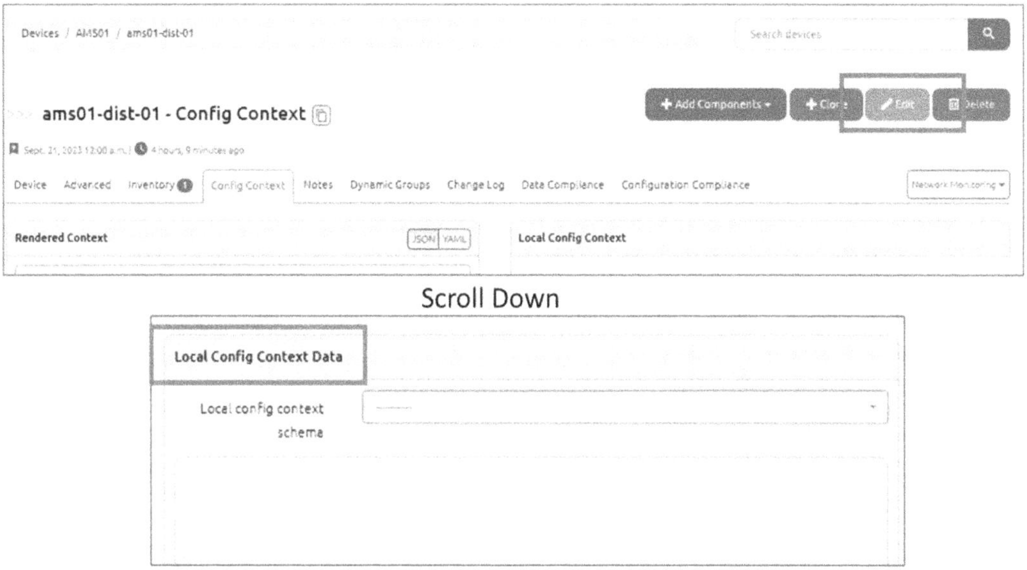

Figure 6.39 – Adding a local config context (device-specific)

The data is automatically applied to the device and virtual machine model instance and is visible at rendering time. When viewing a device or virtual machine model instance on the **Config Context** tab, you can view both the hierarchy and the final rendering of the data, as shown previously.

Let's showcase one more example of config contexts to highlight how they can be used to manage snowflakes in the network design. If you've been in networking long enough, you'll know that no matter how much you know about device configuration standards, there is always one or more devices that need to deviate. In this case, one approach would be to use tags to apply a given context.

Let's walk through an example that maps some YAML that contains data that's required to generate route maps for edge devices when they deviate from the Enterprise standard.

First, we must create a new tag. We'll call this tag `snowflake_route_maps`:

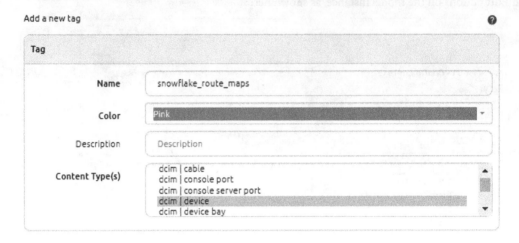

Figure 6.40 – Creating a tag to drive config contexts

Next, we'll create a new config context called `snowflake_route_maps`, giving it a weight of **2000** so that it has the highest priority. Then, we'll select the new tag in the **Tags** form field so that this config context is used whenever the tag is assigned to a device. Finally, we'll add the JSON data that will be used. You can use any data for testing purposes:

Figure 6.41 – Creating a config context for tagged devices

Finally, navigate to a device in the Nautobot demo instance, such as ams01-dist-01. Edit the device and at the bottom of the **Edit** form, choose the new snowflake_route_maps tag. After saving the config context, navigate to the detailed page of the device and click the **Config Contexts** tab. In the **Rendered Context** area in the left pane, you'll see the data that was part of the newly created config context!

This highlights how you can combine tags with config contexts for a very powerful solution to manage deviations in a very methodical way.

Use cases for config contexts

Let's consider some common use cases for config contexts:

- Managing complex nested data, such as BGP or ACLs
- Managing global configurations, such as NTP, DNS, and AAA servers
- Managing services, such as toggling CDP, LLDP, and HTTP on/off

Config context schemas

One of the differences between Nautobot's data model and a traditional **Infrastructure as Code (IaC)** model defined in structured data, such as YAML or JSON, is that the models are stored in databases, and thus naturally enforce a schema. This ensures things such as "If a device is added, it must have a Location" or "If the config context is active, it must be a Boolean value."

However, config contexts act more like IaC, in that anyone can technically type in any value they want. So, instead of typing in "true" for active, you can type in "sure" or type in an IP address of "300.1.1.1" as one of your NTP servers. This is problematic because any programmatic access likely expects a valid Boolean and a valid IP, respectively.

Fortunately, a spec exists called json-schemas (`https://json-schema.org/`) that allows you to describe your data model and enforce that the data defined adheres to it. The data model schema can be used to validate this.

Nautobot config contexts can have an associated JSON schema definition to ensure your data stays clean.

> **Note**
> A config context is tied to a single schema object, and thus they are meant to model individual units of the overall config context. In this way, they validate each config context object, not the fully rendered context, as viewed on a particular device or virtual machine.

Let's add a **Config Context Schema** property that requires two – and only two – IPv4 addresses to define NTP servers:

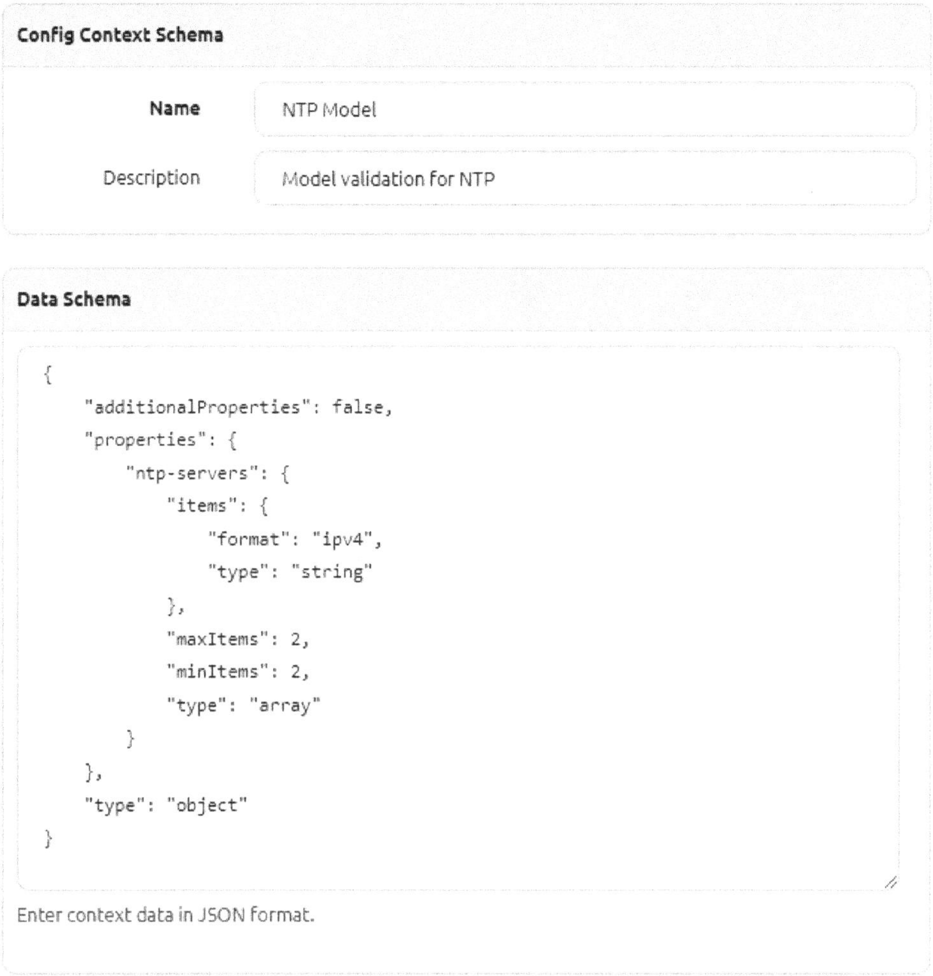

Config Context Schema

Name	NTP Model
Description	Model validation for NTP

Data Schema

```
{
    "additionalProperties": false,
    "properties": {
        "ntp-servers": {
            "items": {
                "format": "ipv4",
                "type": "string"
            },
            "maxItems": 2,
            "minItems": 2,
            "type": "array"
        }
    },
    "type": "object"
}
```

Enter context data in JSON format.

Figure 6.42 – Adding a JSON schema for a config context

Note that this schema would *not* allow any of the following examples to be saved.

Here's example 1:

```
{
    "ntp-servers": [
        "172.16.10.22"
    ]
}
```

Here's example 2:

```
    "ntp": "172.16.10.22,172.16.10.22"
}
```

Here's example 3:

```
{
    "ntp-servers": [
        "172.16.10.22",
        "172.16.10.33",
        "5.5.4"
    ]
}
```

However, it *would* allow the following to be saved and used:

```
{
    "ntp-servers": [
        "172.16.10.22",
        "10.1.5.10"
    ]
}
```

Once the schema is added, you can assign it to a given config context in the **Create or Edit Config Context** form:

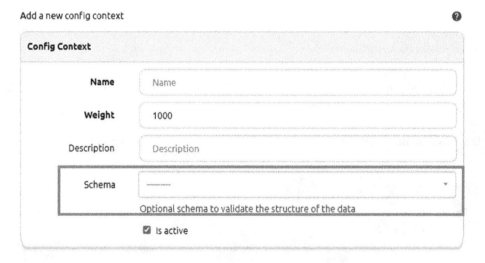

Figure 6.43 – Mapping a JSON schema to a config context

Feel free to try adding the three example schemas to the sandbox and test it out!

Git as a data source

Data sources provide a mechanism to synchronize data and files into Nautobot, specifically using Git. This allows you to leverage the feature set of Git while natively integrating with Nautobot. Git is arguably the most important DevOps tool for any modern automation. The capabilities it provides allow you to provide feedback on changes through pull/merge request reviews before implementations, tracking changes, and much more.

The basic concept is that you manage some data or file artifacts in your Git repository and then synchronize to Nautobot as an extension of the application. From a Nautobot perspective, you can synchronize Nautobot jobs, export templates, config contexts, config context schemas, and more. Jobs will be covered in *Chapter 12*; the other items are covered in this chapter. This means you should not be managing Jinja templates and YAML files in the Nautobot UI (for production) in the long term. While it is great to get started and do testing, it is much more maintainable through proper versioning and peer review.

Managing and applying Git data sources

Nautobot allows you to manage Git data sources directly from the UI. It allows you to sync one or more Git repositories directly into Nautobot. The following inputs are supported when adding a new Git repository to sync:

- **Name**: The logical name to use to reference the Git repository
- **Slug**: The unique Django slug of the model
- **Remote URL**: The HTTP-based URL of the repository
- **Branch**: The branch with which you will synchronize
- **Secrets Group**: The secret that will be used to connect to the repository as required
- **Provides**: This maps up which feature the repository provides, such as config contexts

You can view the current Git repositories by navigating to **Extensibility > Git Repositories**:

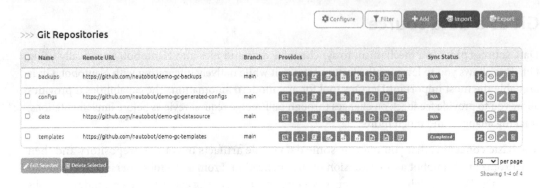

Figure 6.44 – Viewing Git Repositories

From there, you have the standard buttons to add, delete, and edit.

> **Note**
>
> You won't have permission to add more Git repos in the sandbox. So, if you want to try this, use your local installation, something we showcased in *Chapter 3*.

You can also see the icons that show which features the repo *provides*; this is indicated by the icon highlighted in green. You can hover over each icon to see what it is:

Figure 6.45 – Hovering over icons to see what the repository provides to Nautobot

Additionally, you can sync each repository by using the Git icon shown here:

Figure 6.46 – Syncing a repository from the table view

There are also requirements for the directory structure, typically something that just requires a pre-defined directory name based on what the use case(s) is for a given repository. For example, if you want to provide a repo with config context data, you need to create a directory called `config_contexts`, as shown here:

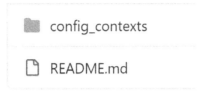

Figure 6.47 – The high-level directory structure required to sync config contexts

Regarding the flexibility of having config contexts based on roles, platforms, and more, as seen in the create form of a config context, you can also create group-based and host-based config contexts. This can be seen in the following screenshot. Take note of the global contexts, as well as the device, platform, and role-specific config contexts:

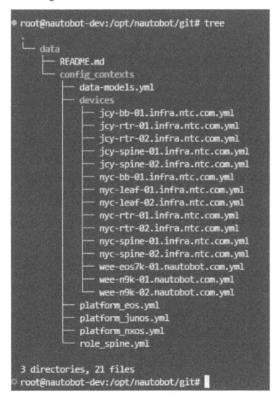

Figure 6.48 – Sync config contexts based on attributes with a defined structure

The previous figure shows a filesystem after a Git sync is performed for the repo that contained config contexts.

Similarly, when using a Git data source, Nautobot jobs should be stored in a `jobs` directory and export templates in an `export_templates` directory. If any Nautobot app adds Git functionality, the app will define any requirements required within the Git repository.

Whenever you synchronize a repo, you will be redirected to the job view, which shows the status of the synchronization.

Within the detail view for a specific Git repository, you can also synchronize and dry-run the data. The dry run will tell you *what would change* if you were to run the synchronize process. You can also see the last known Git commit history. Here is an example of the detail view:

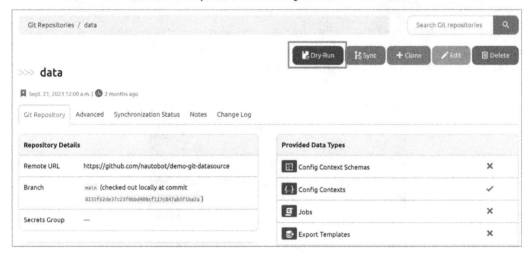

Figure 6.49 – Performing a dry run of a Git repo to see if there would be any updates

Use cases for data sources

There are many use cases for Git as a data source:

- Jobs
- Export templates
- Config contexts
- Config context schemas
- Data compliance rules (part of the *Data Validation Engine* app)

As another example, Nautobot Apps, which will be covered in *Chapters 14* and *15*, can also take advantage of Git as a data source. The Golden Config app extensively uses Git to store and fetch backups and intended config files. Any Nautobot app can add to its own data source and leverage the existing framework for managing the URLs, secrets, synchronization process, UI, API, and underlying

Python functions and classes provided. In other words, the technology can be easily extended to handle future use cases.

Best practices for data sources

It is important to understand the underlying concepts of the filesystem and how it relates to Git as a data source:

- When data is synchronized, it is synchronized to the local filesystem on Nautobot.

- The latest synchronization will update the latest commit hash of the Git repo into the database. You can see this in the detailed view shown previously.

- The synchronization only happens when instructed, such as by a job, a user in the UI, or an API.

When adding a secret group to be used for Git, be sure to use a secret **Access type** of **HTTP(S)** with **Secret Type** set to **Username** and **Token** (this may vary based on your Git provider, such as GitHub or Bitbucket). For GitHub, the secret group should contain an HTTP username and token to access the given repository if it is not a public repository.

Relationships

Nautobot's Relationships feature provides a way for you to associate objects together as if they were model Relationships in Nautobot, allowing you to reflect your network design.

Use cases for relationships

Out of the box, Nautobot has a default network data model, as reviewed in *Chapter 2*. Let's recall a few of those built-in relationships:

- An Interface has a relationship to a Device

- An IP address has a relationship to an Interface

- A Device has a relationship to a Location

- A Circuit has a relationship to an Interface

- A Prefix has a relationship to a VLAN

- A VLAN has a relationship to a Location

These relationships are generally universal; as such, Nautobot provides them to all users as part of the opinionated data model Nautobot ships with.

However, other relationships are not universal. We all know there are different types of network designs and relationships that may make sense for you, including (but not limited to) the following:

- Mapping a Prefix to a Circuit – for example, having awareness of what prefix subnet is associated with a circuit

- Mapping a VLAN to a rack – for example, Layer 3 leaf/spine designs

- Mapping a device to a rack – for example, mapping an OOB server to a rack

- Mapping two devices together – for example, creating a relationship between active/standby or A/B devices

- Mapping a device to multiple VLANs – for example, understanding which VLANs are on a given device

To adapt the Nautobot data model, you can create Nautobot Relationships to accomplish any of these designs.

Managing and applying relationships

You can view Nautobot Relationships by navigating to **Extensibility** > **Relationships**:

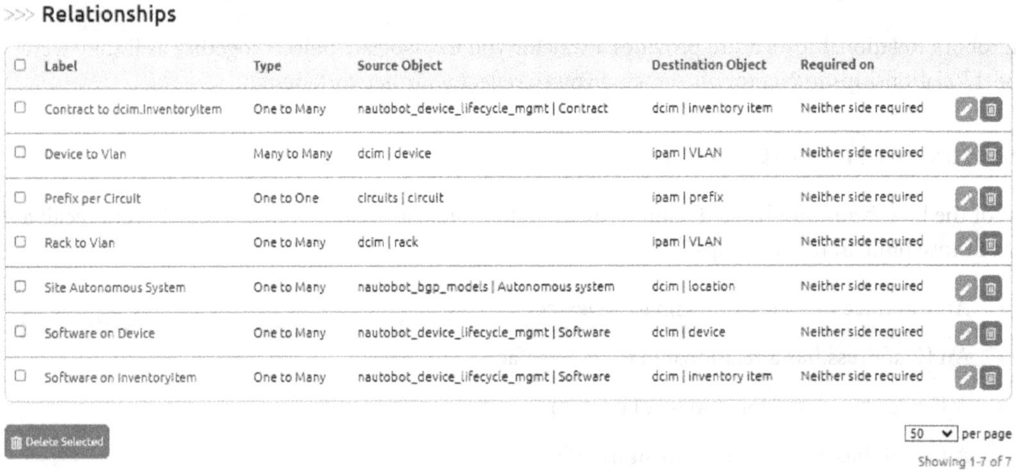

>>> Relationships

	Label	Type	Source Object	Destination Object	Required on		
☐	Contract to dcim.InventoryItem	One to Many	nautobot_device_lifecycle_mgmt | Contract	dcim | inventory item	Neither side required		
☐	Device to Vlan	Many to Many	dcim | device	ipam | VLAN	Neither side required		
☐	Prefix per Circuit	One to One	circuits | circuit	ipam | prefix	Neither side required		
☐	Rack to Vlan	One to Many	dcim | rack	ipam | VLAN	Neither side required		
☐	Site Autonomous System	One to Many	nautobot_bgp_models | Autonomous system	dcim | location	Neither side required		
☐	Software on Device	One to Many	nautobot_device_lifecycle_mgmt | Software	dcim | device	Neither side required		
☐	Software on InventoryItem	One to Many	nautobot_device_lifecycle_mgmt | Software	dcim | inventory item	Neither side required		

🗑 Delete Selected   50 ☑ per page

Showing 1-7 of 7

Figure 6.50 – Viewing Relationships

As shown in the Relationships table, there are a few columns (Type, Source Object, and Destination) that we need to understand before creating Relationships in Nautobot. The best way to understand them is by looking at the **Add a new relationship** form:

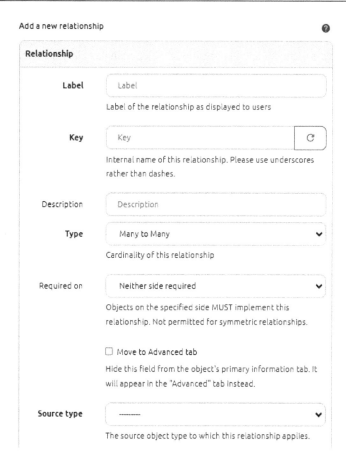

Figure 6.51 – Adding a new relationship

It makes sense to look at the form here first because of a few of the required fields that are unique to Relationships, such as **Type**, **Source type**, and **Destination type**.

Type has five (5) options to consider when creating a new relationship:

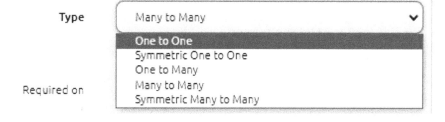

Figure 6.52 – Types of Relationships

These types are described here and have been mapped to a few use cases that were mentioned previously:

- **One to One**: This is a directional relationship in which a single object is associated with another single object. This can often be looked at as an extension of one object. An example use case would be a *prefix to a circuit*.

- **Symmetric One to One**: This is similar to a one-to-one relationship, but the direction does not matter. For instance, in pairing an active/active redundant device, each device is paired precisely with one other device and there are no considerations for the primary or backup.

- **One to Many**: This is a relationship in which a single object can be associated with multiple other objects. An example use case would be a single device to multiple VLANs.

- **Many to Many**: This involves a relationship where both sides can be linked to multiple objects. An example use case would be a device with multiple circuits and a circuit that connects to multiple devices.

- **Symmetric Many to Many**: This is similar to a many-to-many relationship, but the direction doesn't matter. For example, this might be relevant to a set of devices within a routing network, where each device connects to multiple peers, without distinct source or destination differentiation.

Source type and **Destination type** both have the same options, and they allow you to choose the model (or content type) for each side of the relationship:

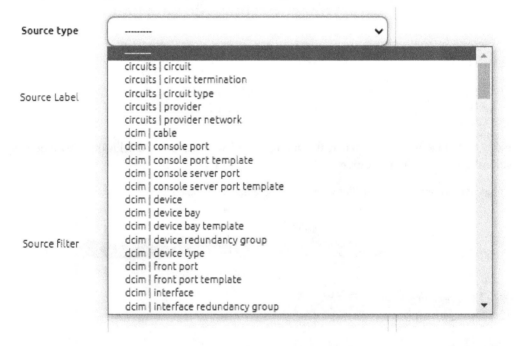

Figure 6.53 – Viewing the source/destination types of Relationships

The following is a complete list and description of the form fields that are shown when creating Relationships:

- **Label**: The label that will be displayed on the UI

- **Key**: The internal name of this relationship; this will be automatically generated but can be specified

- **Description**: Standard description field

- **Type**: One of one-to-one, one-to-many, many-to-many, symmetrical-one-to-one, or symmetrical-many-to-many

- **Required on**: Whether or not this is required on source, destination, or neither side

- **Move to Advanced tab**: A Boolean to move the details of this relationship to the **Advanced** tab

- **Source Type**: The model that will be used as the source of the relationship

- **Source Label**: The label or name that will appear on the source side

- **Hide for source object**: A Boolean to not show on the source side of the relationship

- **Source Filter**: A FilterSet that defines which source objects will have this relationship – for example, if you only want firewalls to have a backup device

- **Destination Type**: The model that will be used as the destination of the relationship

- **Destination Label**: The label or name that will appear on the destination side

- **Hide for destination object**: A Boolean to not show on the destination side of the relationship

- **Destination Filter**: A FilterSet that defines which destination objects will have this relationship

Filters are outside the scope of this book. For details on creating FilterSets, please refer to the Feature Guide in the official Nautobot docs, which includes an example JSON-based FilterSet: `https://docs.nautobot.com/projects/core/en/stable/user-guide/feature-guides/relationships/#extensibility-of-a-relationship`.

> **Pro tip**
>
> Relationships use a Django ORM feature called **Generic Relations** (`https://docs.djangoproject.com/en/5.0/ref/contrib/contenttypes/#generic-relations`), a term that is often used with generic foreign keys. The high-level concept is that there is an additional through table that allows this mapping to happen. However, this is not as efficient as more native methods. Take, for example, a one-to-one relationship. From a practical perspective, it is the same as extending an existing model with more fields – meaning the parent object must exist. So, you can often simply add additional fields on that parent model. Now, in practicality, the models already exist, so this is not usually possible. However, queries do become less efficient at using generic relationships as the queries do not happen in the database but in Python. Occasionally, it could seem like a good idea to add a relationship to get data that is related quicker, but the computational impact may be too high. Let's consider an example of associating a Circuit with a Prefix. This has several benefits, such as making large queries easier for the user. However, it could be more computationally efficient to obtain this information from a computed field as the primary use case would be to understand this data one model instance at a time. As always, there is no one-size-fits-all approach, but consideration for the computational impact should be considered. Using custom fields or computed fields may meet all of your requirements for a more scalable approach.

Creating a relationship

Let's walk through how to create a one-to-one relationship for *Prefix to Circuit* using the public sandbox at https://demo.sandbox.com. Add a new Relationship with the following inputs (all fields not shown are set to their defaults):

- **Label**: `Prefix per Circuit`
- **Key**: `prefix_circuit`
- **Type: One to One**
- **Required on: Neither side required**
- **Source type: circuits | circuit**
- **Destination type: ipam | prefix**

This can be seen in the following screenshot:

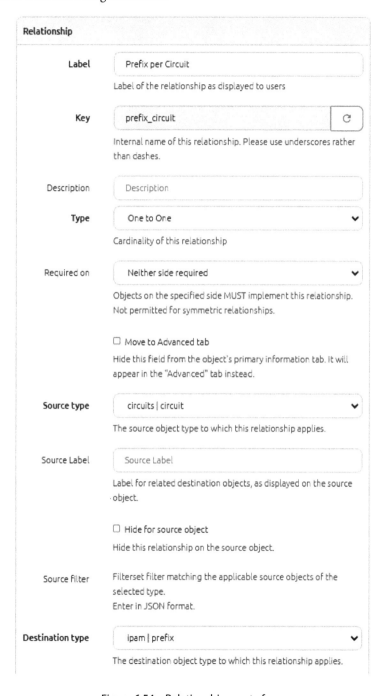

Figure 6.54 – Relationship create form

Because we didn't choose **Hide for source object**, we can view the relationship on the detailed page of the source object. In this case, this is the circuit. This can be seen in the following screenshot:

Figure 6.55 – Viewing the relationship in a detailed object view

As you can see, Relationships can be quite powerful in adapting the data model to your requirements. Leveraging Relationships is often a good way to start tailoring Nautobot to your requirements and can be used before looking at more advanced implementations such as creating new models, something that will be covered in *Chapter 15*.

Dynamic groups

Dynamic groups provide a way to manage collections of objects of the same Nautobot model (content type) that match certain user-defined filters. A dynamic group is used to create groups of objects matching a given filter, such as Devices for a specific location or set of locations that match a given device type, role, and status (as just one example). Dynamic groups update in real time as new or existing objects are created, updated, or deleted from Nautobot.

The following models can currently use dynamic groups:

- Cluster
- Device
- Device redundancy group
- IP address
- IP namespace
- Prefix
- Rack
- Virtual machines

Use cases for dynamic groups

Dynamic groups can and should be leveraged as the centralized definitions of which devices belong to which groups:

- Device grouping to be applied to your network management infrastructure based on regional-like groupings, such as clusters
- Device grouping for managing which devices to use in Golden Config
- IP address definitions and their associations with firewall definitions
- Prefix definitions and their associations with firewall zone definitions
- Rack definitions tied to their role or tag for defining where devices can be correctly racked
- Virtual machines based on tenant or tag to understand the scope of the ownership

Managing and applying dynamic groups

You can view your current dynamic groups by navigating to **Organization** > **Dynamic Groups**, where you can see a list view of them. We'll use the public sandbox for testing purposes, which is available at https://demo.nautobot.com:

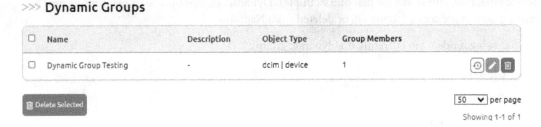

Figure 6.56 – Viewing Dynamic Groups

Let's walk through creating a dynamic group that contains all devices that are active and are data center spine devices. First, click the **Add** button above the table view; you'll be prompted with the **Add a dynamic group** form:

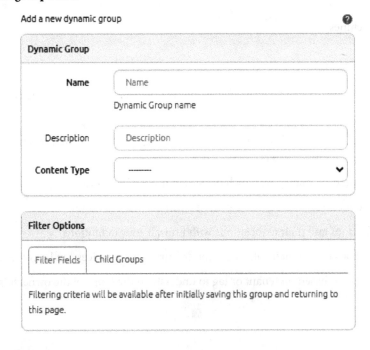

Figure 6.57 – Add a new dynamic group

We'll use the following inputs in the form:

- **Name**: Active Spine Devices
- **Description**: Any Spine Device that is Active
- **Content Type**: dcim | device

At this point, we need to create the group before applying the filter. You can see that via the text, **Filtering criteria will be available after initially saving this group and returning to this page.** within the **Filter Options** section.

So, we must create the group and then open it to edit the filter. Once it is open, set the following two filter options:

Figure 6.58 – Creating a filter for a dynamic group

After defining the filter, update the dynamic group and return to the table view:

Dynamic Groups

Name	Description	Object Type	Group Members
Active Spine Devices	Any Spine Device that is Active	dcim	cevice	4

Figure 6.59 – Viewing the members of a dynamic group

You will see the number of group members that are part of the group. This is not only helpful to understand the data in Nautobot, but this data can be queried via the API to streamline your automation tooling. It would eliminate client-side (tooling) filtering and mangling of data.

You can now click on your new group and see the detailed page for the dynamic group:

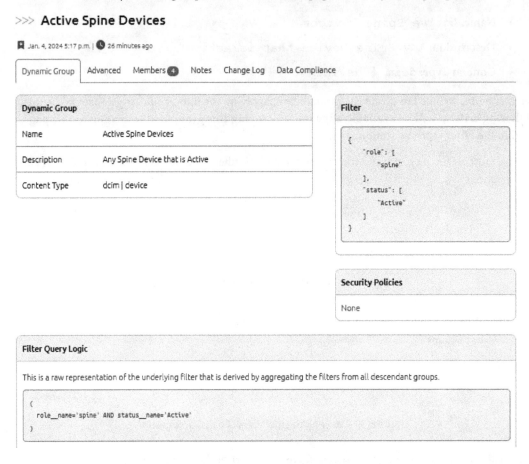

Figure 6.60 – Viewing a dynamic group's detailed page

In addition to the field information you would expect within the detail view, such as all the inputs from the create form, you also will see a few key pieces of information:

- **Filter**, which is in the JSON format, is similar to the JSON FilterSet used in permissions, which will be covered in the next chapter, as well as similar to the FilterSets used in the filters when using Nautobot Relationships

- **Filter Query Logic** is the raw underlying filter that is being applied

- **Related Groups** will show any nesting that the group applies to – that is, both the descendants (where this group is used) and the ancestors (groups that the current dynamic group is using)

- The **Members** tab will look similar to the standard list view of the content type that is applied for the current dynamic group

Best practices for dynamic groups

There is a computational cost associated with dynamic groups, especially when it comes to their nested nature, so consider structures that would flatten the query.

One consideration is to use tags to describe your group and thus keep the queries smaller. This would work well when combined with verifying the data.

Be careful of the truth table logic. A key thing to understand here is that within a single dynamic group, if there are additional values for the same filter field, the results of the query will broaden the group so that it includes additional objects that match those additional values (equivalent to an "or"). Specifying values for additional filter fields will narrow the query and group to only the objects that match the additional filter (equivalent to an "and").

While the implementation of extending FilterForm is outside the scope of this chapter, it is worth mentioning that you can extend what can be filtered on. Nautobot documents Filter Extensions in the official docs for further clarification.

Summary

Nautobot has a significant amount of extensibility features that enable a vast amount of configurability catering to any network design. In this chapter, we learned how to create custom statuses that can better reflect the status of a device or other objects, how tags can be used to better identify services or configurations to an object, how custom fields can be used to extend the data model, how computed fields can dynamically create a new field pulling from existing data sources already in Nautobot, how custom links can enable Nautobot to be the first pane of glass and allow users to jump to any other system, how export templates enable reporting, how config contexts can allow any YAML/JSON data to be tied to a device, how you can easily store YAML data and files in Git and auto-sync repos to Nautobot, and finally, how Relationships and dynamicg groups can be used together or separately to better define groups and associations of objects in Nautobot.

In the next chapter, we will learn about the system and platform administration aspect of Nautobot by diving into the `nautobot-server` command, as well as many of the optional configuration settings in `nautobot_config.py`.

Best practices for dynamic groups

Managing and Administering Nautobot

Over the last few chapters, you learned how to install and configure Nautobot, something that started with *Chapter 3*, went on to learn how to start adding data in Nautobot in *Chapter 5*, and then learned all about extensibility in *Chapter 6*. Through those chapters, the majority of the focus was on using Nautobot, primarily in the UI, so there has been limited focus thus far on Nautobot from a platform administration perspective. We'll shift gears in this chapter and delve into numerous settings of the Nautobot Config file. We'll demonstrate how to enhance and secure a Nautobot deployment by focusing on platform administration tasks and features.

In this chapter, we will cover the following topics:

- Using the Admin UI to set up users, groups, and role-based permissions
- Managing Nautobot settings to better fit your organization
- Developing familiarity with the Nautobot Server commands, and the Nautobot Shell
- Upgrading Nautobot and its apps to new versions
- Troubleshooting Nautobot itself as well as the data within

Administration with the Admin UI

The Nautobot Admin UI provides easy access to tasks such as local user and group management, configuring settings that don't require restarting Nautobot services, viewing audit logs, as well as reviewing and cleaning up old job results and files:

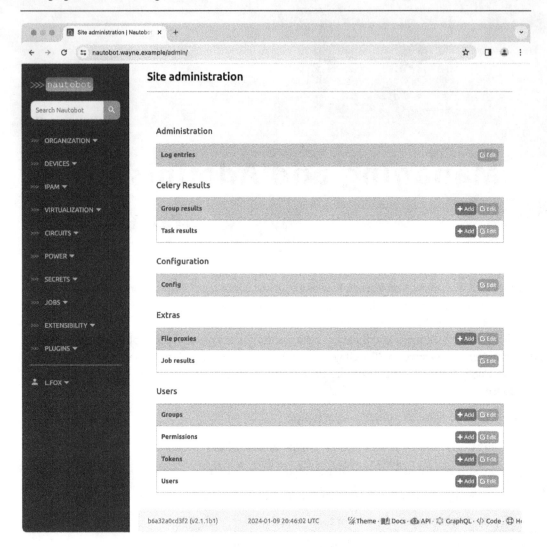

Figure 7.1 – Nautobot Admin UI

The superuser you created in *Chapter 3* will have access to this section, which we'll refer to as the Admin UI. We will cover granting other users access later in this chapter. To visit this section, you can select your username from the menu bar and select **Admin** or navigate to the /admin/ path of your Nautobot installation (for example, http://10.x.y.x:8001/admin/ from our example in *Chapter 3*):

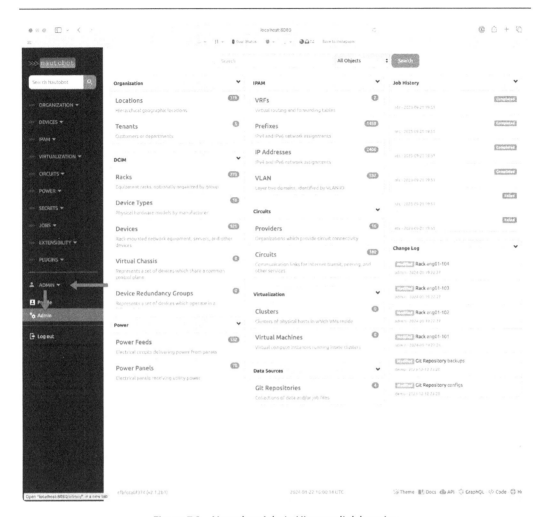

Figure 7.2 – Nautobot Admin UI menu link location

We will reference or use the Admin UI in several sections of this chapter, so we won't dive into each of the pages yet.

Some settings that are defined in `nautobot_config.py` can also be configured via the Nautobot Admin UI. To do so, these settings must not be defined in your `nautobot_config.py` file as any settings defined there will take precedence over any values defined in the Admin UI.

For example, if BANNER_LOGIN is already set in the nautobot_config.py config file, this is what you'd see in the Admin UI, reminding you that the config file takes higher priority over the UI:

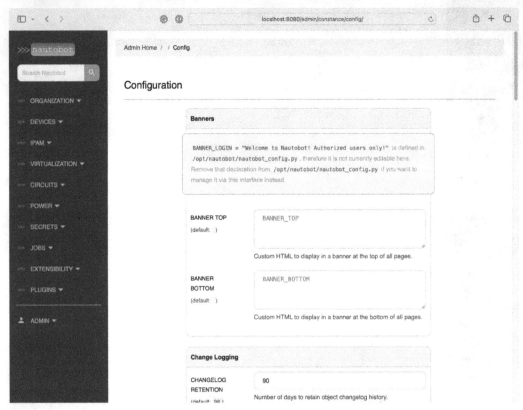

Figure 7.3 – nautobot_config.py and Admin UI precedence

User, group, and permissions management

Nautobot allows for an extensive role-based access system. Roles are managed by assigning permissions either to groups and assigning users to be members of those groups, or by assigning permissions directly to the users. All of the permissions that are granted to the user's groups and directly to the user will be used when determining authorization to perform operations, such as viewing or editing, on objects or views.

Before we work with the power of the permissions framework, let's start by creating users and groups beyond what was set up when we installed Nautobot in *Chapter 3*.

Groups

How you choose to define groups will largely come down to how you plan to group the activities your users will perform. Do you have an operations team who are allowed to view data and run diagnostics but cannot change data directly? Or do you have separate teams managing DCIM versus IPAM-centric data? Ultimately, the choice is up to you. Nautobot allows you to tailor your source of truth to suit your needs.

To create groups, visit **Admin UI** > **Users** > **Groups**. Then, click the **Add** button in the top-right corner. You will be presented with a form for entering the group name:

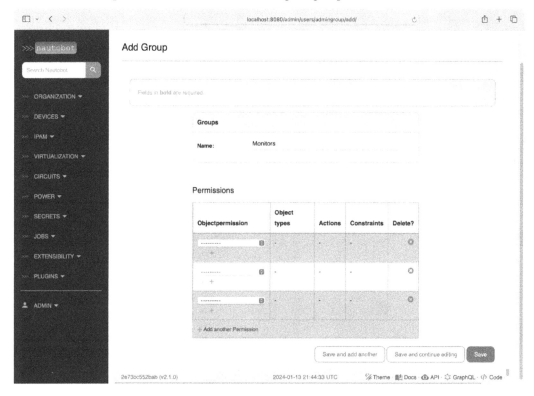

Figure 7.4 – Add group form

You can also set the permissions for the group when creating it but we'll revisit this after we create a user. So, let's create a new group.

Enter the group's name as `Monitors` and click **Save** at the bottom of the form. Next, we will create a user to add to the group.

Users

To create users, visit **Admin UI** > **Users** > **Users**. Then, click the **Add** button in the top-right corner, similar to the list view for groups. You will be presented with a form for entering the user's name and password:

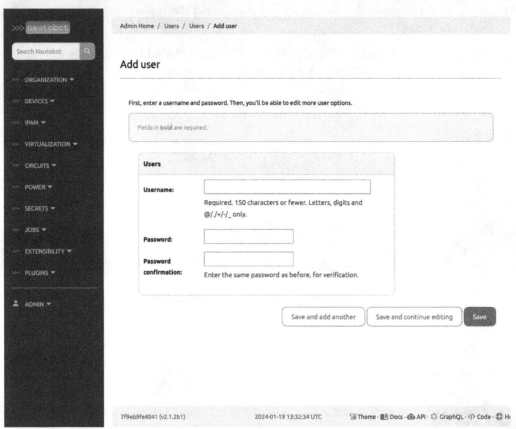

Figure 7.5 – Add user form

Enter j.doe as the username and use a password of your choosing.

We want to add this new user to the group we just created. To do that, we must continue editing the user after we create it. Click **Save and continue editing**:

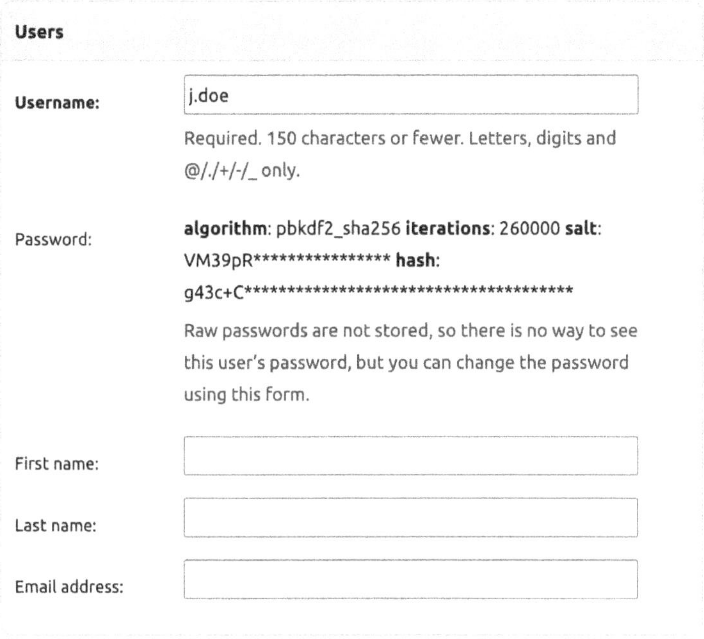

Figure 7.6 – Users form

You will then be presented with the **Change user** form, which contains a lot more information we can modify about the user. We'll review each section down the page. The first section allows you to provide a name and email address for the user you just created. There's also a link to reset the password if necessary:

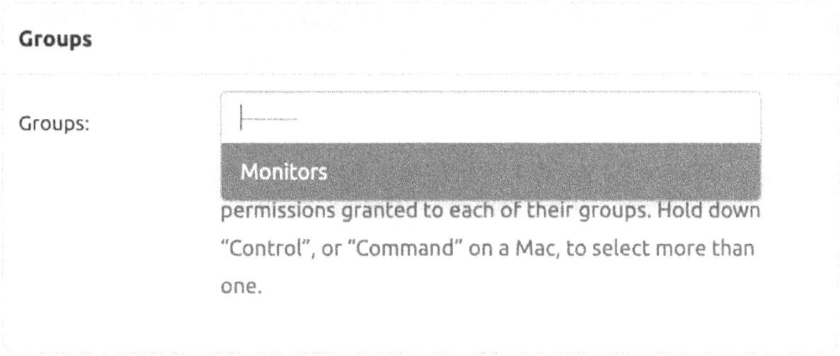

Figure 7.7 – Groups form

The next section will allow you to assign the user to any number of groups. You can click the box and select multiple groups by selecting the group names from the list while holding down the *Ctrl* or *Command* key while clicking.

To test our object permissions, you can go ahead and select the group you created previously:

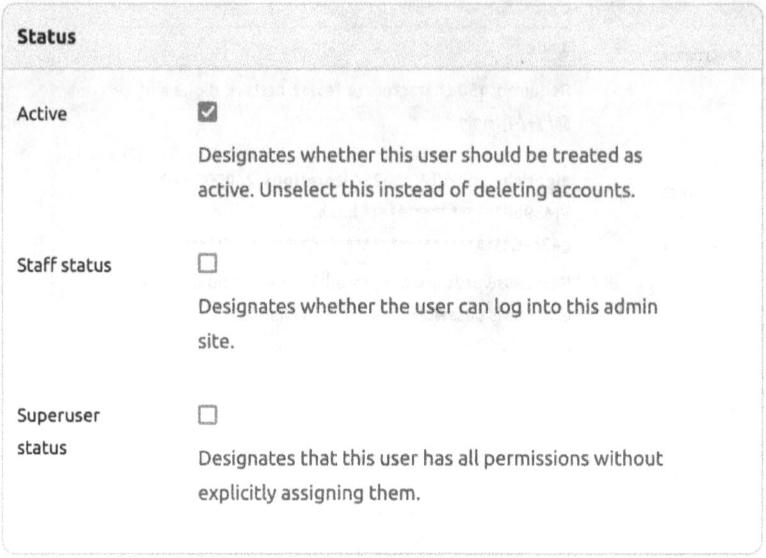

Status	
Active	☑
	Designates whether this user should be treated as active. Unselect this instead of deleting accounts.
Staff status	☐
	Designates whether the user can log into this admin site.
Superuser status	☐
	Designates that this user has all permissions without explicitly assigning them.

Figure 7.8 – Status form

The last section we'll cover in this chapter is the **Status** section of the form. The status controls are as follows:

- **Active**: By default, this is checked. This grants your user the ability to log into Nautobot. You can use this flag to temporarily disable an account or to keep related records (such as **Object Changelog**, **Job Results**, and more) associated with and searchable by their user.

- **Staff**: If you would like to grant the user the ability to navigate to the Admin UI that we have been using and discussing in this chapter, this checkbox controls that. By default, it is unchecked, which prevents their access. Note that this doesn't provide them with unlimited access to the Admin UI. They will still need explicit permissions associated with their associated groups or directly applied to their user account to see or make changes within the Admin UI.

- **Superuser**: If you would like a user to have unlimited access to all objects and settings in Nautobot, you can check this box. This is also disabled by default.

Finally, click the **Save** button to ensure j.doe is a member of the **Monitors** group.

Other sections on the **Change User** form are outside the scope of this book. We'll cover permissions enforcement in the next section.

> **Note**
>
> If you intend to use an external authentication provider via SSO or LDAP (such as Google, Okta, or Active Directory) with Nautobot, you should avoid creating their user profiles in the Admin UI. Their profile will be created automatically once they log into Nautobot for the first time. It is also recommended to manage membership to groups via the external authentication provider. You can create the groups that are managed externally in Nautobot and assign permissions to them. This allows your external authentication system to give a better picture of what a user has access to.

Permissions enforcement

Nautobot provides an object-based permissions framework (which extends Django's built-in permissions model). Object-based permissions enable an administrator to grant users or groups the ability to act on arbitrary subsets of objects in Nautobot, rather than all objects of a certain type. For example, it is possible to grant a user permission to view only sites within a particular location or to modify only VLANs with a numeric ID within a certain range.

It is common for administrators to set permission constraints on a group and never directly apply permissions to a single user, essentially defining a "role" for a user. Note that there is not a formal role concept in Nautobot; rather, it is the resulting permissions applied to a group that results in what you may call a role.

Let's start by creating a Read Only permission that will only allow users to view the data in Nautobot. For that, we will visit **Admin UI** > **Users** > **Permissions**. Once again, like on the **Groups** and **Users** list views, click the **Add** button in the top-right corner.

Let's break down the form sections on this **Permissions** screen:

Figure 7.9 – Permissions form

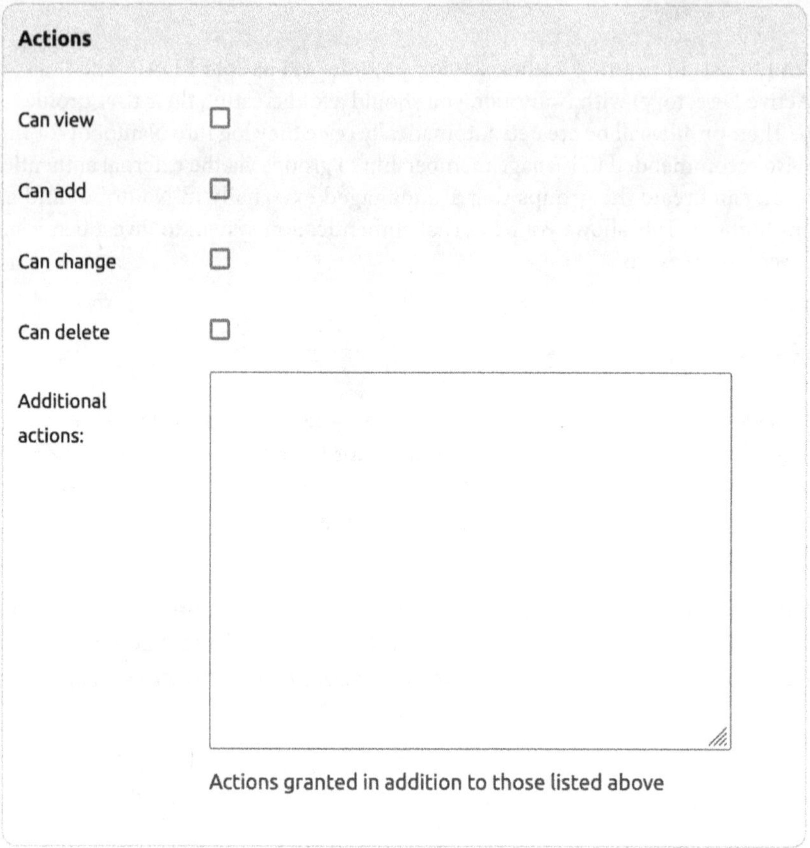

Actions	
Can view	☐
Can add	☐
Can change	☐
Can delete	☐
Additional actions:	

Actions granted in addition to those listed above

Figure 7.10 – Actions form

- **Permissions**: This form section is where you will provide a name and description for what this permission rule provides, and whether or not it is enabled.

- **Actions**: This section sets what users can do with the object types we will be associating with it in the next section. Most actions will be covered by the pre-provided **View**, **Add**, **Change**, or **Delete** options. However, some functions, such as the ability to run a job, are covered by additional actions. We will cover one later in this section:

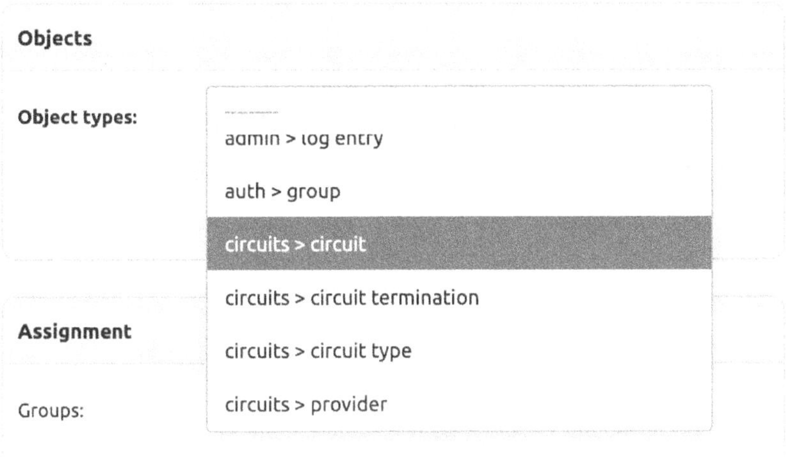

Objects

Object types:

----
admin > log entry

auth > group

circuits > circuit

circuits > circuit termination

Assignment
circuits > circuit type

Groups:
circuits > provider

Figure 7.11 – Objects form

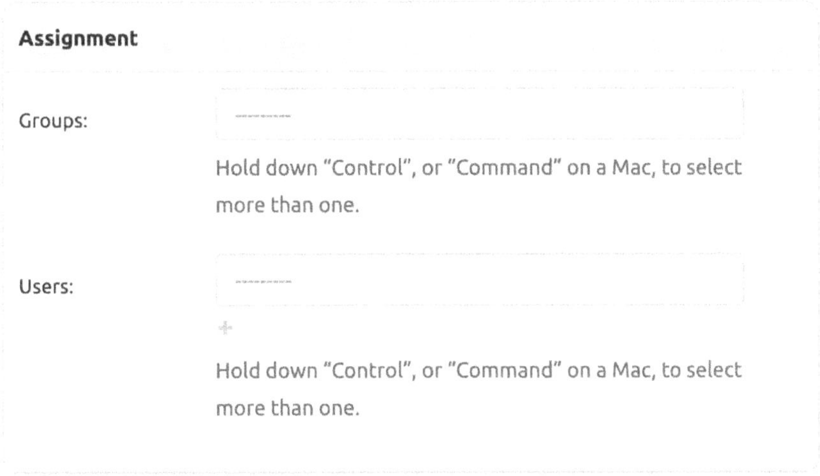

Assignment

Groups:

Hold down "Control", or "Command" on a Mac, to select more than one.

Users:

+
Hold down "Control", or "Command" on a Mac, to select more than one.

Figure 7.12 – Assignment form

- **Objects**: This section sets which objects the users can perform the aforementioned actions on. This is a multi-select section, similar to the one we used to assign users to groups in the previous section. They are organized by model, just like you've seen in many of the extensibility features in *Chapter 6*:

  - *Circuits*: Circuit provider and more

  - *DCIM*: Devices, interfaces, and so on

  - *Extras*: Jobs, roles, and so on

- *IPAM*: IP addresses, prefixes, and so on

- *Tenancy*: Tenants and more

- *Virtualization*: Virtual machines, virtual machine interfaces, and so on

There may be others, depending on which additional Nautobot apps you have installed.

- **Assignment**: This section allows you to quickly associate the permission you are creating to groups and users once it is created. If you intend to assign users to groups, you only need to associate the permissions to the group; the users will use the permissions indirectly.

- **Constraints**: This section allows you to provide limitations beyond just the object types a user is allowed to perform actions on. This is an extremely powerful feature of Nautobot that we will cover in the next section:

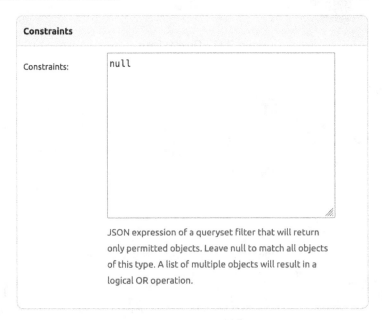

Figure 7.13 – Constraints form

For a simple **Read Only** role, we will use the group we've already created and fill out the form as follows:

- **Name**: **Read Only All**

- **Enabled**: (checked)

- **Actions**:

- **Can view**: (checked)

- **All others**: (unchecked/empty)

- **Object Types**: All object types in the **Circuits, DCIM, Extras, IPAM, Tenancy**, and **Virtualization** areas

- **Assignment > Groups**: **Monitors** (the group we created earlier)

- **Assignment > Users**: **Empty**

- **Constraints**: `null` (the default)

Next, click **Save** at the bottom of the form. We can now log in as the user we created earlier and view the resulting permissions. It may look like nothing has changed, but all the create/edit/delete links, buttons, and forms have disappeared. Here is a side-by-side comparison of the superuser we created in *Chapter 3* and the user we just created:

Figure 7.14 – List view permission comparison; left – admin, right – read-only

It is worth noting that without any permissions, users have no permissions. Also, multiple permissions combine to add to what users can do, instead of taking away. Nautobot jobs require an additional action, beyond **View**, to be able to run them. Let's go ahead and create a permission to do that. You can create a permission similar to the previous one but for this one, we'll set the following:

- **Actions**:

- **Can view**: (checked)

- **Additional actions**: **run**

- **Object Types**:

- **extras > jobs**

- **extras > scheduled jobs**

- **extras > job result**

Users who can run jobs can choose to run them immediately or schedule them for a future time, once or reoccurring. If you would also like to allow users to cancel or change a previously scheduled job, you should also grant them the **Delete** action, but you may want to use a constraint (discussed later) to only allow them to modify the jobs they scheduled:

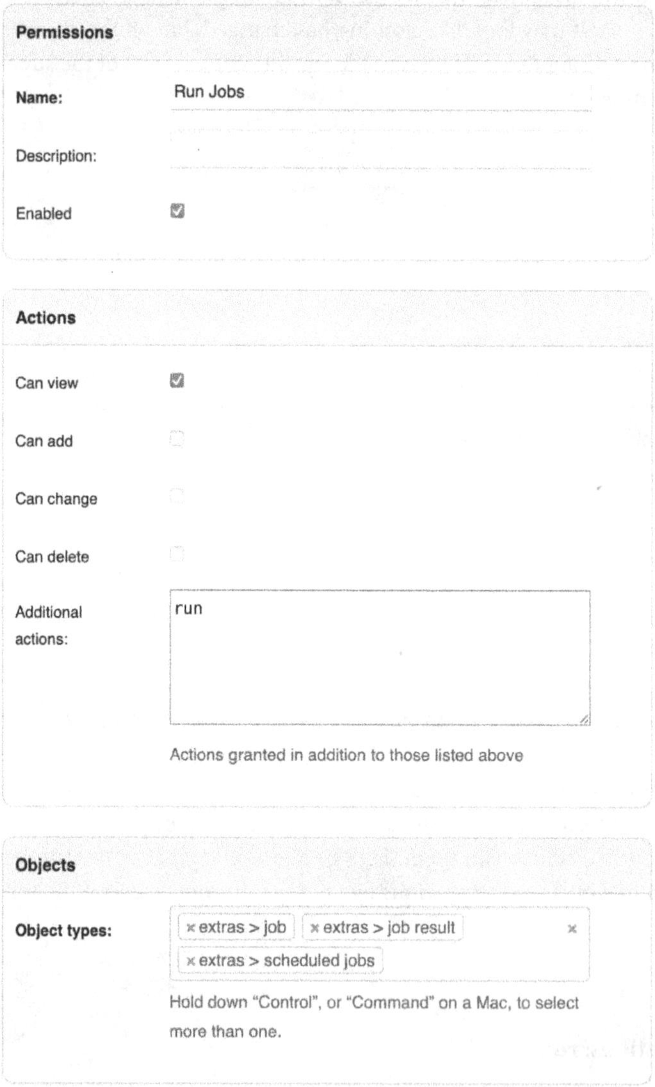

Permissions

Name: Run Jobs

Description:

Enabled   ☑

Actions

Can view   ☑

Can add   ☐

Can change   ☐

Can delete   ☐

Additional   run
actions:

Actions granted in addition to those listed above

Objects

Object types: × extras > job   × extras > job result   ×
                 × extras > scheduled jobs

Hold down "Control", or "Command" on a Mac, to select
more than one.

Figure 7.15 – Run job permissions

We will explain more about building and running jobs in *Chapter 12*.

Let's dive into permission constraints.

Defining permission constraints

Permission constraints afford you the ability to get very granular with what your users, or even automation, can do to data. Here are some examples:

- Can only see devices in a certain location

- Can only edit VLANs between a certain VLAN ID range

- Can only manage IP addresses in a given namespace and with the **Planned** status

Let's use the first example: we want to create a group that can only access devices from a certain location. In our example, we have an existing location called **ANG01**.

First, create a user group called `Rack Managers ANG01`. Now, we will create a new permission with the following settings:

- **Name**: RW Racks ANG01

- **Enabled**: (checked)

- **Actions**:

- **View**, **Add**, **Change**, **Delete** (checked)

- **Object Types**:

- **dcim > device**

- **dcim > rack**

- **Assignment**: `Rack Managers ANG01`

- **Constraints**: `[{"location__name": "ANG01"}]`:

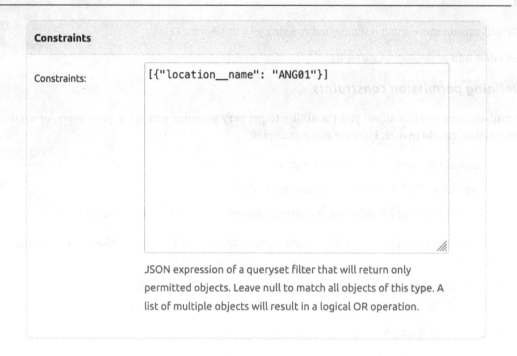

Constraints

Constraints:

```
[{"location__name": "ANG01"}]
```

JSON expression of a queryset filter that will return only permitted objects. Leave null to match all objects of this type. A list of multiple objects will result in a logical OR operation.

Figure 7.16 – Constraint for object permission

As discussed earlier, permissions are additive, so we will ensure the user is only a part of this new group so that they do not get access from the earlier read-only permission. Once we log in with the user, we will see a stark difference between what our original superuser account could view and what our constrained permissions provide:

Figure 7.17 – Permission constraint comparison – home page

Figure 7.18 – Permission constraint comparison – device list

Even though the user was only granted access to device and rack objects, useful information about those objects is displayed so that you don't get a fragmented or incomplete view. This means you'll see the device type, tenant, primary IP address, and more, even if you have not given explicit access to them. However, the user will not be able to access the details of those related objects, instead being presented with an **Access Denied** page:

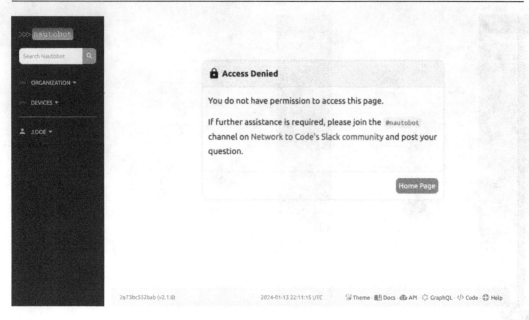

Figure 7.19 – Access Denied

When creating or editing objects and looking to associate objects with permission constraints, selections will be limited to those objects that the user can view. To continue our device and rack example, the user can only assign a device to a rack they can view:

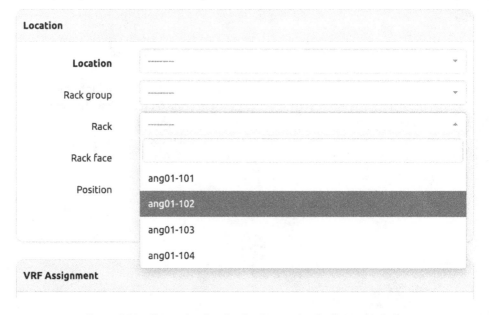

Figure 7.20 – Constrained rack selection on the device create form

It is worth noting that many objects should be left unconstrained as you will be unable to use them when creating related objects. For example, because we never granted the user the ability to view device types, they will be unable to select one when they try to create a device, which is a required property:

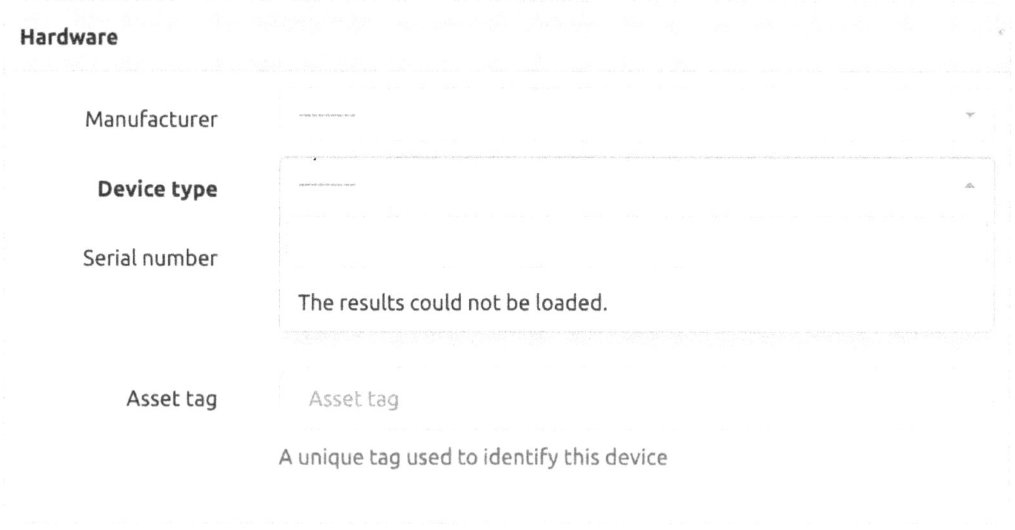

Hardware

Manufacturer

Device type

Serial number

The results could not be loaded.

Asset tag   Asset tag

A unique tag used to identify this device

Figure 7.21 – Unusable device type selection on the device create form

Here are some examples of constraints and what they mean:

Constraints	Description
{"status": "active"}	Status is active
{"status__in": ["planned", "reserved"]}	Status is active OR reserved
{"status": "active", "role": "testing"}	Status is active OR role is testing
{"name__startswith": "Foo"}	The name starts with "Foo" (case-sensitive)
{"vid__gte": 100, "vid__lt": 200}	VLAN ID is greater than or equal to 100 AND less than 200

Table 6.1 – Permission constraints

When more than one constraint is applied and they are both constraining different attributes (such as **Status** and **Role**), the result is a logical "OR." When more than one constraint is used and they are applied to the same model, such as VLAN ID (for example, greater than and less than), the result is a logical "AND."

How constraints are applied

Object-based permissions work by filtering the database query that's generated by a user's request to restrict the set of objects that are returned. When a request is received, Nautobot first determines whether the user is authenticated and has been granted to perform the requested action. If the user has not been assigned this permission (either directly or via a group assignment), Nautobot will return a 403 (forbidden) HTTP response. If the permission *has* been granted, Nautobot will compile any specified constraints for the model and action. For example, suppose two permissions have been assigned to the user granting view access to the device model, with the following constraints:

```
[
    {"location__name__in":  ["NYC1", "NYC2"]},
    {"status":  "offline", "tenant__isnull":  true}
]
```

This grants the user access to view any device that is assigned to a location named NYC1 or NYC2, or which has a status of **offline** and has no tenant assigned. This means that users would be able to see devices outside of NYC1 and NYC2 that are **offline** and not assigned to a tenant, so be sure to test if your constraints are applied.

This is nearly identical to what is executed programmatically inside Nautobot, jobs, or the Nautobot Shell (which we will cover later in this chapter). These constraints are equivalent to the following Nautobot Shell query:

```
Device.objects.filter(
    Q(location__name__in=['NYC1', 'NYC2']),
    Q(status='offline', tenant__isnull=True)
)
```

The same sort of logic is in play when a user attempts to create or modify an object in Nautobot but with a twist. Once validation has been completed, Nautobot starts an atomic database transaction to facilitate the change, and the object is created or saved normally. Next, while still within the transaction, Nautobot issues a second query to retrieve the newly created/updated object, filtering the restricted query set with the object's primary key. If this query fails to return the object, Nautobot knows that the new revision does not match the constraints imposed by the permission. The transaction is then rolled back, leaving the database in its original state before the change, and the user is informed of the violation. Here is what the preceding constraint would allow or revert:

- A device in the **NYC1** location with a status of **Active** and an associated tenant: **ALLOWED**

- A device in the **London** location with a status of **Active** and an associated tenant: **REVERTED**

- A device in the **London** location with a status of **Offline** and no tenant associated: **ALLOWED**

Setting up external authentication (SSO/LDAP)

Nautobot supports several external authentication mechanisms, including OAuth (1 and 2), OpenID, SAML, LDAP, and others. By default, many authentication backends are supported without additional packages, including Google, Microsoft Azure Active Directory, Okta, and more. Others may require additional Python and/or system-level packages to be installed so that they can be used.

The online documentation (`https://docs.nautobot.com/projects/core/en/stable/user-guide/administration/configuration/authentication/sso/`) is the best place to go for the most accurate and up-to-date information on configuring external authentication.

You can configure Nautobot to automatically add users to groups and set default permissions upon setting up external authentication.

Exploring Nautobot's settings

Nautobot is a highly customizable application that can be tailored to the requirements of small businesses and enterprises alike. In this section, we are going to highlight several settings that are common to modify to tailor a Nautobot deployment for any given environment.

Understanding setting precedence

Unless otherwise noted, all settings covered in this section can be set in the `nautobot_config.py` file we've been working with since *Chapter 3*. As mentioned earlier in this chapter, some settings can be set using the Admin UI (also covered in this section). Some settings can also be configured by environment variables.

> **Note**
>
> The version-specific documentation is the best place to learn how to specify these settings. Please check out the Nautobot documentation provided with your installation or online, and be sure to make sure you reference the specific version you have installed.

Here are some rules of thumb on precedence:

- `nautobot_config.py`
- Environment variables (if not set in the config file)
- Admin UI (if not set previously)
- Defaults (if not set previously)

If you do set a setting in `nautobot_config.py`, you will not be able to change or override it in the Admin UI if it's available. Instead, you will see a warning message to that effect:

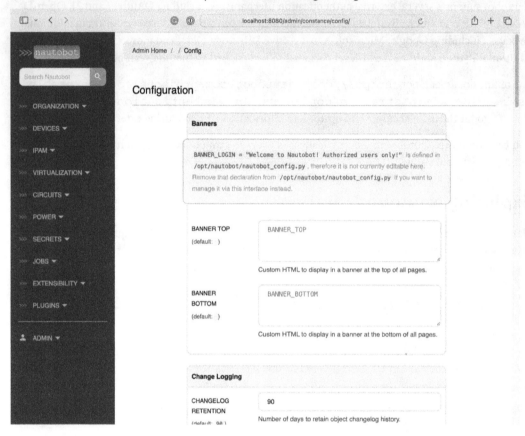

Figure 7.22 – Nautobot Admin UI providing a config conflict warning

Setting banner and support messages

Banner messages provide a simple and easy way to customize your Nautobot deployment to indicate special instructions, maintenance notifications, links to external systems, and more. Setting these variables will display custom content in a banner at the top and/or bottom of the page, login page, or when a user receives an error. These settings can be set via the Nautobot config file:

```
# nautobot_config.py
BANNER_TOP = 'Welcome to Wayne Enterprises Nautobot!'
BANNER_BOTTOM = 'For support, please reach out to <a href="https://
it.wayne.example">WE-IT Helpdesk 📧</a>.'
BANNER_LOGIN = 'Authorized Users Only'
```

Banners support HTML, are applied to all pages except for the login page, and are only visible to logged-in users. By default, all messages are not set/blank.

When making file changes, make the changes as the `nautobot` user.

Once these settings are updated, restart Nautobot's services:

```
root@nautobot-dev~# systemctl restart nautobot nautobot-worker
nautobot-scheduler
```

Here is what the preceding config would change in the UI, both before and after:

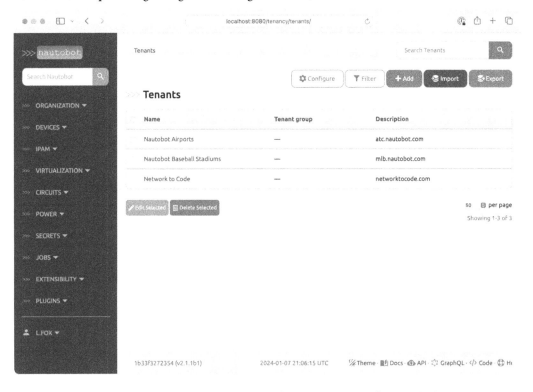

Figure 7.23 – Nautobot UI list view before the banner config is applied

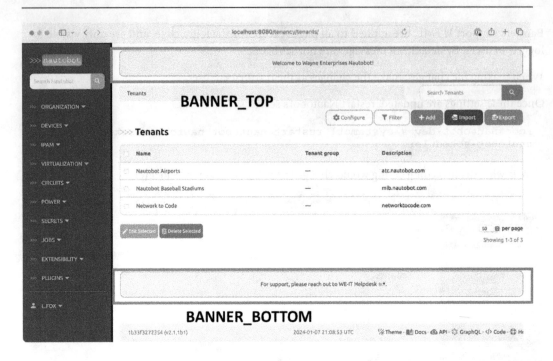

Figure 7.24 – Nautobot UI list view after the banner config is applied

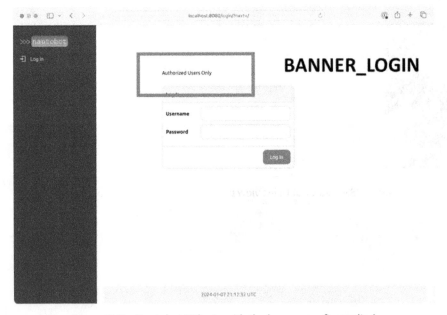

Figure 7.25 – Nautobot UI login with the banner config applied

These settings can also be set via the Nautobot Admin UI, as shown in the following figure (to navigate here, go to **User** dropdown (bottom left) > **Admin** > **Users** > **Groups**):

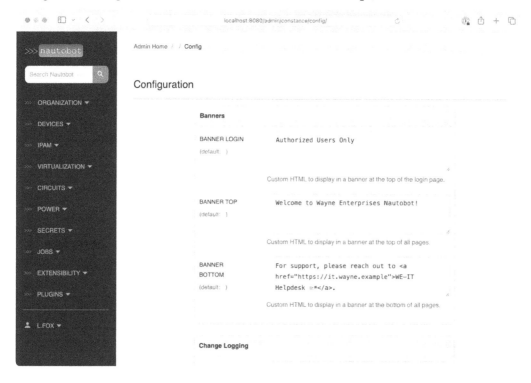

Figure 7.26 – Nautobot Admin UI for banner config settings

Nautobot also presents users with a support message when receiving an error message – for example, if the page is not found or an unexpected error has occurred. By default, Nautobot will instruct users to join the Network to Code #nautobot Slack channel for assistance, but this might not be ideal in many Enterprise or corporate environments. Instead, for example, you may want to instruct users to go to an internal helpdesk to open a ticket with the team supporting your Nautobot installation.

This can be configured by setting the SUPPORT_MESSAGE setting in your Nautobot config:

```
# nautobot_config.py
SUPPORT_MESSAGE = 'For support, please reach out to [WE-IT Helpdesk
🖥️](https://it.wayne.example).'
```

> **Note**
>
> This SUPPORT_MESSAGE setting does not support HTML directly, but many HTML elements may be used via Markdown. This setting can also be set via the Admin UI.

When making the file changes, make the changes as the `nautobot` user.

You can use the following code to restart services:

```
root@nautobot-dev~# systemctl restart nautobot nautobot-worker
nautobot-scheduler
```

Here's an example of an error page showing the customized support message from this example:

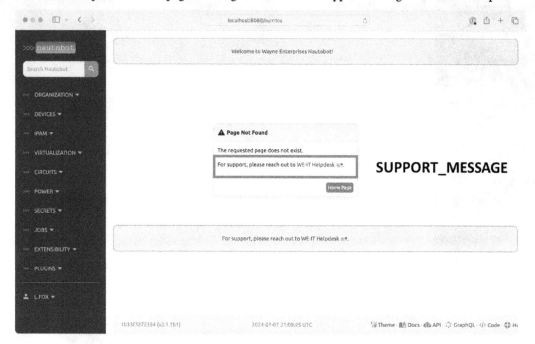

Figure 7.27 – Nautobot UI error page showing a customized support message

Next, we'll customize the branding elements of Nautobot.

Adding your company's logos and branding

Nautobot provides necessary settings to transition your installation of Nautobot to being something powered by Nautobot, without the Nautobot branding. This is common when there is a broad and diverse set of users even outside of IT.

Logos and Icons

By default, you'll see Nautobot on page titles, prefixed on file downloads, as well as via logos and icons. To customize these, we'll be setting configuration items prefixed with `BRANDING_`.

Let's start with the most common ones: the logos and icons. These can be set via BRANDING_FILEPATHS. This is a dictionary that supports multiple paths for different images used within the application. Paths are relative to MEDIA_ROOT (another setting) where these can be retrieved.

To start, we'll set the logo that is shown in the top left of the screen by setting the "logo" dictionary item:

```
# nautobot_config.py
BRANDING_FILEPATHS = {
"logo": "nautobot-logo.png"
}
```

This means we placed an image at MEDIA_ROOT/nautobot-logo.png. The default MEDIA_ROOT is /opt/nautobot/media.

When making these file changes, make them as the nautobot user.

After making this change, you'll need to start the Nautobot application, as we've been doing thus far.

Use the following code to restart services:

```
root@nautobot-dev~# systemctl restart nautobot nautobot-worker
nautobot-scheduler
```

After that, you'll see the new logo in all of the places you would have originally seen the full-sized Nautobot logo!

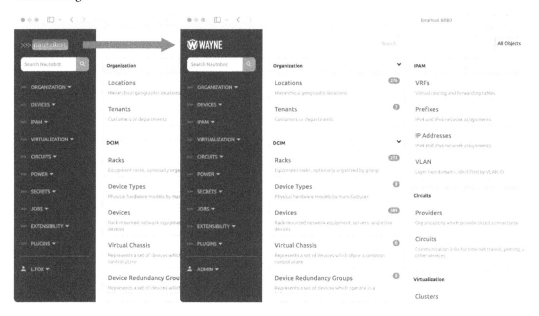

Figure 7.28 – Nautobot UI with a customized logo

You'll likely also want to set the favicon (the icon that appears in the tab bar or bookmarks). This is set with the same `BRANDING_FILEPATHS` setting but with different dictionary keys for different elements:

- `favicon`: Browser favicon
- `icon_16`: 16x16 px icon
- `icon_32`: 32x32 px icon
- `icon_180`: 180x180 px icon – used for the `apple-touch-icon` header
- `icon_192`: 192x192 px icon
- `icon_mask`: `mono-chrome` icon used for the `mask-icon` header
- `header_bullet`: Image used for various page headers (introduced in 2.1)
- `nav_bullet`: Image used for various nav menu headers (introduced in 2.1)

A fully configured example would look like this:

```
# nautobot_config.py
BRANDING_FILEPATHS = {
"logo": "nautobot-logo.png",
"favicon": "favicon.ico",
"icon_16": "favicon-16x16.png",
"icon_32": "favicon-32x32.png",
"icon_180": "apple-touch-icon.png",
"icon_192": "android-chrome-192x192.png",
"icon_mask": "safari-pinned-tab.svg",
"header_bullet": "header-icon.svg",
"nav_bullet": "folder-icon.svg",
}
```

Using those branding paths will render the following in the UI:

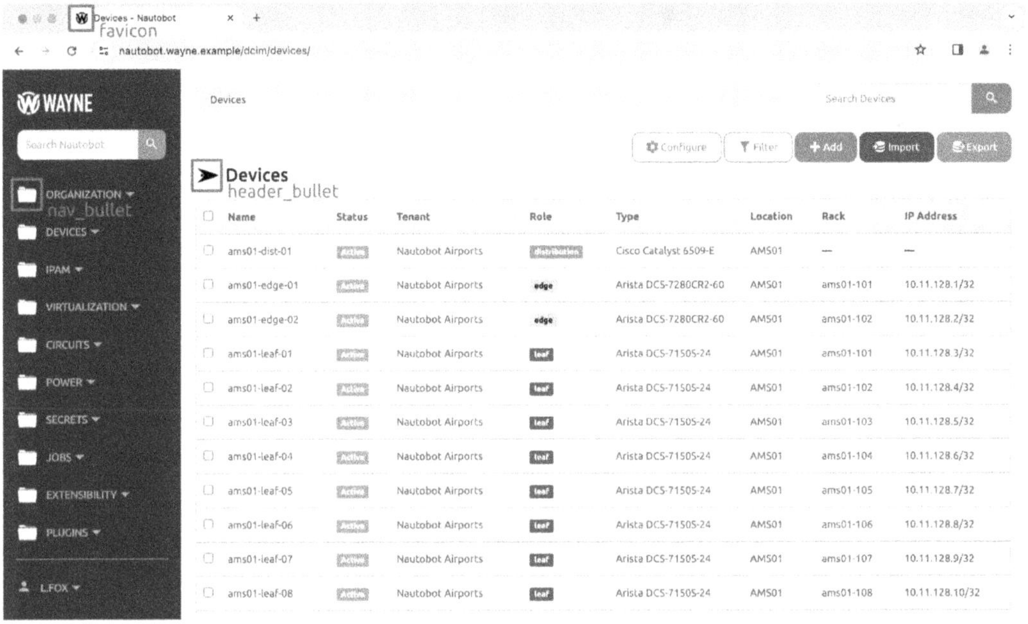

Figure 7.29 – Nautobot UI with fully customized images (labeled for clarity)

It is recommended to keep these files to within 50 Kb each. Anything larger can significantly slow down render performance. The `logo`, `nav_bullet`, and `header_bullet` images also support SVGs, which can improve performance and appearance further. Any branding file path not provided will fall back to the provided defaults.

Titles and filenames

Next, we'll move on to the page titles and the overall name for your Nautobot instance. This can be configured via the `BRANDING_TITLE` setting. The value that's set here is used anywhere the word `Nautobot` would be used for display purposes in a default installation:

```
# nautobot_config.py
BRANDING_TITLE = "W.E. Oracle"
```

This will result in changes seen in the search box and title bar:

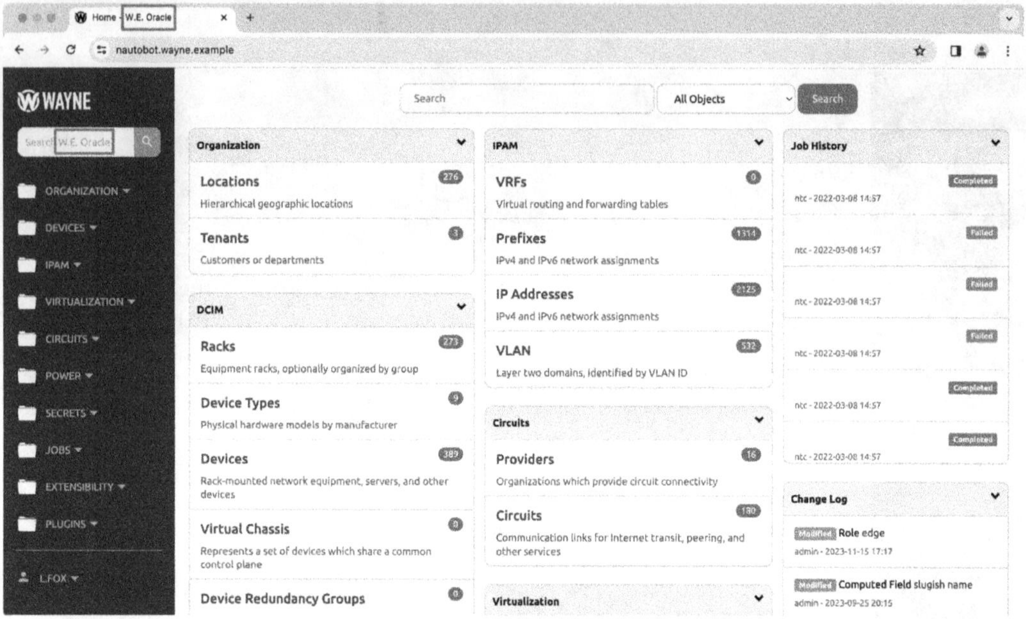

Figure 7.30 – Nautobot UI with an updated page title

By default, `nautobot_` is prepended to files generated by Nautobot to facilitate knowing the origin of file downloads. You probably saw this in export templates in *Chapter 6*. It is possible to change `nautobot_` using the `BRANDING_PREPENDED_FILENAME` setting:

```
# nautobot_config.py
BRANDING_PREPENDED_FILENAME = "weoracle_"
```

When making file changes, make the changes as the `nautobot` user.

After making this change, you'll need to start the Nautobot application, as we've been doing thus far.

Use the following code to restart services:

```
root@nautobot-dev~# systemctl restart nautobot nautobot-worker
nautobot-scheduler
```

You can see the change by triggering an export from any table view:

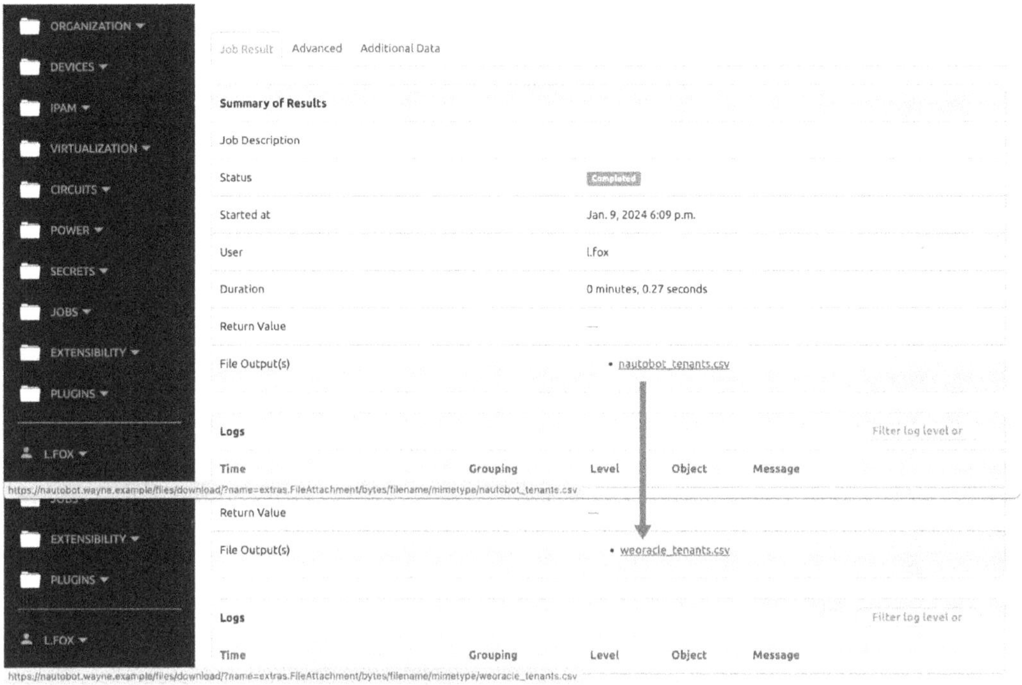

Figure 7.31 – Nautobot UI showing filename prepending changes

Nautobot doesn't prepend filenames with any additional characters, so if you don't end the setting with a delimiter, such as an underscore (_) in this example, the prepended name will be combined with the intended filename outputs.

Lastly, Nautobot provides links in the footer to the Nautobot source code, documentation, and where to find help:

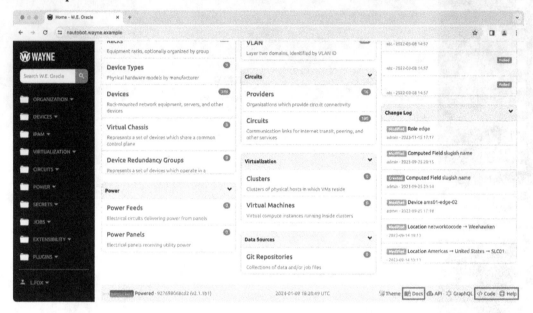

Figure 7.32 – Nautobot UI configurable footer links location

If you would like to guide users to internal resources for this type of information, you can configure them with the BRANDING_URLS setting.

The following URLs are supported:

- code: Code link in the footer (default: https://github.com/nautobot/nautobot)

- docs: Docs link in the footer (default: docs/index.html)

- help: Help link in the footer (default: https://github.com/nautobot/nautobot/wiki):

```
# nautobot_config.py
BRANDING_URLS = {
    "code": "https://github.com/mycompany/our-nautobot",
    "docs": "https://docs.mycompany.com/nautobot",
    "help": "https://help.mycompany.com/nautobot",
}
```

All BRANDING_ settings referenced here can be set either in the config file or via environment variables. Nautobot may still be present in the pre-rendered developer documentation, but for all end user use cases, the application presents as your provided name.

Customizing pagination

Nautobot allows you to customize the number of results returned in any table (or list) view. By default, list views will return up to 50 items, with a maximum size that can be requested of 1,000. Note that this affects both the UI and API. You'll get to test this in the API in *Chapter 9* as well. Changing the maximum size can have effects on the performance of these pages: smaller numbers should result in a faster response time but with the trade-off of having to iterate over more pages; larger numbers will result in a slower response and may cause timeouts or size limitations.

The most common setting to be changed is the PAGINATE_COUNT setting. This setting sets the default page size for results if no limit is specified in the request. Users with a large number of devices or IP addresses may find it annoying to switch to, say, 100 items every time they log into the UI:

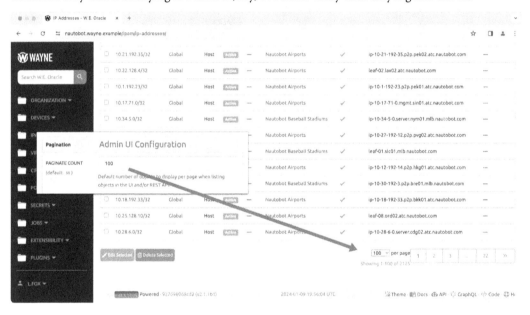

Figure 7.33 – Nautobot UI with a custom PAGINATE_COUNT setting

If you have a significantly large dataset in Nautobot, it can be tempting to increase the maximum number of items you get in a single response to cut down on the number of pages you must iterate over. This can be configured via the MAX_PAGE_SIZE setting. A setting of 0 removes any limitation on page size.

> **Caution**
>
> Setting MAX_PAGE_SIZE to 0 can introduce negative side effects. Nautobot fetches and computes a large amount of information when rendering a single list, which can quickly exceed the memory or time limits of a single request. We recommend testing your set maximum size after changing it before leaving it for general use.

If you choose to modify either setting mentioned previously or find yourself looking for a different set of common page sizes in the dropdown of the UI, you can configure that with the PER_PAGE_ DEFAULTS setting. When setting this in the Admin UI, it is expected to be a comma-separated list of desired common page sizes. If you're setting this in the config file, it should be a list of integers:

```
# nautobot_config.py
PER_PAGE_DEFAULTS = [50, 100, 200, 300]
```

It is recommended that whatever you set as PAGINATE_COUNT be one of the choices in PER_ PAGE_DEFAULTS.

Preferred primary IP version

When displaying a device or virtual machine's primary IP address, as well as connecting to it with the built-in NAPALM integration, which we will cover later in this chapter, Nautobot will default to preferring its associated primary IPv6 address if it is set. If you prefer using the IPv4 address, you can set PREFER_IPV4:

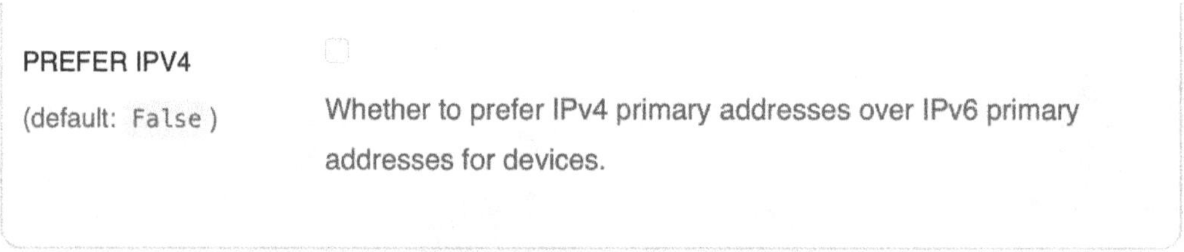

PREFER IPV4

(default: False)   Whether to prefer IPv4 primary addresses over IPv6 primary
                          addresses for devices.

Figure 7.34 – The PREFER IPV4 setting in the Admin UI

You can see the difference when this setting is enabled on this list view:

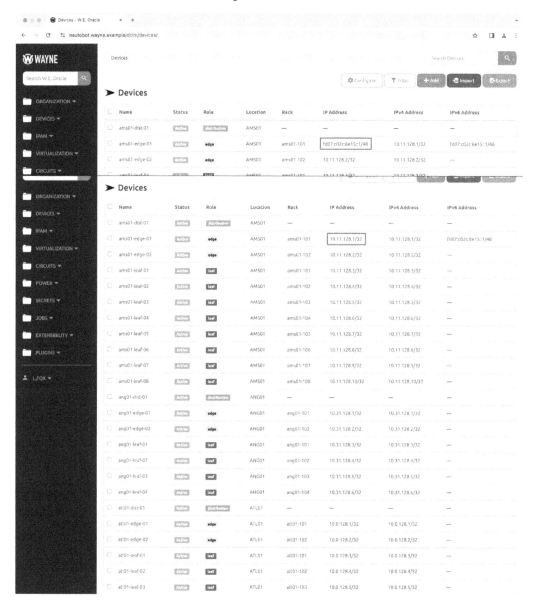

Figure 7.35 – Nautobot UI – top: PREFER_IPV4 set to false (default), bottom: PREFER_IPV4 set to true

Handling logs

Nautobot's logging framework allows you to track when users log into Nautobot, any uncaught error messages, and more. By default, Nautobot will log to the console. Since we configured Nautobot to run as a system service in *Chapter 3*, you will find these console messages emitted to /var/log/ syslog. You can view them as the root user.

These log messages can range in severity from DEBUG, which are messages that are mainly helpful for development or bug-fix reproduction information, to CRITICAL, which means Nautobot was unable to complete a given task. Lower-severity messages can often be less useful in a production environment and more noisy. Therefore, the default configuration of Nautobot excludes DEBUG messages from being emitted.

One common configuration is to have Nautobot log to a non-system log file path. This can then be configured with tools such as Rsyslog to forward those log files to a central server. To enable file-output logging in Nautobot, you can add this snippet to the nautobot_config.py file:

```
# nautobot_config.py
LOGGING["handlers"]["file"] = {
    "level": "INFO",
    "class": "logging.FileHandler",
    "filename": "/tmp/nautobot.log",
    "formatter": "normal",
}

LOGGING["loggers"]["django"]["handlers"] += ["file"]
LOGGING["loggers"]["nautobot"]["handlers"] += ["file"]
```

This tells Nautobot to send all log messages of INFO and above (up to CRITICAL) to the /tmp/ nautobot.log files. This is set up for both Nautobot-specific log messages as well as Django-provided log messages. We will cover using logging for troubleshooting later in this chapter. You will want to change the file path to something the nautobot service user can write to and something that makes sense for your environment.

There are additional affordances for creating much more nuanced configurations, such as splitting out messages based on their context (authentication versus error messages), changing the output format of the log messages, or streaming object change records to an external service. You can read more about these in the online documentation (https://docs.nautobot.com/projects/ core/en/stable/user-guide/administration/configuration/optional- settings/?h=loggin#logging).

Customizing sanitizer patterns

Sometimes, third-party libraries or event messages may be logged and contain sensitive information. These messages might be about failed attempts to connect to devices or tokens contained within a Git repository URL. To overcome the security risk this could impose, Nautobot filters these messages through a sanitizer. This sanitizer contains a collection of configurable find-and-replace patterns to send messages through sequentially. Here are some examples of strings Nautobot will find and what it will replace them with:

- `https://username:password@domain.example` -> `https://(redacted)@domain.example`

- `http://token@domain.example` -> `http://(redacted)@domain.example`

- `"username: admin"` -> `"username: (redacted)"`

- `"USERNAME: local, PASSWD: aBc12e"` -> `"USERNAME: (redacted), PASSWD: (redacted)"`

Again, these are strictly examples and as Nautobot doesn't have complete knowledge of what might be considered sensitive strings for your organization, this is aimed at being a useful but limited baseline. Thankfully, these can be customized for your organization.

Let's say that all service accounts in your organization begin with `svc`. You can add a filter pattern to Nautobot's sanitizer in your `nautobot_config.py` file to find these strings. These *find-and-replace* patterns are implemented via Regex:

```
# nautobot_config.py
import re
from nautobot.core.settings import SANITIZER_PATTERNS as DEFAULT_
SANITIZER_PATTERNS

SANITIZER_PATTERS = DEFAULT_SANITIZER_PATTERNS + [
    (re.compile(r"svc[^\s]*", re.IGNORECASE), r"{replacement}")
]
```

The format of a pattern is a tuple – for example, `(find Regex, replace Regex)`. If you would like to adopt the standard Nautobot behavior where all matches are replaced with `"(redacted)"`, then you can have your replacement Regex be `r"{replacement}"`.

Setting this overrides the default patterns, so be sure to include the default patterns if you want to retain them, as we have done here.

Common settings

Going beyond the examples shown here, which cover many settings that can be changed in `nautobot_config.py`, we wanted to mention a few more settings that are good to be aware of as you start your Nautobot journey:

- `DEBUG`: We will cover this setting in the *Troubleshooting Nautobot* section later in this chapter, but this setting will turn on more verbose logging.

- `TIME_ZONE`: If you prefer to run Nautobot in a timezone other than UTC (the default), you can alter this setting. Nautobot will store all times in the database with UTC but will use this timezone choice when it comes to displaying times. You can find a list of timezone choices here.

- `JOBS_ROOT`: The path at which Nautobot will look for job code. This defaults to a folder called `jobs` in the `NAUTOBOT_ROOT` directory – that is, `/opt/nautobot/jobs` from our setup in *Chapter 3*. This path should be the same on all deployed Nautobot instances and workers. We'll cover jobs in more detail in *Chapter 12*.

- `GIT_ROOT`: Similar to `JOBS_ROOT`, this is the path Nautobot will use to store the files from the Git repository syncs to process locally. This defaults to a folder called `git` in the `NAUTOBOT_ROOT` directory – that is, `/opt/nautobot/git` from our setup in *Chapter 3*.

- `PLUGINS`: Covered in greater detail in *Chapter 14*, this setting provides the list of Nautobot apps you wish Nautobot to register on startup. It is a list of the package names for the apps.

- `PLUGINS_CONFIG`: Covered in *Chapter 13*, some Nautobot apps require their own configuration on startup, as we have covered in this chapter for Nautobot itself. This is where those configurations are applied. `PLUGINS_CONFIG` is a dictionary, with the top-level keys being the package names of the apps for which the configuration is for.

Advanced settings

We've only covered some of the most common settings you can customize in Nautobot. These settings can vary from version to version, and new ones are introduced over time.

Let's say you're looking to change the way Nautobot does any of the following:

- Determines a natural key for a device or location

- Stores files (for example, an S3 bucket on AWS instead of locally)

- Connects to the internet (for example, connecting to a Git repository over an HTTP proxy)

- Accesses Celery or Redis (for example, a distributed workload environment)

- Sets the maximum setting for job file size, timeouts, and more

Please be sure to check out the documentation (`https://docs.nautobot.com/projects/core/en/stable/user-guide/administration/configuration/optional-settings/?h=loggin#optional-configuration-settings`) that's included with your installation of Nautobot or online.

Setting up and using NAPALM integration

Nautobot includes an integration with NAPALM to directly access information from devices. Such information includes uptime, CPU/memory stats, power status, LLDP neighbors, and the running, startup, and candidate configurations:

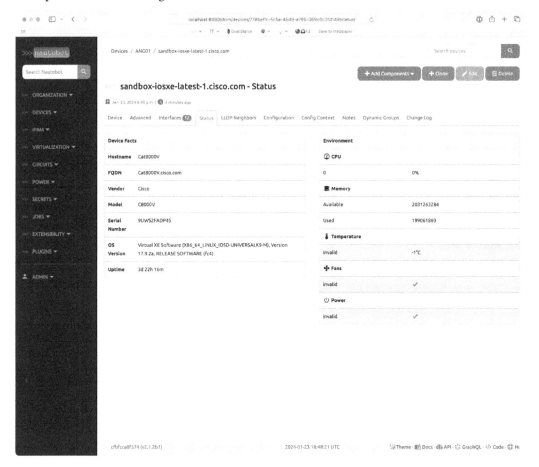

Figure 7.36 – NAPALM integration – device status

You may see **invalid** under **Temperature**, **Fans**, and **Power** if you're using virtual devices. The following screenshot shows how to show neighbors using NAPALM:

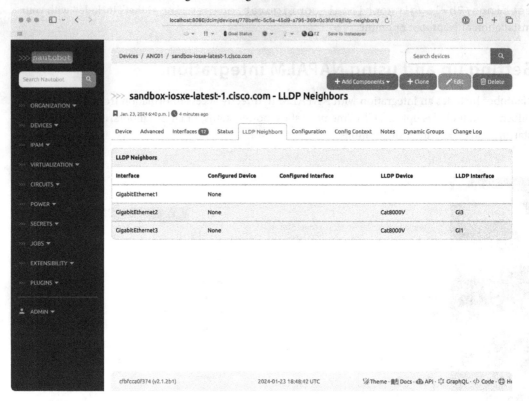

Figure 7.37 – NAPALM integration – LLDP neighbors

Here, we can see the running configuration of a Cisco device:

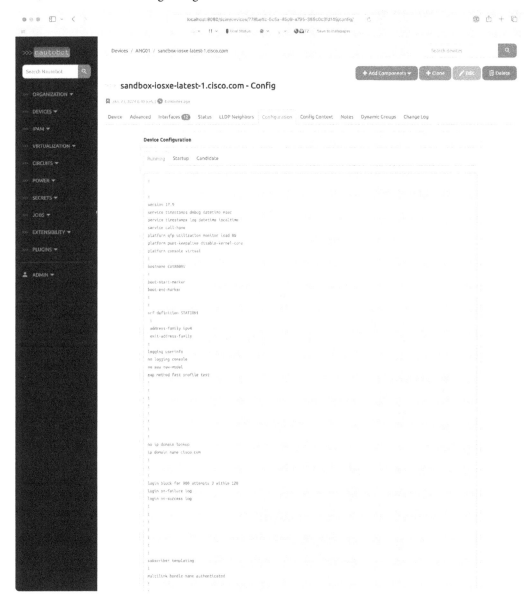

Figure 7.38 – NAPALM integration – device configuration

To do this, Nautobot must have credentials to be able to connect to the device. These credentials will need to be entitled, either via TACACS or some other means, to perform the operations NAPALM uses to fetch the information. You can learn more about NAPALM by reading their documentation.

Nautobot will prefer a secret group associated with the device to determine the username and password it should attempt to connect with:

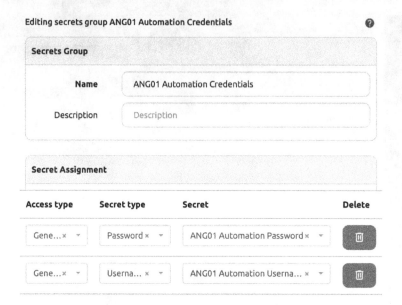

Figure 7.39 – Secrets Group for NAPALM integration

Figure 7.40 – The Secrets group setting on the device form

If no secret group is associated with the device, or you would like to have a single set of credentials for Nautobot to use to access all devices, you can set the NAPALM_USERNAME and NAPALM_PASSWORD settings. To keep these credentials out of your config file, you should use the NAUTOBOT_NAPALM_ USERNAME and NAUTOBOT_NAPALM_PASSWORD environment variables.

If you do this, you can update the Nautobot and worker `systemd` service files:

```
# cat /etc/systemd/system/nautobot.service | grep Environment
Environment="NAUTOBOT_ROOT=/opt/nautobot"
Environment="NAUTOBOT_NAPALM_USERNAME=ntc"
Environment="NAUTOBOT_NAPALM_PASSWORD=ntc123"
#
# cat /etc/systemd/system/nautobot-worker.service | grep Environment
Environment="NAUTOBOT_ROOT=/opt/nautobot"
Environment="NAUTOBOT_NAPALM_USERNAME=ntc"
Environment="NAUTOBOT_NAPALM_PASSWORD=ntc123"
#
```

The default NAPALM arguments may not be all that is needed to enable this integration – for example, if you need to provide an enable password, extended timeout, or any driver-specific settings. The online documentation for NAPALM_ARGS (https://docs.nautobot.com/projects/ core/en/stable/user-guide/administration/configuration/optional- settings/?h=loggin#napalm_args) and NETWORK_DRIVERS (https://docs. nautobot.com/projects/core/en/stable/user-guide/administration/ configuration/authentication/sso/) provides more information.

Exploring nautobot-server CLI commands

Nautobot includes a **command-line interface** (**CLI**) management utility called `nautobot-server` that is used as the entry point for common administrative tasks. For those familiar with Django applications, this CLI utility works exactly as a project's `manage.py` script would, except that it comes bundled with the Nautobot code and, therefore, it's automatically installed in the `bin` directory of your application environment.

Since Nautobot is a Django application, several other built-in management commands are provided by Django itself. We'll only cover some of the commands provided by either Nautobot or Django, so please read the Nautobot documentation to learn more about the available commands.

Creating a superuser account

Whether you're creating your first superuser (as you did in *Chapter 3*) or you need to create another one (if you don't have access to the Admin UI), you can use the `createsuperuser` command. You will be prompted for the desired username, email address, and password. Once created, you should be able to log into the UI.

Here's an example:

```
nautobot@nautobot-dev:~$ nautobot-server createsuperuser
Username: l.fox
Email address: l.fox@wayne.example
Password:
Password (again):
Superuser created successfully.
```

Exporting and importing data

While we recommend that you use the backup tools provided by your database platform (PostgreSQL or MySQL) to perform full database backups and restores, Nautobot provides a built-in way for exporting and importing the data that permits portability between platforms or database versions. These tools are the dumpdata and loaddata commands.

To export data, run the following command:

```
nautobot@nautobot-dev:~$ nautobot-server dumpdata --exclude auth.
permission --format json
--indent 2 --traceback > nautobot_dump.json
```

Importing the data from a previous export is as simple as calling the loaddata command and providing the path to the previously dumped file:

```
nautobot@nautobot-dev:~$ nautobot-server loaddata   --traceback
nautobot_dump.json
```

Cleaning up old scheduled jobs

Nautobot creates objects called **scheduled jobs** to store arguments and details for jobs intended to be run in the future, whether it's once or regularly. This includes jobs that require approvals to run and aren't inherently scheduled to be run on a particular date in the future.

Nautobot does not remove any scheduled job entry proactively, so some of these can proliferate if you are frequently using the one-off future or job approval features. There are no performance concerns for letting these entries linger. However, as Nautobot needs to temporarily store the inputs to the job before they are run, those inputs persist in the database after the job has run. If you would like to remove them from the database, you can run the remove_stale_scheduled_jobs command with a specified "greater than" age:

```
nautobot-server remove_stale_scheduled_jobs 30
```

Running the preceding command will delete scheduled job entries that had their last run more than 30 days ago, and are not enabled or scheduled to run again in the future.

Retracing corrupted/missing cable paths

Nautobot caches cable paths to improve the performance of displaying and relating objects. This cache can sometimes, albeit infrequently, become corrupted. After upgrading the database or working with cables, circuits, or other related objects, there may be a need to rebuild cached cable paths. The `trace_paths` command will help identify and update the paths:

```
nautobot-server trace_paths
```

This command is safe to run at any time. If it does detect any changes, it will exit cleanly. By default, the command will only update paths of cables that have no cache – for example, if you have created a large number of new cables. If you would like to ensure all cable paths have been updated, you can provide the `--force` flag to recalculate all cable paths.

Getting help

To see all available management commands as the Nautobot user, run the following command:

```
nautobot@nautobot-dev:~$ nautobot-server help
```

All management commands have a `-h` or `--help` flag to list all available arguments for that command; here's an example:

```
nautobot@nautobot-dev:~$ nautobot-server migrate --help
```

Some commands are built-in Django commands. Please see the official documentation (`https://docs.djangoproject.com/en/stable/ref/django-admin/#migrate`) for more information.

Exploring the Nautobot Shell

Nautobot includes a Python management shell within which objects can be directly queried, created, modified, and deleted. To enter the shell, run the following command:

```
nautobot@nautobot-dev:~$ nautobot-server nbshell
```

This will launch a lightly customized version of the built-in Django shell with all relevant Nautobot models pre-loaded.

Here's an example output:

```
nautobot@nautobot-dev:~$ nautobot-server nbshell
# Shell Plus Model Imports
from constance.backends.database.models import Constance
# ... A long list of auto-imported modules
from django.db.models import Exists, OuterRef, Subquery
```

```
# Django version 3.2.23
# Nautobot version 2.1.1
Python 3.8.18 (default, Dec 19 2023, 11:04:18)
[GCC 12.2.0] on linux
Type "help", "copyright", "credits" or "license" for more information.
(InteractiveConsole)
>>>
```

> **Note**
> The Nautobot Shell provides direct access to Nautobot data and functions with minimal validation. Consequently, it is vital to restrict access to authorized and knowledgeable users exclusively. Always refrain from executing any actions in the management shell without first ensuring a comprehensive backup is in place.

Working with objects

You may need to validate or manipulate the object data directly. Consider the shell as though it were interpreting Python code live:

```
>>> Device.objects.all()
<QuerySet [<Device: TestDevice1>, <Device: TestDevice2>, <Device:
TestDevice3>,
<Device: TestDevice4>, <Device: TestDevice5>, '...(remaining elements
truncated)...']>
```

To modify an existing object, we must retrieve it, update the desired field(s), and call `validated_save()` again:

```
>>> vlan = VLAN.objects.get(pk="b4b4344f-f6bb-4ceb-85bc-4f169c753157")
>>> vlan.name
'MyNewVLAN'
>>> vlan.name = 'BetterName'
>>> vlan.validated_save()
>>> VLAN.objects.get(pk="b4b4344f-f6bb-4ceb-85bc-4f169c753157").name
'BetterName'
```

As a reminder, because you are modifying objects directly with superuser-level permissions and without authentication, changes will not be reflected in change logs.

Monitoring Nautobot metrics

Nautobot supports optionally exposing native Prometheus metrics from the application. Prometheus (`https://prometheus.io/`) is a popular time series metric platform that's used for monitoring. Metrics are not exposed by default but can be enabled with the `METRICS_ENABLED` configuration setting, which exposes metrics at the `/metrics` HTTP endpoint – for example, `https://<your-nautobot-server>/metrics`. In addition to the `METRICS_ENABLED` setting, database and/or caching metrics can also be enabled by changing the database engine and/or caching backends:

```python
# nautobot_config.py

METRICS_ENABLED = True
DATABASES = {
    "default": {
        # Other settings...
        "ENGINE": "django_prometheus.db.backends.postgresql",
    }
}

CACHES = {
    "default": {
        # Other settings...
        "BACKEND": "django_prometheus.cache.backends.redis.
RedisCache",
    }
}
```

For more information, see the django-prometheus docs at `https://github.com/korfuri/django-prometheus`.

Metrics types

Nautobot makes use of the `django-prometheus` library to export several different types of metrics, including the following:

- Per-model insert, update, and delete counters
- Per-view request counters
- Per-view request latency histograms
- Request body size histograms
- Response body size histograms
- Response code counters

- Database connection, execution, and error counters
- Cache hit, miss, and invalidation counters
- Django middleware latency histograms
- Other Django-related metadata metrics

For the exhaustive list of exposed metrics, visit the `/metrics` endpoint on your Nautobot instance.

Upgrading Nautobot

The Nautobot development teams release updates to Nautobot and open source apps regularly, addressing things such as bug fixes, critical vulnerabilities, as well as new features. It is good to proactively update your Nautobot deployment. Thankfully, Nautobot makes it easy to perform these upgrades.

Before we go any further, it is recommended that you perform a database backup should anything go awry.

You can run the `nautobot-server dumpdata` command we discussed previously in this chapter. You can run the `nautobot-server dumpdata > nautobot_dump.json` command to save the database as JSON.

If you created a Postgres database, as we did in *Chapter 3*, you can also leverage the `pg_dump` utility, which provides a seamless restore option:

```
root@nautobot-dev:~# sudo -iu postgres pg_dump --clean --if-exists
--dbname=nautobot > /tmp/nautobot.sql
```

Should you wish to roll back an upgrade at any point, it is recommended to restore the database as any release may cause changes to the database that are not backward compatible:

```
root@nautobot-dev:~# sudo -u postgres psql postgres -f /tmp/nautobot.
sql
```

In *Chapter 3*, we installed Nautobot by issuing the `pip3 install nautobot` command. This installed the latest stable release at the time of execution. To see which version of Nautobot you installed, you can run the `nautobot-server --version` command from the Nautobot user environment:

```
nautobot@nautobot-dev:~$ nautobot-server --version
2.1.0
```

> **Note**
>
> `nautobot-server version` (without `--`) is a Django-provided command that will give you the Django version, not the Nautobot version. We are concerned with the Nautobot version in this section.

To be able to control the Nautobot upgrade process, you may want to avoid major updates being automatically applied if, for example, a breaking change is introduced. To do this, along with applying updates to Nautobot apps you may install in the future, we will create and maintain a requirements file for our installation:

```
nautobot@nautobot-dev:~$ vi $NAUTOBOT_ROOT/local_requirements.txt
```

You will be presented with a blank screen. We will start by specifying the exact version of Nautobot we have installed (refer to the output of the version command we ran previously and update 2.1.0 in the following code):

```
# /opt/nautobot/local_requirements.txt

nautobot == 2.1.0
```

Then, we can run the pip3 install command, providing this file path. You should be presented with messages about all the requirements already being specified:

```
nautobot@nautobot-dev:~$ pip3 install --upgrade -r $NAUTOBOT_ROOT/
local_requirements.txt
Requirement already satisfied: nautobot==2.1.0 in ./lib/python3.8/
site-packages (from nautobot==2.1.0->-r /opt/nautobot/local_
requirements.txt (line 1)) (2.1.0)
Requirement already satisfied: Django<3.3.0,>=3.2.23 in ./lib/
python3.8/site-packages (from nautobot==2.1.0->nautobot[all]==2.1.0-
>-r /opt/nautobot/local_requirements.txt (line 1)) (3.2.23)
Requirement already satisfied: GitPython<3.2.0,>=3.1.36 in ./lib/
python3.8/site-packages (from nautobot==2.1.0->nautobot[all]==2.1.0-
>-r /opt/nautobot/local_requirements.txt (line 1)) (3.1.40)
# ... output continues
```

Performing this operation will not cause any updates to your Nautobot installation as we've specified in the requirements file that we want the exact version we already have installed to be installed. To allow updates to be installed, we must change the local_requirements.txt file:

```
nautobot@nautobot-dev:~$ vi $NAUTOBOT_ROOT/local_requirements.txt
#
# /opt/nautobot/local_requirements.txt

nautobot ~= 2.1
```

This version specifier (~= 2.1) will allow all updates to Nautobot within the v2 release train. This means you will get updates through a hypothetical Nautobot 2.99, but without modifying this file, Nautobot 3.0 will not be installed.

Once again, we'll run the `pip3 install --upgrade` command from before with the provided file. We'll see that Nautobot has been upgraded:

```
nautobot@nautobot-dev:~$ pip3 install --upgrade -r \ $NAUTOBOT_ROOT/
local_requirements.txt
Requirement already satisfied: nautobot~=2.1 in ./lib/python3.8/
site-packages (from -r /opt/nautobot/local_requirements.txt (line 1))
(2.1.0)
Collecting nautobot~=2.1 (from -r /opt/nautobot/local_requirements.txt
(line 1))
  Downloading nautobot-2.1.2-py3-none-any.whl.metadata (9.9 kB)
Requirement already satisfied: Django<3.3.0,>=3.2.23 in ./lib/
python3.8/site-packages (from nautobot~=2.1->-r /opt/nautobot/local_
requirements.txt (line 1)) (3.2.23)
Collecting GitPython<3.2.0,>=3.1.41 (from nautobot~=2.1->-r /opt/
nautobot/local_requirements.txt (line 1))
# ... output abbreviated
Successfully installed GitPython-3.1.41 Jinja2-3.1.3 nautobot-2.1.2
nh3-0.2.15
nautobot@nautobot-dev:~$ nautobot-server --version
2.1.2
```

Before you can use the new version of Nautobot, you need to run the `post_upgrade` management command, which does the following:

- Database migrations
- Retraces cable paths
- Cleans up stale caches and sessions
- Updates any static files in $NAUTOBOT_ROOT

Upon running the command, you will see an output summary of these operations:

```
nautobot@nautobot-dev:~$ nautobot-server post_upgrade
Performing database migrations...
Operations to perform:
  Apply all migrations: admin, auth, circuits, contenttypes, database,
dcim, django_celery_beat, django_celery_results, extras, ipam,
sessions, social_django, taggit, tenancy, users, virtualization
Running migrations:
  Applying users.0008_make_object_permission_a_changelogged_model...
OK

# ... output abbreviated

Refreshing _content_type cache
CONTENT_TYPE_CACHE_TIMEOUT is set to 0; skipping cache refresh
```

```
Refreshing dynamic group member caches...
DYNAMIC_GROUPS_MEMBER_CACHE_TIMEOUT is set to 0; skipping cache
refresh
```

Once the operation is complete, you will need to restart the services to see the new version reflected in the UI:

```
root@nautobot-dev~# systemctl restart nautobot nautobot-worker
nautobot-scheduler
```

Here's the output:

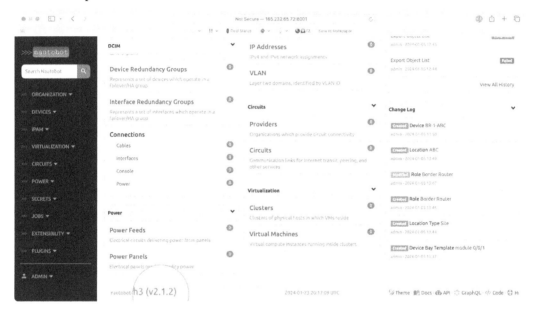

Figure 7.41 – Nautobot UI footer showing the updated version

Congratulations! You have successfully upgraded your Nautobot instance.

Troubleshooting Nautobot

Let's look at a few ways to troubleshoot, starting with performing a health check.

Performing a health check

A quick and easy way to see if Nautobot can connect to its necessary services is to run the `health_check` command. The command will exit successfully (0 status) if everything goes well:

```
nautobot-server health_check
```

Here's an example output:

```
DatabaseBackend            ... working
DefaultFileStorageHealthCheck ... working
RedisBackend               ... working
```

Please see the health check documentation for more information.

Troubleshooting the configuration

To facilitate troubleshooting and debugging settings, you can use the shell to inspect what Nautobot has parsed for them. For example, if you want to check the file Nautobot is looking for in the configuration, check the SETTINGS_PATH property:

```
nautobot@nautobot-dev:~$ nautobot-server nbshell
# ...
>>> from django.conf import settings
>>> settings.SETTINGS_PATH
'/opt/nautobot/nautobot_config.py'
```

Debugging Nautobot

Logging can be another helpful way of digging into an issue with Nautobot. When we covered logs earlier, we discussed that Nautobot keeps log entries concise for a running production environment. Here's an example output you might see:

```
nautobot@nautobot-dev:~$ tail -f /var/log/syslog
Jan 23 21:55:58 nautobot-dev nautobot-server[393844]: 21:55:58.872
INFO    nautobot.auth.login :
Jan 23 21:55:58 nautobot-dev nautobot-server[393844]: User admin
successfully authenticated
Jan 23 21:56:26 nautobot-dev nautobot-server[393844]: 21:56:26.228
WARNING django.request :
Jan 23 21:56:26 nautobot-dev nautobot-server[393844]: Not Found: /
dcim/devices/add2/
Jan 23 21:56:26 nautobot-dev nautobot-server[393844]: [pid:
393844|app: 0|req: 19/25] 104.28.216.170 () {38 vars in 797 bytes}
[Tue Jan 23 21:56:26 2024] GET /dcim/devices/add2/ => generated 148355
bytes in 47 msecs (HTTP/1.1 404) 7 headers in 366 bytes (1 switches on
core 0)
```

In this example, a user successfully logged into Nautobot. Here, we can see an attempt to visit a page in Nautobot that doesn't exist. You won't see log entries for successful page loads or built-in debug information. To do that, you must turn DEBUG on in the nautobot_config.py script:

```
# nautobot_config.py

DEBUG = True
```

You will need to restart Nautobot for the changes to take effect:

```
root@nautobot-dev:~# systemctl restart nautobot.service # Web Service
root@nautobot-dev:~# systemctl restart nautobot-worker.service #
Worker
```

All of a sudden, startup outputs become much more verbose:

```
22:08:37.204 DEBUG    nautobot.core.filters filters.
py                get_filter_predicate() :
  UUID detected: Filtering using field name
22:08:37.205 DEBUG    nautobot.core.filters filters.
py                get_filter_predicate() :
  UUID detected: Filtering using field name


⏳ Running initial systems check...
22:08:38.846 DEBUG    nautobot.core.celery __init__.py        import_
jobs_as_celery_tasks() :
  Importing system Jobs
22:08:38.854 DEBUG    nautobot.core.celery __init__.
py                    register_jobs() :
  Registering job nautobot.core.jobs.ExportObjectList
```

Additionally, requests to pages that log debugging information, such as some esoteric API call outputs, can be captured:

```
22:12:19.819 DEBUG    nautobot.extras.models.groups groups.
py                _map_filter_fields() :
  Added asset_tag (MultiValueCharField) to filter fields
22:12:19.820 DEBUG    nautobot.extras.models.groups groups.
py                _map_filter_fields() :
   22:12:19.821 DEBUG   nautobot.extras.models.groups groups.
py                _map_filter_fields() :
  Deleting local_config_context_data from filterform: has a filter
method
22:12:19.821 DEBUG    nautobot.extras.models.groups groups.
py                _map_filter_fields() :
  Keeping has_primary_ip for filterform: has a `generate_query_`
filter method
22:12:19.821 DEBUG    nautobot.extras.models.groups groups.
py                _map_filter_fields() :
  Skipping removed filterset field: cr_device_soft__source
22:12:20.133 INFO    django.server :
  "GET /extras/dynamic-groups/18e56ae6-df4f-463a-923b-e736ead36417/
edit/ HTTP/1.1" 200 186421
```

It is not recommended to leave this setting on in production environments. Error messages that are displayed to the user become much more verbose and potentially expose system information:

Figure 7.42 – Nautobot 404 page with DEBUG enabled

You may need to leverage this feature when opening a bug report or developing your own Nautobot apps, and then turning `DEBUG` back to `False` once you are done.

Summary

Nautobot is an extremely powerful application with even more powerful configurations and tools under the hood. This chapter is but a dip into an ever-evolving set of what can be done when administering and managing Nautobot as a platform administrator. Please be sure to check out the Nautobot documentation (`https://docs.nautobot.com/projects/core/en/stable/user-guide/administration/configuration/`) provided with your installation or online, and be sure to reference the specific version you have installed.

As you mature your network automation platform powered by Nautobot, you may find yourself needing to scale Nautobot beyond the single-server instance we've been working on so far. You should take a look at our documentation in *Appendix 1*, which covers the Nautobot platform's architecture in more depth. There, you can decide which portions of Nautobot you need to scale or separate, depending on your needs.

Some libraries and tools are published in the Nautobot GitHub organization for container-based deployments, including OpenShift and Kubernetes.

In the next chapter, we will learn about Nautobot's programmatic interfaces, including its REST and GraphQL APIs and its support for event-driven webhooks.

Part 3:
Network Automation
with Nautobot

This part focuses on how Nautobot plays an integral role in network automation solutions. From being a Network Source of Truth to directly powering automation through Nautobot Jobs, you will learn all about Nautobot APIs and how Nautobot integrates with the NetDevOps ecosystem. Then, there is a deep dive into Nautobot Jobs, which allows users to easily write Python scripts and applications and expose them as self-service automations. Finally, you will gain an understanding of how Nautobot fits into an overall enterprise network automation architecture using many of its core and extensibility features.

This part consists of the following chapters:

- *Chapter 8, Learning about Nautobot APIs – REST, GraphQL, and Webhooks*

- *Chapter 9, Understanding Nautobot Integrations for NetDevOps Pipelines*

- *Chapter 10, Embracing Infrastructure as Code with Nautobot, Git, and Ansible*

- *Chapter 11, Automating Networks with Nautobot Jobs*

- *Chapter 12, Data-Driven Network Automation Architecture*

8

Learning about Nautobot APIs – REST, GraphQL, and Webhooks

In the realm of network automation, APIs are critical to enable two or more systems to programmatically interact with each other. APIs are how we can fetch data, make a change, or know when an event has occurred on a remote system. As a network source of truth and network automation platform, Nautobot has robust APIs to enable and power enterprise network automation architectures.

By understanding the APIs Nautobot has to offer, you'll be able to integrate Nautobot into your network automation workflows. For example, you may want to change the status of a circuit or device in a given change window. Going one step further, you may want to pull all data on a given device to build a configuration for that device. On the other hand, when data is changed in Nautobot, you may want to automatically trigger an update on the network or notify a given team.

In this chapter, we'll review various Nautobot APIs. These include REST APIs, GraphQL, Webhooks, and Job Hooks. We will cover specific examples, arming you with the steps needed to easily get started with Nautobot APIs.

> **Note**
> The next chapter introduces **pynautobot**, a layer of abstraction over the Nautobot REST API. If you're looking at accelerating the development of Nautobot API integration, you should read this chapter to understand the API but use **pynautobot** in your Python development.

The following main topics will be covered in this chapter:

- Nautobot REST APIs
- GraphQL with Nautobot
- Webhooks
- Job Hooks

> **Information**
>
> This chapter assumes you have a basic understanding of APIs. Rather than teach the fundamentals of APIs, our focus will be on how to implement APIs in Nautobot. Thus, this chapter requires existing knowledge of using `curl` and the Python `requests` library.

We'll start by exploring Nautobot REST APIs.

Technical requirements

Code files and other resources for this chapter are available at: `https://github.com/PacktPublishing/Network-Automation-with-Nautobot/tree/main/chapter-08`

Nautobot REST APIs

Nautobot has a robust implementation of REST APIs. The best place to dive deep into Nautobot's REST APIs is through the API docs that exist within an installation of Nautobot. We are going to explore these throughout this chapter.

For the examples in this chapter, we are going to be using the Nautobot sandbox located at `https://demo.nautobot.com`. Make sure you can log in to the sandbox if you are going to follow along (the credentials are demo/nautobot).

Nautobot's interactive API documentation

After logging into the sandbox, click the link in the bottom right of the footer that says **API**. To ease going back and forth to the main Nautobot UI, you should open this link in a new browser window or tab. This will open `/api/docs/` for you:

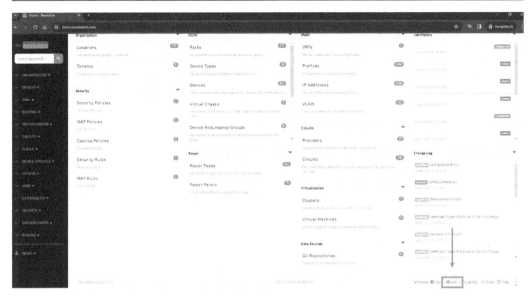

Figure 8.1 – API docs in the Nautobot UI footer

In the new browser tab, you will see the following screen:

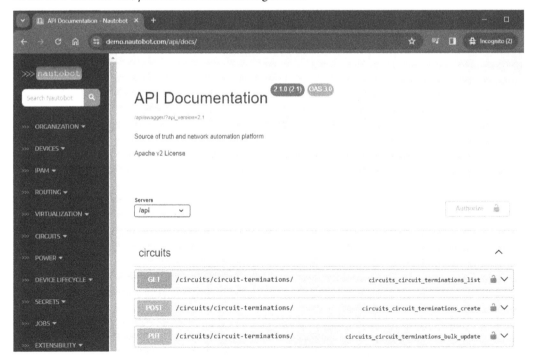

Figure 8.2 – Viewing the Nautobot REST API documentation

At this point, you can scroll up and down this page to see all the different API endpoints that exist within Nautobot. You'll see references to devices, circuits, IP addresses, and other models and APIs. Based on what you are trying to accomplish on a certain object in Nautobot, this is the logical place to start interacting with the API. This page is interactive documentation. Let's try to use the API from the documentation.

Authentication for interactive API documentation

To authenticate, we'll need an API token that'll be used to authenticate to Nautobot. In your original browser tab, do the following:

1. Click the demo username in the bottom left corner of the Nautobot sidebar navigation.
2. Click **Profile**.
3. Click **API Tokens**.
4. You should see the following screen. Once you do, click the green **Copy** button:

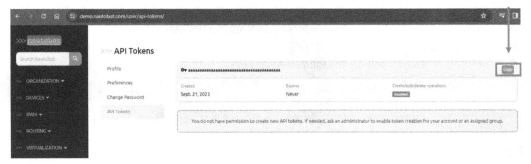

Figure 8.3 – Viewing and copying the API token

5. Navigate back to the API documentation and click the **Authorize** button:

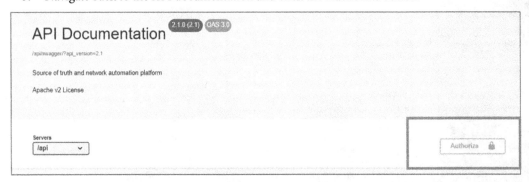

Figure 8.4 – Performing authorization in the API documentation

6.   After clicking **Authorize**, you'll see a few options to authenticate to the API. We are going to use *token-based* authentication. In the **Value** textbox, type the word Token, followed by a space, and then copy the token you copied from the demo user profile. It'll look like this:

Figure 8.5 – Adding an API token to authenticate to API docs

7.   Finally, click **Authorize** and close the **Authorization** popup if it's still open.

Now that we have authentication set up, we are ready to start using the API directly from the documentation.

Example – retrieving all devices

In the first example, we are going to GET all devices from Nautobot. Scroll down on the page and find the /dcim/devices/ API. Once you do, click on the expand arrow, and then click the **Try it out** button:

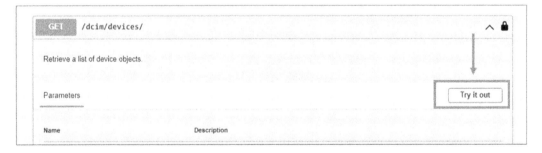

Figure 8.6 – The Try it out button allows you to test the API in the docs

After you click the button, scroll down, taking a look at all of the optional inputs (that's right, they are *all* optional). You'll eventually see a big blue button that says **Execute**:

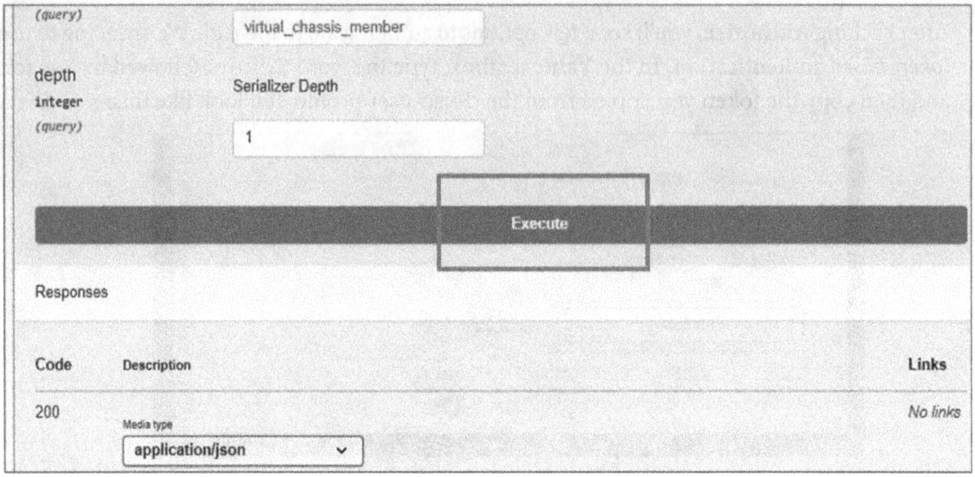

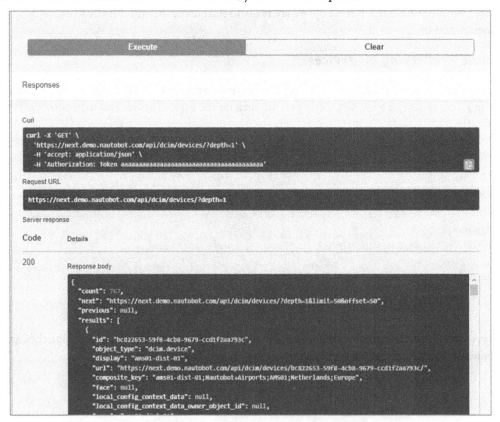

Figure 8.7 – Executing the API call in the API docs

Click the **Execute** button. After a second or two, you'll see the response:

Figure 8.8 – Viewing the API response in the API docs

Let's break down a few things in this response.

Using curl to interact with the Nautobot API

The first is the `curl` command, which you can use with the copy command on the right. Click copy and try it out on a Linux Terminal. It'll just work!

```
root@nautobot-dev:~# curl -X 'GET' \
>    'https://demo.nautobot.com/api/dcim/devices/?depth=1' \
>    -H 'accept: application/json' \
>    -H 'Authorization: Token aaaaaaaaaaaaaaaaaaaaaaaaaaaaaaaaaaaaaaaa'
```

Don't forget to press *Enter* after copying it over.

The next item is the request URL.

It is what you'd expect after seeing the URL in the docs, but it also has `?depth=1`. Don't worry about depth yet. We'll cover this later.

Exploring an API response

Finally, you can see the actual API response. We are going to spend plenty of time going through Nautobot API responses in this chapter, but for now, go ahead and compare the count on the first line of the response body with the number of devices in Nautobot by looking at the home page of Nautobot:

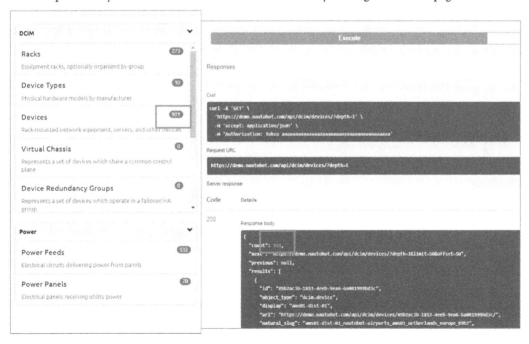

Figure 8.9 – Comparing the number of devices in the UI and API

This is a good test to ensure everything is working. We can see the same number in both, so everything's looking good!

Example – retrieving one device

If you want to get data from one device, you need to use the device's UUID in the API request while using the /dcim/devices/{id}/ URL:

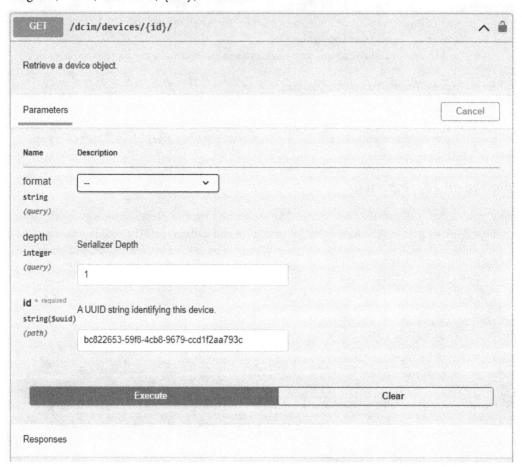

Figure 8.10 – Viewing a device's detailed API endpoint

You can get the UUID in a few different ways:

Figure 8.11 – Getting the UUID of a device (or any object)

Here are the steps shown in the preceding figure:

1. Look in the /dcim/devices/ API response.

2. Browse to a given device and get the ID (UUID) from the URL in the browser's address bar.

3. Browse to a given device, click the **Advanced** tab, and copy the UUID.

Now that we've taken a quick look at the Nautobot interactive documentation, we are going to dive deeper into understanding API responses. We'll continue to dive deeper into the API throughout this chapter.

Understanding Nautobot APIs

You've probably noticed that there is a bit more to understand before using the Nautobot APIs. There are numerous endpoints (URLs), numerous default query parameters, and a few response keys to review before getting started.

The List and Detailed endpoints

As you can see in the interactive API documentation, there are always two endpoints that look nearly identical. These are the *List* and *Detailed* endpoints.

We already saw /dcim/devices/ and /dcim/devices/{id}/. The one without the {id} value is referred to as the List endpoint. The List endpoint is the endpoint or API that will retrieve all objects of a given type (or model). The first example is retrieved devices. More examples will show that /dcim/device-types/ retrieves device types and /circuits/circuits/ retrieves all circuits.

The API that includes {id} is referred to as the Detailed endpoint. Using the Detailed endpoint, you can retrieve a single instance – for example, a single device, device type, or circuit (based on the model). Example APIs include /dcim/device-types/{id} and /circuits/circuits/ {id}, respectively. Similar to the example in the previous section, the ID is the UUID of the instance you want to query. Remember, you can obtain the UUID from the list view, API response, or from the UI in Nautobot by navigating to the object and then clicking the **Advanced** tab. Remember that the UUID is also visible in the web UI browser.

Pagination

In the API response we saw when using the interactive documentation, you may have seen a query parameter called limit with a default request URL that included &limit=50. This tells us that the default request and response is for 50 objects or elements. Like most other parameters, users making API calls can change this value. For now, just know that if you are querying for all devices and see a count of 921, as we have, that initial response (by default) only has 50 of those 921 in the first API response. This is how Nautobot implements *pagination*. If you only have a few hundred total objects in Nautobot, you may be wondering why this is needed. However, think about those larger environments that have tens of thousands of devices or millions of IP addresses. This is to create a predictable manner in which you can use the API without overloading the client or server with a single API call.

The maximum number of objects that can be returned is limited by the MAX_PAGE_SIZE configuration parameter that was covered in *Chapter 7*. This is 1000 by default:

```
nautobot_config.py
379
380    # Maximum number of objects that the UI and API will retrieve in a single request.
381    #
382    # MAX_PAGE_SIZE = 1000
383
```

Figure 8.12 – MAX_PAGE_SIZE in the Nautobot config file

Setting this to 0 or None will remove the maximum limit. An API consumer can then pass ?limit=0 to retrieve all matching objects with a single request. Disabling the page size limit introduces a potential for very resource-intensive requests and is not recommended, since one API request can effectively retrieve an entire table from the database.

Depth

When we queried the API for a single device earlier, you may have noticed that there was an optional query parameter called **depth** that had a default value of 1. Let's explore this further:

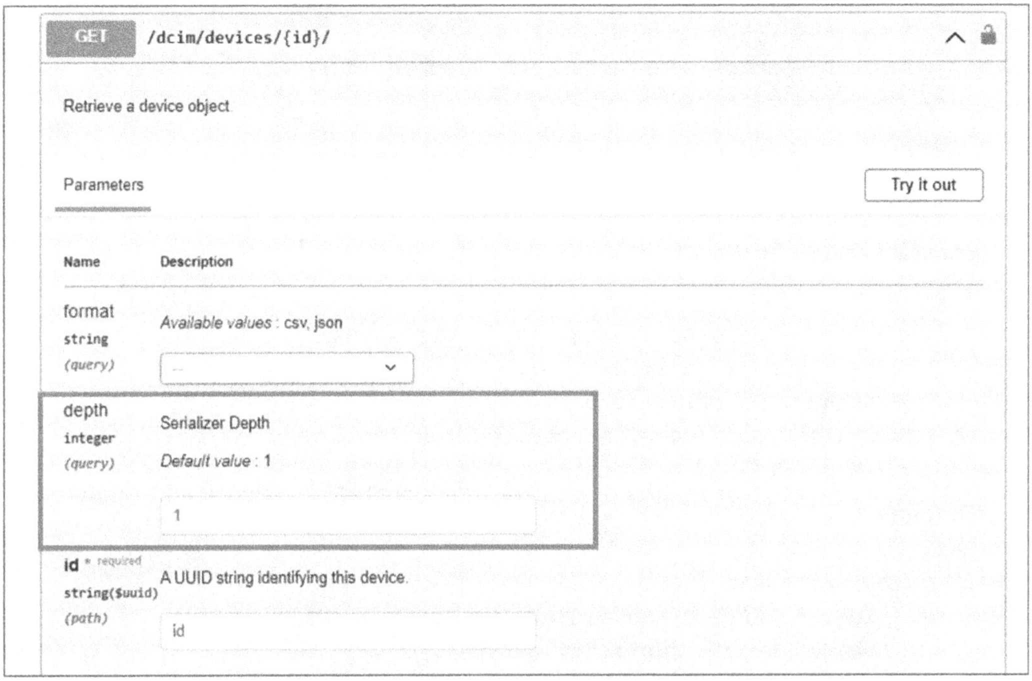

Figure 8.13 – Viewing the default API depth

We are going to execute this API call a few times to better understand what the depth does, starting at a depth of 0 and going to a depth of 3.

The API being requested is a GET request: `/dcim/devices/bc822653-59f8-4cb8-9679-ccd1f2aa7`.

The following are the four API responses using the four different depths (0, 1, 2, 3):

- *Example – device with depth 0*:

```
{
    "id": "bc822653-59f8-4cb8-9679-ccd1f2aa793c",
    "object_type": "dcim.device",
    "display": "ams01-dist-01",
    "url": "https://demo.nautobot.com/api/dcim/devices/bc822653-
59f8-4cb8-9679-ccd1f2aa793c/",
```

```
  "composite_key": "ams01-dist-01;Nautobot+Airports;AMS01;Nether
lands;Europe",
  ...
  "device_type": {
    "id": "87f85340-6fd7-4d3e-b509-893eb0aa3633",
    "object_type": "dcim.devicetype",
    "url": "https://demo.nautobot.com/api/dcim/device-
types/87f85340-6fd7-4d3e-b509-893eb0aa3633/"
  },
  ...
}
```

- *Example – device with depth 1*:

```
{
  "id": "bc822653-59f8-4cb8-9679-ccd1f2aa793c",
  "object_type": "dcim.device",
  "display": "ams01-dist-01",
  "url": "https://demo.nautobot.com/api/dcim/devices/bc822653-
59f8-4cb8-9679-ccd1f2aa793c/",
  "composite_key": "ams01-dist-01;Nautobot+Airports;AMS01;Nether
lands;Europe",
  ...
  "device_type": {
    "id": "87f85340-6fd7-4d3e-b509-893eb0aa3633",
    "object_type": "dcim.devicetype",
    "display": "Cisco Catalyst 6509-E",
    "url": "https://demo.nautobot.com/api/dcim/device-
types/87f85340-6fd7-4d3e-b509-893eb0aa3633/",
    "composite_key": "Cisco;Catalyst+6509-E",
    "subdevice_role": null,
    "front_image": null,
    "rear_image": null,
    "model": "Catalyst 6509-E",
    "part_number": "WS-C6509-E",
    "u_height": 14,
    "is_full_depth": true,
    "comments": "",
    "manufacturer": {
      "id": "706d537d-1cdb-45e0-aa78-df695826f735",
      "object_type": "dcim.manufacturer",
      "url": "https://demo.nautobot.com/api/dcim/
manufacturers/706d537d-1cdb-45e0-aa78-df695826f735/"
    },
    "created": "2023-05-11T00:00:00Z",
```

```
      "last_updated": "2023-05-11T11:26:14.480607Z",
      "notes_url": "https://demo.nautobot.com/api/dcim/device-
  types/87f85340-6fd7-4d3e-b509-893eb0aa3633/notes/",
      "custom_fields": {}
    },
    ...
    ...
  }
```

- *Example – device with depth 2 (and depth 3)*:

For this example with these objects, depths 2 and 3 are the same:

```
  {
    "id": "bc822653-59f8-4cb8-9679-ccd1f2aa793c",
    "object_type": "dcim.device",
    "display": "ams01-dist-01",
    "url": "https://demo.nautobot.com/api/dcim/devices/bc822653-
  59f8-4cb8-9679-ccd1f2aa793c/",
    "composite_key": "ams01-dist-01;Nautobot+Airports;AMS01;Nether
  lands;Europe",
    ...
    "device_type": {
      "id": "87f85340-6fd7-4d3e-b509-893eb0aa3633",
      "object_type": "dcim.devicetype",
      "display": "Cisco Catalyst 6509-E",
      "url": "https://demo.nautobot.com/api/dcim/device-
  types/87f85340-6fd7-4d3e-b509-893eb0aa3633/",
      "composite_key": "Cisco;Catalyst+6509-E",
      "subdevice_role": null,
      "front_image": null,
      "rear_image": null,
      "model": "Catalyst 6509-E",
      "part_number": "WS-C6509-E",
      "u_height": 14,
      "is_full_depth": true,
      "comments": "",
      "manufacturer": {
        "id": "706d537d-1cdb-45e0-aa78-df695826f735",
        "object_type": "dcim.manufacturer",
        "display": "Cisco",
        "url": "https://demo.nautobot.com/api/dcim/
  manufacturers/706d537d-1cdb-45e0-aa78-df695826f735/",
        "composite_key": "Cisco",
        "name": "Cisco",
```

```
            "description": "",
            "created": "2023-05-11T00:00:00Z",
            "last_updated": "2023-05-11T11:17:16.760091Z",
            "notes_url": "https://demo.nautobot.com/api/dcim/
manufacturers/706d537d-1cdb-45e0-aa78-df695826f735/notes/",
            "custom_fields": {}
          },
          "created": "2023-05-11T00:00:00Z",
          "last_updated": "2023-05-11T11:26:14.480607Z",
          "notes_url": "https://demo.nautobot.com/api/dcim/device-
types/87f85340-6fd7-4d3e-b509-893eb0aa3633/notes/",
          "custom_fields": {}
        },
        ...
    }
```

What you can see after examining these responses is that by using the depth query parameter, you can retrieve more information in a single API call. In our example, we are seeing more data on our device's device type. In other words, we are getting more data on *related objects* to the object being queried. Depth is a convenient feature for retrieving more information on related objects. You can also use the URL that is returned in the url key if you need more data than what's returned.

> **Note**
> Nautobot supports depths of 0 to 10, with the default being 0. The higher the depth, the more processing that Nautobot needs to perform, so API responses will be slower. It may be more efficient to query with a lower depth initially and then follow the URL values that the REST API response provides for specific related objects so that you can query those objects directly as a more narrowly focused query approach.

API authentication

The Nautobot REST API uses token-based authentication. To authenticate with a token, you need to generate one. You can do that by logging into Nautobot and navigating to your user profile, clicking your username in the top right, and then clicking **Profile**:

> **Warning**
> You will not have access to create a new token in the Nautobot sandbox.

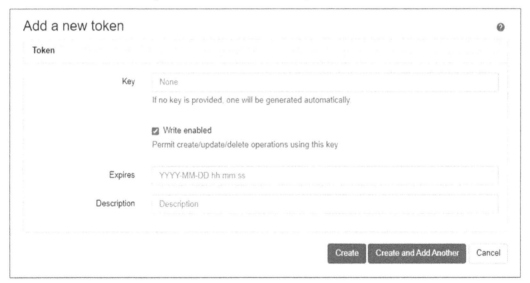

Figure 8.14 – Adding an API token

This is the same screen you saw earlier, where you copied the token to use with the interactive documentation. The difference in the preceding figure is that the permissions were changed to showcase what the UI looks like when you have permission to create a token. After you click + **Add a token**, you'll see that some options allow you to create a read-only or a read-write token, as well as provide a date for when the token expires, and then a short description for the token:

Add a new token

Token

Key	None	
	If no key is provided, one will be generated automatically	
	☑ Write enabled	
	Permit create/update/delete operations using this key	
Expires	YYYY-MM-DD hh:mm:ss	
Description	Description	

Create   Create and Add Another   Cancel

Figure 8.15 – Add a new token

All of these configurable options are helpful when third-party systems are integrating with Nautobot. Specifically, because Nautobot is a source of truth, there may be multiple systems that only fetch data (and do not make any changes). In those instances, you should uncheck **Write enabled**, making it a read-only token.

An authentication token can be sent in the API request by setting the HTTP authorization header to the Token string, followed by a space and then the token. If the token was helloworld, the header would be set to Token helloworld. You might have noticed this when you saw the curl command that was in the interactive API documentation, but here is another example using curl:

```
curl -X 'GET' \
  'https://demo.nautobot.com/api/dcim/devices/' \
  -H 'accept: application/json' \
  -H 'Authorization: Token aaaaaaaaaaaaaaaaaaaaaaaaaaaaaaaaaaaaaaaa'
```

Don't forget the word Token in your header!

Using the API with Python

It's time to start using the API using the Python requests library. We'll walk through a few examples of using the API that show how to retrieve data and make changes while using Python.

The starting Python script

We're going to start with a basic script called nautobot-apis.py:

```
root@nautobot-dev:~# cat nautobot-apis.py
import requests
import json

payload = {}

headers = {
    "Content-Type": "application/json",
    "Authorization": "Token aaaaaaaaaaaaaaaaaaaaaaaaaaaaaaaaaaaaaaaa",
}

devices_url = "https://demo.nautobot.com/api/dcim/devices/"

session = requests.Session()
session.headers.update(headers)

response = session.get(devices_url)
```

Working interactively with the Nautobot API and Python

We'll run this script with the -i flag (to run it in interactive mode). This automatically puts us in the Python shell after the script runs, giving us full access to the objects and variables that are created:

```
root@nautobot-dev:~# python3 -i nautobot-apis.py
>>>
>>>
```

You can view all the objects by issuing the `dir()` function:

```
>>> dir()
['__annotations__', '__builtins__', '__doc__', '__loader__', '__
name__', '__package__', '__spec__', 'devices_url', 'headers', 'json',
'payload', 'requests', 'response', 'session']
>>>
```

Let's start exploring the `response` variable, just as we did visually in the API interactive documentation, starting with the dictionary keys available to us:

```
>>> response.json().keys()
dict_keys(['count', 'next', 'previous', 'results'])
>>>
```

Each key is summarized as follows:

- `count`: The total number of all objects matching the query
- `next`: The API to the next query of results (if applicable)
- `previous`: The API to the previous page of results (if applicable)
- `results`: The list of objects on the current API query

Next, verify how many devices (objects) are available:

```
>>> response.json()['count']
767
>>>
```

We discussed pagination previously, so let's see how to verify the default page size of 50. The results of the API response are stored in the `results` key, so let's check the length of that object:

```
>>> len(response.json()['results'])
50
>>>
```

If there was no pagination, `results` would contain all 921 devices.

Let's view the device name of the first device in the list:

```
>>> response.json()['results'][0]['name']
'ams01-dist-01'
>>>
```

Now, let's view the last device in the list:

```
>>> response.json()['results'][-1]['name']
'bcn01-leaf-07'
>>>
```

To get the next page (or batch) of devices, we can use the URL from the next key:

```
>>> response.json()['next']
'https://demo.nautobot.com/api/dcim/devices/?limit=50&offset=50'
>>>
```

We can see two query parameters of limit and offset. Their use is as follows:

- limit: The number of elements we want in the response limited by MAX_PAGE_SIZE

- offset: The number of elements to exclude in the request – for example, all previous elements since you are paging through all objects

Let's set our new URL so that we can make another API request to get the next page:

```
>>> devices_url = response.json()['next']
>>>
>>> response = session.get(devices_url)
>>>
>>> response.json()['results'][0]['name']
'bcn01-leaf-08'
>>>
```

As you can see, bcn01-leaf-08 is the next device in the list, continuing from the initial request that ended with bcn01-leaf-07.

Let's look at one more example to get the last page using an offset of 905. With a total device count of 921 and excluding the first 905 devices, we should expect 16 in the response:

```
>>> devices_url = 'https://demo.nautobot.com/api/dcim/
devices/?limit=50&offset=905'
>>>
>>> response = session.get(devices_url)
>>>
>>> len(response.json()['results'])
16
>>>
```

Retrieving IP addresses on devices

So far, we've learned how to parse the device's API response, but if you went deeper, you probably realized that you need to access other APIs to understand a device or get all of its related data. An obvious example would be to gather IP addresses on a given device. Let's say you want to gather all IP addresses for all devices for a given site.

> **Tip**
>
> You can use query parameters to filter the API response. API filtering will be covered in more detail later in this chapter, but we'll be using query parameter filters to streamline this example and reduce the lines of code needed.

Here's a script called `restapis.py` that we'll use to print all IP addresses for all devices on the **Jersey City** site:

```python
import requests
import json


payload = {}

headers = {
    "Content-Type": "application/json",
    "Authorization": "Token aaaaaaaaaaaaaaaaaaaaaaaaaaaaaaaaaaaaaaaaaaaa",
}

devices_url = "https://demo.nautobot.com/api/dcim/devices/?location=Jersey%20City"

session = requests.Session()
session.headers.update(headers)


all_devices = []

while devices_url is not None:
    response = session.get(devices_url)
    devices_url = response.json()["next"]
    all_devices.extend(response.json()["results"])

for device in all_devices:
    device["our_interfaces"] = []
    ip_url = f"https://demo.nautobot.com/api/ipam/ip-addresses?device={device['name']}"

    while ip_url is not None:
        ip_url_response = session.get(ip_url)
        ip_url = ip_url_response.json()["next"]
        device["our_interfaces"].extend(ip_url_response.json()["results"])

    for interface in device["our_interfaces"]:
        print(f"Device: {device['name']} has an IP Address {interface['address']}")
```

Figure 8.16 – The restapis.py Python script

Take note of the introduction of the location and device query parameters to filter the API response, but more importantly of the addition of the `/ipam/ip-addresses/` API requests. In both examples, we kept the defaults for pagination.

Executing this script gives us the following output:

```
root@nautobot-dev:~# python3 -i restapis.py
Device: jcy-bb-01.infra.ntc.com has an IP Address 10.0.10.3/32
Device: jcy-bb-01.infra.ntc.com has an IP Address 10.10.0.6/30
# omitted for brevity
>>>
```

While we only wanted the IP addresses for each interface, we had to receive each interface object, which had 25 dictionary keys. In reality, that's 24 more than we needed. This will come in handy when we compare this method to GraphQL later in this chapter:

```
>>> device["our_interfaces"][0].keys()
dict_keys(['id', 'object_type', 'display', 'url', 'composite_key',
'address', 'host', 'mask_length', 'type', 'ip_version', 'dns_name',
'description', 'status', 'role', 'parent', 'tenant', 'nat_inside',
'created', 'last_updated', 'tags', 'notes_url', 'custom_fields', 'nat_
outside_list', 'interfaces', 'vm_interfaces'])
>>>
```

So, when it comes to retrieving information via the API, you'll want to use the interactive documentation as much as possible to learn how to access exactly what you need. Generally speaking, the URLs will closely match what you see in the UI to the API.

For example, if you pick an IP address from the output shown earlier, you'll see `/ipam/ip-addresses/` in the URL, giving you guidance on where to look in the API docs:

Figure 8.17 – Detailed object view of an IP address

Example – onboarding devices

This section will introduce a Python script showing how to add a new device, create a loopback interface on that device, create an IP address, and finally assign that IP address to the newly created loopback interface.

> **Note**
>
> For production environments, you should explore using the Nautobot Device Onboarding App, which abstracts away and automates device onboarding. This example showcases how to use the various Nautobot APIs.

Start of the script

The following code snippet is similar to the previous example in that it is the starting point for the script we are going to walk through. You can find the complete script in this book's GitHub repository: `https://github.com/PacktPublishing/Network-Automation-with-Nautobot/blob/main/chapter-08/start-of-script.py`

```
import requests
import json

payload = {}

headers = {
    "Content-Type": "application/json",
    "Accept":  "application/json",
    "Authorization": "Token aaaaaaaaaaaaaaaaaaaaaaaaaaaaaaaaaaaaaaaa"
}

session = requests.Session()
session.headers.update(headers)
```

Note that the UUIDs used in the following examples are from the Nautobot sandbox. Your instance and maybe even the sandbox UUIDs will be different. Please verify them if you're following along.

Creating a device

To create and add a device to Nautobot, you must make an HTTP POST API call and include particular required fields.

> **Tip**
>
> You can add a device to the UI to see what fields you must streamline using the API. For creating a device, you are required to have **Device Type**, **Status**, **Role**, and **Location**.

In our example, we are also including **Platform** and **Name**. We're doing this to mimic other devices that are already in the site.

The payload is pre-built with the following key-value pairs:

- **Location**: The UUID of the location we want to add the device to. At the time of writing, when using the Nautobot sandbox, this UUID represents the **AMS01** location.

- **Status**: The UUID used represents the **Active** status.

- **Role**: The UUID of the **leaf** role.

- **Device Type**: The UUID of the **Arista DCS-7150S-24** device type.

- **Name**: ams01-leaf-11. **AMS01** already has several leaf switches. This is adding another leaf switch. We chose to use **11** for this switch.

- **Platform**: While optional, we used the **Arista EOS** platform because all other leaf switches in AMS01 are using **Arista EOS**.

- **Tenant**: While optional, we included it to be under the **Nautobot Airports** tenant.

Append the following code to the *Start of script* section:

```
payload = {
    "name": "ams01-leaf-11",
    "device_type": "74cf95a8-4233-46b9-a740-fba4f5dc88d3",
    "status": "9f38bab4-4b47-4e77-b50c-fda62817b2db",
    "role": "869267d8-7d75-4bd3-8a9e-5e6adcf200f6",
    "tenant": "1f7fbd07-111a-4091-81d0-f34db26d961d",
    "platform": "f48fd9e2-45c5-4c2f-aa54-28964edb3e1e",
    "location": "9e39051b-e968-4016-b0cf-63a5607375de"
}

devices_url = "https://demo.nautobot.com/api/dcim/devices/"

# adds ams01-leaf-11 to the location AMS01
r = session.post(devices_url, data=json.dumps(payload))

# the UUID of the device will be saved for the next API call
device_id = r.json()["id"]
```

Creating an interface on the device

To create and add an interface to Nautobot, you must make an HTTP POST API call and include particular required fields.

The following payload is pre-built with the following key-value pairs:

- **Device**: The UUID of the device created in the previous step

- **Name**: The name of the interface we are creating

- **Status**: The UUID used represents the "**Active**" status

- **Type**: Signify that this is a virtual interface from the list of supported interfaces in Nautobot:

```
interface_url = "https://demo.nautobot.com/api/dcim/interfaces/"

payload = {
    "device": device_id,
    "name": "Loopback0",
    "status": "9f38bab4-4b47-4e77-b50c-fda62817b2db",
    "type": "virtual"
}

# adds Loopback0 to ams01-leaf-11
r = session.post(interface_url, data=json.dumps(payload))

interface_id = r.json()["id"]
```

Creating an IP address

The next step for us is to create an IP address that will be assigned to `Loopback0` on `ams01-leaf-11`. There are four required fields we must fill to create an IP address (**Address**, **Type**, **Status**, and **Namespace**). We will also include **Tenant** and **DNS Name** to match the standards that are also in use by the other leaf switches:

```
ip_address_url = "https://demo.nautobot.com/api/ipam/ip-addresses/"

payload = {
    "address": "10.11.128.11/32",
    "namespace": "733d191a-4067-4215-90a8-814bcfe28f03",
    "tenant": "1f7fbd07-111a-4091-81d0-f34db26d961d",
    "type": "host",
    "status": "9f38bab4-4b47-4e77-b50c-fda62817b2db",
    "dns_name": "leaf-11.ams01.atc.nautobot.com"
}

# creates IP
r = session.post(ip_address_url, data=json.dumps(payload))
ipaddr_id = r.json()["id"]
```

Assigning an IP address to the interface

Finally, we must assign 10.11.128.11/32 to Loopback0 on ams01-leaf-11. We can do this using the /ipam/ip-address-to-interface/ API route. This is a simple one that requires two fields: interface and ip address. Both are self-explanatory:

```
ipam_url = "https://demo.nautobot.com/api/ipam/ip-address-to-
interface/"

payload = {
        "interface": interface_id,
        "ip_address": ipaddr_id
    }

# assigns IP to Loopback0 on ams01-leaf-11
r = session.post(ipam_url, data=json.dumps(payload))
```

As a reminder, you can view this complete script in this book's GitHub repository.

Once the script is built, you can validate that everything worked as expected by going to view the new device. Make sure you view the **Interfaces** tab (scroll to the bottom) to ensure the IP address has been assigned to the loop interface:

Figure 8.18 – Viewing the newly created device from the Python script

Now that we've learned how to add objects via the API, we are going to learn how to execute a job from the API.

Executing a job

If you aren't familiar with Nautobot jobs yet, you may want to read *Chapter 11* first; however, it is not a requirement. This section shows how to execute a job via the REST API. The short story is that a job is arbitrary Python code to perform network automation tasks or to analyze the data in Nautobot. It's quite simple to convert scripts into self-service forms, as you'll see in *Chapter 11*. For now, the assumption is that you are at least familiar with the basic concepts of jobs. To see jobs in the demo instance of Nautobot (`demo.nautobot.com`), navigate to **Jobs** > **Jobs** using the sidebar navigation menu.

You'll see this screen:

Figure 8.19 – Viewing Nautobot jobs (list view)

We'll see two examples of executing jobs from the API. Both examples will execute the job *by name*. This is a much more streamlined approach (and easier to understand) than using the UUID of the job.

Executing the Device Software Validation Report job

If you scroll down in the jobs UI, you'll see **Device Software Validation Report**. That is the job's name.

> **Note**
> This particular job is installed automatically when the Nautobot Device Lifecycle Management App is installed.

The API to execute jobs by name is POST: `/api/extras/{JOB NAME}/run/`.

Don't forget that trailing `/`.

Before using the API, let's browse the UI. Click the **Execute** button for the **Device Software Validation Report** job by clicking the **Play** icon:

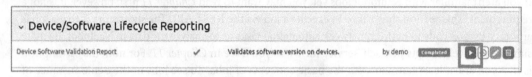

Figure 8.20 – Executing a job

You'll see this screen:

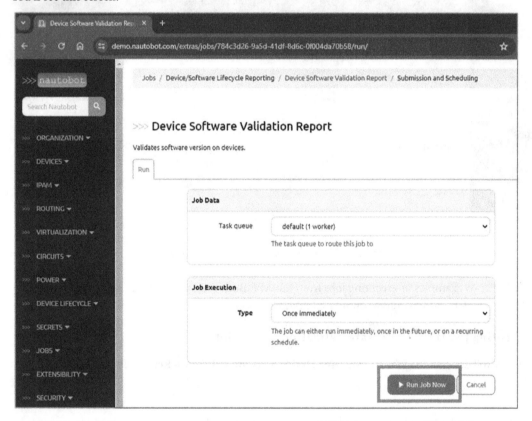

Figure 8.21 – Clicking Run Job Now

> **Note**
> It isn't as important to dive into these specific jobs as it is to navigate them to understand how to collect the data that's needed to execute them from the API.

Click the **Run Job Now** button.

You'll be redirected to the **Job Results** tab on the **Job Result** page. Click the **Additional Data** tab and take note of the **Job Keyword Arguments** section:

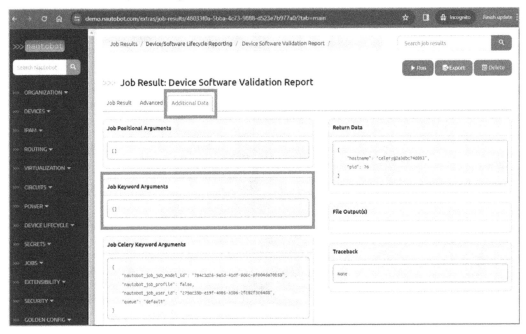

Figure 8.22 – Job Results – viewing keyword arguments

There are no additional arguments. Remember, we ran the job without sending any inputs. This is one of the simplest jobs to execute from an API.

> **Note**
>
> This script uses the same script from the *Start of script* section for imports and headers.

Here's how you'd execute this job from an API:

```
payload = {
    "data": {}
}

run_job_url = "https://demo.nautobot.com/api/extras/jobs/Device
Software Validation Report/run/"
r = session.post(run_job_url, data=json.dumps(payload))
print(r.status_code)
```

Having an empty dictionary helps set the stage for sending keyword arguments to a job. We'll see that in the next example.

Place the *Start of script* code, plus the preceding code block, into a single script called `run-job.py`:

```
run-job.py
 1    import requests
 2    import json
 3
 4    payload = {}
 5
 6    headers = {
 7    "Content-Type": "application/json",
 8    "Accept":  "application/json",
 9    "Authorization": "Token aaaaaaaaaaaaaaaaaaaaaaaaaaaaaaaaaaaaaaaa"
10    }
11
12    session = requests.Session()
13    session.headers.update(headers)
14
15    payload = {
16        "data": {}
17    }
18
19    run_job_url = "https://demo.nautobot.com/api/extras/jobs/Device Software Validation Report/run/"
20    r = session.post(run_job_url, data=json.dumps(payload))
21
22    print(r.status_code)
23
```

Figure 8.23 – Executing a job from an API

Run the script:

```
root@nautobot-dev:~# python3 run-job.py
201
root@nautobot-dev:~#
```

As soon as you run the script, navigate to **Jobs** > **Job Results**; you'll see that the job is running:

Figure 8.24 – Viewing Job Results, ensuring that the job is running

Executing the Generate Vulnerabilities job

In the jobs UI, you'll see another job called **Generate Vulnerabilities**. Execute that job without changing any inputs and navigate to the **Additional Data** tab in **Job Results**.

You should see this screen:

Figure 8.25 – Verifying job inputs as keyword arguments

You probably noticed that there was a date as an input in the **CVEs Published After** form field. You can infer that the published_after keyword argument is the key that's used for that input. Let's make an API call to execute the job while changing the date from **1970** to **1975**. We'll do this in a new script called run-job-gen-vuln.py:

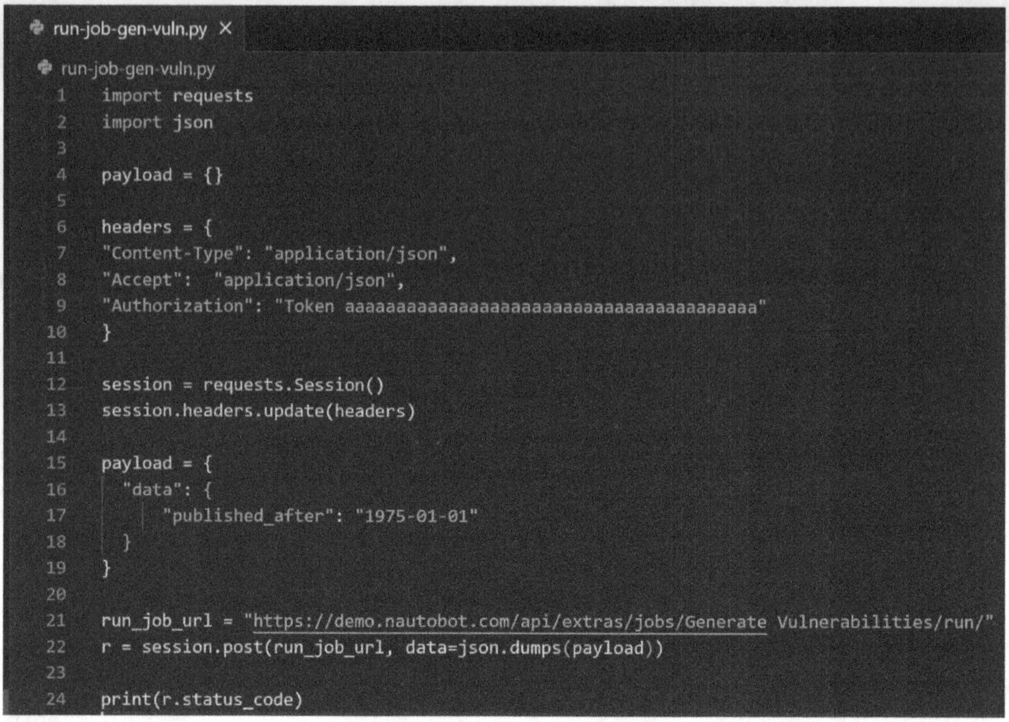

```
run-job-gen-vuln.py  ×

run-job-gen-vuln.py
  1   import requests
  2   import json
  3
  4   payload = {}
  5
  6   headers = {
  7   "Content-Type": "application/json",
  8   "Accept":  "application/json",
  9   "Authorization": "Token aaaaaaaaaaaaaaaaaaaaaaaaaaaaaaaaaaaaaaaa"
 10   }
 11
 12   session = requests.Session()
 13   session.headers.update(headers)
 14
 15   payload = {
 16     "data": {
 17         "published_after": "1975-01-01"
 18     }
 19   }
 20
 21   run_job_url = "https://demo.nautobot.com/api/extras/jobs/Generate Vulnerabilities/run/"
 22   r = session.post(run_job_url, data=json.dumps(payload))
 23
 24   print(r.status_code)
```

Figure 8.26 – Executing a job and passing arguments into the job

After executing the code snippet, you can navigate to **Job Results** and see it:

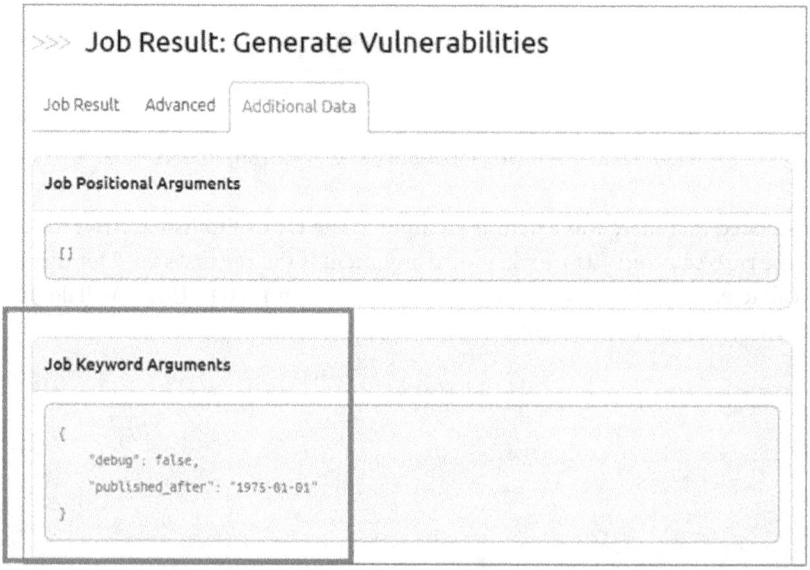

>>> Job Result: Generate Vulnerabilities

Job Result   Advanced   Additional Data

Job Positional Arguments

[]

Job Keyword Arguments

```
{
    "debug": false,
    "published_after": "1975-01-01"
}
```

Figure 8.27 – Verifying the updated keyword arguments

API tips

Now, let's cover a few tips when using the Nautobot REST API.

Filtering

In prior examples, we subtly introduced the concept of filtering when we were using the GET devices API. If we look at the API docs again – that is, the interactive documentation – we'll see several query parameters denoted with (query):

Figure 8.28 – Viewing query parameters in the API docs

These query parameters can be used to filter the API response. It is always a good idea to have the server (in this case, Nautobot) do the heavy processing. If not, you must filter your own code!

One example shown previously was `/api/dcim/devices/?location=Jersey%20City`, which filters the response to only return the Jersey City location.

What if we wanted all `spine` devices in Jersey City? We can add `role` to the URL, like so:

```
/api/dcim/devices/?location=Jersey%20City&role=spine
```

It is also possible to use a query parameter more than once in the same API call. If you do this, the response will return all spine devices and router devices in Jersey City, making it a logical OR operation:

```
/api/dcim/devices/?location=Jersey%20City&role=spine&role=Router
```

Feel free to try these in the demo sandbox and verify them based on the number of devices returned in the API while cross-checking them in the UI (if you have any doubts!). To use the following code, reference the script we created in the *The Starting Python script* section. Run that script interactively by using `python3 -i nautobot-apis.py`. Then, add the following lines in the Python interactive shell (ensure you keep this session open, as we will build on it throughout this chapter):

```
>>> devices_url = "https://demo.nautobot.com/api/dcim/
devices/?location=Jersey%20City"
>>>
>>> response = session.get(devices_url)
>>>
>>> response.json()["count"]
6
>>>
>>> devices_url = "https://demo.nautobot.com/api/dcim/
devices/?location=Jersey%20City&role=spine"
>>>
>>> response = session.get(devices_url)
>>>
>>> response.json()["count"]
2
>>>
>>> devices_url = "https://demo.nautobot.com/api/dcim/
devices/?location=Jersey%20City&role=spine&role=Router"
>>>
>>> response = session.get(devices_url)
>>>
>>> response.json()["count"]
4
>>>
```

Includes

There are two scenarios to call out to assist users in understanding why certain data isn't being returned in API responses to improve response time and the most common API requests. They are querying the API for the following:

- Config contexts
- Relationships
- Computed fields

Config contexts and relationships were both covered in *Chapter 6* when we explored Nautobot's extensibility features. Both config contexts and relationships are computationally heavy and they are not returned in API responses by default. To return them, use the `include` keyword argument in the URL, as follows:

- `?include=config_context`
- `?include=relationships`
- `?include=computed_fields`

Let's look at the config contexts on a given device. We'll query for any given device using a UUID and continue in the Python interactive shell we started previously:

```
>>> device_url = "https://demo.nautobot.com/api/dcim/devices/d200ff18-
ef93-449b-b75e-301e948aaf50/"
>>>
>>> response = session.get(device_url)
>>>
>>> response.json().keys()
dict_keys(['id', 'object_type', 'display', 'url', 'natural_slug',
'face', 'local_config_context_data', 'local_config_context_data_
owner_object_id', 'name', 'serial', 'asset_tag', 'position', 'device_
redundancy_group_priority', 'vc_position', 'vc_priority', 'comments',
'local_config_context_schema', 'local_config_context_data_owner_
content_type', 'device_type', 'status', 'role', 'tenant', 'platform',
'location', 'rack', 'primary_ip4', 'primary_ip6', 'cluster', 'virtual_
chassis', 'device_redundancy_group', 'secrets_group', 'created',
'last_updated', 'tags', 'notes_url', 'custom_fields', 'parent_bay'])
>>>
>>> len(response.json().keys())
37
>>>
```

Now, let's add `?include=config_context`:

```
>>> device_url = "https://demo.nautobot.com/api/dcim/devices/d200ff18-
ef93-449b-b75e-301e948aaf50?include=config_context"
>>>
>>> response = session.get(device_url)
>>>
>>> response.json().keys()
dict_keys(['id', 'object_type', 'display', 'url', 'natural_slug',
# omitted for brevity
'device_redundancy_group', 'secrets_group', 'created', 'last_updated',
'tags', 'notes_url', 'custom_fields', 'parent_bay', 'config_context'])
>>>
>>> len(response.json().keys())
38
>>>
```

You likely saw the new and last key that was added: `config_context`.

Following the same pattern, we can also query for relationships:

```
>>> device_url = "https://demo.nautobot.com/api/dcim/
devices/d200ff18-ef93-449b-b75e-301e948aaf50?include=config_
context&include=relationships"
>>>
>>> response = session.get(device_url)
>>>
>>> response.json().keys()
dict_keys(['id', 'object_type', 'display', 'url', 'natural_slug',
'face', 'local_config_context_data',
#omitted for brevity
'device_redundancy_group', 'secrets_group', 'created', 'last_updated',
'tags', 'notes_url', 'custom_fields', 'relationships', 'parent_bay',
'config_context'])
>>>
>>> len(response.json().keys())
39
```

Now, let's look at API versioning in Nautobot.

API versioning

Nautobot supports API versioning, allowing users to choose the REST API version they want to use. This comes in handy and ensures you know exactly what to expect in API responses. For companies looking to build production-grade automation around Nautobot, having API versioning is a lifesaver.

There are two ways to use API versioning in Nautobot:

- Set the version within the `Accept` and `Content-Type` headers using the version keyword – for example, `version=2.0`

- Set the version as a URL query parameter – for example, `?api_version=2.0`

Let's try it out and verify the version being used using the Python shell. The sandbox is currently using the 2.1 API by default, and we are going to test using the 2.0 API. Again, we'll be continuing in the Python shell we used previously:

```
>>> headers = {
... "Content-Type": "application/json; version=2.0",
... "Accept":  "application/json; version=2.0",
... "Authorization": "Token aaaaaaaaaaaaaaaaaaaaaaaaaaaaaaaaaaaaaaaa"
... }
>>>
>>> session = requests.Session()
>>> session.headers.update(headers)
>>>
>>> payload = {
...     "name": "ams01-leaf-16",
...     "device_type": "74cf95a8-4233-46b9-a740-fba4f5dc88d3",
...     "status": "9f38bab4-4b47-4e77-b50c-fda62817b2db",
...     "role": "869267d8-7d75-4bd3-8a9e-5e6adcf200f6",
...     "tenant": "1f7fbd07-111a-4091-81d0-f34db26d961d",
...     "platform": "f48fd9e2-45c5-4c2f-aa54-28964edb3e1e",
...     "location": "9e39051b-e968-4016-b0cf-63a5607375de"
... }
>>>
>>> devices_url = "https://demo.nautobot.com/api/dcim/devices/"
>>>
>>> r = session.post(devices_url, data=json.dumps(payload))
>>>
>>> r
<Response [201]>
>>>
>>> r.headers
{'Allow': 'GET, POST, PUT, PATCH, DELETE, HEAD, OPTIONS', 'Alt-Svc':
'h3=":443"; ma=2592000', 'Api-Version': '2.0', 'Content-Encoding':
'gzip', 'Content-Type': 'application/json', 'Location': 'https://demo.
nautobot.com/api/dcim/devices/08d97be9-5cbf-4b12-8938-53e9ad9a8118/',
#omitted for brevity
>>>
```

Take note of the `'Api-Version'` header that's returned, validating it was set successfully.

It is recommended to set the API version to ensure your automation is going to work as expected. In newer releases of Nautobot, Nautobot can add response data as backward-compatible changes in the API. This is usually never an issue, but there is also the possibility that Nautobot introduces an API endpoint or changes an existing one (if it's not using the version, this could break your automation). So, the benefit here is that Nautobot can release new features in the API while providing stability and backward compatibility using API versioning. Because Nautobot follows SemVer, there are never breaking API changes between minor versions (for example, 2.1 to 2.2), but there could be with major versions, such as 2.x to 3.0. For this reason, it is recommended to use the API version in your API calls.

pynautobot

Because this chapter is reviewing the REST API for Nautobot, we thought it would be beneficial to point out that it is likely you may never need to use it because of the existence of `pynautobot`, a Python library and wrapper for the Nautobot REST API. Rather than deal with Python requests and build some lower-level logic for your scripts, you can take advantage of all the objects and methods you'd expect with a Python library. However, even when using `pynautobot`, it is helpful to understand concepts such as API versioning, pagination, what methods are required to make changes, how to browse the API docs, and so much more. Be ready to dive into `pynautobot` in *Chapter 9*.

GraphQL with Nautobot

Before we get into the details of using GraphQL with Nautobot, let's review a short primer.

GraphQL primer

GraphQL serves as a query language designed for APIs, enabling users to programmatically request and retrieve data from an API endpoint or server. Using GraphQL to fetch data can transform how you think about APIs. If you remember one thing about GraphQL, it should be that GraphQL empowers you to precisely define the data you want to receive, even when it resides in multiple data silos (or in different data models, in Nautobot parlance). In contrast to REST API queries, a GraphQL query ensures that only the explicitly requested data is returned, even if it involves multiple objects.

A single GraphQL query provides the exact data that would otherwise necessitate multiple REST API calls and extensive filtering.

Note that Nautobot support for GraphQL is limited to read-only queries. Other GraphQL operations such as mutations and subscriptions are not supported.

Consider this scenario: suppose you wish to obtain the IP address for a specific interface and the remote devices that are connected through that interface. This is possible with GraphQL!

It's time to see this in action.

GraphiQL

We are going to start by building GraphQL queries directly inside Nautobot. Click the **GraphQL** link in the bottom right of the footer inside the public sandbox:

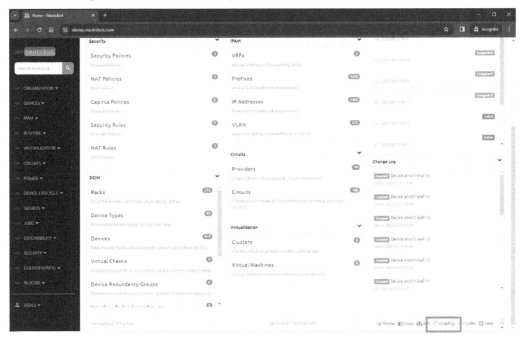

Figure 8.29 – Navigating to GraphiQL from the GraphQL link

You will be presented with GraphiQL, an in-browser tool for building, testing, and exploring GraphQL:

Figure 8.30 – GraphiQL embedded in Nautobot

GraphiQL example 1 – devices with UUID

Let's start with a basic query to return all devices, but only return the UUID of each device. We will take the following query, enter it on the left-hand side of the screen, and click the **Play** button:

```
query {
  devices {
    id
  }
}
```

This can be seen in the following screenshot:

Figure 8.31 – GraphQL query for devices with the UUID

GraphiQL example 2 – devices with UUID and name

As you can see on the right portion of the screen, the response was all the UUIDs of all devices. You may be thinking, *what good is this without more data?* Let's add name to the query:

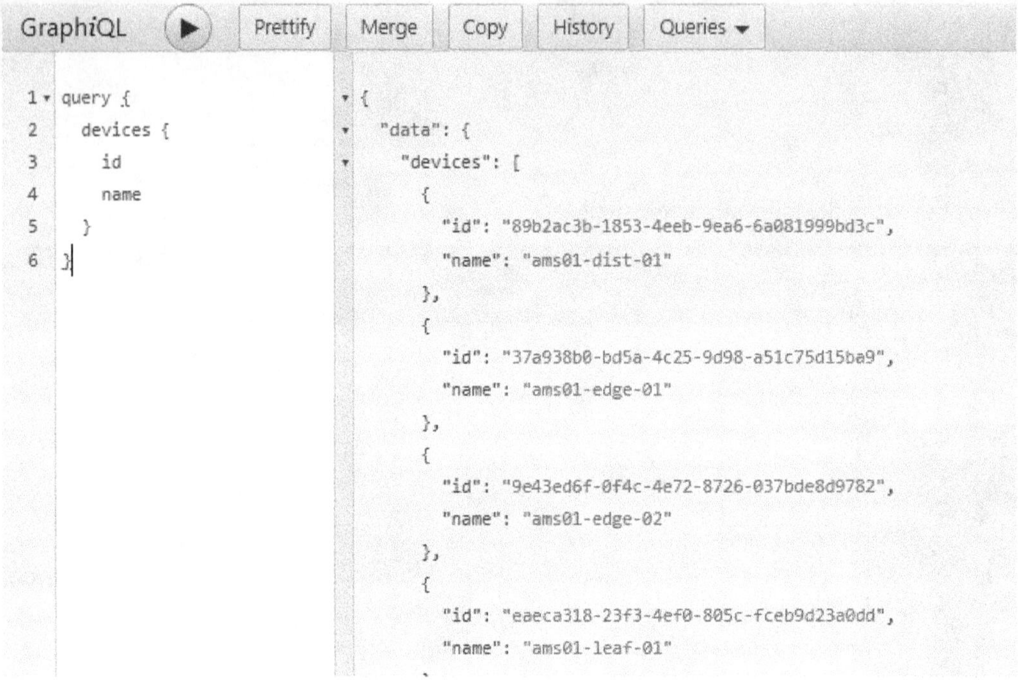

```
GraphiQL  ▶  Prettify   Merge   Copy   History   Queries ▼

1 ▼ query {                  ▼ {
2     devices {              ▼   "data": {
3       id                   ▼     "devices": [
4       name                         {
5     }                                "id": "89b2ac3b-1853-4eeb-9ea6-6a081999bd3c",
6   }                                  "name": "ams01-dist-01"
                                     },
                                     {
                                       "id": "37a938b0-bd5a-4c25-9d98-a51c75d15ba9",
                                       "name": "ams01-edge-01"
                                     },
                                     {
                                       "id": "9e43ed6f-0f4c-4e72-8726-037bde8d9782",
                                       "name": "ams01-edge-02"
                                     },
                                     {
                                       "id": "eaeca318-23f3-4ef0-805c-fceb9d23a0dd",
                                       "name": "ams01-leaf-01"
```

Figure 8.32 – GraphQL query for devices response

We are now getting back name in every device object. In contrast to REST APIs, such as doing a GET on /dcim/devices/, which we did earlier in this chapter, something that returns numerous fields per device, GraphQL allows us to define exactly the data we want to be returned.

GraphiQL example 3 – devices with name and location

You may have been inclined to just type location under name, like so:

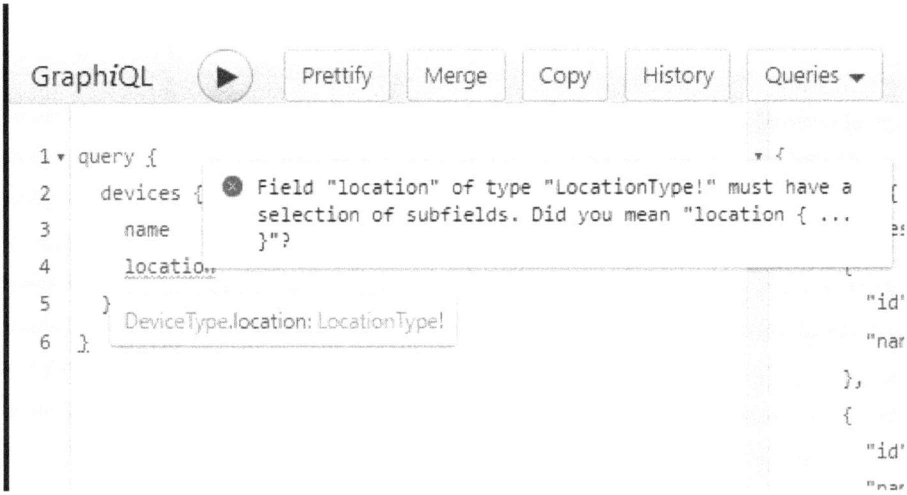

Figure 8.33 – Schema Validation with GraphQL

But if we hover over it, we'll get an error message, telling us we need a subfield or some key value to return for each location, be it a UUID, name, and so on.

Let's return each location's name:

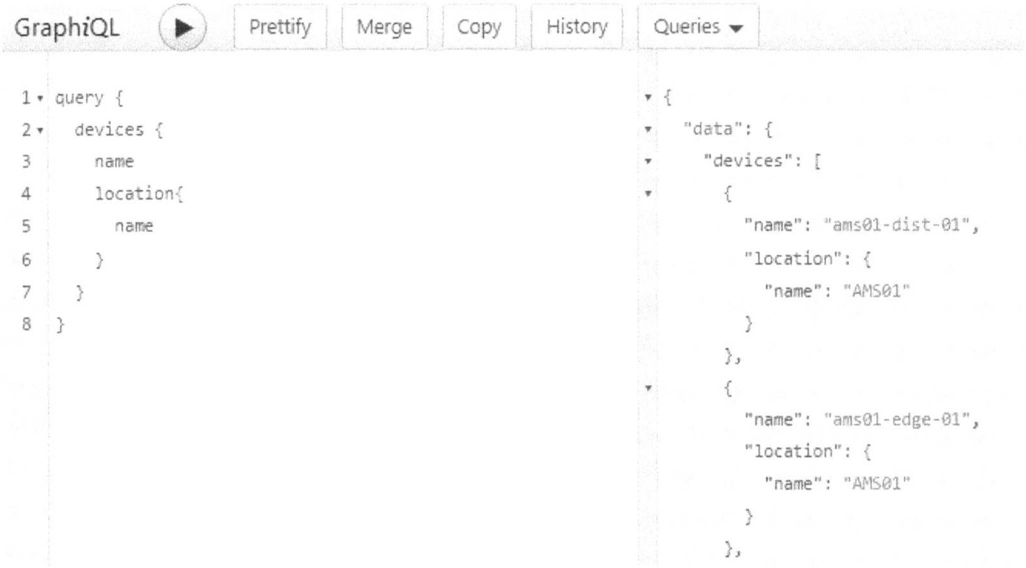

Figure 8.34 – GraphQL query using GraphiQL for devices with name and location

How great is that? We can easily get the location's name for each device.

GraphiQL example 4 – devices with name and location for one location

In most of the examples in this chapter, we were using the **AMS01** location. Let's make the same query from example 3, but now, just for that location:

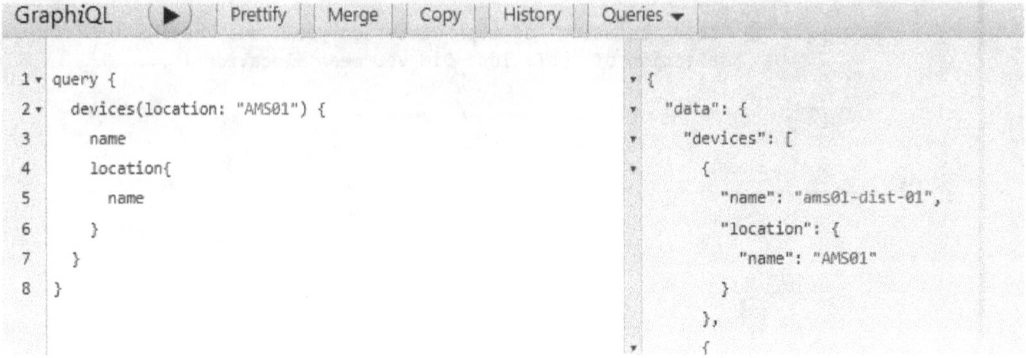

```
GraphiQL   (▶)   Prettify   Merge   Copy   History   Queries ▾

1 ▾ query {                                    ▾ {
2 ▾   devices(location: "AMS01") {             ▾   "data": {
3       name                                   ▾     "devices": [
4       location{                              ▾       {
5         name                                           "name": "ams01-dist-01",
6       }                                                "location": {
7     }                                                    "name": "AMS01"
8   }                                                    }
                                                       },
                                               ▾       {
```

Figure 8.35 – GraphQL query using GraphiQL for devices for a given location

GraphiQL example 5 – devices with name and IP addresses for one location

Let's also ask for interfaces and IP addresses:

```
GraphiQL   (▶)   Prettify   Merge   Copy   History   Queries ▾

 1 ▾ query {                                   ▾ {
 2 ▾   devices(location: "AMS01") {            ▾   "data": {
 3       name                                  ▾     "devices": [
 4 ▾     interfaces {                                  {
 5         name                                          "name": "ams01-dist-01",
 6         ip_addresses {                                "interfaces": []
 7           address                                   },
 8         }                                  ▾        {
 9       }                                               "name": "ams01-edge-01",
10     }                                      ▾          "interfaces": [
11   }.                                       ▾            {
                                                             "name": "Ethernet1/1",
                                              ▾              "ip_addresses": [
                                                               {
                                                                 "address": "10.11.192.0/32"
```

Figure 8.36 – GraphQL query using GraphiQL for device interfaces and IP addresses

GraphiQL example 6 – devices with interfaces and connected interfaces for one location

Now, let's query for the connected interfaces for each local interface per device:

```
GraphiQL  (▶)   Prettify   Merge   Copy   History   Queries ▼

 1 ▾ query {                                    ▾ {
 2 ▾   devices(location: "AMS01") {             ▾   "data": {
 3       name                                   ▾     "devices": [
 4 ▾     interfaces {                                   {
 5         name                                           "name": "ams01-dist-01",
 6 ▾       connected_interface {                          "interfaces": []
 7           device{                                    },
 8             name                             ▾       {
 9           }                                            "name": "ams01-edge-01",
10         }                                    ▾         "interfaces": [
11       }                                      ▾           {
12     }                                                      "name": "Ethernet1/1",
13   }                                          ▾             "connected_interface": {
                                                                "device": {
                                                                  "name": "ams01-edge-02"
                                                                }
```

Figure 8.37 – GraphQL query using GraphiQL for devices with connected interfaces

GraphiQL example 7 – devices with interfaces, connected interfaces, and their names

In this last example, let's look at an example to get the connected interfaces with the name of the remote connected interface:

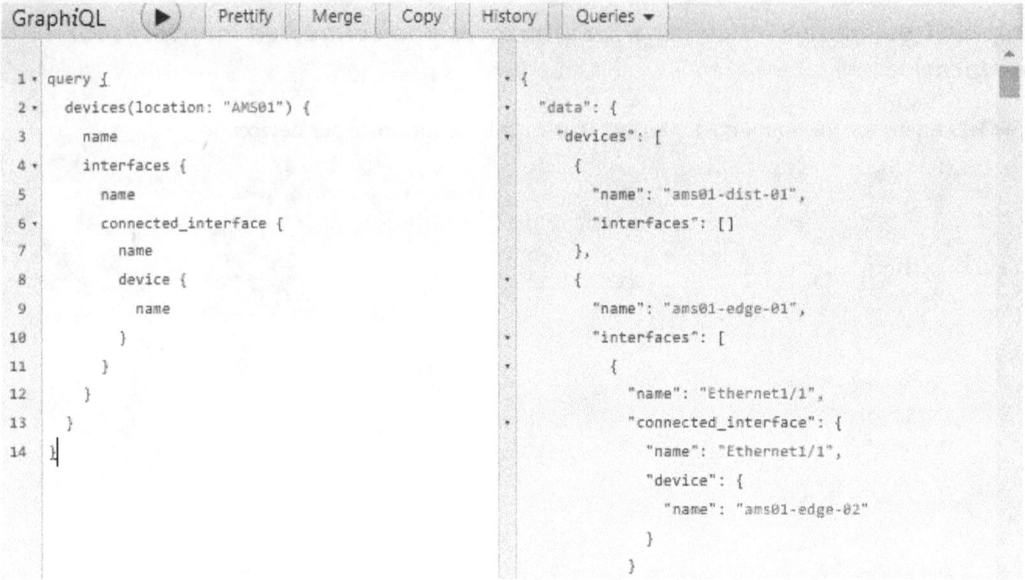

```
GraphiQL    ( ▶ )    Prettify    Merge    Copy    History    Queries ▾

 1 • query {                                      • {
 2 •   devices(location: "AMS01") {               •   "data": {
 3       name                                      •     "devices": [
 4 •     interfaces {                                      {
 5         name                                              "name": "ams01-dist-01",
 6 •       connected_interface {                             "interfaces": []
 7           name                                          },
 8           device {                              •       {
 9             name                                          "name": "ams01-edge-01",
10           }                                     •         "interfaces": [
11         }                                       •           {
12       }                                                       "name": "Ethernet1/1",
13     }                                           •             "connected_interface": {
14   }                                                             "name": "Ethernet1/1",
                                                                   "device": {
                                                                     "name": "ams01-edge-02"
                                                                   }
                                                                 }
                                                               }
```

Figure 8.38 – GraphQL query using GraphiQL for devices with connected interfaces and tags

As you can see, GraphQL is powerful and allows you to make fewer API calls to return a significant amount of data. You could get the same data through the REST API as well, but it would take more requests and additional data transformation in your code to achieve the same result.

GraphiQL example 8 – GraphQL aliasing

What if you wanted to standardize or have more structure on the actual JSON response? For example, in the previous response, Nautobot returns a key called name (for the device name) and interfaces. Maybe you'd prefer to see hostname instead of name in your results, and intfs instead of interfaces (for whatever reason). This is possible with GraphQL aliasing. You can do this by using the <desired_key_name>: <existing_key_name> format.

Let's see this in action:

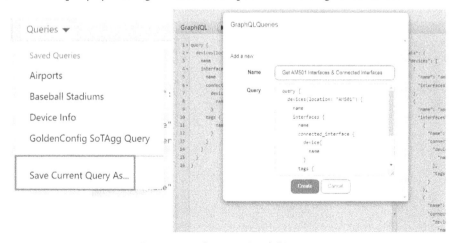

Figure 8.39 – Changing the key's name using GraphQL aliasing

Next, we'll look at how GraphQL queries can be saved in Nautobot.

Saving queries

Going beyond testing and validating queries, you can also save your GraphQL queries. Once they are saved, you can programmatically execute the query from a REST API endpoint. We'll see that in an upcoming example.

Let's save the last query by clicking the **Queries** dropdown and then give it a name:

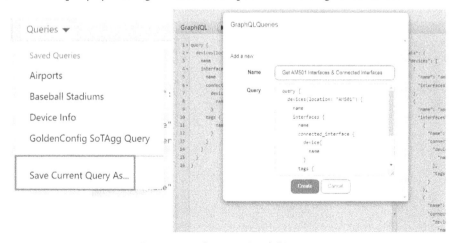

Figure 8.40 – Saving a GraphQL query

As soon as you click **Create** when saving the query, you'll be redirected to the saved queries page for the new query just saved. This is a typical detailed object page in Nautobot that allows you to view the query, date, and time created in the **Advanced** tab, the **Notes** tab, and the changelog.

You can later edit the query from this page too:

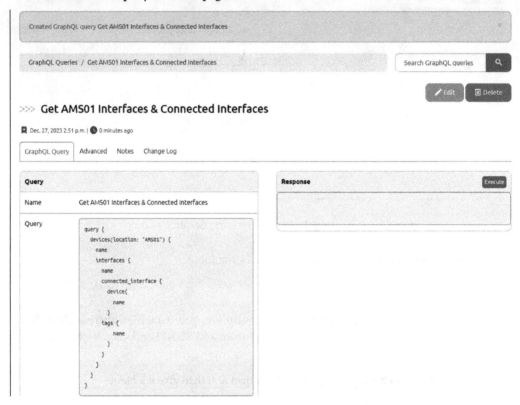

Figure 8.41 – Viewing a saved GraphQL query

To navigate back to this page in the future, you can go to **EXTENSIBILITY | GraphQL Queries** from the navigation sidebar:

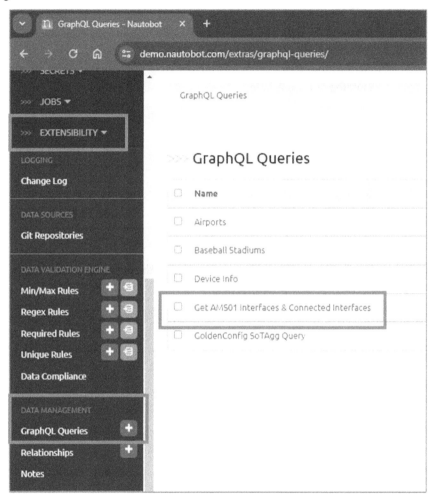

Figure 8.42 – Navigating to GraphQL Queries

> **Tips**
>
> If you were following along in the Nautobot sandbox, you probably already saw the autocomplete feature. Feel free to continue to see what else you can add to the GraphQL query using GraphiQL.
>
> You can also click **< Docs** on the right-hand side of the GraphiQL page to navigate the GraphQL data model so that you can construct queries.
>
> You may have also noticed the **History** tab. Click **History** to view the history of all of your queries. If you are building or testing, being able to recall history is quite helpful.

GraphQL queries with Python

Continuing to build on the previous queries we tested and built with GraphiQL, we'll now review how to execute those through an HTTP API endpoint.

Python example 1 – devices with name and location for one location

We'll look at the same examples we looked at in the previous section.

The API endpoint we need to use is called `/api/graphql/`. The following is an example of how to execute it in Python:

```
query = """
query {
  devices(location: "AMS01") {
    name
    location {
      name
    }
  }
}
"""

payload = {
  "query": query
}

r = session.post(graphql_url, data=json.dumps(payload))
```

You can see a full script here:

```
graphql.py  ×

chapter-08 >  graphql.py
   1    import requests
   2    import json
   3
   4    payload = {}
   5
   6    headers = {
   7        "Content-Type": "application/json",
   8        "Accept": "application/json",
   9        "Authorization": "Token aaaaaaaaaaaaaaaaaaaaaaaaaaaaaaaaaaaaaaaaaaaaa",
  10    }
  11
  12    session = requests.Session()
  13    session.headers.update(headers)
  14
  15    graphql_url = "https://demo.nautobot.com/api/graphql/"
  16
  17    query = """
  18    query {
  19      devices(location: "AMS01") {
  20        name
  21        location {
  22          name
  23        }
  24      }
  25    }
  26    """
  27
  28    payload = {"query": query}
  29
  30    r = session.post(graphql_url, data=json.dumps(payload))
  31
```

Figure 8.43 – Executing a GraphQL query in Python

Execute this script with the `-i` flag so that you can explore the response data. Here is an example to try out:

```
>>> # print(r.json())
>>>
>>> for device in r.json()['data']['devices']:
...      print(f"Device: {device['name']} in Location
{device['location']['name']}")
...
Device: ams01-dist-01 in Location AMS01
Device: ams01-edge-01 in Location AMS01
# output removed for brevity
```

As you can see, it is quite straightforward to execute queries via API. There's no need to show all examples because they'd all be very similar – the key is to pass the query as a string in the `query` parameter. However, we'll show a few more that'll show off a few other features of GraphQL.

Python example 2 – devices with name and location for a variable location

This example introduces how you can use GraphQL variables. To do this, you need to account for a few things:

- Swap out the data – **AMS01**, in our case – for a variable. That is `$nautobot_location` in this example.

- Declare your variable at the top of the query – for example, `$nautobot_location` – along with its data type – for example, `String` (the auto-complete and data validation feature will help you here).

- Pass your variables into the query using the `variables` key in the JSON payload:

```
query = """
query ($nautobot_location: [String]){
  devices(location: $nautobot_location) {
    name
    location {
      name
    }
  }
}
"""

payload = {
    "query": query,
    "variables": {
```

```
          "nautobot_location": "AMS01"
     }
}

r = session.post(graphql_url, data=json.dumps(payload))
```

Python example 3 – querying for a variable device

When using variables, it's always good to validate in GraphiQL too. Here is one example where the data type is incorrect, and using the automatic schema validation helps us correct the type to ID:

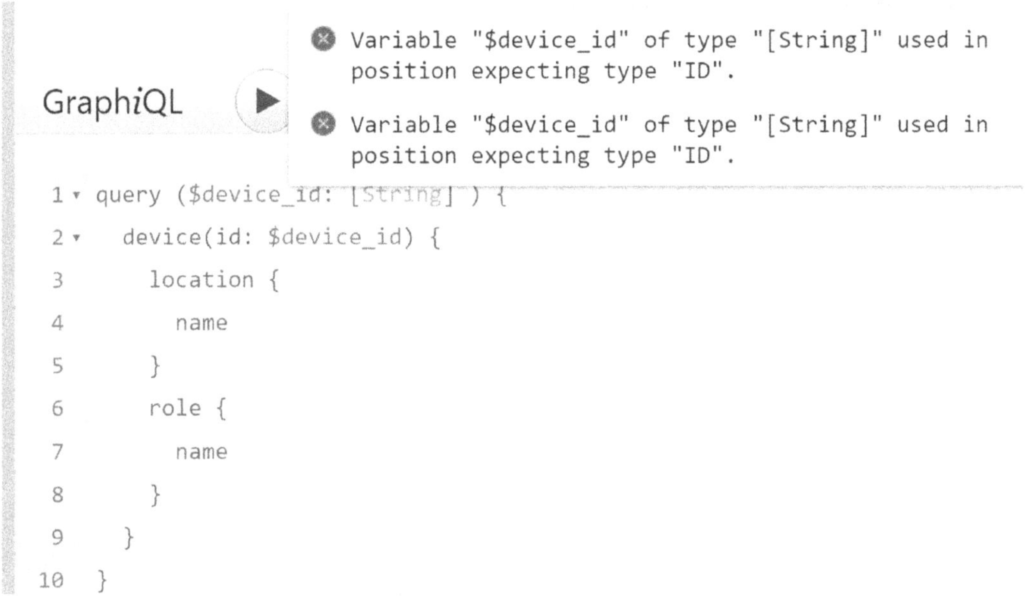

Figure 8.44 – GraphQL schema validation for variables

After fixing it, we'll see the following:

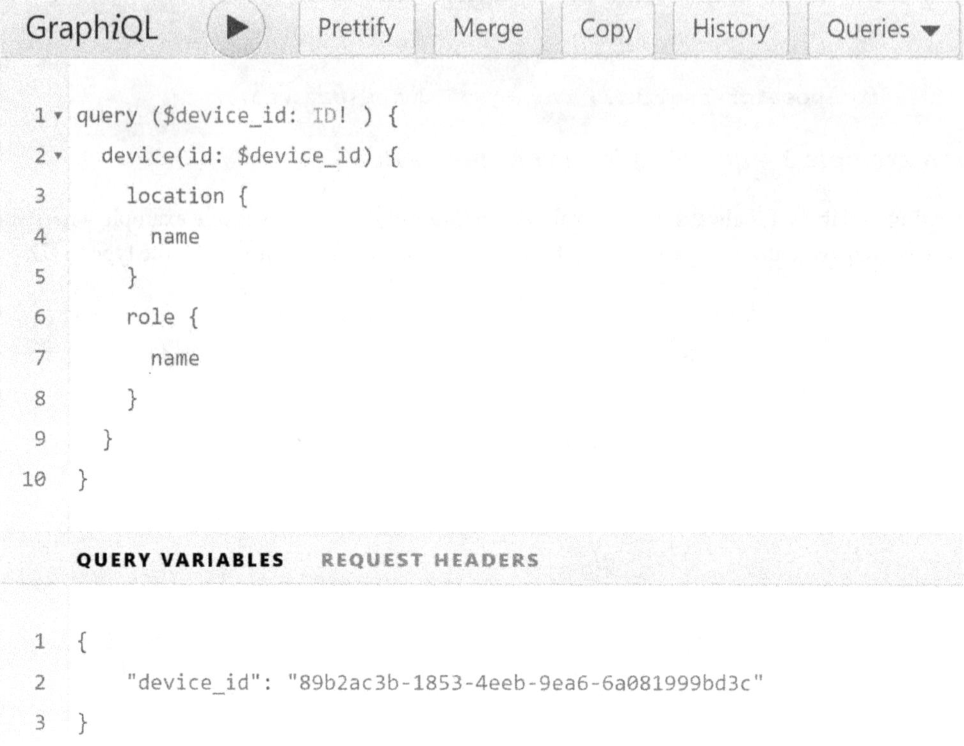

```
GraphiQL    ▶    Prettify    Merge    Copy    History    Queries ▼

1 ▾ query ($device_id: ID! ) {
2 ▾   device(id: $device_id) {
3       location {
4         name
5       }
6       role {
7         name
8       }
9     }
10  }


QUERY VARIABLES    REQUEST HEADERS

1  {
2      "device_id": "89b2ac3b-1853-4eeb-9ea6-6a081999bd3c"
3  }
```

Figure 8.45 – GraphQL schema validation for variables (fixed)

This example also introduced adding an exclamation mark (!) after the type, which means that the variable is required. As you'd expect, this is the code that would be associated with this request.

> **Note**
> If you are following along, be sure to use a valid UUID as the variable data.

Python example 4 – executing a saved GraphQL query

First, create a saved GraphQL query. In the navbar, navigate to **EXTENSIBILITY | GraphQL Queries** and add a new query. Name it **AMS01 Devices** and use the following query (that we've been using):

```
query {
  devices(location: "AMS01") {
    name
    location {
```

```
        name
      }
    }
  }
}
```

Save the query. Then, navigate to the **Advanced** tab of the new query and copy its UUID.

From executing a prior script and entering the Python shell, you can make the following changes:

```
saved_gql_uuid = "<enter the UUID of your new query"
saved_gql_url = f"https://demo.nautobot.com/api/extras/graphql-
queries/{saved_gql_uuid}/run/"
r = session.post(saved_gql_url)
r.json() will have the data you're looking for.
```

You can also pass query variables into the POST request as key-value pairs. If your query had one variable of device_id, your payload would have just one key-value pair (dictionary) of { "device_id": "<device uuid"}.

GraphQL versus REST

It may already be evident the value GraphQL has over REST APIs for querying data. GraphQL allows you to craft the API response you desire for the data you desire. Looking at the response for each of the GraphQL examples, the value is significant and is described here:

- Even if the data you require is in a single REST API GET response, the GraphQL response is cleaner and has only the key-value pairs you asked for.

- In more advanced examples, such as gathering interfaces and IP addresses for devices, interfaces, and IP addresses for their neighbor (or remotely connected), as shown in the GraphiQL examples, that would be numerous GET requests instead of *one* GraphQL request. One is simpler than numerous.

- Building on the previous statement, not only is one request simpler to construct, but the actual data processing (taking responses from one API call, such as the UUID, as an input to the next query) adds more complexity to code that is no longer required, thus simplifying the code even more.

- GraphQL in Nautobot allows you to save GraphQL queries, allowing the same query to be constructed by one person (or team) and allowing others to use it as a trusted source to gather the data they need from Nautobot. Again, this simplifies data consumption for anyone looking to query data from Nautobot.

Webhooks

Webhooks enable you to send outbound HTTP API calls from a system when certain events take place. In this case, we are talking about making API calls from Nautobot when data changes in Nautobot. For example, maybe you want to send a chat notification when there is any change on a particular device; maybe you want to execute an automation workflow on a third-party automation system when a new is device added; maybe you want to remove a device from your monitoring tool when it is deleted from Nautobot. There are numerous reasons you may want to use Webhooks in Nautobot. Let's explore how you'd create one.

Exploring webhooks

Webhooks are managed under **EXTENSIBILITY | Webhooks** in the sidebar navigation.

The next step is to click the **+Add** button in the top right to start creating your Webhook:

Figure 8.46 – Webhooks

When creating the Webhook, you'll notice the following fields:

- **Name**: The arbitrary name you want to give the Webhook.

- **Content Type(s)**: These are the models you can choose from so that when an event occurs, the Webhook will be triggered.

- **Enabled**: You can enable or disable the Webhook. If the box is checked, the Webhook is enabled and active.

- **Type create / update / delete**: There is flexibility for when the Webhook is triggered – for example, if the object is new and just created, if an object is just being updated, or if the object is just being deleted.

- **URL**: The target URL that the Webhook request will be sent to.

- **HTTP method**: The method of the request that'll be sent to the previously defined URL.

- **HTTP content type**: This maps to the Content-Type HTTP header and defaults to `application/json`.

- **Additional headers**: This gives you the flexibility to add more headers that your target system may require. One example is **Authorization** if the target system requires authentication. These headers should be added one per line in key-value format (*note: this is not YAML*).

- **Body template**: If you are sending a POST request to a third-party system, you often need to customize the JSON payload, eliminating the need for third-party middleware to manage Webhooks. This is a Jinja template, and you have access to six built-in Jinja variables that Nautobot exposes for you: `event`, `model`, `timestamp`, `username`, `request_id`, and `data`. It is worth noting that **Additional headers** also support Jinja templating and have access to the same variables, as in **Body template**.

- **Secret**: This adds the **X-Hook-Signature** header containing an HMAC hex digest of the payload body using the secret as the key. Note that the secret is not transmitted in the request.

- **SSL verification**: This allows you to enable/disable SSL verification.

- **CA File Path**: The CA certificate file to use for SSL verification. If no path is defined, system defaults are used.

Example – using a Webhook to trigger an Ansible AWX playbook

We are going to showcase an example of creating a Webhook in Nautobot that will trigger an Ansible playbook in AWX whenever there is a change on a device in Nautobot.

The following inputs were used to create the Webhook:

- **Name**: `Trigger Playbook on Device Change`

- **Content Type(s)**: `dcim | device`

- **Enabled**, **Type create**, **Type update**, and **Type delete** should all be checked

- The URL is the full API endpoint needed to execute (or launch) a playbook – for example, `/api/v2/job_templates/$ID/launch/`

  In this example, an ID of 7 was used. This ID *will* be different for your job template. Browse your job template in AWX; you will easily see the ID in your URL.

- **HTTP Method**: POST

- **HTTP content type**: `applicatio/json`

- **Additional headers**: `Authorization: Bearer $your_token:`

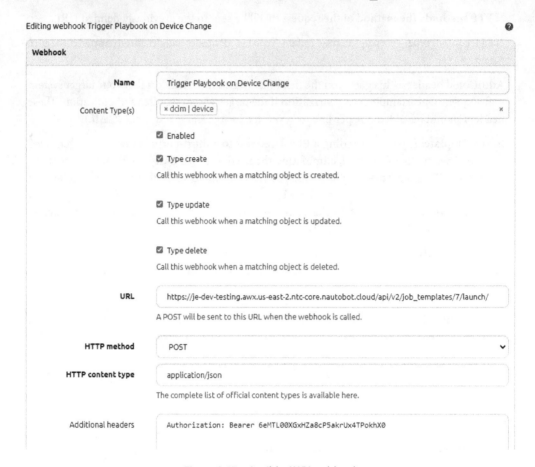

Figure 8.47 – Ansible AWX webhook

Let's see what will be done via Ansible AWX to complete this process.

The following screenshot provides a snapshot of creating the token that will be used for testing. Navigate to your user profile and go to **User Details | Tokens | Add** to add a new Token with a **Write** scope. Copy that token; it will be used as an additional header:

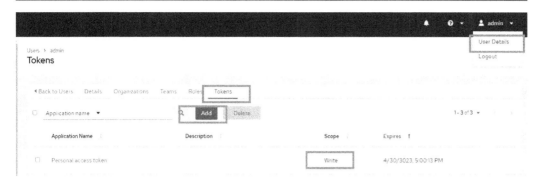

Figure 8.48 – Ansible AWX token creation

Let's continue looking at the rest of the form inputs:

Body template: If you have ever explored the Ansible API, you will see that this is a straightforward JSON payload that only uses the `extra_vars` key. However, it is important to note that we are accessing three of the built-in variables you have access to: `data`, `username`, and `event`. Since they are Jinja variables, you need to use curly brackets.

The rest of the inputs are set to their defaults:

Body template	
| | ```
{
    "extra_vars": {
        "network_device": "{{ data['name'] }}",
        "nautobot_user": "{{ username }}",
        "event": "{{ event }}"
    }
}
``` |

Jinja2 template for a custom request body. If blank, a JSON object representing the change will be included. Available context data includes: `event`, `model`, `timestamp`, `username`, `request_id`, and `data`.

Secret	
	Secret

When provided, the request will include a 'X-Hook-Signature' header containing a HMAC hex digest of the payload body using the secret as the key. The secret is not transmitted in the request.

☑ SSL verification

Enable SSL certificate verification. Disable with caution!

CA File Path	
	CA File Path

The specific CA certificate file to use for SSL verification. Leave blank to use the system defaults.

Figure 8.49 – Ansible AWX Webhook (continued)
```

After saving the Webhook, all you need to do is go to any device in Nautobot and make a change. After that, you should see your playbook execute. After seeing the playbook, you'll see the extra variables rendered as such in AWX:

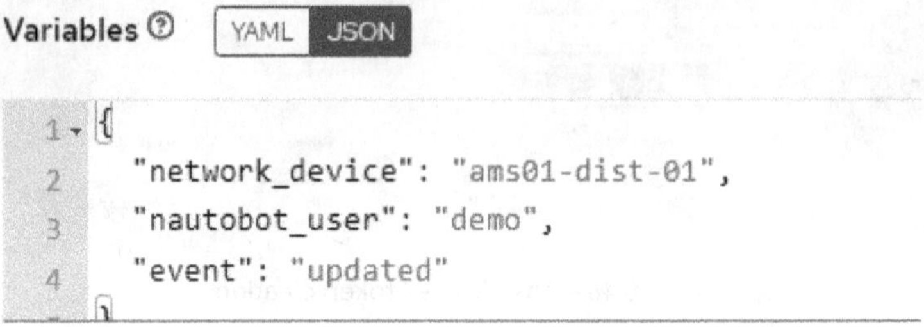

Variables ⓘ   YAML   JSON

```
1 ▾ {
2      "network_device": "ams01-dist-01",
3      "nautobot_user": "demo",
4      "event": "updated"
  }
```

Figure 8.50 – Viewing extra variables in AWX

> **Note**
>
> Be sure to check **Prompt on launch** in the **Variables** section of your AWX job template to allow extra variables to be passed in.

Webhooks are executed as Nautobot jobs (covered in *Chapter 11*) and as such, the status of Webhooks is shown in **Job Results**. Navigate to **Job Results** in the side navigation bar under **JOBS | Job Results**:

Figure 8.51 – Navigating to see the Webhook's results

After clicking **Job Results**, you'll get to see the results of all jobs that include Webhooks:

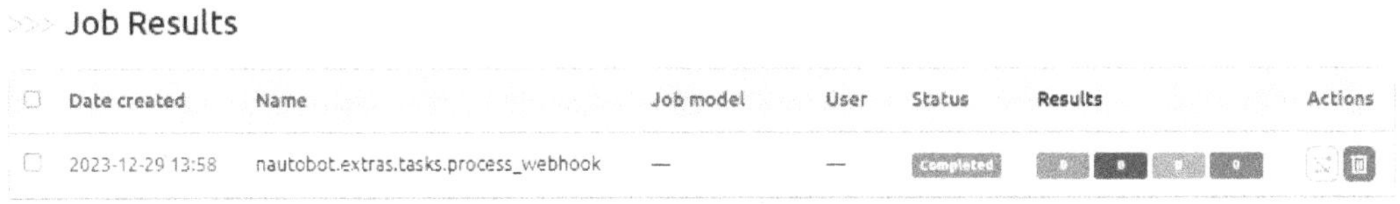

Figure 8.52 – Nautobot Job Results table view

Click the link under **Date created** to see its status:

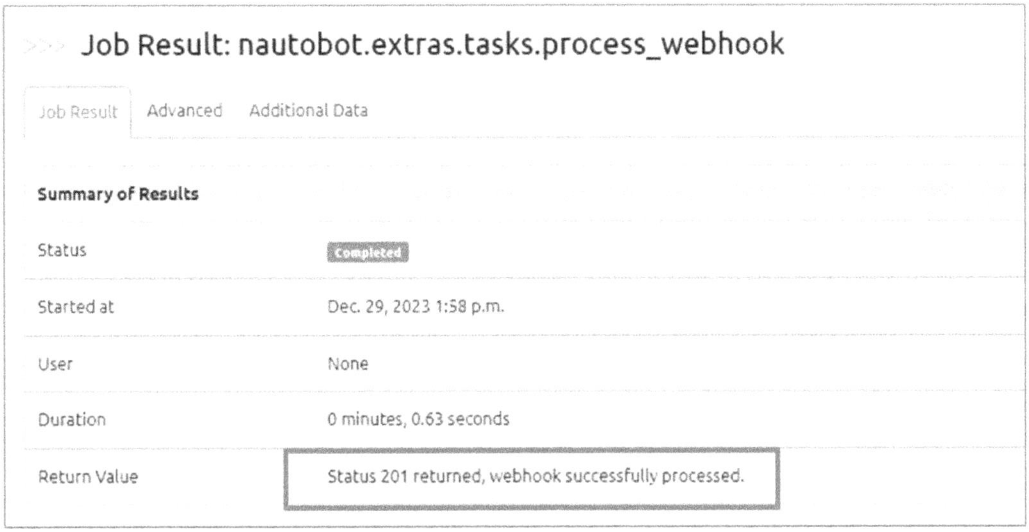

Figure 8.53 – Viewing a Webhook's status

Take note of the return value. If there are errors in the Webhook request, such as typos, invalid authentication, and so on, this is where you'll see them. As an example, if you have a bad token, you'll see an HTTP 401 error stating invalid authentication. So, remember to check **Job Results** when you're testing Webhooks!

> **Note**
>
> In *Chapter 11*, we'll also cover Job Hooks, so they're worth mentioning here. Job Hooks are similar to Webhooks, but instead of triggering a Webhook based on an event (change in data in Nautobot), you can trigger a Nautobot job. Given jobs haven't been covered so far; just know that a Nautobot job is arbitrary Python code to perform some operation on data in Nautobot or any network automation task. Thus, through the use of Job Hooks, you can perform nearly any operation you need within a job when there is a data change in Nautobot. Both jobs and Job Hooks will be covered in *Chapter 11*.

Summary

This chapter provided you with an overview of Nautobot APIs, including its REST APIs, GraphQL, and Webhooks. You saw the power of GraphQL over REST APIs in being able to ask Nautobot for the data you want. You got to work with both APIs in Python, but also saw how to use GraphiQL, a program that's built into Nautobot that accelerates building and testing GraphQL queries. You also saw how Jinja templates embedded with Webhooks can eliminate middleware, which is often needed to transform data into the target destination's format.

In the next chapter, we will explore Nautobot integrations within the NetDevOps ecosystem with a focus on pynautobot and the Nautobot Ansible collection.

9

Understanding Nautobot Integrations for NetDevOps Pipelines

By now, you can probably see that Nautobot as a system is a great platform to have in your arsenal for network automation. You've learned about the data model in Nautobot and installed it, learned about the power of Nautobot's extensibility features, and even learned about its APIs. Going one step further, what integrations does Nautobot have? How do you interact with the data, populate data, and move even quicker once you have Nautobot up and running? There are several tools and integrations available to help along the way.

Nautobot has integrations with Python, Ansible, GoLang, Docker, Kubernetes, and more. We'll be providing a quickstart guide for each integration, ensuring that you know what they are and how they can be used. In this chapter, we will cover the following topics:

- Exploring pynautobot
- Understanding the Nautobot Ansible Collection
- Using Nornir Nautobot
- Exploring Nautobot Docker containers
- Exploring the Nautobot Go library
- Introducing the Nautobot Terraform provider

Technical requirements

Code files and other resources for this chapter are available at: `https://github.com/PacktPublishing/Network-Automation-with-Nautobot/tree/main/chapter-09`

Exploring pynautobot

pynautobot is the recommended method for interacting with Nautobot data when using Python. It is a Python **Software Development Kit** (**SDK**) library for providing object-oriented access to the Nautobot REST API and the capability to make queries against the GraphQL API endpoint discussed in *Chapter 8*. pynautobot is a wrapper on the Python Requests library (`https://requests.readthedocs.io/en/latest/`), which is one of the leading Python libraries for interacting with REST-based APIs. This is the fundamental building block for many other Python applications and may help you with your Nautobot-based pipelines.

If you are looking to build Python scripts or applications working with Nautobot's API, you should be looking to use pynautobot to handle interactions with Nautobot. It will greatly accelerate development—it has predefined patterns for communicating with every API, and it also includes built-in error handling.

Installing pynautobot

The installation of `pynautobot`, like many Python applications, is done via Python PyPI. To complete the installation, you can leverage Python's `pip` or any other Python package managers that you may be using that can integrate with PyPI:

```
pip install pynautobot
```

When running the command, you will see additional Python packages installed as part of the requirements.

For these examples, we'll be using a DigitalOcean Ubuntu server as we have in prior chapters.

```
● root@nautobot-dev:~# pip install pynautobot
  Collecting pynautobot
    Downloading pynautobot-2.0.2-py3-none-any.whl (34 kB)
  Collecting requests<3.0.0,>=2.30.0
    Downloading requests-2.31.0-py3-none-any.whl (62 kB)
                                    ━━━━━━ 62.6/62.6 KB 2.5 MB/s eta 0:00:00
  Collecting packaging<24.0,>=23.2
    Downloading packaging-23.2-py3-none-any.whl (53 kB)
                                    ━━━━━━ 53.0/53.0 KB 6.6 MB/s eta 0:00:00
  Requirement already satisfied: urllib3<1.27,>=1.21.1 in /usr/lib/python3/dist-packages (from pynautobot) (1.26.5)
  Collecting charset-normalizer<4,>=2
    Downloading charset_normalizer-3.3.2-cp310-cp310-manylinux_2_17_x86_64.manylinux2014_x86_64.whl (142 kB)
                                    ━━━━━━ 142.1/142.1 KB 16.2 MB/s eta 0:00:00
  Requirement already satisfied: certifi>=2017.4.17 in /usr/lib/python3/dist-packages (from requests<3.0.0,>=2.30.0->pynautobot) (2020.6.20)
  Requirement already satisfied: idna<4,>=2.5 in /usr/lib/python3/dist-packages (from requests<3.0.0,>=2.30.0->pynautobot) (3.3)
  Installing collected packages: packaging, charset-normalizer, requests, pynautobot
    Attempting uninstall: requests
      Found existing installation: requests 2.25.1
      Not uninstalling requests at /usr/lib/python3/dist-packages, outside environment /usr
      Can't uninstall 'requests'. No files were found to uninstall.
  Successfully installed charset-normalizer-3.3.2 packaging-23.2 pynautobot-2.0.2 requests-2.31.0
  WARNING: Running pip as the 'root' user can result in broken permissions and conflicting behaviour with the system package manager. It is recommended to use
  a virtual environment instead: https://pip.pypa.io/warnings/venv
○ root@nautobot-dev:~# []
```

Figure 9.1 – The pynautobot installation

To verify the installation is complete, you can test importing `pynautobot` from the Python interpreter:

```
root@nautobot-dev:~# python3
Python 3.10.12 (main, Nov 20 2023, 15:14:05) [GCC 11.4.0] on linux
Type "help", "copyright", "credits" or "license" for more information.
>>>
>>> import pynautobot
>>>
```

After we have verified that we can import `pynautobot`, it is time to get started using `pynautobot`!

Getting started

The first part of using the `pynautobot` library is instantiating a new instance of the client that represents a single instance of Nautobot. The minimum requirement for the SDK is to define the URL of the instance that you will be connecting to. Usually, you will also need to have a token created to read/write data to Nautobot. The next most common parameter that is going to be set in the `pynautobot` client object is `api_version`. This is the version of the API that should be leveraged within Nautobot. Nautobot 1.3 introduced the concept of versioned APIs for Jobs, Tags, and IP address objects.

> **Note**
>
> In Nautobot 2.0, the default is to be the latest version of the API, which may bring changes over time.

The versioned API is one more example of the commitment to the stability of the Nautobot project as a whole. This will allow Nautobot to be able to update the REST APIs without needing to have a full major version upgrade. This will also allow for your confidence in the deployment of Nautobot that applications/scripts that are written will continue to function as expected as Nautobot upgrades occur.

> **Tip**
>
> Nautobot follows **Semantic Versioning (SemVer)** ensuring that breaking changes only occur between major version changes, such as v1 to v2 to v3. This means any change in minor or patch version such as from 2.1.2 to 2.1.3 or from 2.1.2 to 2.2.0 will never have breaking changes. Using API versioning allows the API to evolve while providing backward compatibility between major and minor versions. Setting the API version also allows your automated tests to easily run against different API versions.

To instantiate an instance of `pynautobot`, see the following example at the Python interpreter:

```
>>> import pynautobot
>>> nautobot = pynautobot.api(url="https://demo.nautobot.com",
token="aaaaaaaaaaaaaaaaaaaaaaaaaaaaaaaaaaaaaaaa", api_version="2.1")
>>>
```

At this point, we are ready to start interacting with Nautobot data from `pynautobot`.

Retrieving objects

At this point, you created an instance of `nautobot` that has the following objects:

```
>>> dir(nautobot)
[...internal objects removed for brevity…,'_validate_version',
'api_version', 'base_url', 'circuits', 'dcim', 'extras', 'graphql',
'headers', 'http_session', 'ipam', 'max_workers', 'openapi',
'plugins', 'status', 'tenancy', 'threading', 'token', 'users',
'version', 'virtualization']
>>>
```

The `nautobot` object simplifies access to the primary data models that are already present in Nautobot (e.g., devices, circuits, etc.). `pynautobot` will query Nautobot's REST API endpoint in order to construct and execute API calls, so the need to know the exact API endpoint (as you needed in *Chapter 8* when using requests) goes away. Instead, you need to know the object that is essentially the wrapper for a group of objects or API calls. The best way to learn about the objects for `pynautobot` is to continue to understand the hierarchy of API calls when browsing the interactive API documentation that is available through the web UI at `/api/`. For example, you can see from the output of `dir(nautobot)` that there are objects such as circuits, DCIM, extras, virtualization, and so on. Each of these has a corresponding API route that you saw in *Chapter 8* when browsing the interactive API documentation.

Let's continue down the device side of the API within DCIM or `/api/dcim`. In the case of `pynautobot`, that'll be `nautobot.dcim`.

Going to the URL of the public demo instance of Nautobot (`demo.nautobot.com/api/dcim`) using a web browser, you get the following response:

Figure 9.2 – Browsing the DCIM API at /api/

Note that in URLs/APIs, there are a number of URLs that have a – character (hyphen) in them. Since Python objects cannot have hyphens, `pynautobot` replaces all hyphens with underscores (_).

Let's now explore the objects available when executing `dir(nautobot.dcim)` at the Python interpreter:

```
>>> dir(nautobot.dcim)
[...internal objects removed for brevity…', '_choices', '_setmodel',
'api', 'choices', 'config', 'get_custom_field_choices', 'get_custom_
fields', 'model', 'models', 'name']
>>>
```

So, after knowing that hyphens are replaced with underscores, let's look at a quick example of how we'd fetch `/api/dcim/location-types/` from `pynautobot`.

Evaluating `nautobot.dcim.location_types` at the Python interpreter yields the following:

```
>>> nautobot.dcim.location_types
<pynautobot.core.endpoint.Endpoint object at 0x7f4ebd3c0eb0>
>>>
```

As you can see, an `endpoint` object is returned. This is the same type of object that'll be returned when making API calls to fetch different models with `pynautobot`. This `endpoint` object then has several additional attributes and methods available:

```
>>> dir(nautobot.dcim.device_types)
[...internal objects removed for brevity..., 'all', 'api', 'base_url',
'choices', 'count', 'create', 'filter', 'get', 'name', 'return_obj',
'token', 'update', 'url']
>>>
```

As we dive deeper, our focus is going to be on four of the methods that will help us in making HTTP GET requests to Nautobot:

- `all`
- `get`
- `count`
- `filter`

These methods are what will be used to actually see real data from Nautobot.

The all method

The `all` method will retrieve all of the results from Nautobot of a given object type. In this example, you will get all of the location types that are in Nautobot:

```
>>> nautobot.dcim.location_types.all()
[<pynautobot.core.response.Record ('Region') at 0x7f4ebc9fca60>,
<pynautobot.core.response.Record ('Region → Site') at 0x7f4ebc9fc820>,
<pynautobot.core.response.Record ('Region → Site → Data Center Pod')
at 0x7f4ebc9fcc40>]
>>>
```

To get the name of the first location type, you can make the following query:

```
>>> nautobot.dcim.location_types.all()[1].name
'Site'
>>>
```

You can follow the same process for any object within Nautobot. We'll explore more examples in an upcoming section.

The get method

The get method is a method that returns a single value (or record). If you are looking to retrieve data on a single object, you should use get () for the same reason you'd use the /dcim/devices/{id}/ API and not /dcim/devices/.

Let's explore the Region location type that we just saw returned as one of the three location types:

```
>>> nautobot.dcim.location_types.get(name="Site")
<pynautobot.core.response.Record ('Region → Site') at 0x7f3befec5270>
>>>
```

We can see that an object of the Record type is returned. Let's save that into its own variable:

```
>>> location_type = nautobot.dcim.location_types.get(name="Site")
>>>
```

Now, let's explore some of the attributes returned previously:

```
>>> dir(location_type)
[...internal objects removed for brevity..., 'api', 'content_types',
'created', 'custom_fields', 'default_ret', 'delete', 'description',
'display', 'endpoint', 'full_details', 'has_details', 'id', 'last_
updated', 'name', 'natural_slug', 'nestable', 'notes_url', 'object_
type', 'parent', 'save', 'serialize', 'tree_depth', 'update', 'url']
>>>
>>> location_type.description
''  # you can set a description in the UI and see this change!
>>>
>>> location_type.nestable
False
>>>
>>> location_type.parent
<pynautobot.core.response.Record ('Region') at 0x7f3bef482fb0>
>>>
>>> location_type.content_types
['dcim.device', 'dcim.rack', 'ipam.prefix', 'ipam.vlan', 'circuits.
circuittermination', 'dcim.powerpanel', 'dcim.rackgroup', 'ipam.
vlangroup', 'virtualization.cluster']
>>>
```

These are the same attributes you'd configure when adding or viewing a new location type in the Nautobot UI.

All and Get examples

Let's walk through a few examples showing the pattern of retrieving objects of a few different model types.

Here are a few that fall under /dcim/.

This example retrieves all platforms:

```
>>> nautobot.dcim.platforms.all()
[<pynautobot.core.response.Record ('Arista EOS') at 0x7f4ebc1084c0>,
<pynautobot.core.response.Record ('Cisco ASA') at 0x7f4ebc1089a0>,
<pynautobot.core.response.Record ('Cisco IOS') at 0x7f4ebc108490>,
#omitted for brevity<pynautobot.core.response.Record ('Juniper Junos')
at 0x7f4ebc108880>]
>>>
```

This one retrieves only a single platform:

```
>>> nautobot.dcim.platforms.get(name="Arista EOS")
<pynautobot.core.response.Record ('Arista EOS') at 0x7f4ebc7cc5e0>
>>>
>>> nautobot.dcim.platforms.get(name="Arista EOS").name
'Arista EOS'
>>>
```

We can use the dir() function to understand the attributes of a given object:

```
>>> dir(nautobot.dcim.platforms.get(name="Arista EOS"))
# output omitted for brevity
'api', 'created', 'custom_fields', 'default_ret', 'delete',
'description', 'device_count', 'display', 'endpoint', 'full_details',
'has_details', 'id', 'last_updated', 'manufacturer', 'name', 'napalm_
args', 'napalm_driver', 'natural_slug', 'network_driver', 'network_
driver_mappings', 'notes_url', 'object_type', 'save', 'serialize',
'update', 'url', 'virtual_machine_count']
>>>
>>>
```

In this example, we are viewing the network driver for a specific platform:

```
>>> nautobot.dcim.platforms.get(name="Arista EOS").network_driver
'arista_eos'
>>>
```

Here, we are viewing the manufacturer of a specific platform:

```
>>> nautobot.dcim.platforms.get(name="Arista EOS").manufacturer
<pynautobot.core.response.Record ('Arista') at 0x7f4ebc7cc910>
```

```
>>>
>>> nautobot.dcim.platforms.get(name="Arista EOS").manufacturer.name
'Arista'
>>>
```

This example retrieves all device types:

```
>>> nautobot.dcim.device_types.all()
[<pynautobot.models.dcim.DeviceTypes ('APDU9941') at 0x7f4ebc108430>,
<pynautobot.models.dcim.DeviceTypes ('DCS-7048-T') at 0x7f4ebc108dc0>,
<pynautobot.models.dcim.DeviceTypes ('DCS-7150S-24') at
0x7f4ebc108400>, #omitted for brevity
<pynautobot.models.dcim.DeviceTypes ('CSR1000V') at 0x7f4ebc108b20>,
<pynautobot.models.dcim.DeviceTypes ('Nexus 9372TX') at
0x7f4ebc109210>, <pynautobot.models.dcim.DeviceTypes ('Nexus 9Kv')
at 0x7f4ebc109150>, <pynautobot.models.dcim.DeviceTypes ('vMX') at
0x7f4ebc109090>]
>>>
```

This one retrieves all power feeds and then views a single record:

```
>>> nautobot.dcim.power_feeds.all()[0]
<pynautobot.core.response.Record ('ams01-101 Primary Power Feed') at
0x7f4ebc794190>
>>>
```

Here are a few examples that fall under /circuits/:

```
>>> nautobot.circuits.circuit_types.all()
[<pynautobot.core.response.Record ('Dark Fiber') at 0x7f4ebc7cc130>,
<pynautobot.core.response.Record ('IX') at 0x7f4ebc7cc250>,
<pynautobot.core.response.Record ('MPLS') at 0x7f4ebc7cc790>,
<pynautobot.core.response.Record ('Private Peering') at
0x7f4ebc7cc040>, <pynautobot.core.response.Record ('Transit') at
0x7f4ebc7cc070>]
>>>
>>> nautobot.circuits.circuits.all()[0]
<pynautobot.models.circuits.Circuits ('ntt-104265404093023273') at
0x7f4ebc7cc3d0>
>>>
>>> nautobot.circuits.providers.all()[0]
<pynautobot.core.response.Record ('AT&T') at 0x7f4ebc7cc040>
>>>
>>> nautobot.circuits.providers.all()[0].name
'AT&T'
```

The preceding examples show examples of the common patterns of using `all()` and `get()`. Remember to check the API docs to better understand which objects are accessible within each parent, such as DCIM and circuits, while also using the `help()` and `dir()` functions to browse them while in the Python shell.

The count method

The number of devices in the demo instance shows **921** when accessing the `count` method.

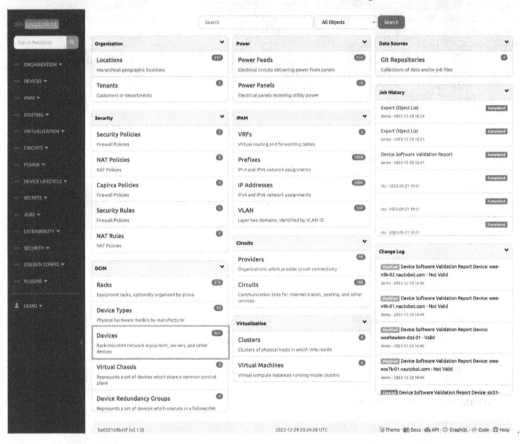

Figure 9.3 – Nautobot home screen with device count highlighted

Let's take a look at the `count` method. It returns the number of devices rapidly, rather than getting all of the devices back and then counting the length of the list:

```
>>> nautobot.dcim.devices.count()
921
>>>
```

> **Note**
>
> Because this is a public demo instance, when you actually test, the number of devices may not be 921, but what you see in the UI should definitely match the output from the `count()` method.

Using `count` is a nice shortcut—better than retrieving all objects (which is more time-consuming) and then doing a `len()` on the response.

The filter method

The `filter` method is similar to the `get` method; however, it returns a list of records instead. This method has the ability to return multiple responses. In order to showcase this, we'll query for devices at the `AMS01` location.

Rather than loop through all objects using the `all()` method, it's cleaner to filter just for the objects required for your query. It'll be a wise choice to leverage the `filter` method to allow Nautobot to handle the filtering rather than having to work through all of the data returned.

Let's retrieve all devices at the `AMS01` location:

```
>>> nautobot.dcim.devices.filter(location="AMS01")
[<pynautobot.models.dcim.Devices ('ams01-dist-01') at 0xffff90755420>,
<pynautobot.models.dcim.Devices ('ams01-edge-01') at 0xffff90755c90>,
<pynautobot.models.dcim.Devices ('ams01-edge-02') at 0xffff90757af0>,
…objects removed for brevity…
<pynautobot.models.dcim.Devices ('ams01-pdu-16') at 0xffff8f405c90>,
<pynautobot.models.dcim.Devices ('ams01-pdu-17') at 0xffff8f406260>,
<pynautobot.models.dcim.Devices ('ams01-pdu-18') at 0xffff8f4067a0>]
>>>
```

It is also possible to filter for a list of locations. Let's query for two locations, `AMS01` and `BRE01`:

```
>>> nautobot.dcim.devices.filter(location=["AMS01", "BRE01"])
[<pynautobot.models.dcim.Devices ('ams01-dist-01') at 0x7f1dcb2f00a0>,
<pynautobot.models.dcim.Devices ('ams01-edge-01') at 0x7f1dcac58070>,
<pynautobot.models.dcim.Devices ('ams01-edge-02') at 0x7f1dcac58130>,
<pynautobot.models.dcim.Devices ('ams01-leaf-01') at 0x7f1dcac582e0>,
<pynautobot.models.dcim.Devices ('ams01-leaf-02') at 0x7f1dcac58670>,
…objects removed for brevity…
```

```
<pynautobot.models.dcim.Devices ('bre01-edge-02') at 0x7f1dca8b9870>,
<pynautobot.models.dcim.Devices ('bre01-leaf-01') at 0x7f1dca8b9ae0>,
<pynautobot.models.dcim.Devices ('bre01-leaf-02') at 0x7f1dca8b9bd0>,
<pynautobot.models.dcim.Devices ('bre01-leaf-03') at 0x7f1dca8ba380>,
<pynautobot.models.dcim.Devices ('bre01-leaf-04') at 0x7f1dca8ba6b0>]
>>>
```

You can also filter on multiple parameters, such as `location` and `role`:

```
>>> nautobot.dcim.devices.filter(location=["AMS01", "BRE01"],
role="leaf")
[<pynautobot.models.dcim.Devices ('ams01-leaf-01') at 0xffff906011e0>,
<pynautobot.models.dcim.Devices ('ams01-leaf-02') at 0xffff90602110>,
<pynautobot.models.dcim.Devices ('ams01-leaf-03') at 0xffff906022c0>,
# omitted for brevity
 <pynautobot.models.dcim.Devices ('bre01-leaf-03') at 0xffff90602ad0>,
<pynautobot.models.dcim.Devices ('bre01-leaf-04') at 0xffff90600f10>]
>>>
```

While the prior example passed in a list of locations, you can use sets, lists, or tuples to search for multiple values.

Filter options

It is also possible to filter beyond just retrieving the list of a given object type. It is possible to have a more narrowly scoped filter, which greatly simplifies the code needed if your focus is always on a subset of devices.

One way to understand which filters are available for objects is to use the API docs again.

Let's navigate to `/api/docs`.

To look at the filters available for devices, scroll down until you get into the `dcim` header. You can click on each of the header items to collapse/expand the selection.

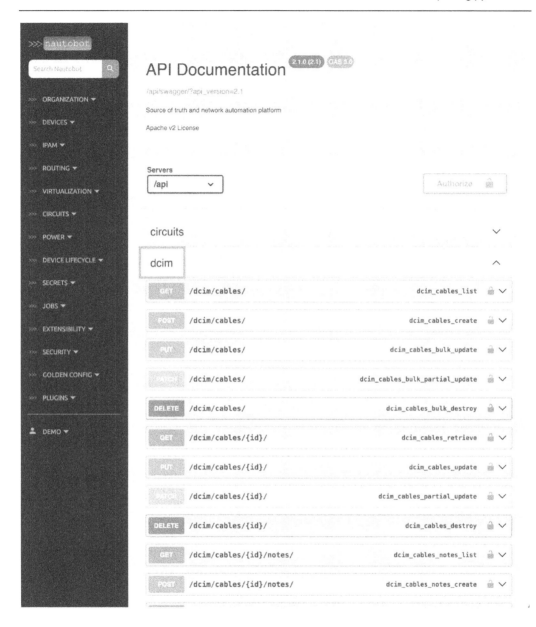

Figure 9.4 – Interactive API docs for DCIM

From this section, you will find all of the particular API-specific details. As you expand the GET request for /dcim/devices/, you will see the filtering options:

Figure 9.5 – Viewing API filters

This is just an example of a few of them. There is a long list available for most object types. These filtering parameters can also be used by pynautobot.

The following table defines what the various filters (using abbreviations) are, as the preceding figure was just an abbreviated list:

Filtering Abbreviation	Field Type	Definition
ic	string	Case-insensitive contains
ie	string	Case-insensitive exact match
iew	string	Case-insensitive ends with
ire	string	Case-insensitive regular expression match
isw	string	Case-insensitive starts with
n	string/numeric	Not equal to (negation)
nic	string	Negated case-insensitive contains
nie	string	Negated case-insensitive exact match
niew	string	Negated case-insensitive ends with
nire	string	Negated case-insensitive regular expression match
nisw	string	Negated case-insensitive starts with
nre	string	Negated case-sensitive regular expression match
re	string	Case-sensitive regular expression match
gt	numeric	Greater than
gte	numeric	Greater than or equal to
lt	numeric	Less than
lte	numeric	Less than or equal to

Table 9.1 – Table of filtering attributes

Let's look at using some of the filters shown.

This filter retrieves all locations that are not the AMS01 location:

```
>>> nautobot.dcim.devices.filter(location__n="AMS01")
>>> # output omitted for brevity
```

This one retrieves all locations that start with a capital A and have some alphanumeric characters after using a RegEx pattern:

```
>>> nautobot.dcim.locations.filter(name__re="A\w+")
>>> # output omitted for brevity
```

Use this one to retrieve devices that have bb in their name (referring to any backbone device):

```
>>> nautobot.dcim.devices.filter(name__re="\S+bb\S+")
>>> # output omitted for brevity
```

This one retrieves manufacturers that are not Cisco:

```
>>> nautobot.dcim.manufacturers.filter(name__n="Cisco")
>>> # output omitted for brevity
```

Keep in mind that if you want to test specific filters, you can use the API docs as well.

Updating an object

Initially, once an object is gathered into the memory of your Python application, it is just that—running in memory on your system. In order to make changes to the record, you need to complete two steps:

1. Assign the new value to the record
2. Use the save method that will save it (or have it persist in Nautobot)

Let's take a look at updating by adding some fields such as **Time Zone** and **Description** for the AMS01 location in the demo Nautobot instance:

```
>>> ams = nautobot.dcim.locations.get(name="AMS01")
>>>
>>> ams.time_zone
>>>
>>> ams.description
''
>>>
```

At this point, the AMS01 time zone and description are empty (or not set). This is also confirmed in the UI of Nautobot:

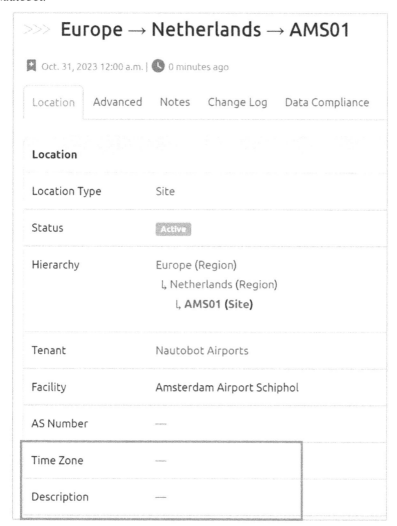

Figure 9.6 – Verifying that Time Zone and Description are not set

Now, let's update the location on the Python object with some values:

```
>>> ams.time_zone = "Europe/Amsterdam"
>>>
>>> ams.description = "Primary site in Europe"
>>>
```

Refreshing the Nautobot UI, you see that this has not changed yet.

The missing piece is to issue the `save` method. This will actually update (or patch) the API so data is reflected in Nautobot:

```
>>> ams.save()
True
>>>
```

Once the object is saved, you can check back in the UI to verify that the object was updated.

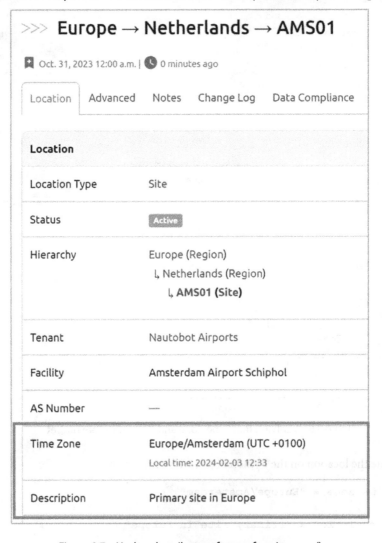

Figure 9.7 – Updated attributes after performing save()

> **Important**
>
> As you are working with `pynautobot` objects, remember to call the `save()` method on them to have Nautobot updated.

Deleting an object

In order to delete an object, first retrieve the object and then use the `delete` method on it:

```
>>> device = nautobot.dcim.devices.get(name="ams01-pdu-01")
>>>
>>> device.delete()
True
>>>
```

If everything works as expected, you'll receive `True` as a response.

If there are dependent objects, you will receive an error message like this:

```
>>> device = nautobot.dcim.devices.get(name="ams01-leaf-01")
>>>
>>> device.delete()
Traceback (most recent call last):
…omitted for brevity…
pynautobot.core.query.RequestError: The request failed with code 409
Conflict: {'detail': 'Unable to delete object. 1 dependent objects
were found: ams01-leaf-01 - AS 4200000000 (84ab4665-f670-4cfc-888b-
fb572810d8ac)'}
>>>
```

In this case, you'd need to first remove the objects depending on the object you're trying to delete.

Creating an object

In order to create an object, you must pass in all required fields just like in the UI or using the native REST API. This is done in pynautobot using the `create()` method on a specific list (or endpoint) object:

```
>>> locations = nautobot.dcim.locations
>>>
>>> type(locations)
<class 'pynautobot.core.endpoint.Endpoint'>
>>>
>>> new_location = locations.create(name="001-new-location", location_
type="Region", status="Active")
>>>
```

You can then log in to the UI and see the new site just created.

>>> Locations

☐	Name	Status	Parent	Tenant
☐	001-new-location	Active	—	—

Figure 9.8 – Viewing the new location created from pynautobot

Note that you could optionally build a dictionary of keyword arguments and pass that into the `create` method using `**kwargs` (or whatever variable you choose to use).

Working with Nautobot Apps

When you take a look at the URLs that are generated by an installed Nautobot App, you'll notice that the URL has `/plugins/` in it. This is due to the original name of Apps being plugins. In the demo instance of Nautobot, there are several Apps already installed. We'll access data from two of them, the BGP and Firewall model Apps.

You'll notice the pattern is identical to access core native models with the exception of needing the word `plugin` in the object path.

It is also worth noting that many of the plugin URLs are singular and not plural (in contrast to core models). For example, the URL for the **Firewall Address Object** page is `/plugins/firewall/address-object/`.

Take note of this in the UI URLs or API docs to construct the right object using pynautobot.

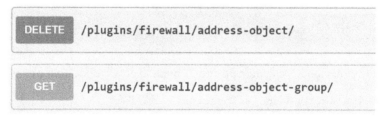

```
DELETE  /plugins/firewall/address-object/
```

```
GET  /plugins/firewall/address-object-group/
```

Figure 9.9 – Viewing the singular URLs of the Firewall App

Let's explore a few examples.

Retrieve the `address` objects from the Nautobot Firewall App like this:

```
>>> nautobot.plugins.firewall.address_object.all()
...output omitted for brevity...
>>>
>>> nautobot.plugins.firewall.address_object.all()[0].name
'Nat-destination'
>>>
```

Retrieve the firewall policies from the Nautobot Firewall App like this:

```
>>> nautobot.plugins.firewall.policy.all()
>>>
```

Retrieve ASNs from the Nautobot BGP App like this:

```
>>> nautobot.plugins.bgp.autonomous_systems.all()
>>> ...output omitted for brevity...
>>> nautobot.plugins.nautobot_device_lifecycle_mgmt.cve.all()
[<pynautobot.core.response.Record ('CVE-2020-3508') at
0x7f5bda540f40>]
>>>
```

Retrieve compliance rules in the Nautobot Golden Config App like so:

```
>>> nautobot.plugins.golden_config.compliance_rule.all()
>>> ...output omitted for brevity...
```

Retrieve details of a compliance rule in the Nautobot Golden Config App like so:

```
>>> nautobot.plugins.golden_config.compliance_rule.all()[0].feature.
name
'aaa'
>>>
>>> nautobot.plugins.golden_config.compliance_rule.all()[0].platform.
name
'Arista EOS'
>>> nautobot.plugins.golden_config.compliance_rule.all()[0].match_
config
'aaa\nno aaa\nmanagement\nusername\nrole\nradius-server'
>>>
>>> nautobot.plugins.golden_config.compliance_rule.all()[0].config_
ordered
True
>>>
```

```
>>> nautobot.plugins.golden_config.compliance_rule.all()[0].config_
type
'cli'
>>>
```

By now, the patterns between Apps should be apparent. As you test these using the public sandbox, be sure to check the URLs and API docs to ensure you are using the right object because some are singular even when requesting the list of objects.

Using GraphQL with pynautobot

In *Chapter 8*, you learned how to work with the GraphQL API. Now, you are going to learn how to use pynautobot to get data from the Nautobot GraphQL API. You are able to leverage all aspects of the GraphQL API, including aliasing, filtering, and providing variables.

The first thing to do is build a query string; this is often done by putting a multiline string into a variable within Python. This is no different from what we showed in *Chapter 8*. We'll use the same examples from the last chapter:

```
>>> query = """
... query {
...     devices {
...         id
...     }
... }
... """
>>>
```

To make the query after having the string identified, pass the query in using the following statement:

```
>>> response = nautobot.graphql.query(query=query)
>>>
>>> response
GraphQLRecord(json={'data': {'devices': [{'id': '95014f3f-c125-4c33-
8b76-472211d1429e'}, {'id': '89b2ac3b-1853-4eeb-9ea6-6a081999bd3c'},
{'id':
...output omitted for brevity...
>>>
```

You can then access it as a Python dictionary using the json attribute:

```
>>> response.json
>>> # output omitted for brevity
```

Within the dictionary, the primary data you'll be interested in will be found within the data key in the dictionary.

`status_code` is also available to ensure you can retrieve the status of the HTTP API call to Nautobot:

```
>>> response.status_code
200
>>>
```

Looking at the last example from *Chapter 8*, let's look at using variables within a GraphQL query too:

```
>>> query = """
... query ($nautobot_location: [String]){
...    devices(location: $nautobot_location) {
...       name
...       location {
...          name
...       }
...    }
... }
... """
>>>
>>> variables = {"nautobot_location": "AMS01"}
>>>
>>> response = nautobot.graphql.query(query=query,
variables=variables)
>>>
>>> response.status_code
200
>>>
>>> response.json
{'data': {'devices': [{'name': 'ams01-dist-01', 'location': {'name':
'AMS01'}}, … omitted for brevity…
>>>
```

Using pynautobot to get the next available IP address

It is quite common to use Nautobot for IPAM too. One common task is to get the next available IP address within a given prefix (or subnet). There are convenience API endpoints that pynautobot is able to access to simplify getting the next available IP address. Let's take a look at the process of getting the next available IP address that can be used for the `servers` role in the `DEN01` location.

The next steps perform the following tasks:

1. Identify the server prefixes (subnets) in `DEN01`.

2. Verify that there are IP addresses available in one or more of the prefixes.

3. Verify what the next available IP address is within the prefix.

Identify the server prefixes (subnets) in DEN01 like so:

```
>>> nautobot.ipam.prefixes.filter(location="DEN01", role="server",
prefix_length="24")
[<pynautobot.models.ipam.Prefixes ('10.15.0.0/24') at 0xffff8f495ea0>,
<pynautobot.models.ipam.Prefixes ('10.15.1.0/24') at 0xffff8f496650>,
<pynautobot.models.ipam.Prefixes ('10.15.2.0/24') at 0xffff8f4967d0>,
<pynautobot.models.ipam.Prefixes ('10.15.3.0/24') at 0xffff8f497700>,
<pynautobot.models.ipam.Prefixes ('10.15.4.0/24') at 0xffff8f4973a0>,
<pynautobot.models.ipam.Prefixes ('10.15.5.0/24') at 0xffff8f497160>,
<pynautobot.models.ipam.Prefixes ('10.15.6.0/24') at 0xffff8f497310>,
<pynautobot.models.ipam.Prefixes ('10.15.7.0/24') at 0xffff8f497af0>]
>>>
```

Verify that there are IP addresses available in one or more of the prefixes, starting with the first prefix:

```
>>> first_prefix = nautobot.ipam.prefixes.filter(location="DEN01",
role="server", prefix_length="24")[0]
>>> len(first_prefix.available_ips.list())
50
>>>
```

Verify what the next available IP address is within the prefix:

```
>>> first_available_address = first_prefix.available_ips.list()[0]
>>> first_available_address
<pynautobot.models.ipam.IpAddresses ('10.15.0.1/24') at
0xffff8f4943a0>
>>> first_available_prefix.address
'10.15.0.1/24'
>>>
```

From here, you have identified what the first available IP address is. To create the IP address object in Nautobot and reserve it, you use the `create` method that we covered previously:

```
>>> ip_address = first_available_prefix.address
>>>
>>> nautobot.ipam.ip_addresses.create(address=ip_address,
status='Active', description="Reserved via pynautobot", parent=first_
prefix.id)
<pynautobot.models.ipam.IpAddresses ('10.15.0.1/24') at
0xffff8f494c70>
>>>
```

In the Nautobot UI, the change log of recent changes shows up on the home page. You can also view the actual object or IP address just created. You will now see the IP address, with the associated description that says that the IP address was created via pynautobot.

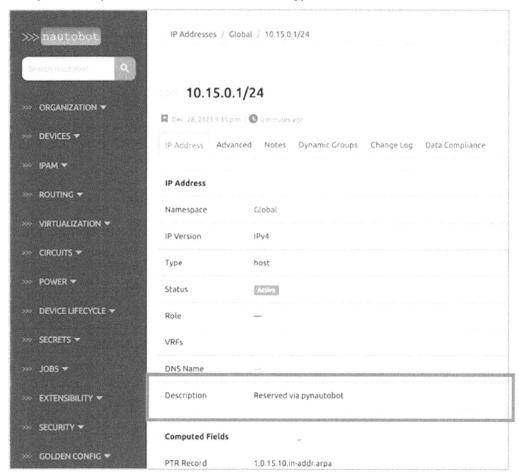

Figure 9.10 – The IP address view that was created via pynautobot

Working with pynautobot (`https://www.google.com/url?q=https://pynautobot.readthedocs.io/en/latest/index.html&sa=D&source=docs&ust=1709588712418702&usg=AOvVaw1Flt1FmNCh0M3dRN4sQZih`) as your Python SDK adds a powerful capability for creating Python objects to interact with Nautobot via the REST and GraphQL APIs. The library uses Python Requests to make the corresponding requests in a consistent fashion with the API itself since the methods are built from the API.

Exploring the Nautobot Ansible Collection

The Nautobot Ansible Content Collection is a collection of modules and inventory sources to assist in **creating, reading, updating, and deleting** (**CRUD** operations) data objects within Nautobot.

The most common use case for using Ansible and Nautobot is to use Nautobot as the primary network data store and Ansible to perform network automation tasks. You'll see this visually when we discuss a reference network automation architecture in *Chapter 12*.

> **Note**
>
> Remember, YAML files you may already have working with Ansible can be synchronized as Nautobot configuration contexts, as you learned in *Chapter 6*. It is possible to execute "tasks" from Nautobot using Nautobot Jobs (with Python), covered in *Chapter 11*, giving you flexibility with how you want to implement network automation, such as with Ansible playbooks, Nautobot Jobs, or a combination of both.

For example, you can fetch dynamic inventory from Nautobot, get the IPs, VLANs, and ASNs from Nautobot, and use that to generate configurations within Ansible. You can also use Ansible to query Nautobot for the next available IP address. Anything you can do from a Nautobot API you can do in Ansible. Another use case that is possible is using Ansible to broker data transformations between two systems (i.e., pull data from one system and push it to Nautobot).

The Nautobot Ansible Collection also provides two inventory plugins that may be used to create an Ansible Inventory from data within Nautobot. We'll explore these later in the chapter.

For a complete list of modules, navigate to `https://nautobot-ansible.readthedocs.io/en/latest/plugins/index.html#plugin-index`.

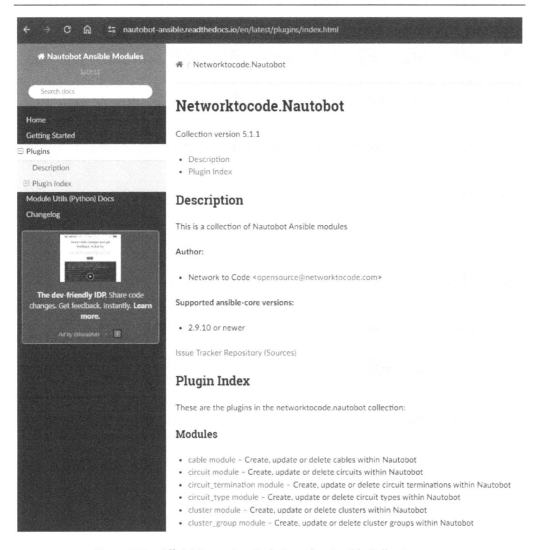

Figure 9.11 – Official Network to Code Nautobot Ansible Collection page

These modules are set up to align with many of the data categories within Nautobot. Just to highlight a few, they provide the capability to handle the CRUD operations for locations, circuits, circuit providers, devices, power connections, prefixes, VLANs, and tags. There are several more, so we encourage you to look at the link that provides a list of all of the modules.

> **Note**
>
> The Ansible playbooks covered in the chapter can also be found in the `https://github.com/PacktPublishing/Network-Automation-with-Nautobot/tree/main/chapter-09` GitHub repository.

Installing the collection

The Nautobot Ansible Content Collection is available on Ansible Galaxy. This can be installed using the Ansible Galaxy command in an environment that has Ansible.

We'll be installing Ansible and the collection as root in the same DigitalOcean Ubuntu server we've been using throughout the book:

```
root@nautobot-dev:~/ansible# pip install ansible
Collecting ansible
…output omitted for brevity…
Installing collected packages: resolvelib, ansible-core, ansible
Successfully installed ansible-9.2.0 ansible-core-2.16.3
resolvelib-1.0.1
WARNING: Running pip as the 'root' user can result in broken
permissions and conflicting behaviour with the system package manager.
It is recommended to use a virtual environment instead: https://pip.
pypa.io/warnings/venv
root@nautobot-dev:~/ansible#
```

With Ansible installed, we can install the collection with `ansible-galaxy`:

```
root@nautobot-dev:~/ansible# ansible-galaxy collection install
networktocode.nautobot
Starting galaxy collection install process
Process install dependency map
Starting collection install process
Downloading https://galaxy.ansible.com/api/v3/plugin/ansible/content/
published/collections/artifacts/networktocode-nautobot-5.1.1.tar.
gz to /root/.ansible/tmp/ansible-local-37476571pu73y4z/tmp4ccnj3nz/
networktocode-nautobot-5.1.1-7u_op7pk
Installing 'networktocode.nautobot:5.1.1' to '/root/.ansible/
collections/ansible_collections/networktocode/nautobot'
networktocode.nautobot:5.1.1 was installed successfully
root@nautobot-dev:~/ansible#
```

The collection is installed and we're ready to start reading Nautobot data with Ansible.

> **Note**
>
> The collection also relies on `pynautobot`, so the assumption is that you already installed pynautobot, as shown in the prior section with `pip install pynautobot`.

Reading data with Ansible

The first operation we'll look at with Ansible will be to read data from Nautobot. We'll use the public demo instance as a starting point for these examples.

There are four ways you can read data from Nautobot:

- Use the lookup plugin that uses the Nautobot REST API
- Use the module that uses the Nautobot GraphQL API
- Use the lookup plugin that uses the Nautobot GraphQL API (not covered in the book)
- Use the `uri` module (not covered in the book)

Our focus is on the first two because they are the most commonly used in production environments.

Using the networktocode.nautobot.lookup plugin

Let's start with a playbook called `01-get-devices.yml` that uses the Nautobot Ansible lookup plugin to query for all devices in Nautobot:

```
---
- name: " GET DEVICES"
  hosts: localhost
  connection: local
  gather_facts: false
  vars:
    NAUTOBOT_URL: https://demo.nautobot.com
    NAUTOBOT_TOKEN: "aaaaaaaaaaaaaaaaaaaaaaaaaaaaaaaaaaaaaaaa"
  tasks:

    - name: "RETRIEVE DEVICES AND SET AS FACT"
      ansible.builtin.set_fact:
        nautobot_devices:
          "{{ query('networktocode.nautobot.lookup', 'devices',
          api_endpoint=NAUTOBOT_URL, token=NAUTOBOT_TOKEN ) }}"

    - name: "LOOP OVER DEVICES AND DEBUG NAME OF EACH DEVICE"
      ansible.builtin.debug:
        msg: "The name of the device is {{ item.value.name }}."
      loop: "{{ nautobot_devices }}"
```

You can optionally (and should) use an environment variable for the token and even the URL, as shown here:

```
root@nautobot-dev:~/ansible# export NAUTOBOT_URL=https://demo.
nautobot.com
```

```
root@nautobot-dev:~/ansible# export NAUTOBOT_
TOKEN=aaaaaaaaaaaaaaaaaaaaaaaaaaaaaaaaaaaaaaaaaaaaa
root@nautobot-dev:~/ansible#
```

> **Note**
>
> These environment variables are not saved to persist user sessions. They should be set in
> `.bashrc` if you want them to persist.

Once the environment variables are set, it is possible to remove those keywords from the lookup query in the playbook. Take note here that we no longer need to specific the URL and token. The lookup plugin automatically looks for those environment variables:

```
- name: "RETRIEVE DEVICES AND SET AS FACT"
  ansible.builtin.set_fact:
    nautobot_devices:
      "{{ query('networktocode.nautobot.lookup', 'devices') }}"
```

You can also use the `api_filter` keyword to filter the query just as you can with an API call. Let's query for one device:

```
- name: "RETRIEVE DEVICES AND SET AS FACT"
  ansible.builtin.set_fact:
    Nautobot_devices:
      "{{ query('networktocode.nautobot.lookup', 'devices', api_
filter='name=ams01-leaf-01') }}"
```

The full playbook execution for this one is shown here:

Figure 9.12 – Viewing playbook execution when using the lookup plugin

Let's now look at using the GraphQL lookup plugin.

Using the networktocode.nautobot.query_graphql module

As you may guess, this lookup plugin lets you use the GraphQL endpoint directly from Ansible. You've already learned how to use it with Python requests in *Chapter 8* and how to use it with pynautobot earlier in this chapter—now, we'll use it in Ansible.

Let's create a playbook that uses this Nautobot GraphQL module and call it 01-get-devices-lookup-gql.yml:

```
---
- name: " GET DEVICES"
  hosts: localhost
  connection: local
  gather_facts: false
  tasks:

    - name: "CREATE QUERY STRING"
      ansible.builtin.set_fact:
        query_string: |
          query {
            devices {
              id
              name
            }
          }

    - name: "RETRIEVE DEVICES"
      networktocode.nautobot.query_graphql:
        query: "{{ query_string }}"
      register: nautobot_devices

    - name: "LOOP OVER DEVICES AND DEBUG NAME OF EACH DEVICE"
      ansible.builtin.debug:
        msg: "The name of the device is {{ item.name }}."
      loop: "{{ nautobot_devices.data.devices }}"
```

Remember to ensure your environment variables are set for the Nautobot URL and token.

Let's look at the second example we've been using that introduces query variables. This requires the use of a query_variables parameter that is passed into the module:

```
---
- name: " GET DEVICES"
```

```
hosts: localhost
connection: local
gather_facts: false
tasks:

  - name: "CREATE QUERY STRING"
    ansible.builtin.set_fact:
      graph_variables:
        nautobot_location: AMS01
      query_string: |
        query ($nautobot_location: [String]){
          devices(location: $nautobot_location) {
            name
            location {
              name
            }
          }
        }

  - name: "RETRIEVE DEVICES"
    networktocode.nautobot.query_graphql:
      query: "{{ query_string }}"
      graph_variables: "{{ graph_variables }}"
    register: devices

  - name: "PRINT DEVICES"
    ansible.builtin.debug:
      var: devices
```

Let's move on to adding or writing data to Nautobot with Ansible.

Ansible write operations

Let's dive into creating new objects within Nautobot. The modules in the collection will allow you to create locations, VLANs, IP addresses, prefixes, and more. All of the modules are idempotent, which means that you can run the same playbook over and over again and expect to get consistent data results.

Let's take a look at what an Ansible Playbook leveraging the modules may look like, one that will create devices at a location.

> **Note**
>
> When using the majority of Nautobot modules, you are required to pass in the URL and token as arguments. In order to facilitate this, we'll define URL and token variables at the top of each playbook that will read the environment variables that were set earlier.

```
1  ---
2  # ch09-ansible/01-add_devices.yml
3  # This demostrates adding devices to Nautobot using the Ansible Nautobot modules
4  - name: "ADD DEVICES TO NAUTOBOT"
5    hosts: localhost
6    connection: local
7    gather_facts: no
8    vars:
9      nautobot_url: "{{ lookup('ansible.builtin.env', 'NAUTOBOT_URL') }}"
10     nautobot_token: "{{ lookup('ansible.builtin.env', 'NAUTOBOT_TOKEN') }}"
11   tasks:
12     - name: "10: CREATE DEVICES"
13       networktocode.nautobot.device:
14         url: "{{ nautobot_url }}"
15         token: "{{ nautobot_token }}"
16         name: "{{ item['name'] }}"
17         device_type: "{{ item['device_type'] }}"
18         status: "{{ item['status'] }}"
19         role: "{{ item['role'] }}"
20         location: "{{ item['location'] }}"
21         tenant: "{{ item['tenant'] }}"
22         state: present
23       loop:
24         - name: ord01-leaf-10
25           device_type: Nexus 9372TX
26           status: "Active"
27           role: "leaf"
28           location: "ORD01"
29           tenant: "Nautobot Airports"
30         - name: ord01-leaf-11
31           device_type: Nexus 9372TX
32           status: "Active"
33           role: "leaf"
34           location: "ORD01"
35           tenant: "Nautobot Airports"
36         - name: ord01-leaf-12
37           device_type: Nexus 9372TX
38           status: "Active"
39           role: "leaf"
40           location: "ORD01"
41           tenant: "Nautobot Airports"
42         - name: ord01-leaf-13
43           device_type: Nexus 9372TX
44           status: "Active"
45           role: "leaf"
46           location: "ORD01"
47           tenant: "Nautobot Airports"
48
```

Figure 9.13 – Viewing a playbook that adds devices to Nautobot

Note that the preceding playbook is not meant to be run by you. We will be diving much deeper into executing playbooks that add data to Nautobot in the next chapter.

On the first execution of a playbook using these modules, you will see that the objects will be created, no different from how Ansible modules should work. The standard messaging from running the Ansible Playbook command will show that the devices were created. If you run a playbook a second time using the Nautobot modules, everything will read ok instead of changed shown at the bottom of your typical playbook output. This is due to the idempotent nature of the Ansible module.

Using the Nautobot Ansible modules is a powerful way to get data into your Nautobot environment from various sources, not only from network devices. This is a recommended method of working with other tools and systems as well to incorporate data into Nautobot.

Exploring Ansible inventory sources

The Nautobot Ansible Content Collection has two inventories available for use with Nautobot.

The first is the REST API version of the inventory, as the Nautobot project roots originally had a REST API only. This inventory continues to provide a familiar experience of working with a network inventory.

The second inventory provided is for using the GraphQL API to gather data, which handles larger amounts of data a bit easier than the REST API inventory. This inventory plugin is used more commonly than the initial REST API version.

With the GraphQL inventory, you specify data arguments for the device that are required for your inventory. And that's all that is requested. There is no need to worry about processing data unnecessarily. Let's take a look at the two inventories and what the output looks like.

> **Tip**
>
> If you are just getting started with your first Nautobot Ansible inventory, we recommend that you use the GraphQL inventory. GraphQL is always recommended, if applicable, for gathering information from Nautobot.

The inventories are configured by creating a YAML file that will allow you to provide additional configuration to pass into the inventory.

Ansible inventory – REST API

Let's explore the first of two inventory plugins. This first one uses the REST API.

> **Important note**
>
> There is a lot of processing of data that is involved with retrieving configuration contexts (we initially talked about this in *Chapter 8*). The default behavior is to not return configuration context data. If there is configuration context data in your Nautobot instance and it is not needed in the inventory, it is still recommended to set this variable to `False` so it is visible to you. If you require it, you should set `config_context` to `True`.

Let's create a dynamic inventory file that will use the REST API inventory plugin. We'll name this file `inventory-rest.yml`. This will contain the instructions on how to use the plugin:

```
# inventory-rest.yml
plugin: networktocode.nautobot.inventory
api_endpoint: https://demo.nautobot.com
validate_certs: true
config_context: false
group_by:
  - device_roles
  - locations
  - platforms
device_query_filters:
  - has_primary_ip: 'true'
```

Once the dynamic inventory file has been created, ensure you set the token for the Nautobot instance and create a playbook called `pb-test-inventory.yml` that'll use the new inventory:

```
root@nautobot-dev:~/ansible# export NAUTOBOT_
TOKEN=aaaaaaaaaaaaaaaaaaaaaaaaaaaaaaaaaaaaaaaa
root@nautobot-dev:~/ansible#

# pb-test-inventory.yml
---

- name: "BASIC PLAY TESTING DYNAMIC INVENTORY"
  hosts: "all"
  gather_facts: false
  connection: "local"

  tasks:

    - name: "PRINT ALL GROUPS"
      debug:
        var: groups
```

As you can see, it is a pretty basic playbook and is just printing the groups to the terminal. The groups we should expect are groups for each role, location, and platform as defined in the inventory file.

Let's run the playbook:

```
root@nautobot-dev:~/ansible# ansible-playbook -i inventory-rest.yml
pb-test-inventory.yml
```

This playbook has significant output, so we trimmed the output showing two of the groups that were returned and debugged:

```
...output omitted for brevity...
        ],
        "locations_weehawken": [
            "wee-eos7k-01.nautobot.com",
            "wee-n9k-01.nautobot.com",
            "wee-n9k-02.nautobot.com"
        ],
        "platforms_arista_eos": [
            "ams01-edge-01",
            "ams01-edge-02",
            "ams01-leaf-01",
...output omitted for brevity...
wee-eos7k-01.nautobot.com    :
ok=1     changed=0    unreachable=0    failed=0    skipped=0
rescued=0    ignored=0
wee-n9k-01.nautobot.com    :
ok=1     changed=0    unreachable=0    failed=0    skipped=0
rescued=0    ignored=0
wee-n9k-02.nautobot.com    :
ok=1     changed=0    unreachable=0    failed=0    skipped=0
rescued=0    ignored=0

root@nautobot-dev:~/ansible#
```

With all groups being printed to the terminal, you can see the patterns of how groups are constructed. Let's use one of those groups in a playbook. We'll use the platforms_arista_eos group and debug all hosts (which is just printing the name of the device) in that group to the terminal in a new playbook called pb-test-inventory-2.yml:

```
# pb-test-inventory-2.yml
---

- name: "BASIC PLAY TESTING DYNAMIC INVENTORY"
  hosts: "platforms_arista_eos"
  gather_facts: false
```

```
    connection: "local"

    tasks:

      - name: "PRINT ALL GROUPS"
        debug:
          var: inventory_hostname
```

You can then run the playbook as you'd expect against all devices that have a platform of Arista EOS:

```
root@nautobot-dev:~/ansible# ansible-playbook -i inventory-rest.yml
pb-test-inventory-2.yml
#output omitted
```

Here is another example showing the power of the dynamic inventory:

```
---
# rest_inv.yml
plugin: networktocode.nautobot.inventory
api_endpoint: https://demo.nautobot.com
validate_certs: true
config_context: false
device_query_filters:
  - has_primary_ip: "true"
  - location: "ORD01"
  - role: "edge"
group_names_raw: yes
group_by:
  - platforms
```

To demonstrate this inventory, we will use the `ansible-inventory` command to call the YAML file defined and get the data for connecting to the ORD01 hosts that are of the edge type for the role. This is a great way to test dynamic inventories without having to create playbooks.

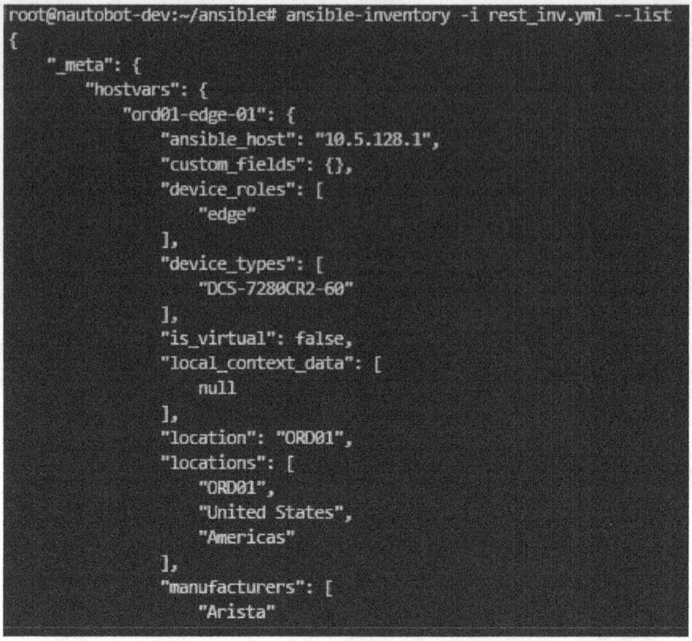

```
root@nautobot-dev:~/ansible# ansible-inventory -i rest_inv.yml --list
{
    "_meta": {
        "hostvars": {
            "ord01-edge-01": {
                "ansible_host": "10.5.128.1",
                "custom_fields": {},
                "device_roles": [
                    "edge"
                ],
                "device_types": [
                    "DCS-7280CR2-60"
                ],
                "is_virtual": false,
                "local_context_data": [
                    null
                ],
                "location": "ORD01",
                "locations": [
                    "ORD01",
                    "United States",
                    "Americas"
                ],
                "manufacturers": [
                    "Arista"
```

Figure 9.14 – Viewing the Nautobot REST API inventory plugin

You will notice that there is quite a bit of information that is processed since the data is coming back from the REST API. There are statuses, tags, tenants, device roles, device types, manufacturers, platforms, primary IP addresses, racks, rack groups, and regions all being provided back to the inventory. You will not need the majority of these items in your inventory, but since the REST API calls are providing this detail, that data is automatically available (although not needed). All that is really needed for an inventory is the hostname, IP address, and the platform for the device. We'll highlight this in the next example when we look at the GraphQL inventory plugin continuing to showcase the power of GraphQL.

Ansible inventory – GraphQL API

The newer Nautobot Ansible inventory plugin leverages the GraphQL API. Keep in mind that the prior example in the chapter was using a GraphQL Ansible module. This is an inventory plugin.

The GraphQL inventory plugin leverages the same type of YAML setup file to construct the necessary query string to use to query the GraphQL interface. Because this is GraphQL, you get only the information that is needed in order to connect to the devices for the inventory. So if the environment has DNS records already set up for name resolution on the hostname, there could be a need to only get the device name (hostname) and the platform in order to connect to the device.

The GraphQL inventory plugin requires the netutils and pynautobot libraries to be installed in order to function. To install netutils, leverage Python's pip as previously completed for other Python libraries.

> **Tip**
>
> `netutils` is a Python library for networking. It has so many helper functions that anyone writing Python and automating networks should check it out at its docs site (`https://netutils.readthedocs.io/en/latest/`).

```
root@nautobot-dev:~/ansible# pip install netutils
Collecting netutils
  Downloading netutils-1.6.0-py3-none-any.whl (481 kB)
```

```
──── 481.3/481.3 KB 9.2 MB/s eta 0:00:00
Installing collected packages: netutils
Successfully installed netutils-1.6.0
WARNING: Running pip as the 'root' user can result in broken
permissions and conflicting behaviour with the system package manager.
It is recommended to use a virtual environment instead: https://pip.
pypa.io/warnings/venv
root@nautobot-dev:~/ansible#
```

The following is an example of a GraphQL inventory file. We have a filter to only query for devices in the `AMS01` location. If you want to expand this for all devices in Nautobot, you can easily just delete the `filters` key and subsequently the location: `AMS01` within that object.

Create a file called `inventory-gql.yml` and save the following in it:

```
# inventory-gql.yml
plugin: networktocode.nautobot.gql_inventory
api_endpoint: https://demo.nautobot.com
query:
  devices:
    filters:
      location: AMS01
    id:
    name:
    location:
      name:
    tenant:
      name:
    platform:
      name:
      manufacturer:
        name:
    device_type:
      model:
    role:
      name:
```

```
group_by:
    - tenant.name
    - platform.name
    - device_type.model
    - role.name
    - location.name
```

Note that, as you look through the GraphQL queries that are set up, in order to use the `group_by` parameter and generate groups via the inventory, you must have the data in the query. If you do not have the data, you cannot group by it.

You should take note that the query is just a normal GraphQL query converting it into a YAML format. For every new indentation or curly bracket in the GraphQL query, it is a new key in YAML. The following would be the same query if we were trying it in GraphiQL (from *Chapter 8*), too:

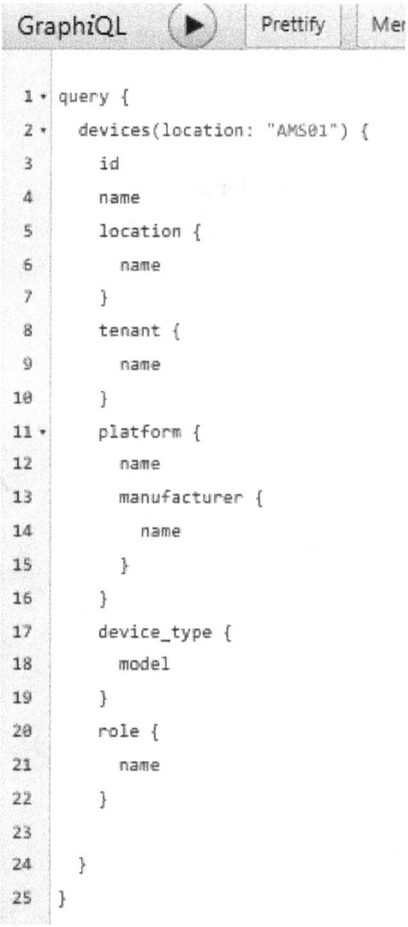

Figure 9.15 – GraphQL inventory plugin query shown in GraphiQL

As a reminder, be sure to set your token as an environment variable. This is used by the dynamic inventory script:

```
root@nautobot-dev:~/ansible# export NAUTOBOT_
TOKEN=aaaaaaaaaaaaaaaaaaaaaaaaaaaaaaaaaaaaaaaaaaaa
root@nautobot-dev:~/ansible#
```

Once the inventory file is built, you can use it. We'll use the `ansible-inventory` utility to test it:

```
root@nautobot-dev:~/ansible# ansible-inventory -i inventory-gql.yml
--list
…output omitted for brevity…

    "distribution": {
        "hosts": [
            "ams01-dist-01"
        ]
    },
    "edge": {
        "hosts": [
            "ams01-edge-01",
            "ams01-edge-02"
        ]
    },
    "leaf": {
        "hosts": [
            "ams01-leaf-01",
            "ams01-leaf-02",
            "ams01-leaf-03",
…output omitted for brevity…
```

The preceding snippet shows a few of the groups created from `role`. Remember, with this example, we created groups so we can automate by tenant, platform, device type (model), role, and location!

You will notice a significant performance increase in using the GraphQL inventory compared to the REST API inventory. That is the nature of the GraphQL interface—the data is provided back quicker as the web application is handling the relationships instead of the data being built with multiple REST API calls.

Querying for too much data in a dynamic inventory can cause a significant load on your Nautobot instance. We will cover some inventory best practices for Ansible and Nautobot in the next chapter.

Using Nornir Nautobot

Nornir is a Python-based automation framework that has grown in popularity for network automation. It provides an alternative to Ansible for those who want an automation framework without a **domain-specific language** (DSL), but it also provides a framework that can be used by other Python applications and projects such as Nautobot! This makes Nornir and Nautobot a great match for their use in Nautobot Jobs and Apps.

This book assumes some background in Nornir already as we aren't covering Nornir basics. Our goal is to introduce Nornir Nautobot, which is a project that provides a set of utilities to simplify using Nautobot with Nornir. Specifically, Nornir Nautobot contains an inventory plugin (similar to what we just covered with Ansible) as well as a series of plugins to simplify doing network automation tasks.

In this section, we will review how to get started with the Nornir inventory plugin so that you can use Nautobot as an inventory source for your Nornir applications. In this example, we will jump right into filtering from the API point of view, so that the inventory matches what you are looking to accomplish.

Installing Nornir Nautobot

The installation of the library is handled through Python's `pip` with any typical Python package hosted on PyPI:

```
root@nautobot-dev:~/ansible# pip install nornir-nautobot
# output omitted for brevity
```

> **Note**
> `nornir-nautobot` also installs Nornir if it wasn't previously installed.

For this example Python script, we'll use environment variables just as we did earlier with Ansible. They should still be set if you were following, but you can set them again to be safe:

```
root@nautobot-dev:~/nornir# export NAUTOBOT_URL=https://demo.nautobot.com
root@nautobot-dev:~/nornir# export NAUTOBOT_TOKEN=aaaaaaaaaaaaaaaaaaaaaaaaaaaaaaaaaaaaaaaaaaaa
root@nautobot-dev:~/nornir#
```

We are going to build a Nornir script that uses the `NautobotInventory` plugin from Nornir Nautobot. We name the script `nornir_example.py`. We'll review the script in sections:

```
# nornir_example.py
import os
from nornir import InitNornir
```

```
from nornir.core.task import Task, Result

from nornir_utils.plugins.functions import print_result
```

The preceding section is just all of the imports from your typical Nornir network automation script.

Next, we have a basic function called `hello_world()` that accepts one argument, which is a Nornir `task` object:

```
def hello_world(task):
  return Result(host=task.host, result=f"{task.host.name} says hello
world!")
```

The function is a Nornir task that will print out a hello message. If you are working with network devices here, this is where you would put the logic for working with the devices. For more details on getting started working with Nornir itself, take a look at the blog post on the *Network to Code* blog (`https://blog.networktocode.com/post/getting-started-with-python-network-libraries-4/`).

```
def main():
    locations = ["AMS01"]

    book_nornir = InitNornir(
        inventory = {
            "plugin": "NautobotInventory",
            "options": {
                "nautobot_url": os.getenv("NAUTOBOT_URL"),
                "nautobot_token": os.getenv("NAUTOBOT_TOKEN"),
                "filter_parameters": {"location": locations },
                "ssl_verify": True,
            },
        },
    )
```

In the preceding block of code, there is a function called `main()`, a list called `locations` that will be used as a filter for the inventory query, and finally, a new instance of `InitNornir` that will be the basis for our automation communication with Nautobot.

```
    print(f"Hosts found: {len(book_nornir.inventory.hosts)}")

    print(book_nornir.inventory.hosts.keys())

    result = book_nornir.run(task=hello_world)
```

```
    print_result(result)

if __name__ == "__main__":
    main()
```

After the Nornir instance is created, we'll start to interact with the objects of our instance, as shown in the preceding code. We'll print the number of hosts found in the inventory. Given each host is a dictionary, we'll show all devices by printing host keys. We'll execute our automation task that calls hello_world(). Finally, the Nornir task result happens using the Nornir utility and function previously imported, print_result, to make the output look nice.

By executing the preceding Python script, you get the following outputs (they are broken up into a snippet and figure to save space):

```
root@nautobot-dev:~/nornir# python3 nornir_example.py
Hosts found: 27
dict_keys(['ams01-dist-01', 'ams01-edge-01', 'ams01-edge-02',
'ams01-leaf-01', 'ams01-leaf-02', 'ams01-leaf-03', 'ams01-leaf-04',
'ams01-leaf-05', 'ams01-leaf-06', 'ams01-leaf-07', 'ams01-leaf-08',
'ams01-pdu-01', 'ams01-pdu-02', 'ams01-pdu-03', 'ams01-pdu-04',
'ams01-pdu-05', 'ams01-pdu-06', 'ams01-pdu-07', 'ams01-pdu-08',
'ams01-pdu-11', 'ams01-pdu-12', 'ams01-pdu-13', 'ams01-pdu-14',
'ams01-pdu-15', 'ams01-pdu-16', 'ams01-pdu-17', 'ams01-pdu-18'])
hello_world*****************************************************
**********
* ams01-dist-01 ** changed : False ********************************
**********
vvvv hello_world ** changed : False vvvvvvvvvvvvvvvvvvvvvvvvvvvvvvvvvvv
vvvvvvvvvv INFO
ams01-dist-01 says hello world!
^^^^ END hello_world ^^^^^^^^^^^^^^^^^^^^^^^^^^^^^^^^^^^^^^^^^^^^^^^^^^
^^^^^^^^^^
* ams01-edge-01 ** changed : False ********************************
**********
vvvv hello_world ** changed : False vvvvvvvvvvvvvvvvvvvvvvvvvvvvvvvvvvv
vvvvvvvvvv INFO
ams01-edge-01 says hello world!
^^^^ END hello_world ^^^^^^^^^^^^^^^^^^^^^^^^^^^^^^^^^^^^^^^^^^^^^^^^^^
^^^^^^^^^^
```

```
* ams01-pdu-15 ** changed : False ******************************************
vvvv hello_world ** changed : False vvvvvvvvvvvvvvvvvvvvvvvvvvvvvvvvvvvvvvvvv INFO
ams01-pdu-15 says hello world!
^^^^ END hello_world ^^^^^^^^^^^^^^^^^^^^^^^^^^^^^^^^^^^^^^^^^^^^^^^^^^^^^^^^
* ams01-pdu-16 ** changed : False ******************************************
vvvv hello_world ** changed : False vvvvvvvvvvvvvvvvvvvvvvvvvvvvvvvvvvvvvvvvv INFO
ams01-pdu-16 says hello world!
^^^^ END hello_world ^^^^^^^^^^^^^^^^^^^^^^^^^^^^^^^^^^^^^^^^^^^^^^^^^^^^^^^^
* ams01-pdu-17 ** changed : False ******************************************
vvvv hello_world ** changed : False vvvvvvvvvvvvvvvvvvvvvvvvvvvvvvvvvvvvvvvvv INFO
ams01-pdu-17 says hello world!
^^^^ END hello_world ^^^^^^^^^^^^^^^^^^^^^^^^^^^^^^^^^^^^^^^^^^^^^^^^^^^^^^^^
* ams01-pdu-18 ** changed : False ******************************************
vvvv hello_world ** changed : False vvvvvvvvvvvvvvvvvvvvvvvvvvvvvvvvvvvvvvvvv INFO
ams01-pdu-18 says hello world!
^^^^ END hello_world ^^^^^^^^^^^^^^^^^^^^^^^^^^^^^^^^^^^^^^^^^^^^^^^^^^^^^^^^
root@nautobot-dev:~/nornir#
```

Figure 9.16 – Output of a Nornir script using Nornir Nautobot

Since we aren't going into great detail in these sections, the important part to remember is that the integration exists, and use official docs or support channels to really dive into the details.

Exploring Nautobot Docker containers

In *Chapter 3*, we introduced how to install Nautobot from scratch, installing every service needed to get Nautobot up and running. There is another option that can be quite helpful, and that is to use Docker containers. This can be helpful in building out NetDevOps pipelines and deployments that are repeatable, immutable, and used for quick testing.

There is a GitHub repository (https://github.com/nautobot/nautobot-docker-compose) that has prebuilt Docker Compose and Docker files for different environments to help you get started even quicker for container environments.

The repository offers three different sets of options to deploy Nautobot onto a Docker Compose environment:

- Standard Nautobot deployment
- LDAP-supported environment
- Nautobot Apps-supported environment

We'll focus on the first option.

Detailed installation instructions for Docker and the Docker Compose plugin can be found on the Docker installation pages starting at https://docs.docker.com.

The following summary of Linux commands can be used to get Docker up and running:

```
root@nautobot-dev:~# apt install -y curl software-properties-common
tree
root@nautobot-dev:~# curl -fsSL https://download.docker.com/linux/
ubuntu/gpg | apt-key add -
root@nautobot-dev:~# add-apt-repository "deb [arch=amd64] https://
download.docker.com/linux/ubuntu focal stable"
root@nautobot-dev:~# apt install -y docker-ce
```

Now, clone the repository to a host that has Docker and the Docker Compose plugin installed:

```
root@nautobot-dev:~# git clone https://github.com/nautobot/nautobot-
docker-compose.git
root@nautobot-dev:~# cd nautobot-docker-compose/
root@nautobot-dev:~/nautobot-docker-compose#
```

Once the repository is cloned down, you should make a copy of the example .env file and place it in the directory. This file will maintain credential information. It will be protected by the Linux authentication that is set up on the machine. This may be a place where you would create a Nautobot user account on the host or other best practice from your organization:

```
root@nautobot-dev:~/nautobot-docker-compose# cp local.env.example
local.env
root@nautobot-dev:~/nautobot-docker-compose#
```

After you copy the file, open the local.env file and change NAUTOBOT_CREATE_SUPERUSER, setting it to true:

```
NAUTOBOT_CREATE_SUPERUSER=true
```

Note that we are following the steps for dev and testing. If you are using this Docker Compose setup in a corporate environment, you should do the following:

1. Update NAUTOBOT_ALLOWED_HOSTS

2. Update any PASSWORD-related entry

3. Update any SECRET_KEY variable

4. Change the permissions of local.env to 0600

As you update passwords, be sure that the database passwords (Postgres or MySQL) match for the NAUTOBOT_DB_PASSWORD setting. If these differ, the Nautobot application will not be able to communicate with the database. The database containers read the environment file to set the credentials for the account. This also applies to Redis as a separate database type. Redis and a choice of either PostgreSQL or MySQL is required. You should *not* be using both MySQL and PostgreSQL in this setup.

To start Nautobot for the first time, run the `docker compose up` command, which will pull all required containers from container registries, including those for Redis and your database.

Use `Control-C` to stop the containers.

Note that you may have to exit and rerun `docker compose up` for the scheduler service to work properly:

```
root@nautobot-dev:~/nautobot-docker-compose# docker compose up
```

Once the log messages stop, you can access the new instance on HTTP port `8080` (e.g., `http://<your-nautobot:8080`). If you didn't change the credentials in the `env` file, they are `admin/admin`.

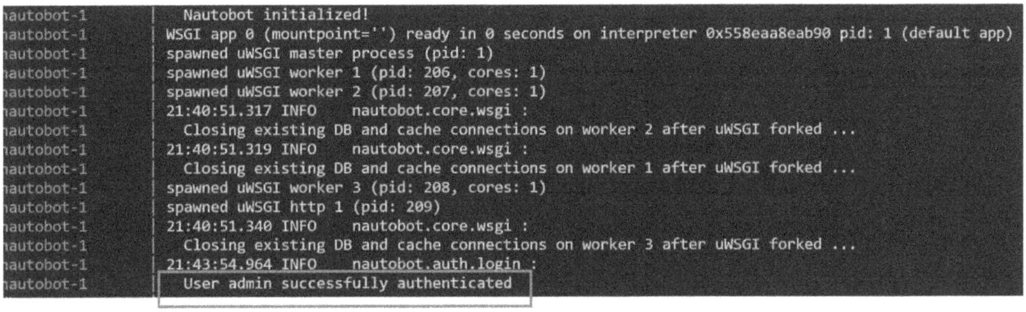

```
nautobot-1          Nautobot initialized!
nautobot-1          WSGI app 0 (mountpoint='') ready in 0 seconds on interpreter 0x558eaa8eab90 pid: 1 (default app)
nautobot-1          spawned uWSGI master process (pid: 1)
nautobot-1          spawned uWSGI worker 1 (pid: 206, cores: 1)
nautobot-1          spawned uWSGI worker 2 (pid: 207, cores: 1)
nautobot-1          21:40:51.317 INFO    nautobot.core.wsgi :
nautobot-1            Closing existing DB and cache connections on worker 2 after uWSGI forked ...
nautobot-1          21:40:51.319 INFO    nautobot.core.wsgi :
nautobot-1            Closing existing DB and cache connections on worker 1 after uWSGI forked ...
nautobot-1          spawned uWSGI worker 3 (pid: 208, cores: 1)
nautobot-1          spawned uWSGI http 1 (pid: 209)
nautobot-1          21:40:51.340 INFO    nautobot.core.wsgi :
nautobot-1            Closing existing DB and cache connections on worker 3 after uWSGI forked ...
nautobot-1          21:43:54.964 INFO    nautobot.auth.login :
nautobot-1            User admin successfully authenticated
```

Figure 9.17 – Viewing stdout starting Docker Compose and showing a user login (admin/admin)

Once you log in, you will see a message in the log that you were able to log in successfully.

Once you use `Control-C` to stop the containers, you'll see all containers being gracefully stopped.

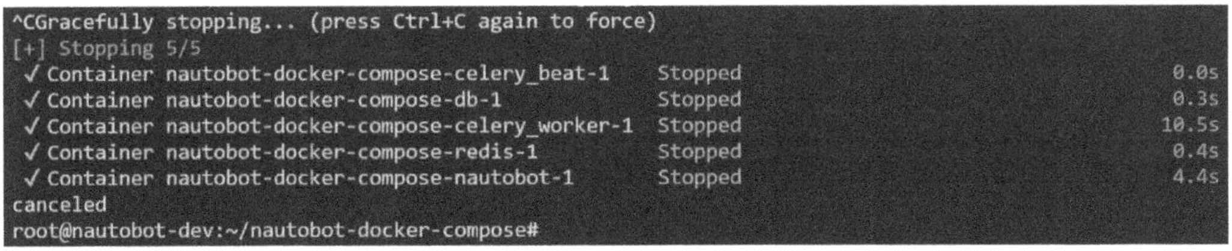

```
^CGracefully stopping... (press Ctrl+C again to force)
[+] Stopping 5/5
 ✓ Container nautobot-docker-compose-celery_beat-1      Stopped                          0.0s
 ✓ Container nautobot-docker-compose-db-1               Stopped                          0.3s
 ✓ Container nautobot-docker-compose-celery_worker-1    Stopped                         10.5s
 ✓ Container nautobot-docker-compose-redis-1            Stopped                          0.4s
 ✓ Container nautobot-docker-compose-nautobot-1         Stopped                          4.4s
canceled
root@nautobot-dev:~/nautobot-docker-compose#
```

Figure 9.18 – Showing all Docker Compose containers in a Stopped state

> **Tip**
>
> To start the containers without seeing the logs, you should use *Ctrl+C* in the command line, then execute the `docker compose up` command with the `-d` flag. This will start the containers in a detached state. The containers themselves have the `restart: unless-stopped` setting by default. So, when the system reboots, the containers will automatically restart.

Exploring the Nautobot Go library

Nautobot's latest companion is an SDK written for GoLang named `go-nautobot`. The library is currently in an alpha state as we look to grow its capabilities as the automation community desires the added features. `go-nautobot` is a library that is autogenerated from the Nautobot Swagger API and provides a clean integration to Nautobot via a GoLang application. Here are a few examples to get started. As part of the design of the SDK for Go, every iteration of Nautobot will produce a tightly coupled SDK for GoLang. So, when a new release of Nautobot comes out, a corresponding release for `go-nautobot` will also be released. The structure of the SDK will be tightly integrated with the Swagger API definition for the release so if something changes in the Nautobot API, the Go library will read in these changes.

Let's start by creating a new Go script in a file called `main.go`. The script will query Nautobot for manufacturers and simply print the received JSON payload with those manufacturers:

```
root@nautobot-dev:~# touch main.go
```

Add the following code inside the newly created `main.go` file:

```
// main.go
package main

import (
        "context"
        "fmt"

        nb "github.com/nautobot/go-nautobot/pkg/nautobot"
)

func check(err error) {
        if err != nil {
                panic(err)
        }
}
```

In this code, there is a method defined within the application to not repeat the error checking that is common practice within GoLang. Next is the `main` function, which you should add directly below the `check` function:

```
func main() {
        token, err := nb.NewSecurityProviderNautobotToken("aaaaaaaaaaaaaa
aaaaaaaaaaaaaaaaaaaaaaaaaa")
        check(err)

        c, err := nb.NewClientWithResponses(
```

```
        "https://demo.nautobot.com/api/",
        nb.WithRequestEditorFn(token.Intercept),
    )
    check(err)

    ctx := context.Background()

    resp, err := c.DcimManufacturersListWithResponse(ctx, &nb.
DcimManufacturersListParams{})
    check(err)

    fmt.Printf("%v", string(resp.Body))

}
```

The following figure shows the full script:

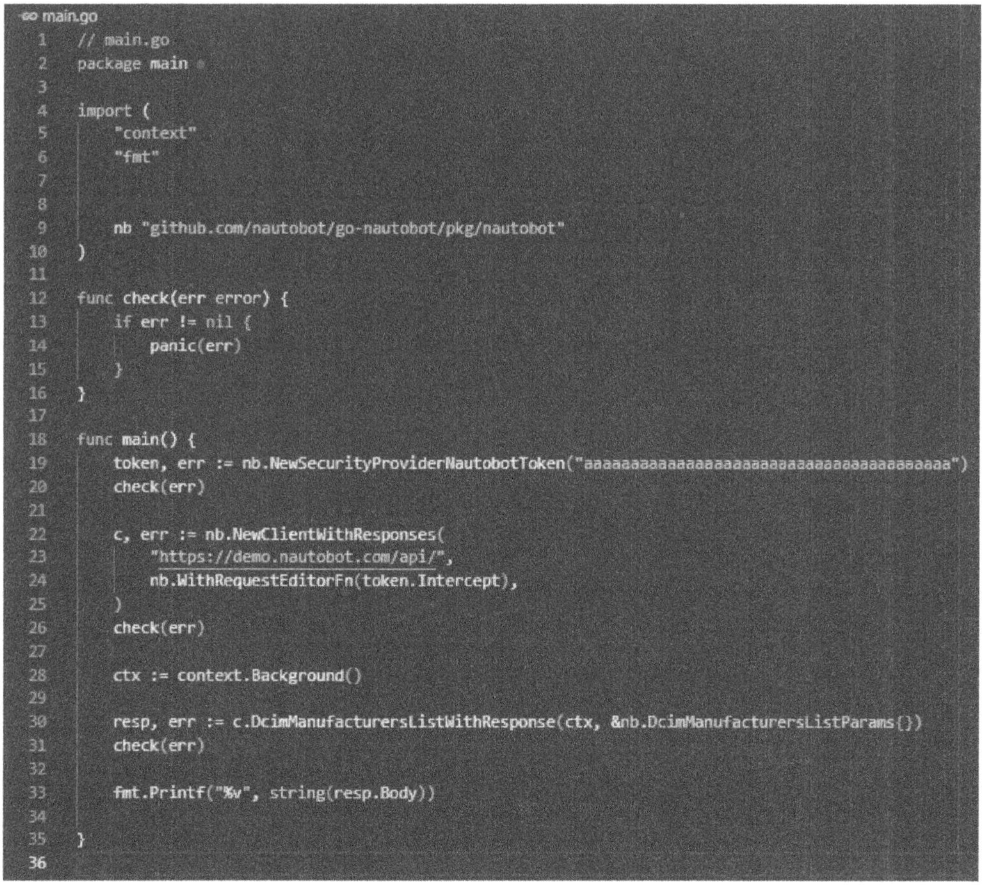

```go
// main.go
package main

import (
    "context"
    "fmt"

    nb "github.com/nautobot/go-nautobot/pkg/nautobot"
)

func check(err error) {
    if err != nil {
        panic(err)
    }
}

func main() {
    token, err := nb.NewSecurityProviderNautobotToken("aaaaaaaaaaaaaaaaaaaaaaaaaaaaaaaaaaaaaaaa")
    check(err)

    c, err := nb.NewClientWithResponses(
        "https://demo.nautobot.com/api/",
        nb.WithRequestEditorFn(token.Intercept),
    )
    check(err)

    ctx := context.Background()

    resp, err := c.DcimManufacturersListWithResponse(ctx, &nb.DcimManufacturersListParams{})
    check(err)

    fmt.Printf("%v", string(resp.Body))

}
```

Figure 9.19 – Go script that queries Nautobot for manufacturers

Let's prepare the system so we can run the script. We will be installing Go, setting up the application, and then fetching the go-nautobot package from GitHub:

```
root@nautobot-dev:~# apt install golang-go
<... output omitted for brevity …>
root@nautobot-dev:~# go mod init nautobot.com/bookExample
go: creating new go.mod: module nautobot.com/bookExample
go: to add module requirements and sums:
        go mod tidy
root@nautobot-dev:~# go mod tidy
go: finding module for package github.com/nautobot/go-nautobot/pkg/
nautobot
go: downloading github.com/nautobot/go-nautobot v1.5.8-beta
go: found github.com/nautobot/go-nautobot/pkg/nautobot in github.com/
nautobot/go-nautobot v1.5.8-beta
go: downloading github.com/deepmap/oapi-codegen v1.10.1
go: downloading github.com/tidwall/gjson v1.14.1
go: downloading github.com/google/uuid v1.3.0
go: downloading github.com/tidwall/pretty v1.2.0
go: downloading github.com/tidwall/match v1.1.1
go: downloading github.com/stretchr/testify v1.7.1
go: downloading github.com/davecgh/go-spew v1.1.1
go: downloading github.com/pmezard/go-difflib v1.0.0
go: downloading gopkg.in/yaml.v3 v3.0.0-20210107192922-496545a6307b
```

Let's install the Nautobot Go library called go-nautobot:

```
root@nautobot-dev:~# go get github.com/nautobot/go-nautobot/pkg/
nautobot
```

Now that the packages are all installed, run go run main.go to get the output from the Nautobot instance:

```
root@nautobot-dev:~# go run main.go
{"count":10,"next":null,"previous":null,"results":[{"id":"
9a195533-0663-44bb-a99f-3e1879e88dc8","object_type":"dcim.
manufacturer","display":"APC","url":"https://demo.nautobot.com/api/
dcim/manufacturers/9a195533-0663-44bb-a99f-3e1879e88dc8/","natural_
slug":"apc_9a19","device_type_count":1,"inventory_item_
count":0,"platform_count":0,"name":"APC","description":""
,"created":"2023-09-21T00:00:00Z","last_updated":"2023-09-
21T19:52:53.167151Z","notes_url":"https://demo.nautobot.com/api/dcim/
manufacturers/9a195533-0663-44bb-a99f-3e1879e88dc8/notes/",
<... output omitted for brevity …>
```

Now, let's take a look at the Terraform that uses the Go library.

Introducing the Nautobot Terraform provider

Along with providing a `go-nautobot` library comes the follow-on project of a Terraform provider to work with Nautobot within Terraform. This module is in the early stages of development today and will continue to see improvements as the community is looking for updates. The Nautobot provider for Terraform provides a methodology to interact with Nautobot to create resources inside of Nautobot and to use Nautobot as a source to complete the Terraform activities.

Summary

This chapter has shown that there are quite a few integrations that can be used to round out a Nautobot installation and even influence how you choose to install Nautobot. We covered introductions for pynautobot, the Nautobot Ansible Collection, Nornir Nautobot, Docker Compose, and the Go Library, and finally, gave mention to the newest project just getting started, which is a Terraform provider for Nautobot. These integrations give you the freedom to develop network automation as it fits into your needs and your organization. They also help to get to your goal of automating networks and your manual workflows much quicker.

In the next chapter, we will dive much deeper into what is possible when integrating Nautobot with Ansible in *Infrastructure as Code* deployments.

Embracing Infrastructure as Code with Nautobot, Git, and Ansible

In the previous chapter, you learned a bit about the Nautobot Ansible collection. This chapter builds on that and dives into using Ansible with Nautobot even more. We'll populate data using Ansible, use the GraphQL dynamic inventory by walking through real use cases, and look at how to integrate Nautobot with Ansible playbooks that do network automation. We'll walk through a multi-vendor network, its inventory, and how to get systems set up, and ultimately deploy and configure interfaces, IP addresses, and BGP across Cisco IOS and NXOS, Arista EOS, and Juniper Junos networks.

The following are the main topics covered in this chapter:

- Setting up the environment
- Adding data to Nautobot with Ansible
- Setting up a dynamic inventory
- Backing up network devices
- Performing a config replace with Nautobot, NAPALM, and Ansible on Arista and Juniper devices
- Performing config changes with Nautobot and Ansible for Cisco IOS devices
- Performing config changes with Nautobot and Ansible for Cisco NX-OS devices
- Managing data with config contexts and using Git
- Nautobot jobs versus Ansible playbooks

Technical requirements

Code files and other resources for this chapter are available at: https://github.com/PacktPublishing/Network-Automation-with-Nautobot/tree/main/chapter-10

Setting up the environment

We are going to walk through how to build the exact environment we'll be using in this chapter. Even though we will be building on the Nautobot and Ansible installations from previous chapters, each step is provided again as a summary, just in case you want a fresh installation. It will also serve as a reference on your Nautobot journey.

Network topology

The network topology is the one area where we will not provide instructions on how to build a lab. We realize some people may prefer containerlab, netlab, Cisco Modeling Labs, GNS3, or just general virtual or physical devices.

However, let's review the network topology that we'll be automating.

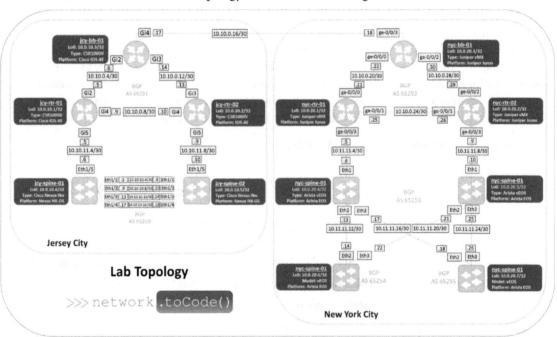

Figure 10.1 – Lab network topology

There are 12 devices (4 device types) split across 2 locations: Jersey City and New York City.

The following is the list of Jersey City devices. This is a Cisco-only site that is using Cisco routers (CSR1KVs) and Nexus switches (N9Kvs).

Device	Location	Role	Manufacturer	Device Type	Platform
jcy-bb-01	Jersey City	backbone	Cisco	CSR1000V	Cisco IOS-XE
jcy-rtr-01	Jersey City	router	Cisco	CSR1000V	Cisco IOS-XE
jcy-rtr-02	Jersey City	router	Cisco	CSR1000V	Cisco IOS-XE
jcy-spine-01	Jersey City	spine	Cisco	Nexus 9kv	Nexus NX-OS
jcy-spine-02	Jersey City	spine	Cisco	Nexus 9kv	Nexus NX-OS

Figure 10.2 – Cisco IOS and NX-OS inventory

The following is the list of New York City devices. This is a Juniper and Arista site that is using Juniper routers (vMXs) and Arista switches (vEOSs).

Device	Location	Role	Manufacturer	Device Type	Platform
nyc-bb-01	New York City	backbone	Juniper	vMX	Juniper Junos
nyc-rtr-01	New York City	router	Juniper	vMX	Juniper Junos
nyc-rtr-02	New York City	router	Juniper	vMX	Juniper Junos
nyc-spine-01	New York City	spine	Arista	vEOS	Arista EOS
nyc-spine-02	New York City	spine	Arista	vEOS	Arista EOS
nyc-leaf-01	New York City	leaf	Arista	vEOS	Arista EOS
nyc-leaf-02	New York City	leaf	Arista	vEOS	Arista EOS

Figure 10.3 – Arista EOS and Juniper Junos inventory

Each device has an FQDN using the `infra.ntc.com` domain; thus, it is possible to ping each host using its FQDN, for example, `ping jcy-bb-01.infra.ntc.com`. This is important because when you're using Nautobot with Ansible, Ansible uses the Nautobot device name `ansible_host` unless you set a Primary IPv4 or IPv6 address for that device.

Let's test one device (trust us – the rest work!):

```
root@nautobot-dev:~# ping jcy-spine-01.infra.ntc.com
PING jcy-spine-01.infra.ntc.com (172.18.0.20) 56(84) bytes of data.
64 bytes from ip-172-18-0-20.us-east-2.compute.internal (172.18.0.20):
icmp_seq=1 ttl=60 time=84.5 ms
```

It is worth noting that you can still follow along with this chapter even without a network to test on. You'll still learn how to perform several new operations and you'll be prepared for when you do have a network.

Linux host

As in other chapters, we're referencing and using a Linux host built from scratch. We chose to use one from DigitalOcean, but feel free to use any Ubuntu 22.04 installation if you want to follow along.

The Linux operations performed in this chapter use the `root` user unless the `nautobot` user is called out, just like we did in *Chapter 3*.

Here is the output of our system to show we're running on an Ubuntu 22.04 LTS machine:

```
root@nautobot-dev:~# lsb_release -a
No LSB modules are available.
Distributor ID: Ubuntu
Description:    Ubuntu 22.04.2 LTS
Release:        22.04
Codename:       jammy
```

Here, we verify the Python version we are using:

```
root@nautobot-dev:~# python3 --version
Python 3.10.12
root@nautobot-dev:~#
```

Let's move on to Ansible.

Ansible

We'll be using the Ubuntu 22.04 machine as our Ansible control host installing Ansible directly from PyPI. This will include installing Ansible, a few Ansible collections, and then their associated dependencies.

As you'll see in this chapter, for many of the things that were done in previous chapters, we are not restating what each command does or how it works, but rather providing a summary in a more concise way so you can reinstall and reconfigure as necessary, or just have it as a reference.

Let's build the Ansible control host using the following steps. Again, for the book, each package was installed as the `root` user. We realize that, for production, you are likely not using the `root` user, but in the same vein, you won't be installing these manually either. You should be using automation or containers:

```
# pip install ansible
# ansible-galaxy collection install networktocode.nautobot
# pip install napalm
# ansible-galaxy collection install napalm.napalm
# pip install pynautobot
# pip install netutils
```

We'll verify the Ansible version:

```
root@nautobot-dev:~# ansible --version
ansible [core 2.16.3]
  config file = None
  configured module search path = ['/root/.ansible/plugins/modules',
'/usr/share/ansible/plugins/modules']
  ansible python module location = /usr/local/lib/python3.10/dist-
packages/ansible
  ansible collection location = /root/.ansible/collections:/usr/share/
ansible/collections
  executable location = /usr/local/bin/ansible
  python version = 3.10.12 (main, Nov 20 2023, 15:14:05) [GCC 11.4.0]
(/usr/bin/python3)
  jinja version = 3.0.3
  libyaml = True
root@nautobot-dev:~#
```

You can also view the collections that were installed using the paths shown above:

```
root@nautobot-dev:~# ls /root/.ansible/collections/ansible_
collections/
```

At this point, Ansible is set up and ready to go!

Time to move on to Nautobot.

Nautobot

We'll now get Nautobot up and running. This section includes exactly the same process as was employed in *Chapter 3*, but in this chapter, we've pinned the Nautobot installation to v.2.1.4.

The following outlines the functional steps to get Nautobot up and running:

1. Update key system dependencies:

    ```
    apt update
    apt install -y git python3 python3-pip python3-venv python3-dev
    ```

2. Install Redis and Postgres:

    ```
    apt install -y redis-server
    apt install -y postgresql
    ```

3. Reboot (if prompted for kernel changes):

    ```
    reboot
    ```

4. Set up Postgres:

```
sudo -u postgres psql
CREATE DATABASE nautobot;
CREATE USER nautobot WITH PASSWORD 'nautobot123';
GRANT ALL PRIVILEGES ON DATABASE nautobot TO nautobot;
\q
```

5. Validate access:

```
psql --username nautobot --password --host localhost nautobot
```

6. Create and verify the Nautobot system user:

```
useradd --system --shell /bin/bash --create-home --home-dir /
opt/nautobot nautobot
eval echo ~nautobot
ls -l /opt/ | grep nautobot
```

7. Set up a Python virtual environment for the Nautobot user:

```
sudo -u nautobot python3 -m venv /opt/nautobot
echo "export NAUTOBOT_ROOT=/opt/nautobot" | sudo tee -a
~nautobot/.bashrc
sudo -iu nautobot
echo $NAUTOBOT_ROOT
```

8. Install Nautobot:

```
pip3 install --upgrade pip wheel
pip3 install nautobot==2.1.4
nautobot-server --version
```

9. Set up and configure the Nautobot settings:

```
nautobot-server init
vi $NAUTOBOT_ROOT/nautobot_config.py
# do not forget to uncomment required settings
echo "export NAUTOBOT_ALLOWED_HOSTS=*" | tee -a ~nautobot/.
bashrc
echo "export NAUTOBOT_DB_USER=nautobot" | tee -a ~nautobot/.
bashrc
echo "export NAUTOBOT_DB_PASSWORD=nautobot123" | tee -a
~nautobot/.bashrc
source ~/.bashrc
env | grep NAUTOBOT
nautobot-server migrate
nautobot-server createsuperuser
```

```
# do not forget to add a new user
nautobot-server collectstatic
nautobot-server check
```

10. Run the web server and log in to Nautobot:

```
nautobot-server runserver 0.0.0.0:8080 --insecure
# time to login to Nautobot on port 8080
```

11. Create a Nautobot celery worker:

```
nautobot-server celery worker
```

12. Set up a Nautobot web service:

```
touch $NAUTOBOT_ROOT/uwsgi.ini
# do not forget to update the uwsgi.ini file
nautobot-server start --ini /opt/nautobot/uwsgi.ini
```

13. Add and configure Linux services for Nautobot:

```
touch /etc/systemd/system/nautobot.service
vi /etc/systemd/system/nautobot.service
# time to update service files
systemctl daemon-reload
systemctl enable --now nautobot
systemctl status nautobot.service
touch /etc/systemd/system/nautobot-worker.service
vi /etc/systemd/system/nautobot-worker.service
# time to update service files
systemctl daemon-reload
systemctl enable --now nautobot-worker
systemctl status nautobot-worker.service
touch /etc/systemd/system/nautobot-scheduler.service
vi /etc/systemd/system/nautobot-scheduler.service
# time to update service files
systemctl daemon-reload
systemctl enable --now nautobot-scheduler
systemctl status nautobot-scheduler.service
```

14. Log in to Nautobot using port 8001:

```
# time to access Nautobot on port 8001
```

The next step is to ensure there is at least one API token that will be used for CRUD operations in the following sections.

Log in to Nautobot, and then do the following:

1. Click **admin** in the lower-left corner (your logged-in user).
2. Click **Profile**.
3. Click **API Tokens**.
4. Click + **Add a token**.

> **Note**
> If you already have a token, you can use that. Otherwise, you can still use a new token for this chapter.

5. Add a new token with the following key:

    `nautobot-book-token-123456789-abcdefghij`

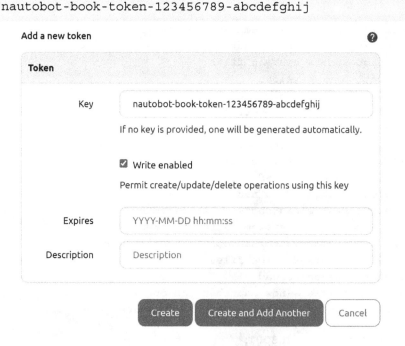

Figure 10.4 – Add API token

After adding it, you'll see it in the **API Tokens** list.

>>> **API Tokens**

Profile			
	⊙┭ nautobot-book-token-123456789-abcdefghij		Copy Edit Delete
Preferences	Created	Expires	Create/edit/delete operations
Change Password	Feb. 17, 2024	Never	Enabled
API Tokens			
Advanced	+ Add a token		
Settings			

Figure 10.5 – View newly created API token

We'll see this token in the next section when we complete the setup of the Ansible environment.

The book's Git repo

This section ensures your environment is set up with all required environment variables and playbooks needed to begin automating with Nautobot and Ansible.

Let's download the Git repo built for this book:

`https://github.com/PacktPublishing/Network-Automation-with-Nautobot`

For this chapter, the repo contains environment variables, config contexts, dynamic inventory, playbooks, and a Jinja template we'll be using.

Clone the repository:

```
root@nautobot-dev:~# git clone https://github.com/PacktPublishing/
Network-Automation-with-Nautobot.git nautobot-book
```

Navigate to the `chapter-10` directory:

```
root@nautobot-dev:~#  cd nautobot-book/chapter-10
```

Copy the provided example creds file and create a creds.env file:

```
root@nautobot-dev:~/nautobot-book/chapter-10#  cp creds.example.env
creds.env
```

Update `creds.env` with the following – ensure you use the right information for your devices:

```
export NAUTOBOT_TOKEN=nautobot-book-token-123456789-abcdefghij
export NAUTOBOT_URL=http://localhost:8001
```

```
export NTC_NET_USERNAME=<enter network device username>
export NTC_NET_PASSWORD=<enter network device password>
```

Now we'll source the file to bring the environment variables into our shell environment:

```
root@nautobot-dev:~/nautobot-book/chapter-10# source creds.env
root@nautobot-dev:~/nautobot-book/chapter-10# cat creds.env   # shows
ours
export NAUTOBOT_TOKEN=nautobot-book-token-123456789-abcdefghij
export NAUTOBOT_URL=http://localhost:8001

export ansible_user=ntc
export NTC_NET_PASSWORD=ntc123

root@nautobot-dev:~/nautobot-book/chapter-10#
Let's show the Nautobot env vars in our environment:
root@nautobot-dev:~/nautobot-book/chapter-10# env | grep NAUTOBOT
NAUTOBOT_TOKEN=nautobot-book-token-123456789-abcdefghij
NAUTOBOT_URL=http://localhost:8001
root@nautobot-dev:~/nautobot-book/chapter-10#
```

Now we're ready to start running Ansible playbooks to add the network devices to Nautobot.

Adding data to Nautobot with Ansible

Based on a fresh install, Nautobot still doesn't have data in it, so it is pretty much useless for us. This section is going to show how Ansible can be used to add devices (and much more data) to Nautobot.

Ensure you are in the chapter-10 directory in the book's repo.

Execute the 01-add-locations.yml playbook:

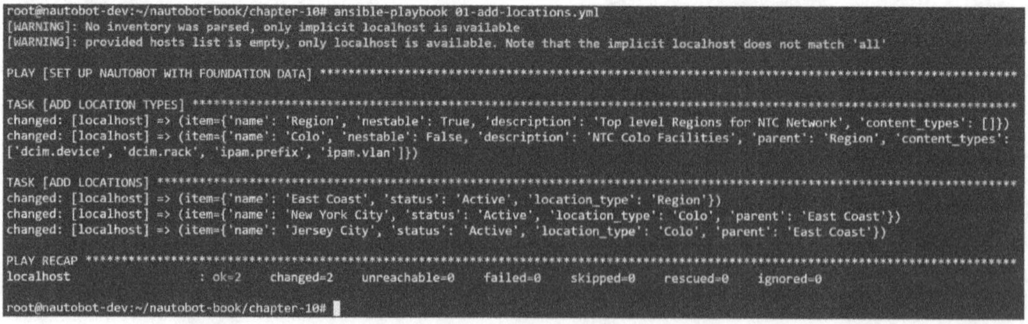

Figure 10.6 – Adding data to Nautobot with an Ansible playbook

This playbook uses two Ansible modules in the Nautobot Ansible collection: `networktocode.nautobot.location_type` and `networktocode.nautobot.location`.

In our network, we are creating two location types – `Region` and `Colo` – and creating three locations: `East Coast` (which is of type `Region`) and then `Jersey City` and `New York City` (which are of type `Colo`).

> **Note**
>
> Given this book assumes prior knowledge of Ansible, we are not going to cover the details of each module. You can use `ansible-doc` or view the Ansible docs to understand the arguments that each module supports. Additionally, the data that these playbooks have embedded is to make them easier to read and learn from. In production, you'd want to use a more holistic data structure to reduce the amount of duplicate data.

Execute the `02-add-device-metadata.yml` playbook.

This playbook uses five Ansible modules in the Nautobot Ansible collection: `networktocode.nautobot.manufacturer`, `networktocode.nautobot.platform`, `networktocode.nautobot.role`, `networktocode.nautobot.device_type`, and `networktocode.nautobot.device_interface_template`.

In our network, we are creating the following:

- Three manufacturers: Cisco, Arista, and Juniper
- Four platforms: Cisco IOS-XE, Nexus NX-OS, Juniper Junos, and Arista EOS
- Four device types: CSR1000V, Nexus 9Kv, vEOS, and vMX
- Interfaces for each device type — these are broken down in the playbook to make the playbook more readable
- Four device roles: spine, leaf, router, and backbone

Execute the `03-add-devices.yml` playbook.

This playbook uses one Ansible module in the Nautobot Ansible collection: `networktocode.nautobot.device`.

This playbook adds all 12 devices to Nautobot using the FQDN as its name and assigns a location, device type, platform, status, and role.

As stated earlier, this playbook does not add a primary IP address per device since the device name is an FQDN. However, if you do not use FQDNs, you can add the `primary_ip4: "a.b.c.d/yz"` (or `primary_ip6`) parameters to your playbook. Just remember to ensure those addresses are added to Nautobot already and assigned to an interface on your device (as we did for the `05-add-prefix-ips.yml` playbook).

Execute the `04-add-cables.yml` playbook.

This playbook uses one Ansible module in the Nautobot Ansible collection: `networktocode.nautobot.device`.

In our network, we are creating 19 physical cables. Each of these cables can be seen in the diagram shown in *Figure 10.1*, earlier in this chapter.

Execute the `05-add-prefix-ips.yml` playbook.

This playbook uses three Ansible modules in the Nautobot Ansible collection: `networktocode.nautobot.prefix`, `networktocode.nautobot.ip_address`, and `networktocode.nautobot.ip_address_to_interface`.

In our network, we are creating five prefixes that are used primarily for the loopback interfaces and transit links and then 62 IP addresses that are all assigned to Layer 3 interfaces on the devices.

> **Note**
> Due to our lab setup, we need to assign the same IP to our `mgmt` interfaces so we'll only see 50 IP addresses in Nautobot.

Execute the `06-add-devices-macs.yml` playbook.

This playbook uses one Ansible module in the Nautobot Ansible collection: `networktocode.nautobot.device_interface`.

This is used to assign a static MAC address for Nexus 9Kv due to the default being the same MAC on routed interfaces. This was a workaround for our lab but still shows how you can assign a MAC address to an interface in Nautobot using Ansible.

By now, you've probably verified objects were being added to Nautobot as you ran playbooks, but after you run all of them, you'll see the following object counts on the Nautobot home screen:

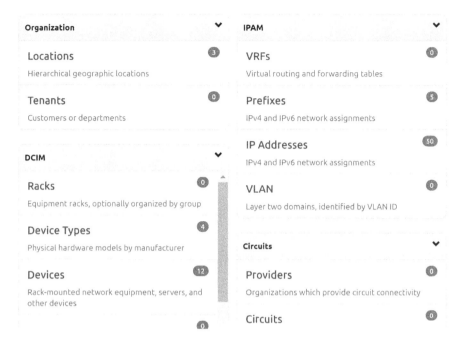

Figure 10.7 – Viewing data populated from Ansible

Since we now have data in Nautobot, we are ready to start doing some network automation powered by Nautobot.

Setting up a dynamic inventory

The `chapter-09` directory also has a few dynamic inventory files we are going to use in this chapter. We are going to start with a stripped-down version to ensure the inventory and connectivity from Ansible to Nautobot are working as expected.

Let's view the `nautobot-inventory-01.yml` inventory file:

```
root@nautobot-dev:~/nautobot-book/chapter-09# cat nautobot-
inventory-01.yml
---
plugin: networktocode.nautobot.gql_inventory
query:
  devices:
    name:

root@nautobot-dev:~/nautobot-book/chapter-09#
```

This is a basic inventory file that will perform a GraphQL query to Nautobot only asking for device hostnames. This uses `networktocode.nautobot.gql_inventory`, which we introduced in *Chapter 8*.

Let's test it out using `ansible-inventory`:

```
root@nautobot-dev:~/nautobot-book/chapter-09# ansible-inventory -i
nautobot-inventory-01.yml --list
{
    "_meta": {
        "hostvars": {
            "jcy-bb-01.infra.ntc.com": {
                "ansible_host": "jcy-bb-01.infra.ntc.com",
                "ansible_network_os": "cisco.ios.ios",
                "name": "jcy-bb-01.infra.ntc.com",
                "platform": {
                    "napalm_driver": "ios"
                # all all other hosts
                #  omitted for brevity
    "all": {
        "children": [
            "ungrouped"
        ]
    },
    "ungrouped": {
        "hosts": [
            "jcy-bb-01.infra.ntc.com",
            "jcy-rtr-01.infra.ntc.com",
            # all other hosts
        # removed for brevity
root@nautobot-dev:~/nautobot-book/chapter-09#
```

So, we've tested that Ansible can retrieve the inventory from Nautobot. There are no groups for us to use in our playbooks other than `"all"` and we can also see there are four "hostvars" returned for each host. We'll be sure to use these in upcoming playbooks.

> **Tip**
>
> When you start out with Nautobot and Ansible, use separate, small inventories whenever possible. The inventories should be defined as one for production devices that have many groups, one for `stage/UAT/sub-prod` (or whatever the name of the environments are before production). There may be several development inventories, which will be a bit more individualized according to what devices are used by each network automation engineer. It is not recommended to have inventories per workflow, for example, there should not be an inventory for code upgrades and a separate inventory for switchport management. You should put production devices into the production inventory and leverage groups to differentiate between the device types, the locations, and so on.
>
> An inventory should also be as lightweight as possible. That said, the inventory should only contain the information that is needed to get connected to the network device. Then, the first set of tasks within the first play of the playbook should be to gather additional data from Nautobot as variables and attach them to the inventory. This allows the inventory to be light and nimble for all playbooks, and only gather what you need. We'll talk more about this when we cover config contexts later in the chapter.
>
> For example, in a playbook where you are looking to do code upgrades on a device, you do not need to know about each interface, their associated VLANs, and IP addresses to complete a code upgrade. You may need some of this information for doing pre/post verification within the playbook, but the act of doing just the code upgrade does not need this information.

Backing up network devices

The next step is to use the inventory file we just tested (`nautobot-inventory-01.yml`) and verified with `ansible-inventory` to connect to those devices and perform a backup. Make sure to start small and expand from there. Starting small here is just performing configuration backup.

The nice thing about this is we can use a single multi-vendor Ansible core module, as shown here:

```
- name: "BACKUP ALL CONFIGURATIONS"
  ansible.netcommon.cli_config:
    backup: true
    backup_options:
      filename: "{{ ansible_host }}.cfg"
      dir_path: "{{ backups_dir }}/"
```

The `backups_dir` directory is a directory defined earlier in the playbook. Be sure to look at it on your filesystem or in the Git repo.

Execute the `backups` playbook:

```
# ansible-playbook -i nautobot-inventory-01.yml pb_backup_network.yml
```

Once the playbook has finished, you can view your new backups:

```
#  ls network_backups/
jcy-bb-01.infra.ntc.com.cfg    jcy-spine-01.infra.ntc.com.cfg   nyc-
leaf-01.infra.ntc.com.cfg
jcy-rtr-01.infra.ntc.com.cfg   jcy-spine-02.infra.ntc.com.cfg   nyc-
leaf-02.infra.ntc.com.cfg
# some files removed for brevity
```

As you can see, once you have a functional Ansible and Nautobot installation, there isn't too much to do to start using an Ansible dynamic inventory with Nautobot as its source.

We'll continue to build on this by showing more network automation powered by Nautobot.

> **Tip**
> If you're looking for a functioning backup solution, be sure to check out the Nautobot Golden Config App. It'll likely save you some cycles instead of building something custom for a task as common as backups!

Performing a config replace with Nautobot, NAPALM, and Ansible on Arista and Juniper devices

Because different devices support different ways of interacting with their operating systems, this section focuses on only automating Arista EOS and Juniper Junos devices.

We are going to do what is called a full configuration replacement on these devices, but not using built-in Ansible modules. We are going to do this using the NAPALM Ansible collection. We'll explain this more when we get to the playbook.

Before we get to the playbook, let's look at an updated Nautobot inventory:

```
# cat nautobot-inventory-02.yml
---
plugin: networktocode.nautobot.gql_inventory
query:
  devices:
    name:
    location:
      name:
group_by:
  - location.name
  - platform.napalm_driver
```

Take note of the changes:

- There is now a location and its associated name per device

- There is now a group_by section and there are groups added by location name and the associated NAPALM driver

As you're going to see in the next playbook (pb_deploy_network_eos_junos.yml), we use these new variables and groups like this:

```
---

- name: "DEPLOY EOS & JUNOS CONFIGURATIONS"
  hosts: "eos,junos"
  gather_facts: false
  connection: "local"

  vars:
    project_path: "{{ playbook_dir }}"
    backups_dir: "{{ project_path }}/network_backups"
    diffs_dir: "{{ project_path }}/diffs"
    config_path: "{{ project_path }}/baseline/{{ location['name'] }}"
```

Because we added a group_by section for platform.napalm_driver, we are able to automate just the eos and junos devices as shown in the play declaration. The repo also has a predefined location where "baseline" configurations are stored. There is a subdirectory per location. Thus in order to dynamically pull the correct config file based on location, we added the location and its name to the GraphQL query.

> **Note**
>
> This book does not provide an overview of NAPALM. If you are not familiar with NAPALM, you should check out its documentation at https://napalm.readthedocs.io/en/latest/. If you recall, we also mentioned NAPALM in *Chapter 6* due to Nautobot's built-in integration that fetches device and LLDP information from network devices using NAPALM.

The remainder of the NAPALM config replace playbook (pb_deploy_network_eos_junos.yml) looks like this:

```
tasks:
  - name: "BUILD DIFFS DIRECTORY"
    ansible.builtin.file:
      path: "{{ diffs_dir }}/"
      state: "directory"
```

```
      run_once: true

  - name: "DEPLOY AND REPLACE JUNOS-EOS CONFIGURATIONS"
    napalm_install_config:
      provider:
        username: "{{ lookup('env', 'ansible_user') }}"
        password: "{{ lookup('env', 'NTC_NET_PASSWORD') }}"
        host: "{{ ansible_host }}"
      config_file: "{{ config_path }}/{{ ansible_host }}.cfg"
      diff_file: "{{ project_path }}/diffs/{{ ansible_host }}.diffs"
      replace_config: true
      commit_changes: true
      dev_os: "{{ platform['napalm_driver'] }}"
```

This playbook does not care what is currently configured on the device. It is a declarative way to manage the network. Many organizations seek to manage their network this way, but it is a journey and, more realistically, network teams manage features declaratively before they get to manage full configs this way. However, it is still good to understand what is possible with the automation readily available today.

The NAPALM module references configs that are pre-built in the chapter-09 directory. In reality, you'd want to build out the required Jinja templates, which would generate the configs you'd be deploying to the network.

Let's run the playbook:

```
# ansible-playbook -i nautobot-inventory-02.yml pb_deploy_network_eos_
junos.yml
```

Feel free to make changes to the devices or the config files and analyze the diffs directory that is automatically created during the deployment. This allows you to keep track of the changes you're overwriting when you're doing a full config replacement.

Performing config changes with Nautobot and Ansible for Cisco IOS devices

Now we'll shift gears from using multi-vendor modules to a vendor-specific module. We are going to update the dynamic and deploy configurations to the Cisco IOS devices.

Let's look at the updated inventory file we'll be using:

```
root@nautobot-dev:~/nautobot-book/chapter-10# cat nautobot-
inventory-03.yml
---
plugin: networktocode.nautobot.gql_inventory
```

```
query:
  devices:
    name:
    location:
      name:
    config_context:
    interfaces:
      name:
      ip_addresses:
        address:
      mac_address:
  group_by:
    - location.name
    - platform.napalm_driver
```

This new inventory query adds two new fields to the response we're going to receive from Nautobot. It'll now return config contexts and interfaces.

Remember that config contexts were first introduced in *Chapter 5*. We also highlighted that config contexts can be configured manually in Nautobot via the UI or added using Git as a data source. In this chapter, we are going to add them using a Git data source.

Browse to this book's Git repo and look at the `config_contexts` directory (you can also do this on your filesystem if preferred):

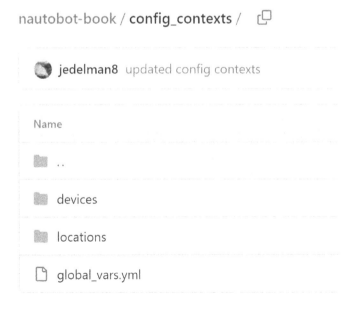

Figure 10.8 – Browsing the Git repo viewing config contexts

Click through to the `devices` directory and view the config context for `jcy-bb-01.infra.ntc.com.yml`.

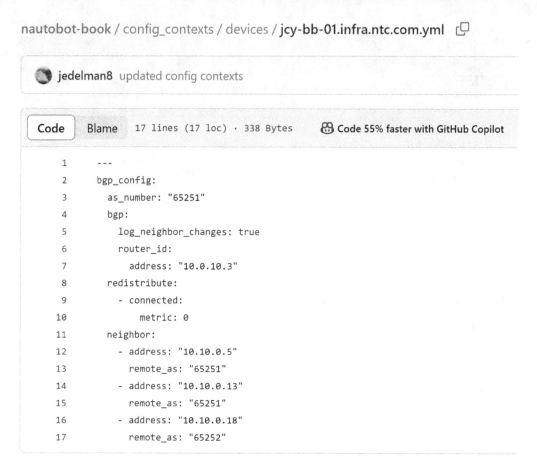

nautobot-book / config_contexts / devices / **jcy-bb-01.infra.ntc.com.yml**

jedelman8 updated config contexts

Code   Blame   17 lines (17 loc) · 338 Bytes   Code 55% faster with GitHub Copilot

```
 1    ---
 2    bgp_config:
 3      as_number: "65251"
 4      bgp:
 5        log_neighbor_changes: true
 6        router_id:
 7          address: "10.0.10.3"
 8      redistribute:
 9        - connected:
10            metric: 0
11      neighbor:
12        - address: "10.10.0.5"
13          remote_as: "65251"
14        - address: "10.10.0.13"
15          remote_as: "65251"
16        - address: "10.10.0.18"
17          remote_as: "65252"
```

Figure 10.9 – Viewing a single (local) device's config context

The data you see here for the Nexus switches will be used by the devices we are about to automate.

It is worth noting that the data structure built for these exercises was built to match the default data structure required by the `ios_bgp_global` and `ios_l3_interfaces` Ansible modules. In reality, you may prefer to model your data first, independent of any module, which would then require more data manipulation in Ansible.

Let's add these config contexts to our Nautobot instance.

Navigate to the Nautobot UI and then **Extensibility** > **Git Repositories**.

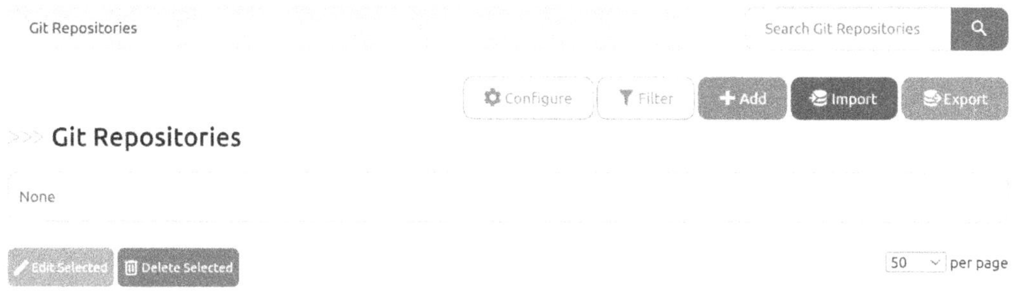

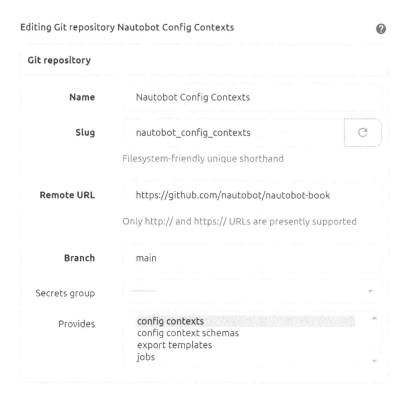

Figure 10.10 – Git Repositories table

Once you see the **Git Repositories** table, click +**Add**.

Ensure you use the `https` URL for the book's Git repository, **Branch** is set to **main**, and **config contexts** is selected in the **Provides** list.

Figure 10.11 – Adding a Git data source for config contexts

When done, click **Create & Sync**. Once you do, you'll see the status of the job that is syncing the repo to Nautobot. If something doesn't work, check your URL and branch.

> **Note**
> If you choose to use a different Git repo that is not public, you will need to create secrets and a secret group first and then reference it when you're adding the Git data source to Nautobot.

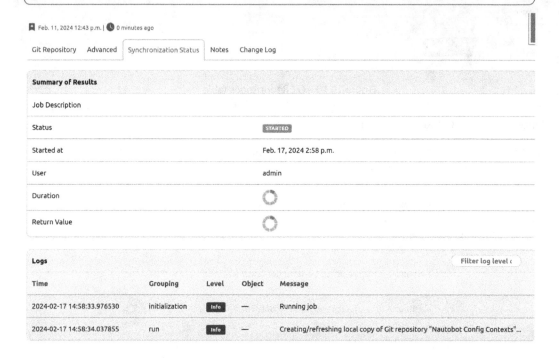

Figure 10.12 – Viewing the Git sync process (it is a Nautobot job)

Once done, you can view your new data source in the **Git Repositories** list:

Figure 10.13 – View Git Repositories list and Sync Status

Going one step further, you can also view your new config contexts by clicking **EXTENSIBILITY |
Config Contexts**. This will show all contexts that are broader than a single device:

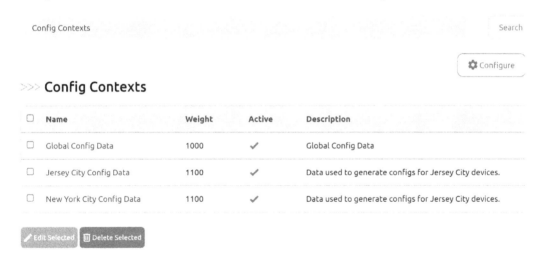

Figure 10.14 – Viewing Nautobot Config Contexts

Let's also view a local (single-device) config context:

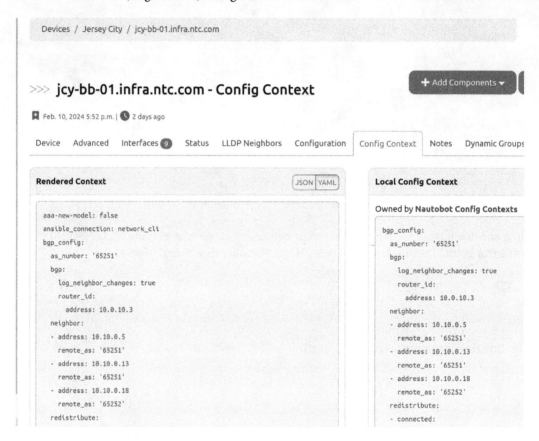

Figure 10.15 – Viewing a device-specific local config context

We now have config context data added to Nautobot – and remember, we also added `config_context` to the dynamic inventory GraphQL query. Before we automate devices with this data, let's examine the data being returned with `ansible-inventory`.

Let's run the following command as in previous examples:

```
# ansible-inventory -i nautobot-inventory-03.yml --list
```

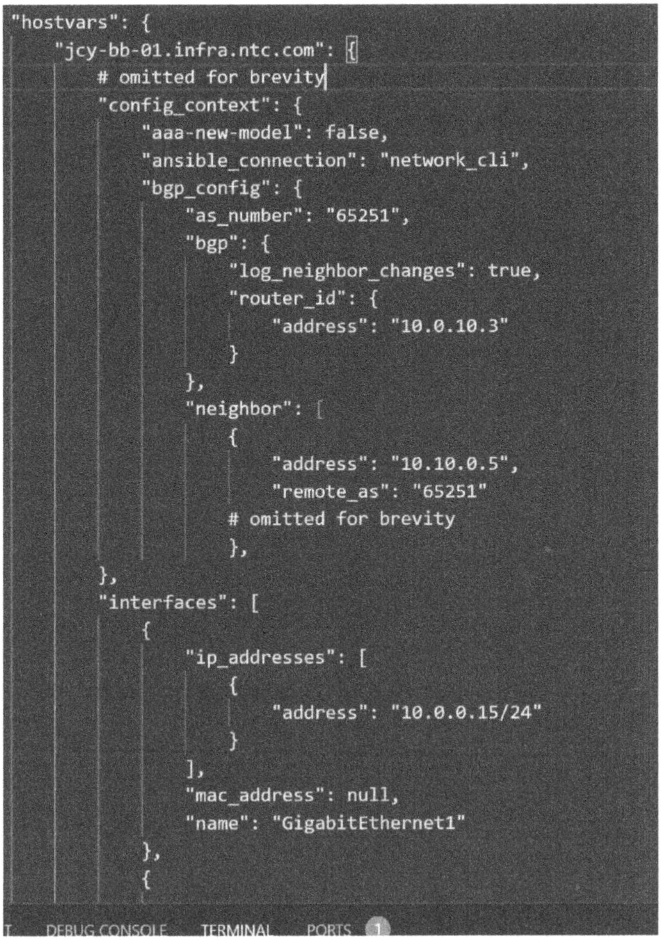

```
"hostvars": {
    "jcy-bb-01.infra.ntc.com": {
        # omitted for brevity
        "config_context": {
            "aaa-new-model": false,
            "ansible_connection": "network_cli",
            "bgp_config": {
                "as_number": "65251",
                "bgp": {
                    "log_neighbor_changes": true,
                    "router_id": {
                        "address": "10.0.10.3"
                    }
                },
                "neighbor": [
                    {
                        "address": "10.10.0.5",
                        "remote_as": "65251"
                        # omitted for brevity
                    },
            },
            "interfaces": [
                {
                    "ip_addresses": [
                        {
                            "address": "10.0.0.15/24"
                        }
                    ],
                    "mac_address": null,
                    "name": "GigabitEthernet1"
                },
                {
```

DEBUG CONSOLE   TERMINAL   PORTS   1

Figure 10.16 – Viewing hostvars with interfaces and config context(s)

You'll notice that `interfaces` is a hostvar and that `bgp_config` is stored under the `config_context` key. We need to know this to use the data in our playbooks and templates.

Let's look at the `pb_deploy_ios_cfg.yml` playbook that'll be using the config contexts and interfaces.

We will first see the standard play definition:

```
---

- name: "DEPLOY DEVICE CONFIGURATIONS"
  hosts: "ios"
  gather_facts: false
  connection: "ansible.netcommon.network_cli"

  vars:
    ansible_password: "{{ lookup('env', 'NTC_NET_PASSWORD') }}"
```

The play declaration is followed by two tasks.

The first uses the `ios_l3_interfaces` module to add and configure the IP addresses from Nautobot on the devices. It is worth noting that we are iterating over interfaces even though it is possible to use a `list` object for the `config` parameter. This was chosen to focus on using the interfaces from Nautobot in Ansible instead of shifting the focus to what would be required to get the data exactly as we'd need it. Instead, we chained together a few Jinja filters. These filters do not return `GigabitEthernet1` (so we don't change management) and do not return any interface that does not have an IP address in Nautobot, keeping things nice and simple.

```
tasks:

  - name: "CONFIGURE L3 INTERFACES"
    cisco.ios.ios_l3_interfaces:
      config:
        - name: "{{ item['name'] }}"
          ipv4:
            - address: "{{ item['ip_addresses'][0]['address'] }}"
      state: replaced
    loop: "{{ interfaces | rejectattr('name', 'equalto', 'GigabitEthernet1') | selectattr('ip_addresses', 'ne', []) | list }}"

  - name: "DEPLOY BGP CONFIGURATION"
    cisco.ios.ios_bgp_global:
      config:  "{{ config_context['bgp_config'] }}"
      state: replaced
```

Figure 10.17 – Interfaces and BGP configuration for IOS devices

You can now run the playbook:

```
# ansible-playbook -i nautobot-inventory-03.yml pb_deploy_ios_cfg.yml
```

> **Note**
>
> Config contexts can add significant load when used in dynamic inventories and are run at the beginning of the playbook as we've been showing. Consider hundreds or thousands of devices. It would not be recommended to query for config contexts in the dynamic inventory configuration. The following snippet shows what is not recommended for medium to large production deployments:
>
> ```
> ---
> plugin: networktocode.nautobot.gql_inventory
> query:
>   devices:
>     name:
>     location:
>       name:
>     config_context:  # remove this line for production deployments
> ```

Instead, another, more scalable, option than gathering config contexts in dynamic inventory is using the `networktocode.nautobot.query_graphql` module in the Ansible Nautobot collection. This is more scalable because the data is gathered per device when that device is automated and is in contrast to a dynamic inventory, which queries Nautobot for *all* devices at playbook runtime.

You can then call the module when you need it, also taking advantage of Ansible's built-in concurrency for devices:

```
- name: "GET DEVICE INFO FROM GRAPHQL"
  networktocode.nautobot.query_graphql:
    url: "{{ nautobot_url }}"
    token: "{{ nautobot_token }}"
    validate_certs: false
    query: "{{ query_string }}"
    update_hostvars: true # bypass using the "config_context" key to
access config context data
```

Performing config changes with Nautobot and Ansible for Cisco NX-OS devices

Next, we are going to shift gears again and do some automation for Cisco Nexus switches. We already covered many of the concepts used, so we are going to jump right into the playbook.

The play declaration is just as we've seen in previous playbooks, but now we'll automate the Nexus devices by using the `nxos` group that was built by creating groups from the NAPALM platform driver:

```
---

- name: "DEPLOY DEVICE CONFIGURATIONS"
  hosts: "nxos"
  gather_facts: false
  connection: "ansible.netcommon.network_cli"

  vars:
    ansible_ssh_password: "{{ lookup('env', 'NTC_NET_PASSWORD') }}"
```

This time, we are going to generate configs from a Jinja template. The next task ensures there is a directory created per location. For production use cases, you'd want to do this task in a different play so we didn't need to run it for every device:

```
  tasks:

    - name: "BUILD CONFIGS DIRECTORY"
      ansible.builtin.file:
```

```
        path: "configs/{{ location['name'] }}/"
        state: "directory"
```

The next two tasks generate the desired configs using the `ansible.builtin.template` and `cisco.nxos.nxos_config` modules:

```
    - name: "GENERATE BGP CONFIGS"
      ansible.builtin.template:
        src: "nxos_config.j2"
        dest: "configs/{{ location['name'] }}/{{ ansible_host }}.cfg"

    - name: "ENSURE INTERFACES ARE L3"
      cisco.nxos.nxos_config:
        src: "configs/{{ location['name'] }}/{{ ansible_host }}.cfg"
```

If you explore the `nxos_config.j2` template (found in the `chapter-09/templates/` folder), you'll notice that it is similar to what we previously integrated into the playbook, but with some logic added for some required Nexus-specific requirements.

Let's run this final playbook:

```
# ansible-playbook -i nautobot-inventory-03.yml pb_deploy_nxos_cfg.yml
```

If you didn't have Nexus or other platforms to automate during this chapter, we still encourage you to review all files used in this chapter.

Take note that there is also a `nautobot-inventory-final.yml` file in the directory that shows even more data being returned, along with several more groups. This creates more flexibility based on your Ansible automation requirements.

Managing data with config contexts and using Git

It may be evident by now, but let's look at common scenarios where you would choose to use config contexts with Nautobot and Git:

- Adding a network change locally, for example, adding a new BGP peer in the right config context YAML file:

  - Consider creating a JSON Schema definition to ensure the data adheres to your standard data model

  - You should follow good Git practices as well, for example, create a new branch that holds the new changes

- Opening a pull request to the repo.

- Allowing your team to review, comment on, and then finally merge your pull request.

- Syncing a Nautobot Git repo (on a schedule, in the UI, or using the API, and possibly even from a webhook upon merge).

- Executing automation such as an Ansible playbook that'll then use the updated data, such as adding a new BGP peer.

For more use cases, please do not forget to consider how to add a specific feature, remove a feature, or simply add to it. Of course, always do ample testing to account for your specific design and configurations.

For BGP specifically, you can also explore using the Nautobot BGP Model App (`https://docs.nautobot.com/projects/bgp-models/en/latest/`) in contrast to YAML data. This allows users to easily update the data in the UI with predefined opinionated BGP models.

Nautobot jobs versus Ansible playbooks

This book has its focus on Nautobot, but we must all recognize the role Ansible has played and continues to play in the world of network automation. We thought it was good to call out some reasons why you may opt to use Ansible playbooks versus Nautobot jobs (which are covered in the next chapter). Here is what we've generally seen in the industry over the last few years:

- Ansible has a lower barrier to entry because it does not require coding but Ansible is more difficult to use for more complex use cases

- Python/Jobs have a higher barrier to entry because users have to learn coding, but it allows for more complex use cases further down the automation path

- Ansible has been the de facto standard for getting started with network automation

- Python and the use of Netmiko, NAPALM, and Nornir is a close second to Ansible

- Due to the lower barrier of entry of Ansible and enterprises adopting Ansible AWX or **Ansible Automation Platform** (**AAP**), there are usually many more engineers creating playbooks than there are creating Python scripts

- The more Ansible playbooks a team writes, the more Python they end up creating (for custom plugins, filters, modules, etc.), which naturally improves a team's Python skills

- As Python skills improve, there is a natural tendency to write more Python

- These Python scripts end up being converted to Nautobot jobs because jobs have direct access to the data (you'll see this in the next chapter)

So, where does this leave us? What should be the path forward if this sounds like your team?

- For the most common automation requirements, such as config backups, generating intended configurations, doing compliance, and even remediation, Nautobot Golden Config has them built atop Python using Nornir. Our belief is that you shouldn't reinvent the wheel.

- Once broad solutions are built, look for what scales across the team or organization.

- Continue to use Ansible playbooks for automation that crosses different teams and organizations if you've already seen success with it.

- Use Nautobot jobs for automation when you need to interact with data in Nautobot heavily, when it is a smaller team doing the automation (or when there are inter-team software standards), or when you're finding it difficult to create playbooks that need heavy data manipulation, loops, and so on that are forcing you to break into Python anyway.

Going beyond Nautobot jobs (Python) versus Ansible playbooks, you must consider all the features that enterprises typically require as well, such as logging, inventory integrations, fine-grained role-based access control, secret integration, and workflow management and orchestration.

Overall, remember it is a journey and no two teams are alike. While goals are often similar between teams, skills, people, and culture are not.

Summary

In this chapter, we explored a few different integration points in Nautobot and Ansible. We looked at a few different ways to manage configurations, exploring multi-vendor modules (the NAPALM and Core modules) that both can send and replace full configurations and perform backups; perform specific configuration changes using Ansible resource modules; and finally, we looked at template configs and sending them with a platform-specific config module.

In the next chapter, we'll take a deep dive into Nautobot jobs, exploring the power of Nautobot data combined with Python for network automation.

11

Automating Networks with Nautobot Jobs

In the previous chapter, we focused on using Nautobot with Ansible. We also touched on the value of Nautobot Jobs and how they may compare to Ansible playbooks. In this chapter, we'll dive deep into Nautobot Jobs. Nautobot Jobs are a great opportunity to bring automation into your network while using Nautobot as a full-blown automation platform. By bringing together the power of direct database access to your Nautobot data with the ability to run Python code, you can iterate very quickly and create powerful automation solutions.

This chapter begins with an overview and an introduction to the Django **Object-Relational Mapper** (**ORM**). We'll walk through how to create Jobs using the ORM, migrate Python scripts to Nautobot and run them as Jobs, and orchestrate Jobs together. After, we'll cover permissions, logging, scheduling, and approvals before covering how to add and distribute Jobs to Nautobot.

In this chapter, we will cover the following topics:

- Nautobot Jobs overview
- Introduction to the Django ORM
- Learning about the Nautobot Shell
- Adding Jobs to Nautobot
- Creating your first Nautobot Job
- Using Jobs to populate data in Nautobot
- Working with Job buttons, Job Hooks, APIs, and more

Technical requirements

Code files and other resources for this chapter are available at: `https://github.com/PacktPublishing/Network-Automation-with-Nautobot/tree/main/chapter-11`

Nautobot Jobs overview

Nautobot Jobs are the primary automation engines within Nautobot and can automate anything you can imagine. Because Nautobot Jobs are Python scripts or applications, if you can do it in Python, you can do it in a Nautobot Job. Nautobot Jobs have access to the Nautobot Django ORM, which means you can get direct access to the data in the database. This, in turn, means you can read, update, create, and delete objects with advanced ORM filter and query capabilities within Nautobot from a Job.

The primary use cases for Jobs are as follows:

- **Validating and verifying data in Nautobot**: For example, are all devices that should be in a specific rack in the rack? Do all devices have a primary IPv4 and IPv6 address? Do all devices have a serial number?

- **Generating files (and reports)**: Ensure these can be downloaded later. The reports can be for data hygiene or the results of show commands after connecting to devices.

- **Performing actual network automation tasks**: It is quite common to use common network automation libraries such as Netmiko, NAPALM, and Nornir inside Jobs to do automation directly from Nautobot.

- **Self-service network automation**: With predefined patterns, users can easily create self-service forms that have direct access to the data in Nautobot, improving overall accessibility for network automation. This makes it much easier for teams to rally around someone's scripts instead of using static user inputs or something such as command-line arguments in a Python script.

Before we dive into Jobs, because Jobs have access to the Django ORM, we thought we'd start by providing an introduction to the ORM.

Introduction to the Django ORM

The Django ORM acts as a middleware between Python and the data in the database. It is a tool that comes built into Django projects, and Nautobot is a Python Django project. The ORM provides an object-oriented method to interact with relational databases. The two supported database types for Nautobot are Postgres and MySQL. Nautobot packages ORM utilities to interact with both Postgres and MySQL in the same fashion.

The Python API for the ORM allows you to perform all of the actions that would be associated with **create, read, update, delete (CRUD)** operations on the data in Nautobot.

The ORM is different from what we've seen so far. Let's compare it with `pynautobot`, which we looked at in *Chapter 9*.

When creating a location with `pynautobot`, we must use the following code:

```
import pynautobot

nautobot = pynautobot.api(
    url="https://demo.nautobot.com",
    token="aaaaaaaaaaaaaaaaaaaaaaaaaaaaaaaaaaaaaaaa"
    )

nautobot.dcim.locations.create(name="ATW01", location_type="Site",
parent="Global Region", status="Active")
```

When working with the Django ORM, we'd be using the following code (while also creating the location type):

```
from nautobot.dcim.models.locations import Location, LocationType
from nautobot.extras.models.statuses import Status

status_active = Status.objects.get(name="Active")
location_type = LocationType.objects.get(name="Site")
location = Location.objects.create(name="ATW01", location_
type=location_type, status=status_active)
```

Don't worry – we'll show you how to start using the ORM in the next section. For now, we are showing it as a comparison to `pynautobot`.

To access objects within the ORM, the first thing to do is import the models into the Python environment. In the previous example, the `Status` model and the `Location` model were imported, so they can be referenced. The ORM doesn't do any lookups or have any shortcuts built in like the REST API does, so we need to be able to access the **Active** status object that will be used when creating the location, and then pass it as a parameter when creating a location. Finally, we can create the location with the `Location.objects.create()` method.

When working to develop the various object changes, it is common practice to either use the Nautobot Shell or Nautobot Shell Plus environments.

Nautobot Shell (`nautobot-server nbshell`) brings you into a Python interpreter session. You will need to import each item that you are using.

Shell Plus (`nautobot-server shell_plus`) brings you into a similar environment but looks more like an iPython environment, and it automatically imports all the Nautobot models that are in your environment.

Let's dive into the Nautobot Shell and the ORM so that you can start your Nautobot Job journey.

Learning about the Nautobot Shell and ORM

To access Nautobot Shell on your Nautobot instance, you will need to access the terminal of your Nautobot instance. While at the command line, as the Nautobot user, run the `nautobot-server shell_plus` command. If you have IPython installed, you will get a similar IPython experience.

Based on our setup, we'll be keeping two terminal sessions open: one to manage working with Jobs and restarting services as the `root` user, and the other for `nautobot-server` commands as the `nautobot` user.

To install IPython, do so via `pip`. We'll do this as the `nautobot` user:

```
nautobot@nautobot-dev:~$ pip install ipython
Collecting ipython
#output omitted for brevity
nautobot@nautobot-dev:~$
```

> **Note**
>
> One of the nice references that you get when using the Enhanced Shell is the list of many of the Nautobot models that will be used in working with the ORM. Use this as a reference for working with Jobs.

Enter the Shell Plus environment:

```
nautobot@nautobot-dev:~$ nautobot-server shell_plus
```

You'll see the output shown in the following figure:

```
# Shell Plus Model Imports
from constance.backends.database.models import Constance
from django.contrib.admin.models import LogEntry
from django.contrib.auth.models import Group, Permission
from django.contrib.contenttypes.models import ContentType
from django.contrib.sessions.models import Session
from django_celery_beat.models import ClockedSchedule, CrontabSchedule, IntervalSchedule, PeriodicTask,
PeriodicTasks, SolarSchedule
from django_celery_results.models import ChordCounter, GroupResult, TaskResult
from nautobot.circuits.models import Circuit, CircuitTermination, CircuitType, Provider, ProviderNetwork
from nautobot.dcim.models.cables import Cable, CablePath
from nautobot.dcim.models.device_component_templates import ConsolePortTemplate, ConsoleServerPortTemplate,
DeviceBayTemplate, FrontPortTemplate, InterfaceTemplate, PowerOutletTemplate, PowerPortTemplate, RearPortTemplate
from nautobot.dcim.models.device_components import ConsolePort, ConsoleServerPort, DeviceBay, FrontPort, Interface,
InterfaceRedundancyGroup, InterfaceRedundancyGroupAssociation, InventoryItem, PowerOutlet, PowerPort, RearPort
from nautobot.dcim.models.devices import Device, DeviceRedundancyGroup, DeviceType, Manufacturer, Platform,
VirtualChassis
from nautobot.dcim.models.locations import Location, LocationType
from nautobot.dcim.models.power import PowerFeed, PowerPanel
from nautobot.dcim.models.racks import Rack, RackGroup, RackReservation
from nautobot.extras.models.change_logging import ObjectChange
from nautobot.extras.models.customfields import ComputedField, CustomField, CustomFieldChoice
from nautobot.extras.models.datasources import GitRepository
from nautobot.extras.models.groups import DynamicGroup, DynamicGroupMembership
from nautobot.extras.models.jobs import Job, JobButton, JobHook, JobLogEntry, JobResult, ScheduledJob,
ScheduledJobs from nautobot.extras.models.models import ConfigContext, ConfigContextSchema, CustomLink, ExportTemplate,
ExternalIntegration, FileAttachment, FileProxy, GraphQLQuery, HealthCheckTestModel, ImageAttachment, Note, Webhook
from nautobot.extras.models.relationships import Relationship, RelationshipAssociation
from nautobot.extras.models.roles import Role
from nautobot.extras.models.secrets import Secret, SecretsGroup, SecretsGroupAssociation
from nautobot.extras.models.statuses import Status
from nautobot.extras.models.tags import Tag, TaggedItem
from nautobot.ipam.models import IPAddress, IPAddressToInterface, Namespace, Prefix, RIR, RouteTarget, Service,
VLAN, VLANGroup, VRF, VRFDeviceAssignment, VRFPrefixAssignment
from nautobot.tenancy.models import Tenant, TenantGroup
from nautobot.users.models import AdminGroup, ObjectPermission, Token, User
from nautobot.virtualization.models import Cluster, ClusterGroup, ClusterType, VMInterface, VirtualMachine
from social_django.models import Association, Code, Nonce, Partial, UserSocialAuth
# Shell Plus Django Imports
from django.core.cache import cache
from django.conf import settings
from django.contrib.auth import get_user_model
from django.db import transaction
from django.db.models import Avg, Case, Count, F, Max, Min, Prefetch, Q, Sum, When
from django.utils import timezone
from django.urls import reverse
from django.db.models import Exists, OuterRef, Subquery
```

Figure 11.1 – Nautobot Shell Plus importing all data models

At this point, you're ready to start interacting with the shell.

Reading data

Before you can read the data, make sure that there is data in your Nautobot instance! For this example, we'll add LocationType and Location from the UI, even though you should have all the data from *Chapters 4, 5*, and/or *Chapter 10* too:

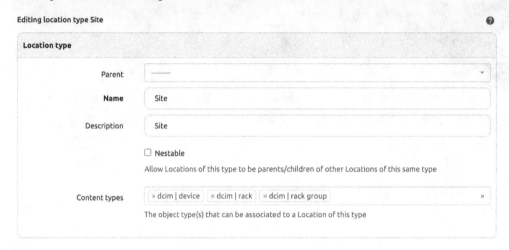

Figure 11.2 – Example location type

Now, add a new location using that location type:

> **Note**
>
> If you are using the instance that we created in *Chapter 4*, your site instance will also need a parent site specified.

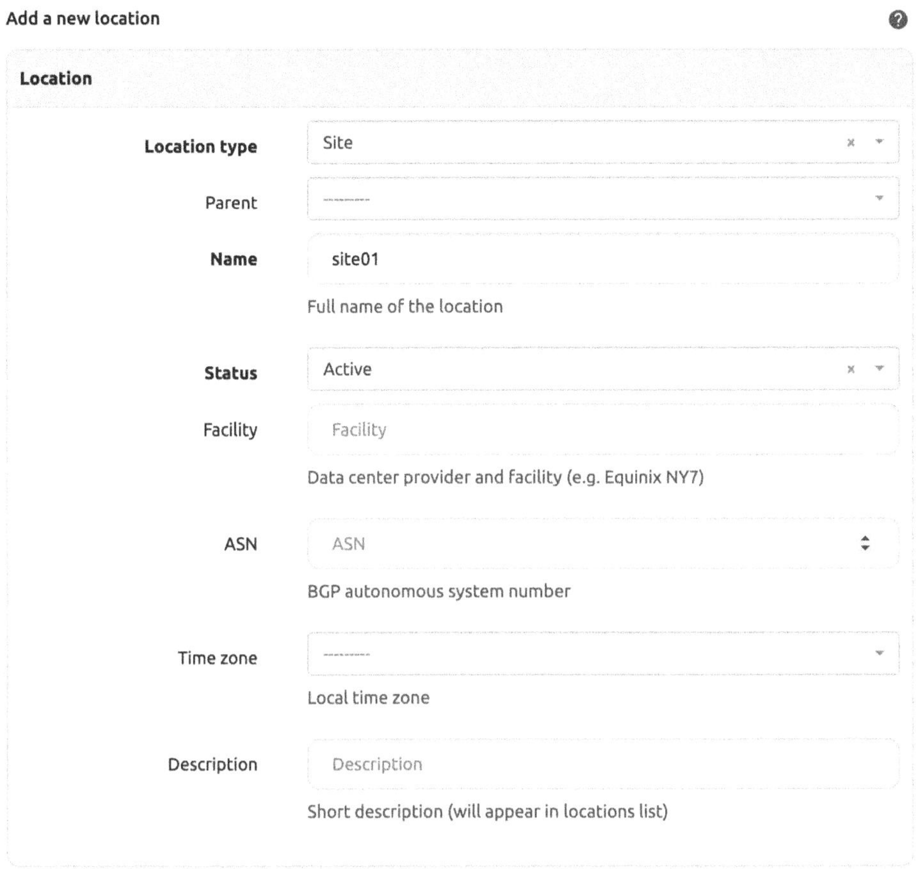

Add a new location

Location

Location type	Site
Parent	---------
Name	site01

Full name of the location

Status	Active
Facility	Facility

Data center provider and facility (e.g. Equinix NY7)

ASN	ASN

BGP autonomous system number

Time zone	---------

Local time zone

Description	Description

Short description (will appear in locations list)

Figure 11.3 – Example location of site01 that is of the Site type

The data can now be accessed using the ORM. Let's take a look at a few examples to access LocationType and the Site location that we created previously.

The first query we'll look at is getting all of the LocationType values.

We can do this with the LocationType.objects.all() query.

This provides all of the LocationType values. The following response gives you a QuerySet back, but we aren't able to do much with it. The second stanza makes more sense here by assigning QuerySet to a variable that you can work with:

```
In [1]: LocationType.objects.all()
Out[1]: <TreeQuerySet [<LocationType: Region>, <LocationType: Colo>,
<LocationType: Site>]>

In [2]: location_types = LocationType.objects.all()
```

As we saw with `pynautobot`, if you need to query all of a particular object type that is quite expansive, it is better to work with the `filter()` method to filter the response data.

Let's filter for all locations that are of the `Site` type:

```
In [3]: sites = Location.objects.filter(location_type__name="Site")
```

When you work with the `filter` method, you can use the double underscore on a particular attribute to help with filtering. In the preceding example, if you were to just use the `location_type` filter, then you would need to provide the UUID for the location type. Then, you would need to query for the location type to be able to use the `id` attribute. The double underscore makes it a bit easier to search for the data that you are looking for. The filtering returns an iterable `QuerySet` that can be used within a *for loop*:

```
In [4]: type(sites)
Out[4]: nautobot.dcim.models.locations.LocationQuerySet
```

If there is an expected singular response, then there will be another method that has some error handling to perform. That is what the `get()` method does. Getting a particular device or site would be the perfect example of this:

```
In [5]: site01 = Location.objects.get(name="site01")

In [6]: type(site01)
Out[6]: nautobot.dcim.models.locations.Location
```

As you can see, the type of the object is a singular location in the `Out[6]` output. However, in `Out[4]`, we had `LocationQuerySet`. If there are multiple responses in this output within the ORM, an error will be generated.

Another commonly used method in doing Job development within the ORM is to utilize the `first()` method. This method will get the first response within the database so that a single object is returned, but it doesn't matter what the location type is:

```
In [7]: location_type = LocationType.objects.first()

In [8]: type(location_type)
Out[8]: nautobot.dcim.models.locations.LocationType

In [9]: location_type.name
Out[9]: 'Site'
```

Adding and updating data

Now that we've read the data, it's time to look at adding and updating data via the ORM. Two methods are generally used in Django to create objects: `create()` and `get_or_create()`.

The `create` method will create the object and return exactly that object, while the `get_or_create()` method will return two objects – the object that was requested or created, as well as a Boolean response regarding whether or not the object was created.

Note that you can't see if you are creating a duplicate object when rerunning the command multiple times, which is why it is important to run `validated_save()`, something we'll look at next.

With the `get_or_create()` method, if you create a site, such as `Site10`, with one set of data, and then create another site with different data but the same name (`Site10`), the method will create a new site. Let's take a look.

Let's ensure `location_type` is set to `Site` for our examples:

```
In [10]: location_type = LocationType.objects.get(name="Site")
In [11]: status_active = Status.objects.get(name="Active")
In [12]: location, created = Location.objects.get_or_
create(name="Site10", location_type=location_type, status=status_
active, description="Hi")
```

With that, you've created a new site called `Site10`.

Go into the UI and change the description to something else, such as `Hello`.

Then, rerun the statement – two sites will be created:

```
In [13]: location, created = Location.objects.get_or_
create(name="Site10", location_type=location_type, status=status_
active, description="Hi")
```

| ☐ | Site10 | Active | — | — | Hi | — | ⟳ ✎ 🗑 |
| ☐ | Site10 | Active | — | — | Hello | — | ⟳ ✎ 🗑 |

Figure 11.4 – UI of the site created via the ORM

This is invalid, however. If you attempt to go in and edit one of the sites in the UI, you will get a validation error stating that there cannot be two sites with the same name:

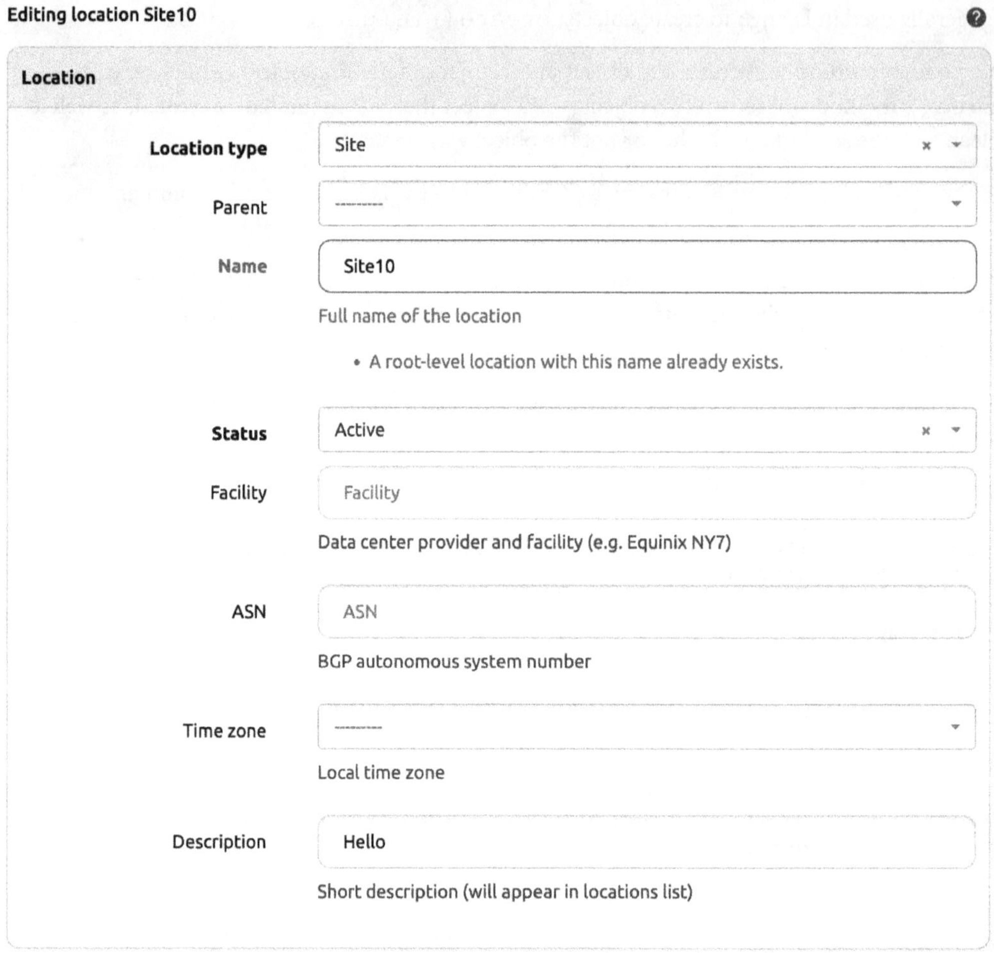

Editing location Site10

Location	
Location type	Site
Parent	--------
Name	Site10
	Full name of the location
	• A root-level location with this name already exists.
Status	Active
Facility	Facility
	Data center provider and facility (e.g. Equinix NY7)
ASN	ASN
	BGP autonomous system number
Time zone	--------
	Local time zone
Description	Hello
	Short description (will appear in locations list)

Figure 11.5 – Editing a duplicated site

The method to verify this in the ORM is to run the `validated_save()` method within a `try/except` block. If there is a `ValidationError` except, then you need to remove the object and handle the error appropriately for the process you are working through.

Let's delete the last location that was just created with the `delete()` method:

```
In [14]: location.delete()
Out[14]: (1, {'dcim.Location': 1})
```

It is good practice to run `validated_save()` so that you don't run into issues like this:

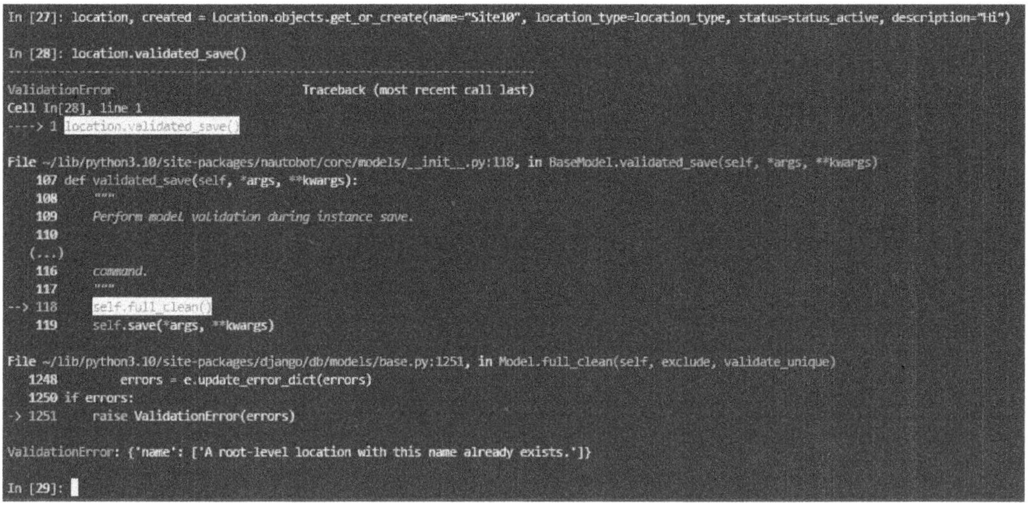

```
In [27]: location, created = Location.objects.get_or_create(name="Site10", location_type=location_type, status=status_active, description="Hi")

In [28]: location.validated_save()
```
```
ValidationError                         Traceback (most recent call last)
Cell In[28], line 1
----> 1 location.validated_save()

File ~/lib/python3.10/site-packages/nautobot/core/models/__init__.py:118, in BaseModel.validated_save(self, *args, **kwargs)
    107 def validated_save(self, *args, **kwargs):
    108     """
    109     Perform model validation during instance save.
    110
    (...)
    116     command.
    117     """
--> 118     self.full_clean()
    119     self.save(*args, **kwargs)

File ~/lib/python3.10/site-packages/django/db/models/base.py:1251, in Model.full_clean(self, exclude, validate_unique)
    1248         errors = e.update_error_dict(errors)
    1250 if errors:
-> 1251     raise ValidationError(errors)

ValidationError: {'name': ['A root-level location with this name already exists.']}

In [29]:
```

Figure 11.6 – Performing validated_save() after creating a new location

With the power of Python, you can iterate through a list of sites using the ORM to create the sites pretty easily:

```
In [15]: site_list_to_add = ["Site02", "Site03", "Site04"]
In [16]: location_type = LocationType.objects.get(name="Site")
In [17]: status_active = Status.objects.get(name="Active")
In [18]: for site_addition in site_list_to_add:
   ...:     try:
   ...:         location, _new = Location.objects.get_or_
create(name=site_addition, location_type=location_type, status=status_
active)
   ...:         location.validated_save()
   ...:     except ValidationError:
   ...:         print("Looks like there is an error. This location may
exist already.")
   ...:
```

With this code, you will create three new sites in the for loop when a validated save is performed. The first two items must be gathered first for the creation method. Then, you need to provide the Nautobot object to the `get_or_create()` method for the `LocationType` and `Status` values that you want to assign to the site. `Status` and `LocationType` are the two required items when creating a location.

Now, we must update the data. We can do this by assigning the attributes of the object. So, let's walk through the process of getting `Site02` and changing the physical address on the site to a fictitious address of 1 NTC Path. Note that to see all of the objects, it is common practice to use the Python dir() function on an object. This allows you to see what methods and attributes are available:

```
In [19]: site_to_update = Location.objects.get(name="Site02")
In [20]: site_to_update.name
Out[20]: 'Site02'
In [21]: site_to_update.physical_address
Out[21]: ''
In [22]: site_to_update.physical_address = "1 NTC Path"
```

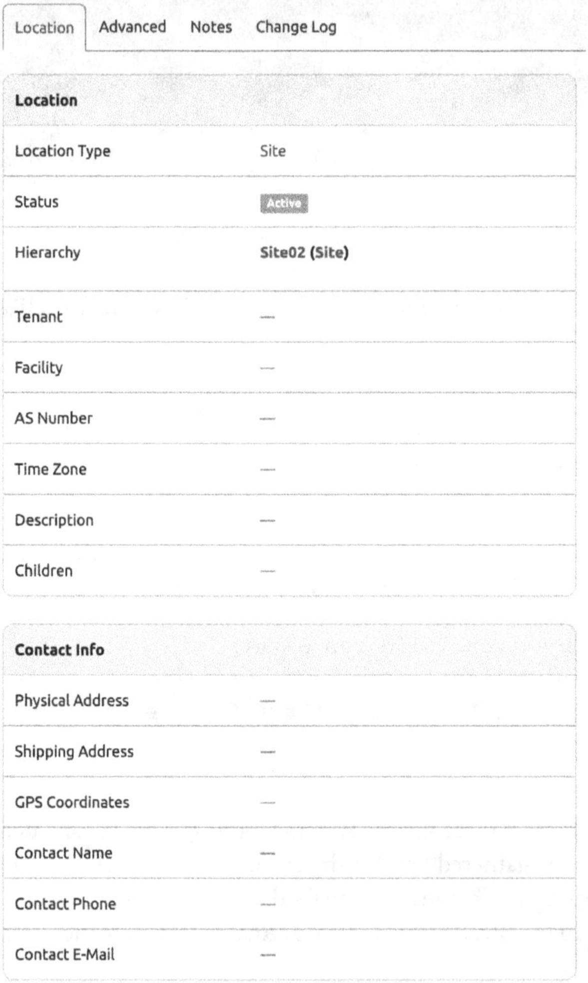

| Location | Advanced | Notes | Change Log |

Location	
Location Type	Site
Status	Active
Hierarchy	Site02 (Site)
Tenant	—
Facility	—
AS Number	—
Time Zone	—
Description	—
Children	—

Contact Info	
Physical Address	—
Shipping Address	—
GPS Coordinates	—
Contact Name	—
Contact Phone	—
Contact E-Mail	—

Figure 11.6 – Site not updated yet

Note that at this point, even though you have assigned a new address, the object hasn't been updated because it hasn't been saved:

```
In [23]: site_to_update.validated_save()
```

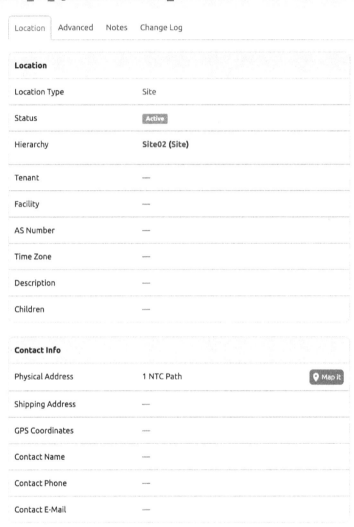

Figure 11.7 – Site with an updated physical address

Remember, changes that are made in your local Python interpreter need to be saved into the database for them to be visible in the web UI and become permanent!

Let's move on to deleting data.

Deleting data

Deleting data from Nautobot can be done at different levels. If you're just looking to delete an attribute of the object, such as the address that was just created, we can do this by using the methodology in Python on the object. So, assigning an empty string to the `physical_address` attribute will empty the data:

```
In [24]: site_to_update.physical_address = ""

In [25]: site_to_update.validated_save()
```

If you're looking to delete an object that has been created, you can get the object and then run the `delete()` method on the object. This can be done inline in one statement:

```
In [26]: Location.objects.get(name="Site04").delete()
Out[26]: (1, {'dcim.Location': 1})
```

> **Important**
>
> When working directly in the shells, you don't get a changelog associated with the object that is created/modified/deleted here. When working with Jobs, the Job execution is wrapped within a user context, which will then associate the appropriate changes with the user account that runs the Job.

Now that we've covered some of the basics of the ORM, we are going to learn more about Jobs themselves. However, keep in mind that we'll be using the ORM code in Jobs throughout this chapter.

Adding Jobs to Nautobot

One of Nautobot's goals is to make it easy to be able to make your Nautobot environment extensible and flexible. With Nautobot's flexibility, three methods can be used to add (or distribute) Jobs into an instance of Nautobot:

- Synchronizing Jobs into Nautobot from a Git repository
- Distributing Jobs as part of a Nautobot app
- Mounting or placing Jobs directly in the `JOBS_ROOT` path (the default is `/opt/nautobot/jobs`)

Having these three methods available to get your Jobs into Nautobot allows you to quickly iterate and have Jobs do what you want to do so that you can automate your environment.

For Jobs that are part of either a Git repository or mounted into the `JOBS_ROOT` path, there is no support for cross-file dependencies, meaning you cannot do relative imports. If you are looking to have a common library module for common code, you should package Jobs within a Nautobot App, which you can reference via Python paths. More on Nautobot Apps will be covered in *Part 4* of this book.

Synchronizing Jobs into Nautobot from a Git repository

As covered in *Chapter 6*, you can add Jobs to Nautobot by using the Git data source feature.

For any Jobs that you wish to import from a Git repository, you must have a few things set up with the repository. First, you should have a directory at the root of the repository named `jobs`. This is where the jobs will exist. Within this directory, you should have an empty `__init__.py` file. The last requirement for the Git repository is to have an actual Job (`jobs_file.py` in the following output) file inside the repository. See the following example regarding the directory's layout:

```
> tree
.
├── jobs
│   ├── __init__.py
│   └── jobs_file.py
└── README.md
```

Distributing Jobs as part of a Nautobot app

Nautobot Apps can act as a distribution point for Nautobot Jobs. This allows Jobs to be developed in a self-contained development environment. The Nautobot Apps route for Jobs also allows for relative imports so that you can build a utilities library package and then reference that file. This is not available via the Git Sync or file placed in the Jobs Root directory. This path is also helpful if you have dependencies for your Jobs.

More details on this will be discussed in *Chapter 16*.

Mounting or placing Jobs directly in JOBS_ROOT

JOBS_ROOT was first mentioned in *Chapter 7*. This is the file path where Jobs need to be placed so that you can use them. Our focus for the remainder of this chapter will be using JOBS_ROOT to simplify working with Jobs.

Creating your first Nautobot Job

In this section, you will be walked through two Nautobot Jobs to understand how to get started with Jobs. The first will be a Hello World Job that will introduce you to the structure and how to log to the **Job Result** page in the UI. The second example will provide a bit more context on building out Job form fields that enable self-service in the UI for users of your Jobs.

"Hello World" Nautobot Job

Remember your first Python script? It probably had a single `print()` statement. We are going to start with the same for a Nautobot Job. There is only one difference: we are going to use Nautobot Job logging instead of print statements.

Our examples will use the `JOBS_ROOT` method of placing a Job Python file into the Nautobot file structure.

Configure `JOBS_ROOT` open `nautobot_config.py` as the `nautobot` user and uncomment the following line:

```
JOBS_ROOT = os.getenv("NAUTOBOT_JOBS_ROOT", os.path.join(NAUTOBOT_
ROOT, "jobs").rstrip("/"))
```

This will use the default path of `/opt/nautobot/jobs` based on our prior configurations of `NAUTOBOT_ROOT`.

Let's create our first Nautobot Job file:

```
root@nautobot-dev:/opt/nautobot/jobs# touch my_job_module.py
root@nautobot-dev:/opt/nautobot/jobs#
```

That should be the only file you see in the jobs directory:

```
root@nautobot-dev:/opt/nautobot/jobs# ls
my_job_module.py
root@nautobot-dev:/opt/nautobot/jobs#
```

Add the following code to your new Job file. We already added the code to the file and are showing it printed to the Terminal through the use of the `cat` command:

```
root@nautobot-dev:/opt/nautobot/jobs# cat my_job_module.py
from nautobot.apps.jobs import Job, register_jobs

class HelloWorld(Job):

    def run(self):
        self.logger.debug("Hello, this is my first Nautobot Job.")

register_jobs(HelloWorld)
root@nautobot-dev:/opt/nautobot/jobs#
```

As you can see, it is a very basic and minimal Job that is equivalent to a `Hello World` script.

In this definition, the minimum requirements to build the Nautobot Job are as follows:

- Import the `Job` class and the `register_jobs` function that's on the last line from `nautobot.apps.jobs`. These are standard objects that need to be used in all Nautobot Jobs.

- A Python class is used to define the Nautobot Job.

- The `run` method is a required method of your Job class. This is the actual execution of the Job.

- Call the `register_jobs()` function to register the Job within Nautobot.

Without the last line that registers the Job, Nautobot will not know about the job that was created. This function takes classes as arguments. You may see the syntax of `*jobs_list`, which will expand the list to their individual items as they are passed in.

With the file saved in the directory, the next step is to run the Nautobot server post upgrade command (`nautobot-server post_upgrade`) as the Nautobot user account. This will take care of registering the job with Nautobot since it is installed as part of the JOBS_ROOT path:

```
nautobot@nautobot-dev:~$ nautobot-server post_upgrade
Performing database migrations…
Operations to perform:
  Apply all migrations: admin, auth, circuits, contenttypes, database,
dcim,
<… output omitted for brevity …>
15:20:13.036 INFO    nautobot.extras.utils :
  Created Job "my_job_module: HelloWorld" from <HelloWorld>
<… output omitted for brevity …>
```

Immediately after the running migrations section, you will notice that there is an INFO log that shows `Created Job "Hello World" from <HelloWorld>`. This indicates that Nautobot has successfully registered the Job to Nautobot.

After running the `post_upgrade` command, restart the Nautobot services via `systemctl`:

```
root@nautobot-dev:/opt/nautobot/jobs# systemctl restart nautobot
nautobot-worker nautobot-scheduler
root@nautobot-dev:/opt/nautobot/jobs#
```

Now that the Job is registered, navigate to the **Jobs** page on the Nautobot UI by selecting **Jobs | Jobs**.

Here, you see a list of all the Jobs that are available, including your first Nautobot Job!

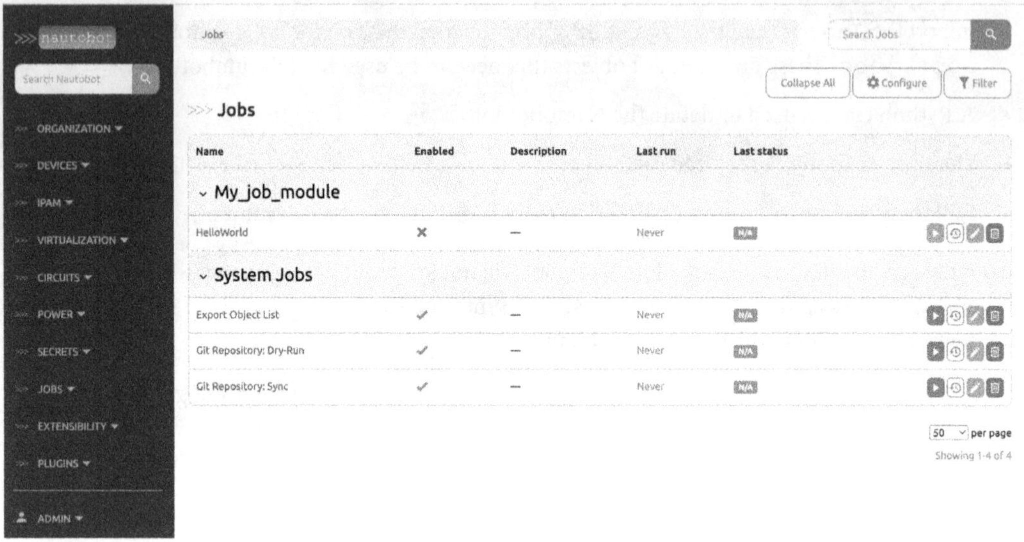

Figure 11.8 – Nautobot Hello World Job showing s disabled

However, there is an **X** icon on the Hello World Job, signifying that it hasn't been **Enabled** yet.

To run the Job, you need to enable the Job. To do this, select the edit button (pencil icon) on the right-hand side of the row. There is a checkbox labeled **Enabled** in the **Job** section that needs to be selected. After checking it, select **Update** at the bottom of the page:

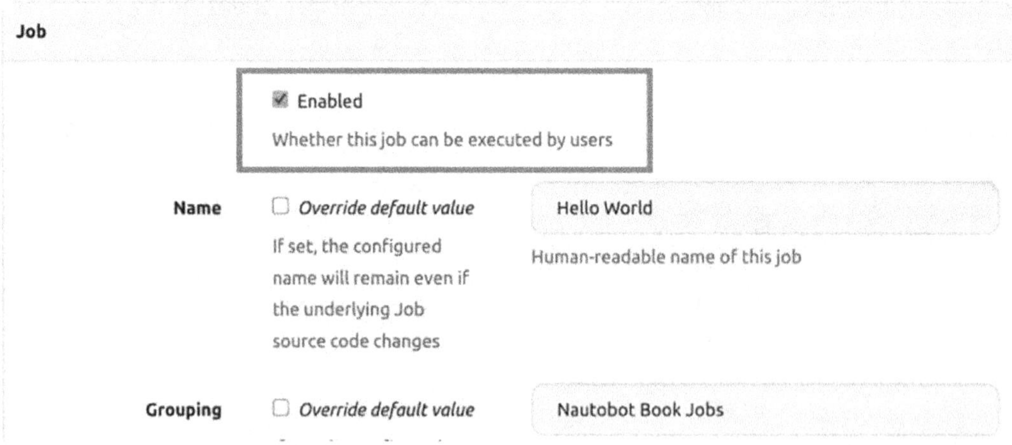

Figure 11.9 – Editing the Hello World Job, with the checkbox for Enabled disabled

Once saved you are taken back to the main Jobs page where you can now see the green checkbox seeing it is now **Enabled**:

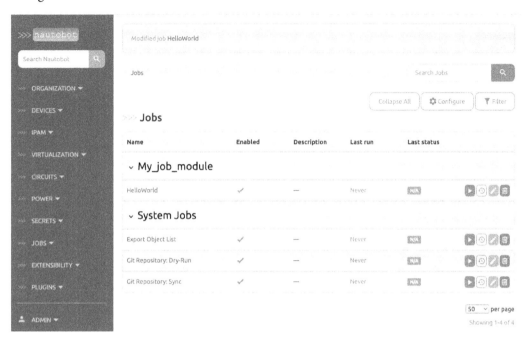

Figure 11.10 – Nautobot Job list with Hello World job enabled

Now that the Job is enabled, the Job can be run. Select the blue **Play** button on the right-hand side of the row to get to the Job run page:

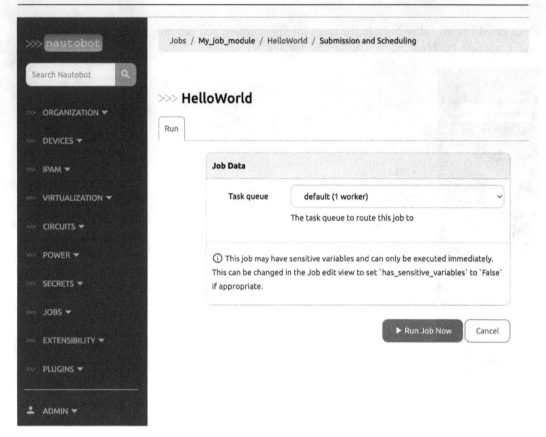

Figure 11.11 – Nautobot Job run page

This page is short at the moment, with a **Task queue** dropdown being the only item presented. The dropdown has one option showing **default (1 worker)**. If you have followed all the instructions in this book regarding setting up Nautobot, you will have only created one celery worker. That's what **default (1 worker)** means in the dropdown. Each worker supports one or more queues. For advanced deployments, you may want to have one or more workers support one or more queues to distribute work and automation. When you're doing an install for testing, like we're doing here, there will be one queue called **default** running on the one worker we created.

Moving on, the next Job shows how to add additional fields to the Job form so that you can get input from the users of your Job. Select the **Run Job Now** button to execute the Job:

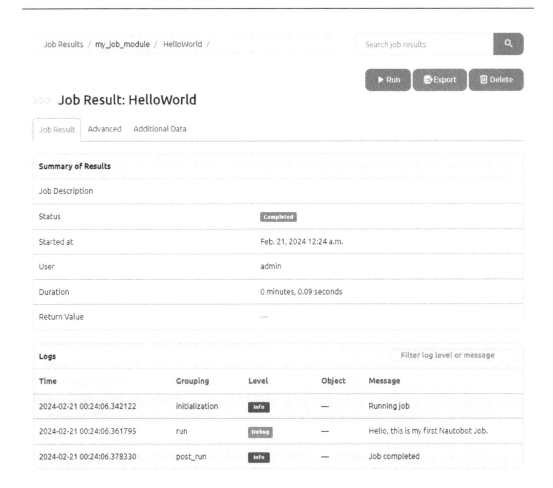

Figure 11.12 – The Job Result: HelloWorld page

Upon executing the Hello World Job, you will have three logs in the **Logs** section below the **Job Results** summary page. The first and last log messages provide information about the start and stop of the Job. In the middle is the log message from the Job – that is, **Hello, this is my Nautobot Job!**.

Let's explore the rest of the log levels that can provide feedback during the Job execution. Update your Job so that it looks like this:

```
root@nautobot-dev:/opt/nautobot/jobs# cat my_job_module.py
from nautobot.apps.jobs import Job, register_jobs

class HelloWorld(Job):
```

```
    def run(self):
        self.logger.debug("Hello, this is my first Nautobot Job.")
        self.logger.debug("Hello World, this is a debug log.")
        self.logger.info("This is an informational log.")
        self.logger.warning("This is a warning log.")
        self.logger.error("This is an error log.")
        self.logger.critical("This is a critical log.")

register_jobs(HelloWorld)
root@nautobot-dev:/opt/nautobot/jobs#
```

Let's run this job (only after we run `nautobot-server post_upgrade` and restart our services again!).

Execute a post upgrade, like so:

```
nautobot@nautobot-dev:~$ nautobot-server post_upgrade
```

Restart the services:

```
root@nautobot-dev:/opt/nautobot/jobs# systemctl restart nautobot
nautobot-worker nautobot-scheduler
```

Logs				Filter log level or message
Time	**Grouping**	**Level**	**Object**	**Message**
2024-02-21 00:47:21.755000	initialization	Info	—	Running job
2024-02-21 00:47:21.776000	run	Debug	—	Hello, this is my first Nautobot Job.
2024-02-21 00:47:21.783000	run	Debug	—	Hello World, this is a debug log.
2024-02-21 00:47:21.788000	run	Info	—	This is an informational log.
2024-02-21 00:47:21.793000	run	Warning	—	This is a warning log.
2024-02-21 00:47:21.797000	run	Error	—	This is an error log.
2024-02-21 00:47:21.802000	run	Critical	—	This is a critical log.
2024-02-21 00:47:21.817000	post_run	Info	—	Job completed

Figure 11.13 – The Job Result: HelloWorld page showing additional logging levels

As you can see, different colors are associated with the row of the log. Here's a mapping of those colors:

Log Level	Background Color
Debug	Clear
Info	Blue
Warning	Yellow
Error	Red
Critical	Red

Table 11.1 – Nautobot Job log levels

Breaking down and building a Nautobot Job

Let's dive into a Nautobot Job that will add additional information to make the Job form look much more user-friendly and work with data within the Nautobot platform. We'll create a new file that will contain the next couple of jobs to be created.

The file will be called my_jobs.py and will contain multiple jobs.

So, create the new my_jobs.py file:

```
root@nautobot-dev:/opt/nautobot/jobs# touch my_jobs.py
root@nautobot-dev:/opt/nautobot/jobs# ls
__pycache__   my_job_module.py   my_jobs.py
root@nautobot-dev:/opt/nautobot/jobs#
```

Add the following lines to the new Job:

```
from nautobot.apps.jobs import Job, ObjectVar, register_jobs
from nautobot.dcim.models.locations import Location
from nautobot.dcim.models.devices import Device

name = "Nautobot Book Jobs"
```

The first new addition to the Job code is the use of name at the global constant level of the file. This name is what is shown on the Jobs page. On the Jobs page in the UI, you'll have a new Jobs section called **Nautobot Book Jobs** when we're finished. name is defined at the same level as the class definitions, with the common practice of having the name defined immediately after the Python imports.

Next, add the following lines below name:

```
class SerialNumberReport(Job):

    location_to_check = ObjectVar(
        model=Location,
    )

    class Meta:
        name = "Check Serial Numbers"
        has_sensitive_variables = False
        description = "First Nautobot Job: Reading and reporting data"
```

Here, we created a new class that declares a new Job called `SerialNumberReport` that must inherit the base `Job` class. Next up is the use of a variable that we'll use as user input to the Job. Using an `ObjectVar` class is going to allow us to have a dropdown of all locations in Nautobot when this Job form renders.

In general, the `ObjectVar` class is used for Nautobot to say that there are going to be objects (data, in Nautobot) that will be used for the form and that the user of the job will need to choose from the objects within the model. The `ObjectVar` class requires a model to be passed in. In this example, this will be the `Location` model. This is why the Python file's `Location` was included in the models imported.

Next up will be a subclass called `Meta`. This is where metadata about the Job is defined, including a name, description, and several other optional attributes. See the following table for the attributes you can use in the `Meta` class.

For the most up-to-date information about what attributes are available for jobs, be sure to check out the Nautobot documentation in the **Developer Guide** section and the Jobs Developer Guide:

Attribute	Default	Description
name	Name of the class	A name to display and a reference for the job.
description		A human-friendly description of the job. It accepts plain text, Markdown, and a limited set of HTML.
approval_required	False	Whether or not a job requires another individual to approve the job before it launches.
dryrun_default	False	Sets the default for a job to be executed without making database changes. Note that dryrun capabilities for interacting with non-Nautobot database-related items require logic to be built into the Job to handle this.

Attribute	Default	Description
field_order	[]	This setting sets the order in which the Job form is rendered; the top is first on the list. This allows for the variables defined at the class level to be in any order. The default order is the order in which the variables are defined.
has_sensitive _variables	True	This tells Nautobot if there are secrets or other variables that should not be displayed. When this is set to True, the Job cannot be scheduled.

Table 11.2 – Table of commonly used Job Meta attributes

Our Job uses name and description and sets has_sensitive_variables to False.

Now that the Job form has been defined, it's time to put the logic into the run method on the Job class:

```
def run(self, location_to_check):
    device_query = Device.objects.filter(
        location=location_to_check
    )

    for device in device_query:
        self.logger.info(
            "Checking the device %s for a serial number.",
            device.name,
            extra={"object": device},
        )
        if device.serial == "":
            self.logger.error(
                "Device %s does not have serial number defined.",
                device.name,
                extra={"object": device},
            )
        else:
            self.logger.debug(
                "Device %s has serial number: %s",
                device.name,
                device.serial,
                extra={"object": device},
            )

register_jobs(SerialNumberReport)
```

For completeness, this is the full Job we've created so far:

```python
my_jobs.py ×

jobs >  my_jobs.py
1    from nautobot.apps.jobs import Job, ObjectVar, register_jobs
2    from nautobot.dcim.models.locations import Location
3    from nautobot.dcim.models.devices import Device
4
5    name = "Nautobot Book Jobs"
6
7    class SerialNumberReport(Job):
8
9        location_to_check = ObjectVar(
10           model=Location,
11       )
12
13       class Meta:
14           name = "Check Serial Numbers"
15           has_sensitive_variables = False
16           description = "First Nautobot Job: Reading and reporting data"
17
18       def run(self, location_to_check):
19           device_query = Device.objects.filter(
20               location=location_to_check
21           )
22
23           for device in device_query:
24               self.logger.info(
25                   "Checking the device %s for a serial number.",
26                   device.name,
27                   extra={"object": device},
28               )
29               if device.serial == "":
30                   self.logger.error(
31                       "Device %s does not have serial number defined.",
32                       device.name,
33                       extra={"object": device},
34                   )
35               else:
36                   self.logger.debug(
37                       "Device %s has serial number: %s",
38                       device.name,
39                       device.serial,
40                       extra={"object": device},
41                   )
42
43   register_jobs(SerialNumberReport)
```

Figure 11.14 – Performing a serial number check on the Job code

The preceding code starts by adding the required run method by passing in the previously defined variable – that is, `location_to_check`. You can probably infer that we are going to be looping over devices in the location the user selects in the Job form.

Regarding the logging shown, the structure follows traditional Python logging syntax, with `%s` handling the variable substitution with the variables listed in order immediately following the log string. The second piece is the `extra={"object": device}` syntax. This provides a link to the Nautobot object inline on the log. This allows the users of the Job to be able to quickly link to the object that contains the log line. So, in this case, having the object on the log line will allow the operator to quickly navigate to any device that is missing a serial number and update it right inside the UI.

Now, we're ready to run the Job, but again, we need to perform `post_upgrade` and restart our services:

```
nautobot@nautobot-dev:~$ nautobot-server post_upgrade
root@nautobot-dev:/opt/nautobot/jobs# systemctl restart nautobot
nautobot-worker nautobot-scheduler
```

Figure 11.15 – The Jobs page showing the Nautobot Book Jobs section and a Job description

Notice that on the main Jobs page, there is now a description and a section for the Job using the `name` and `description` variables. Pretty neat, right?

Now, enable the Job, as shown earlier in this chapter. At this point, we can run the Job and see the output.

The first new item on the Job run page is the form fields that were described, including the **Location to check** dropdown, with a single object selection, meaning that you have to select one of those objects. It is in bold, indicating that this is a required field.

You will also see the sites from *Chapter 10* if you previously ran the data population playbooks. Let's choose **East Coast | Jersey City**. If you didn't complete *Chapter 10*, you can also choose **WayneHQ** from *Chapter 4*:

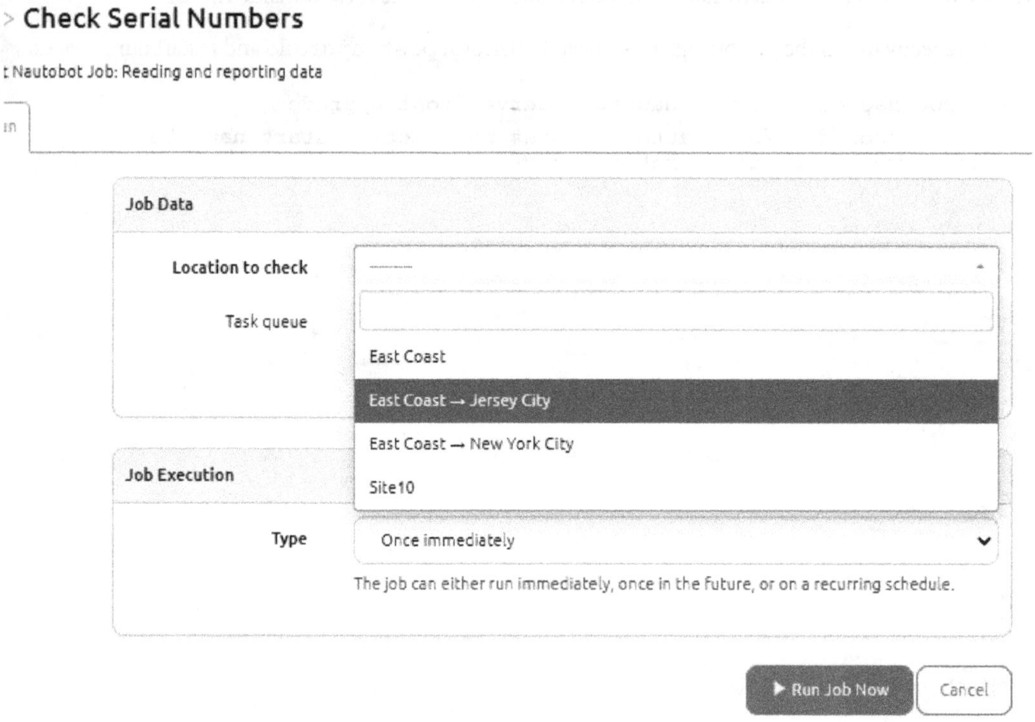

> **Check Serial Numbers**

t Nautobot Job: Reading and reporting data

Figure 11.16 – The Job launch page for Check Serial Numbers with a Location to check dropdown

After running the Job, you'll see the following Job results:

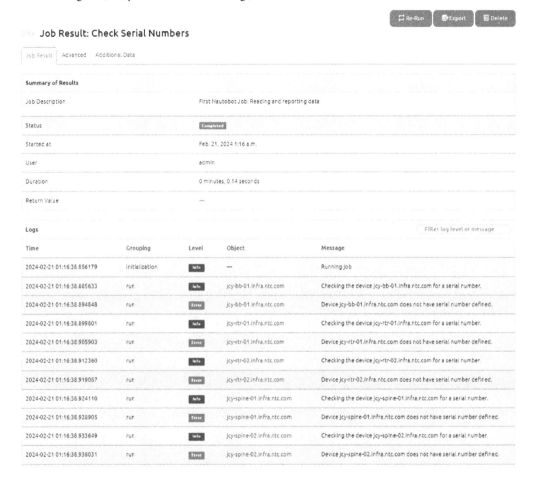

Figure 11.17 – The Job Result page with devices showing an error or success

Let's continue by building an even more robust and more network automation-centric Job.

Adding dynamic dropdowns to your job

In this example, we're going to create a command runner execution Job. This allows a user to choose a location, choose a device in that location, and then choose between one and five show commands to run on the device, followed by generating a file for each command output gathered from the device.

The form will start with having a user select a location that will be the starting filter. The second dropdown will give a filtered list of devices at that location. The third dropdown will provide what command(s) to run. Once the command has run against the network device, the Job will save the output to a file so that the output can be downloaded.

This script is going to use Netmiko (`https://github.com/ktbyers/netmiko`), a common open source SSH library for network automation.

> **Note**
>
> Later in this example, we'll introduce a workaround that bypasses connecting to devices so that you can still understand the overall process and workflow. So, even if you don't have a network to test with, you can follow these steps!

We'll start by installing Netmiko:

```
nautobot@nautobot-dev:~$ pip3 install netmiko
# output omitted.
```

Just so we don't mess up our last Job, let's create a new file using `my_jobs.py` as the source. Copy `my_jobs.py` into `my_jobs_2.py`:

```
root@nautobot-dev:/opt/nautobot/jobs# cp my_jobs.py my_jobs_2.py
root@nautobot-dev:/opt/nautobot/jobs#
```

Let's walk through the changes we are going to make.

Add the following imports and add the `COMMAND_CHOICES` variable:

```
import os

from nautobot.apps.jobs import MultiChoiceVar, Job, ObjectVar,
register_jobs, StringVar
from nautobot.dcim.models.locations import Location
from nautobot.dcim.models.devices import Device
from netmiko import ConnectHandler


name = "Nautobot Book Jobs Part 2" # change in name

COMMAND_CHOICES = (
    ("show ip interface brief", "show ip int bri"),
    ("show ip route", "show ip route"),
    ("show version", "show version"),
    ("show log", "show log"),
```

```
        ("show ip ospf neighbor", "show ip ospf neighbor"),
    )
```

COMMAND_CHOICES (tuple of tuples) will be used in conjunction with MultiChoiceVar in the Job. We'll see this in more detail in the following code block.

Add the following code block below the SerialNumberReport class, but above register_jobs():

```
class CommandRunner(Job):
    device_location = ObjectVar(
        model=Location,
        required=False
    )

    user_name = StringVar(
        description="Input your username",
        required=False
    )

    device = ObjectVar(
        model=Device,
        query_params={
            "location": "$device_location",
        }
    )

    commands = MultiChoiceVar(choices=COMMAND_CHOICES)

    class Meta:
        name = "Command Runner"
        has_sensitive_variables = False
        description = "Command Runner"
```

The SerialNumberReport class remains unchanged and will be omitted for brevity. Here, we are showing how you can have more than one Job (that is, a class) in one Job file. Be sure to check out the final script in this book's GitHub repository (https://github.com/PacktPublishing/ Network-Automation-with-Nautobot/blob/main/chapter-11/my_jobs_2.py) to ensure your script reflects what it should be.

Looking at the class definition for the new CommandRunner Job, there are a few new variable types and a new parameter that will allow us to perform filtering based on the earlier selection.

The `device_location` variable is the same as before, except that it has been changed to be an optional variable. This will assist in filtering for the device, which is required, but not required for the job itself to run. The visual of the job form with the variables from previously looks like what's shown here.

The `user_name` variable is a `StringVar` input variable that takes text input and represents the username required to log into the network device. It is a simple definition with a description and sets the required parameter to `False`. Note that there is also a `TextVar` input variable type that will take in longer text in a paragraph format.

The `device` variable shows how to filter an `ObjectVar` selection based on a previously selected form item. The addition of `query_params={"location": "$device_location"}` is what will complete the filtering for the devices to be only at the location selected in the first dropdown. If no device location is selected, then all of the devices will be listed. `query_params` takes a dictionary of REST API filters to be applied. See *Chapter 8* for more information about the REST API. The dollar sign (`$`) in front of the text string within `$device_location` tells Nautobot that the following text will be a variable rather than the string to be interpreted.

The last new variable command in this section is `MultiChoiceVar`, which takes the input of `COMMAND_CHOICES` that was created near the top of the file. This allows multiple commands to be generated. The first member in the tuple is the response that will be passed in as part of the selection. The second member of the tuple is what is displayed to the user in the job input form. So, regarding the first command choice of "show ip interface brief," the full command will be passed to the form entry. This allows for the creator of the job to have one thing displayed but pass something different.

The `Meta` class in this example doesn't introduce anything new at this point and continues to set the grouping and metadata information for the job.

The following figure showcases what we're in the middle of building to bring it to life:

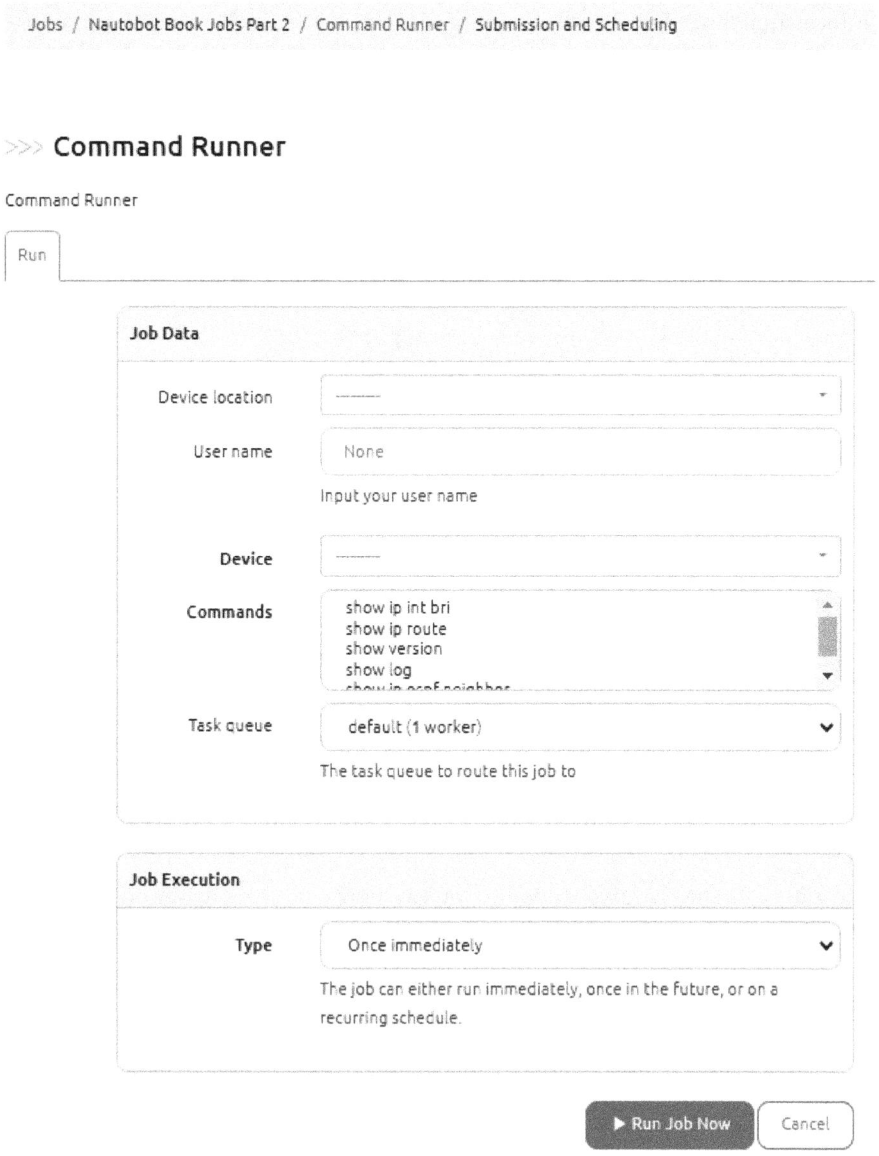

Figure 11.18 – Command Runner

Starting with the `run` method for this class, the Job provides information about what device the Job is being run against. Next, there are a few verification steps that the device performs to ensure it has all of the information it needs to be able to connect to the device, such as making sure that an IP address has been assigned, that a platform has been set for the device, and that the Netmiko mapping is available for the platform so that Netmiko knows what device type it is connecting to:

```python
    def run(self, device_location, user_name, device, commands):
        self.logger.info("Device name: %s", device.name)
        if user_name != "":
            self.logger.debug("User executing the command: %s", user_
name)

        # Verify that the device has a primary IP
        if device.primary_ip is None:
            self.logger.fatal("Device does not have a primary IP
address set.")
            return

        if device.platform is None:
            self.logger.fatal("Device does not have a platform set.")
            return

        if device.platform.network_driver_mappings.get("netmiko") is
None:
            self.logger.fatal("Device mapping for Netmiko is not
present, please set.")
            return
```

> **Note**
>
> To automatically set the Netmiko mapping, you can set the network driver on a platform to the normalized network driver. This will automatically populate all network driver mappings:

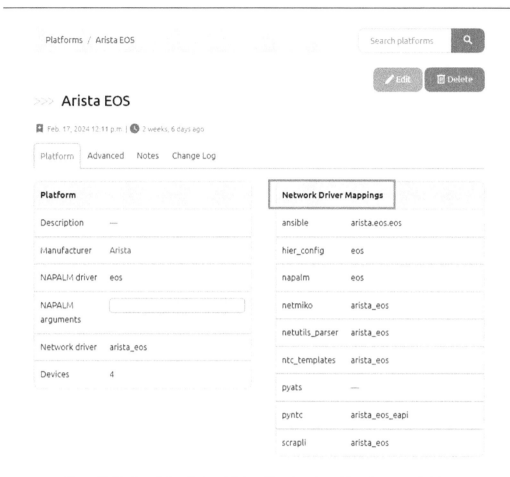

Figure 11.19 - Populating Network Driver Mappings by setting the network driver

The next step involves setting up Netmiko with the data from Nautobot. This script is using environment variables to connect to the device:

```
# Connect to the device, get some output
net_connect = ConnectHandler(
    device_type=device.platform.network_driver_
mappings['netmiko'],
    host=device.primary_ip.host, #or device.name if name is a
FQDN
    username=os.getenv("DEVICE_USERNAME"),
    # you can also set username to user_name if you want to
use the
    # username passed in from the form. We're showing both as
    # options.
    password=os.getenv("DEVICE_PASSWORD")
)
```

To use the environment variables that are being read by the `os.getenv()` method, you need to set the variables in the Nautobot and Nautobot-Worker `systemd` files. Place the `Environment` statement below the existing `Environment` statement:

```
# cat /etc/systemd/system/nautobot.service | grep Environment
Environment="NAUTOBOT_ROOT=/opt/nautobot"
Environment="DEVICE_USERNAME=ntc"
Environment="DEVICE_PASSWORD=ntc123"
#
# cat /etc/systemd/system/nautobot-worker.service | grep Environment
Environment="NAUTOBOT_ROOT=/opt/nautobot"
Environment="DEVICE_USERNAME=ntc"
Environment="DEVICE_PASSWORD=ntc123"
#
```

After modifying both files, rerun the `daemon-reload` command and restart the Nautobot services:

```
# systemctl daemon-reload
# systemctl restart nautobot nautobot-worker
#
```

> **Note**
>
> The use of Nautobot Secrets is recommended when implementing for production. Nautobot itself does not maintain secrets, instead providing a methodology to be able to access secrets through the environment or password management systems such as HashiCorp Vault.

Here is the rest of the `run` method on Command Runner. The last new introduction of a capability of the Job in Nautobot here is the `create_file` method. This will take in a filename and text string and create a file that can be downloaded once the job is completed:

```
        for command in commands:
            output = net_connect.send_command(command)
            self.create_file(f"{device.name}-{command}.txt", output)
register_jobs(SerialNumberReport, CommandRunner)
```

The last line of the file still contains the `register_jobs` method, which is now registering multiple jobs to Nautobot.

Running the job can now be done in the same fashion as before. Make sure that if the job is not showing enabled, you enable the job so that it can be run. Let's run the job with a few of the commands we've selected.

The following are a few figures to visualize the full script for completeness:

```
jobs >  my_jobs_2.py
  1    import os
  2
  3    from nautobot.apps.jobs import MultiChoiceVar, Job, ObjectVar, register_jobs, StringVar
  4    from nautobot.dcim.models.locations import Location
  5    from nautobot.dcim.models.devices import Device
  6    from netmiko import ConnectHandler
  7
  8    name = "Nautobot Book Jobs Part 2"
  9
 10    COMMAND_CHOICES = (
 11        ("show ip interface brief", "show ip int bri"),
 12        ("show ip route", "show ip route"),
 13        ("show version", "show version"),
 14        ("show log", "show log"),
 15        ("show ip ospf neighbor", "show ip ospf neighbor"),
 16    )
 17
 18    class SerialNumberReport(Job):
 19
 20        location_to_check = ObjectVar(
 21            model=Location,
 22        )
 23
 24        class Meta:
 25            name = "Check Serial Numbers"
 26            has_sensitive_variables = False
 27            description = "First Nautobot Job: Reading and reporting data"
 28
 29        def run(self, location_to_check):
 30            device_query = Device.objects.filter(
 31                location=location_to_check
 32            )
 33
 34            for device in device_query:
 35                self.logger.info(
 36                    "Checking the device %s for a serial number.",
 37                    device.name,
 38                    extra={"object": device},
 39                )
 40                if device.serial == "":
 41                    self.logger.error(
 42                        "Device %s does not have serial number defined.",
 43                        device.name,
 44                        extra={"object": device},
 45                    )
```

Figure 11.20 – Network automation Job – part 1

Here, we can see `SerialNumberReport` and the start of the `CommandRunner` Job:

```
46 ⌄            else:
47 ⌄                self.logger.debug(
48                     "Device %s has serial number: %s",
49                     device.name,
50                     device.serial,
51                     extra={"object": device},
52                 )
53
54
55 ⌄ class CommandRunner(Job):
56 ⌄     device_location = ObjectVar(
57             model=Location,
58             required=False
59         )
60
61 ⌄     user_name = StringVar(
62             description="Input your user name",
63             required=False
64         )
65
66 ⌄     device = ObjectVar(
67             model=Device,
68 ⌄         query_params={
69                 "location": "$device_location",
70             }
71         )
72
73         commands = MultiChoiceVar(choices=COMMAND_CHOICES)
74
75 ⌄     class Meta:
76             name = "Command Runner"
77             has_sensitive_variables = False
78             description = "Command Runner"
79
80 ⌄     def run(self, device_location, user_name, device, commands):
81             self.logger.info("Device name: %s", device.name)
82 ⌄         if user_name != "":
83                 self.logger.debug("User executing the command: %s", user_name)
84
85             # Verify that the device has a primary IP
86 ⌄         if device.primary_ip is None:
87                 self.logger.fatal("Device does not have a primary IP address set.")
88                 return
```

Figure 11.21 – Network automation Job - part 2/3

Finally, here are the last 20 lines of the Jobs file:

```
 89
 90        if device.platform is None:
 91            self.logger.fatal("Device does not have a platform set.")
 92            return
 93
 94        if device.platform.network_driver_mappings.get("netmiko") is None:
 95            self.logger.fatal("Device mapping for Netmiko is not present, please set.")
 96            return
 97
 98        # Connect to the device, get some output - comment this out if you are simulating
 99        net_connect = ConnectHandler(
100            device_type=device.platform.network_driver_mappings['netmiko'],
101            host=device.primary_ip.host, #or device.name if name is a FQDN
102            username=os.getenv("DEVICE_USERNAME"),
103            password=os.getenv("DEVICE_PASSWORD")
104        )
105        for command in commands:
106            output = net_connect.send_command(command)
107            self.create_file(f"{device.name}-{command}.txt", output)
108
109  register_jobs(SerialNumberReport, CommandRunner)
```

Figure 11.22 – Network automation Job – part 3/3

Let's run this job (only after we run `nautobot-server post_upgrade` and restart the services again!).

Perform a post upgrade:

```
nautobot@nautobot-dev:~$ nautobot-server post_upgrade
```

Restart the services:

```
root@nautobot-dev:/opt/nautobot/jobs# systemctl restart nautobot
nautobot-worker nautobot-scheduler
```

Don't have a network to test with and still want to run this Job? If that is your setup, update your `run` method so that it looks like this:

```python
class Meta:
    name = "Command Runner"
    has_sensitive_variables = False
    description = "Command Runner"

def run(self, device_location, user_name, device, commands):
    self.logger.info("Device name: %s", device.name)
    if user_name != "":
        self.logger.debug("User executing the command: %s", user_name)

    for command in commands:
        #output = net_connect.send_command(command) # comment this out if you are only testing
        output = "show command response goes here"
        self.create_file(f"{device.name}-{command}.txt", output)

register_jobs(SerialNumberReport, CommandRunner)
```

Figure 11.23 – Updating the run method if you don't have network devices to test with

Navigate to the UI and view the new Jobs list:

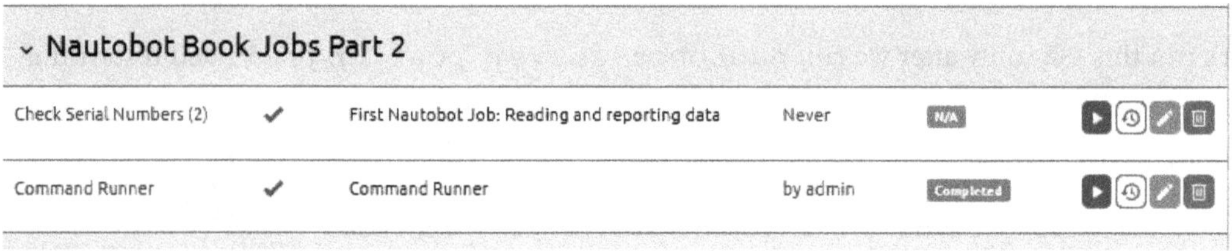

Figure 11.24 – Ensuring the Command Runner Job is shown

Make sure you enable the new Command Runner Job by repeating the process from earlier.

Now, we're finally ready to execute the Job:

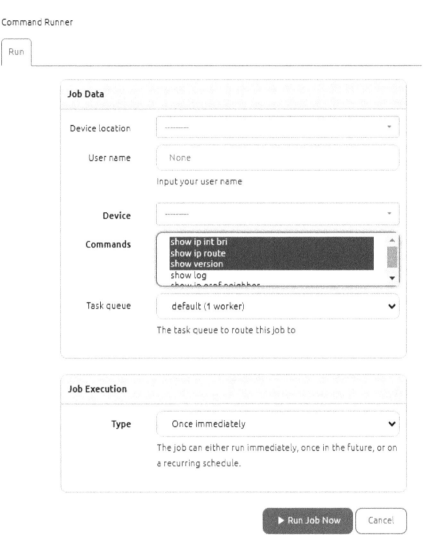

Figure 11.25 – Showing the Job execution page with multiple commands selected

When the Job is executed, new files are created. These appear both in the inline logs and in the **Job Result** summary page:

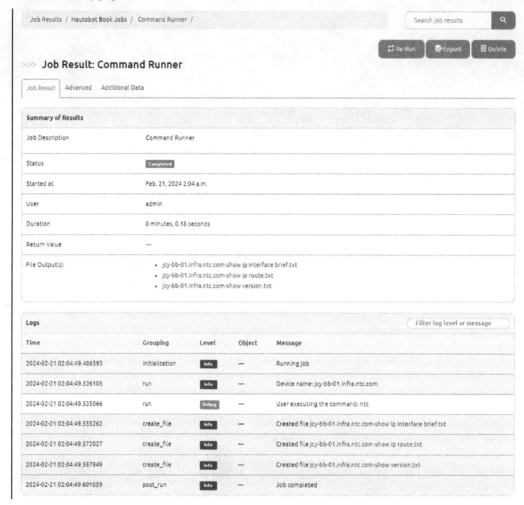

Figure 11.26 – The Command Runner Job with the files available for download

You can click on either the file in the Job log or in the file output whenever you want to download the contents of that particular file to your system.

In this section, different Job variables and attributes were used. Here is a larger table of some of the variable types available at the time of writing. Take a look at the Jobs Developer documentation (`https://docs.nautobot.com/projects/core/en/latest/development/jobs/`) for the most up-to-date variables and documentation:

Prompt Type	Details
BooleanVar	Boolean true/false selection box
ChoiceVar	A tuple of choices, with the value first in the tuple and then in the human-readable prompt
FileVar	Prompt for a file
IntegerVar	Whole number variable input
IPAddressVar	An IP address input
IPAddressWithMaskVar	An IP address input with a CIDR mask accompanying the input
IPNetworkVar	A prefix variable
MultiChoiceVar	From choices, allowing multiple to be selected
MultiObjectVar	Select multiple Nautobot objects
ObjectVar	Select a single selection of a Nautobot object
StringVar	A single line of text input that allows for regex validation
TextVar	Multi-line input text box

Table 11.3 – Job form variable types

That completes this example of doing network automation with Nautobot Jobs. There is a lot of power in being able to easily build self-service forms with built-in object types such as `StringVar` and `ObjectVar`. When you couple self-service with easy access to data, Nautobot Jobs are great tools to have in your toolkit.

Using Jobs to populate data in Nautobot

In *Chapter 13*, we'll highlight a Nautobot app called Design Builder (`https://blog.networktocode.com/post/design-builder/`). It is an app (that uses Nautobot Jobs) that highlights this use case and allows users to simplify adding data to Nautobot through codified network designs. The codified designs might be T-shirt size designs for small, medium, and large branch sites. The design can include every piece of data needed to onboard new sites to Nautobot, from devices and interfaces to IP addressing, VLANs, and circuits.

From a user experience perspective, if a user is trying to add a new device, rack, location, or something else, that can be codified in a Job or framework (such as Design Builder) so that there are minimal inputs in a Job form. Once a user submits the Job, all the data that's needed for the new design is added to Nautobot. This helps significantly in growing organizations.

Converting Python scripts into Nautobot Jobs

Now that you have seen how Nautobot Jobs can be executed, what if you have some Python command-line scripts that you would like to make available via Nautobot Jobs? The first step is to identify what input you need to gather from a user running your scripts. If you need to prompt for devices, sites, or other items (such as an email address), you need to determine if this data is in Nautobot already or if you need to create a variable prompt. If you have the data in Nautobot already, you should do your best to see if it is something that can be gathered without asking for input. An example of this is often a prompt for a network address when building out new addressing. Can you create a parent prefix in Nautobot identified with prefix roles and not ask to just grab the next available one? This is the power of having a network source of truth.

Let's take a look at what a common command runner Python script looks like and what you will need to do to convert it into a Nautobot Job. Two primary things are missing from the Python script that you get when a running script as a Job from Nautobot:

- The ability to dynamically select what commands to run against a device

- The ability to provide a place to download the output natively within a computer operating system

The following Job will have similar capabilities to the command runner on the CLI, similar to the Command Runner Job described earlier:

```
1    """Example Netmiko application to interact with network devices."""
2    import os
3    import logging
4
5    from netmiko import ConnectHandler
6
7  v def setup_logging():
8          """Configure logging settings."""
9  v       logging.basicConfig(
10               filename="netmiko_app.log",
11               level=logging.INFO,
12               format="%(asctime)s - %(levelname)s - %(message)s"
13         )
14
15 v def main():
16       """Execute when running this file."""
17       setup_logging()
18       logging.info("Starting")
19
20 v     device = {
21           "device_type": "cisco_ios",
22           "host": os.getenv("DEVICE_IP") or input("Enter device IP:"),
23           "username": os.getenv("DEVICE_USERNAME"),
24           "password": os.getenv("DEVICE_PASSWORD"),
25       }
26
27 v     with ConnectHandler(**device) as net_conect:
28 v         for command in ["show version", "show ip int brief"]:
29               logging.info(f"Sending command to device: {command}")
30               output = net_conect.send_command(command)
31
32               filename = f"{device['host']}-{command}.txt"
33               logging.info(f"Writing output to file: {filename}")
34 v             with open(filename, 'w') as file:
35                   file.write(output)
36
37       logging.info("Done.")
38
39 v if __name__ == "__main__":
40       main()
```

Figure 11.27 – Python Netmiko application to run commands to a file

The following table outlines the details of what can be eliminated, kept, or moved from the Python application or what can be done in a more user-friendly way. Three sections are eliminated by using Nautobot Jobs, while two sections remain the same – that is, the business logic handling.

Finally, two portions of the code have improved capabilities:

Code Description	Line Numbers	Action on Code	Nautobot Application Support
Logging setup	8-14	Eliminated	Job logging comes natively with Nautobot.
Logging the start	20	Eliminated	Completed as part of the Job within Nautobot.
Setting up the connection to the network device	22-29	Kept	Mostly the same setup but with a few variables, such as the device IP address information, to come from the source of truth.
Looping over commands	30	Improved	In the Nautobot environment, there is now a UI that provides a list of commands to choose from. In the Python CLI, a menu system may be needed, along with much more complexity.
Interactions with network devices over SSH	32	Kept	This is the same regardless of whether you're using the Python CLI or a Nautobot Job.
File handling	36-37	Improved	Nautobot performs file handling and provides a UI to be downloaded by multiple users.
Logging the completion	39	Eliminated	The ability to provide a log is handled without the need for code to be written.

Table 11.4 – Migrating your Python applications to a Nautobot Job

As you work through the table, you'll see that many items have been improved, including providing a UI to the users. This may be even easier for a Level 1 team member who does not need network access to be able to execute and get information; this will help reduce the amount of time needed to get work completed.

So, you will be able to grab many of the important portions of the existing Python scripts that your network engineers have and put them in a centralized location without much effort. There may be a few other places to get information that will change, such as prompts for information. Remember that all prompts or user inputs will come from Nautobot when you're using Jobs.

Diving into even more Job features

Going beyond just creating Jobs that users execute from the Jobs UI, many other features enhance Jobs or make them even more functional in enterprise environments. Let's review some of those features.

Job buttons

Job buttons allow you to launch Nautobot Jobs via *buttons* that can be placed within the Nautobot UI. Specifically, these buttons can be placed on detailed views for specific object types. This allows you to enhance the UI and launch automations right from the page of an object you want to automate, reducing copying and pasting and possible errors going between pages and tools.

For example, you can place a button on a device's page that validates it' in your monitoring tool, connect to a device to compare its data to what's in Nautobot, or generate a compliance report for just that specific device.

Without Job buttons, you end up going to the Jobs page, finding the Job you want to run, filling out a user form, and then finally running the Job. So, using Job buttons helps the overall user experience when you need to run a Job against a single object.

Let's create a Job button that will check a device. Although our Job button will be a Hello World example, it conveys how these buttons can be used in production environments:

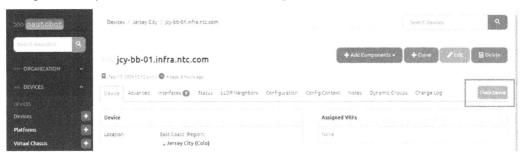

Figure 11.28 – The Check Device button on the right of the Device details page

We'll start the process of building a Job button by creating a new Job file.

So, create a new file called `job_button_test.py`:

```
root@nautobot-dev:/opt/nautobot/jobs# touch job_button_test.py
root@nautobot-dev:/opt/nautobot/jobs#
```

Add the following code to the new Job button file:

```
root@nautobot-dev:/opt/nautobot/jobs# cat job_button_test.py
from nautobot.apps.jobs import Job, register_jobs
```

```
from nautobot.apps.jobs import JobButtonReceiver

name = "Job Button Receivers"

class HelloWorldJobButton(JobButtonReceiver):

    class Meta:
        name = "This is my first JobButton Receiver."

    def receive_job_button(self, obj):
        self.logger.info("Hello, this is my first Nautobot Job
Button.", extra={"object": obj})
        self.logger.info("Hello, this is my first Nautobot Job
Button.", extra={"object": obj.name})
        self.logger.info("Hello, this is my first Nautobot Job
Button.", extra={"object": obj.status})
        self.logger.info("Hello, this is my first Nautobot Job
Button.", extra={"object": obj.role})

register_jobs(HelloWorldJobButton)
```

While this file is a Job file and you are still registering a Job, you are creating what is called a Job button receiver. Take note that the class being inherited and required method names are different from Jobs that users run using the Job form.

Another important piece of information to note is the use of obj – for example, obj.name, obj. status, and so on. obj is the object's data from where the button we're creating was clicked. So, if a Job button was clicked on the NYC page, obj will be the NYC object data. If a Job button was clicked on the nyc-rtr-01 page, obj will be the nyc-rtr-01 data:

```
root@nautobot-dev:/opt/nautobot/jobs#
```

Perform a post upgrade:

```
nautobot@nautobot-dev:~$ nautobot-server post_upgrade
```

Restart the services:

```
root@nautobot-dev:/opt/nautobot/jobs# systemctl restart nautobot
nautobot-worker nautobot-scheduler
```

Once the services have been restarted, navigate to the Jobs page in the UI. You will not see the Job button receiver. Receivers are not shown by default. Click **Filter** at the top of the page, filter for **Is job button receiver**, and set it to **Yes**:

Figure 11.29 – Using a Jobs filter to view Job button receivers

After the filter is applied, we need to edit the Job button receiver. We still need to enable it:

Figure 11.30 – Viewing the Job Button Receivers area

Take note of the section's name (**Job Button Receivers**) and Job name, and ensure you enable the Job receiver by editing the Job file. Once you do, you'll see a green checkmark, signifying it is enabled.

Once it's been enabled, we need to create a button that'll be shown on an object's detailed view pages. Click the + icon next to the **Job Buttons** menu item:

Figure 11.31 – Adding a Job button

When creating the new Job button, use the information shown in the following figure:

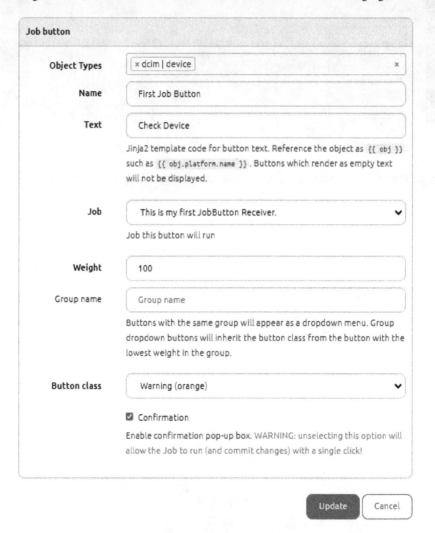

Figure 11.32 – Adding a new Job button

The **Text** field in the form is where you could also add a Jinja template (using a Jinja filter) to help have the button only show up when all of the object's data attributes are matched and have been set appropriately.

Once it is saved, navigate to a device in your inventory. You'll see the new button in the top right of the page, in line with the detailed view tabs:

Figure 11.33 – Viewing the new button on a device's detailed page

Let's execute the Job by clicking the Job button and confirming that this is what we want to do:

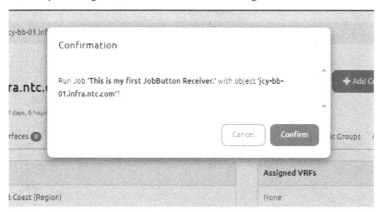

Figure 11.34 - Job button execution confirmation

Once you **Confirm** the Job, click to view the results by clicking the link at the top of the screen:

Figure 11.35 – The Job enqueued message after clicking the Job button

This will take you to the typical **Job Result** page for the Job that just ran:

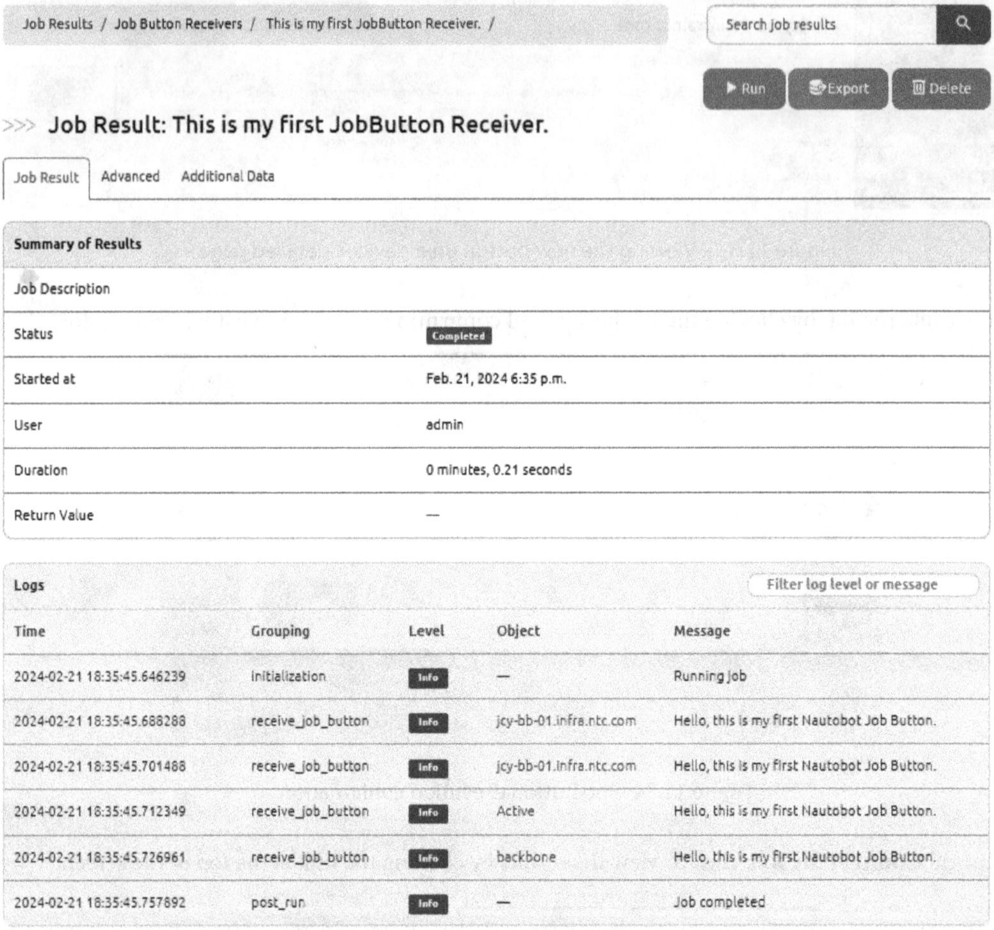

| Job Results / Job Button Receivers / This is my first JobButton Receiver. / | Search job results | 🔍 |

▶ Run   ⇖ Export   🗑 Delete

>>> **Job Result: This is my first JobButton Receiver.**

| Job Result | Advanced | Additional Data |

Summary of Results

Job Description	
Status	Completed
Started at	Feb. 21, 2024 6:35 p.m.
User	admin
Duration	0 minutes, 0.21 seconds
Return Value	—

Logs   Filter log level or message

Time	Grouping	Level	Object	Message
2024-02-21 18:35:45.646239	initialization	Info	—	Running job
2024-02-21 18:35:45.688288	receive_job_button	Info	jcy-bb-01.infra.ntc.com	Hello, this is my first Nautobot Job Button.
2024-02-21 18:35:45.701488	receive_job_button	Info	jcy-bb-01.infra.ntc.com	Hello, this is my first Nautobot Job Button.
2024-02-21 18:35:45.712349	receive_job_button	Info	Active	Hello, this is my first Nautobot Job Button.
2024-02-21 18:35:45.726961	receive_job_button	Info	backbone	Hello, this is my first Nautobot Job Button.
2024-02-21 18:35:45.757892	post_run	Info	—	Job completed

Figure 11.36 – Viewing the Job Result page for Job button automation

The major point to realize is that our Job only logs the attributes of a device. Consider a real Job button that uses those attributes to perform ad hoc network automation (read-only or verification tasks), does a compliance check, or queries a third-party tool using that device's data.

Please check this book's GitHub repository at `https://github.com/PacktPublishing/Network-Automation-with-Nautobot` to find a more detailed Job button that uses Netmiko and updates the serial number in Nautobot after fetching it live from a network device. In the *Chapter 11* directory, you'll find a Job file called `job_button_netmiko.py`.

Job Hooks

Job Hooks are a concept introduced by Nautobot inspired by webhooks. Instead of sending a webhook (or HTTP request), a Nautobot Job is executed. This means that when data changes in Nautobot, you have the option to send a webhook or run a Job. This allows you to still send an API call in the Job but with much more flexibility.

Very similar to Job buttons, Job Hooks also have their own class that is used when creating the Job files. You still need to create a Job file and register the Job, but we're creating what is called a `JobHook` receiver. This `JobHook` receiver is what we build and is executed by a Job Hook. Each Job Hook receiver that's built inherits from a `JobHookReceiver` class. There are no variables to import since all of the data is expected to come from the object being changed.

Let's take a look at an example of a Job Hook that will launch an Ansible AWX inventory update.

We'll start the process of building a Job Hook by creating a new Job file.

So, create a new file called `job_button_test.py`:

```
root@nautobot-dev:/opt/nautobot/jobs# touch jobhook_test.py
root@nautobot-dev:/opt/nautobot/jobs#
```

Add the following code to the new Job button file:

```
root@nautobot-dev:/opt/nautobot/jobs# cat jobhook_test.py
import os
from nautobot.apps.jobs import Job, register_jobs
from nautobot.apps.jobs import JobHookReceiver


class AwxInventoryJobHook(JobHookReceiver):
    """Launch an AWX Inventory Update when data is changed in
Nautobot."""

    def receive_job_hook(self, change, action, changed_object):

        self.logger.info("Launching AWX Inventory Update Job")
        token = os.getenv("AWX_TOKEN")
        url = os.getenv("AWX_INVENTORY_URL")

        if token is None or url is None:
            self.logger.fatal(
                "AWX_TOKEN or AWX_INVENTORY_URL env variable not set."
            )
            Return
```

```
        headers = {
            "Content-Type": "application/json",
            "Authorization": f"Bearer {token}",
        }

        response = requests.post(
            url,
            headers=headers,
        )

        if response.status_code == 200:
            self.logger.info("AWX Inventory Update Job Launched")

register_jobs(AwxInventoryJobHook)
root@nautobot-dev:/opt/nautobot/jobs#
```

You will need to add AWX_TOKEN and AWX_INVENTORY_URL as environment variables, just like we did earlier. Keep in mind that for production, you should use Nautobot Secrets, even if you are still using environment variables:

```
# cat /etc/systemd/system/nautobot.service | grep Environment
Environment="NAUTOBOT_ROOT=/opt/nautobot"
Environment="DEVICE_USERNAME=ntc"
Environment="DEVICE_PASSWORD=ntc123"
Environment="AWX_TOKEN=<>"
Environment="AWX_INVENTORY_URL=<>"
#
# cat /etc/systemd/system/nautobot-worker.service | grep Environment
Environment="NAUTOBOT_ROOT=/opt/nautobot"
Environment="DEVICE_USERNAME=ntc"
Environment="DEVICE_PASSWORD=ntc123"
Environment="AWX_TOKEN=<>"
Environment="AWX_INVENTORY_URL=<>"
#
```

Similar to Jobs and Job buttons, you need to add the new class to register the Job, restart the Nautobot app with the post_upgrade command, and restart your Nautobot services. Once the services have been re-enabled, you can enable the JobHook receiver.

Job Hooks are then created in the same menu under Jobs and Job Hooks, similar to Job buttons. You do not run the Job Hook from the Jobs page. Remember, its purpose is to be triggered when data changes in Nautobot.

On the page where you created the Job Hook, select the content types that, when a create/update/delete is completed, the Job Hook will launch. Then, select the actions that will trigger the Job Hook. At that point, any time an object meets the criteria for the Job Hook, the Job Hook code will be executed. When this happens, a `JobResult` object will be created to log the results of the job:

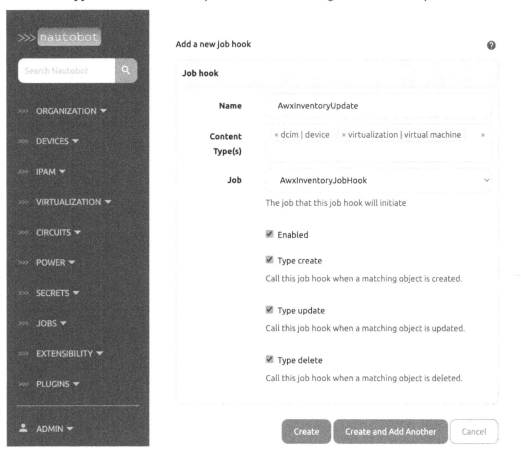

Figure 11.37 – Job Hook create/edit page

Now that you have seen the several different types of Jobs and how you may execute them, let's dive into Job scheduling.

Job scheduling

Job scheduling allows you to either run a Job one time in the future or set up a recurring Job execution. The recurring Job execution is particularly helpful in that of the Golden Config App, where there is a configuration backup Job that can be executed.

There is also an option called **Recurring custom** that allows you to use a **crontab** schedule. When using the powerful `crontab` format to schedule Jobs, you will be able to heavily customize when your Jobs are executed:

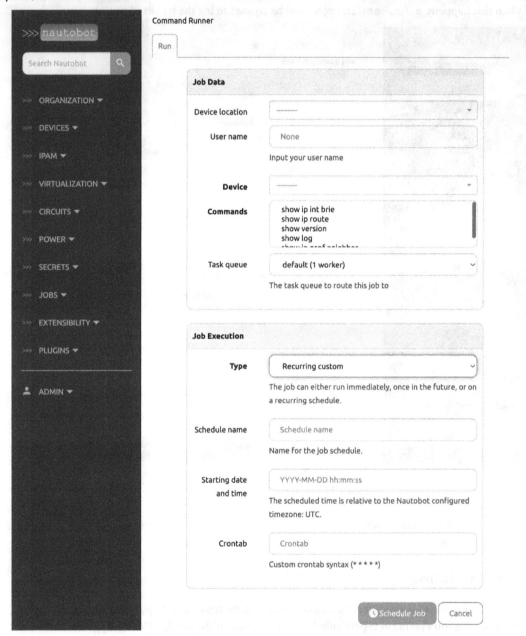

Figure 11.38 – Crontab scheduling

To schedule a Job, you must run a Job; you will see the button change from **Run Now** to **Schedule Job** when **Type** is changed within the **Job Execution** area of the Job form.

Job approvals

Job approvals allow you to create a Job that requires someone other than the person running the Job to approve it before it is executed. This is managed by an attribute in the `Meta` class of the Job. This attribute is called `approval_required`. It is a Boolean field that defaults to **False** and when set to **True** will require that after a Job's execution is initiated, an approval will be required before the Job is executed.

When a user executes a Job that has `approval_required` set to `True`, they'll see the following button:

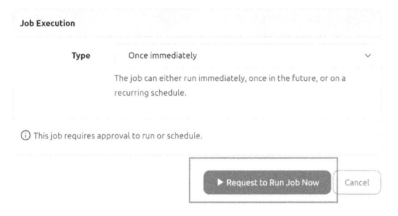

Figure 11.39 – Requesting to run a job that requires approval

The list of Jobs that are in the approval queue can be found by navigating to **Jobs** > **Job Approval Queue**:

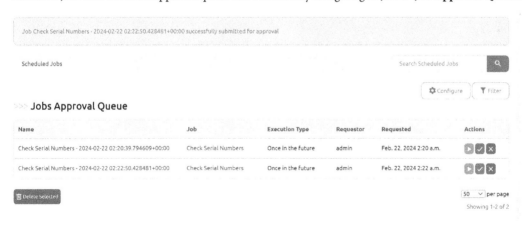

Figure 11.40 – The Jobs Approval Queue page

This is where you'll see all of the Jobs that need to be approved. From this screen, if you have permission to approve Jobs (which can be adjusted in the admin UI), then you will find all of the approval requests and can approve the Job execution.

When looking at a list of approvals, you'll see three buttons that you can execute per Job waiting to be approved:

- Dry run (the blue play button) allows you to execute a dry run without any changes to the database. If the Job connects to other sources, such as an API, you have to design your Job to not execute those actions. The dry run part only applies to Nautobot database changes.

- Approve (the green checkmark button) or Deny (the red X button). Depending on whether the Job is an immediate run or a future run, the Job will be executed at that given time.

Try it out by setting `approval_required` on one of your Jobs. All you need to do is add that one new attribute to `Meta`, like this (and do a post upgrade and restart the services):

```
class Meta:
    name = "Check Serial Numbers"
    approval_required = True
```

The Jobs API

The Jobs API was initially reviewed in *Chapter 8*. We'll look at it again briefly here. The URL for the API has the Job name in the API.

You should now have the URL to provide to others – that is, `https://<nautobot_url>/api/extras/jobs/{JobName}/run/`.

You can provide a JSON payload such as the following to the Command Runner Job:

```
{
  "data": {
    "device": "jcy-bb-01.infra.ntc.com",
    "user_name": "John Doe"
  }
}
```

As you can see, the first-level JSON keys are `data` and `commit`, which aligns with what is being passed into the class method of `run`. The result of calling the Job will be a `JobResult` ID that can then be queried for its status. The following code is being executed using Python requests to demonstrate this:

```
import os
import requests
import json
```

```python
import pynautobot

url = f"{os.getenv('NAUTOBOT_URL')}/api/extras/jobs/Command Runner/
run/"

# Get the UUID of the device name
nautobot = pynautobot.api(
    url=os.getenv("NAUTOBOT_URL"), token=os.getenv("NAUTOBOT_TOKEN"),
verify=False
)

device_id = nautobot.dcim.devices.get(name="jcy-bb-01.infra.ntc.com").
id

payload = json.dumps(
    {
        "data": {
            "device": device_id,
            "user_name": "John Doe",
            "commands": ["show version"],
        }
    }
)


headers = {
    "Authorization": f'Token {os.getenv("NAUTOBOT_TOKEN")}',
    "Content-Type": "application/json",
    "accept": "application/json",
}

response = requests.request("POST", url, headers=headers,
data=payload, verify=False)
print(json.dumps(response.json(), indent=2))
```

Note that it is possible to run a Job with `pynautobot`, but it requires getting the class path from the Job. To simplify the execution, we chose to show it using `requests` using only the Job name. If you want to explore running a Job with `pynautobot`, you can use the `extras.job.run()` method of your `nautobot` instance:

```
root@nautobot-dev:~/nautobot-book/chapter-10# export NAUTOBOT_
TOKEN=nautobot-book-token-123456789-abcdefghij
root@nautobot-dev:~/nautobot-book/chapter-10# export NAUTOBOT_
URL=http://<your-ip-address>:8001
root@nautobot-dev:~/nautobot-book/chapter-10# python3 pyjob-run.py
```

As you can see, you get several keys in the response, including the `result` key, which will give you information about the Job result object that gets created:

```
{
  "scheduled_job": null,
  "job_result": {
    "id": "ada0b1ac-93b9-49de-a5eb-7581b58a56e5",
    "object_type": "extras.jobresult",
    "display": "Command Runner started at 2024-03-12
17:49:43.172512+00:00 (PENDING)",
    "url": "http://198.22.11.88:8001/api/extras/job-results/ada0b1ac-
93b9-49de-a5eb-7581b58a56e5/",

    ... OUTPUT OMITTED FOR BREVITY ...
    "computed_fields": {}
    "files": []
  }
}
```

With the Job result ID, you can now query that endpoint to see where the Job is and, once completed, the result:

```
job_result_url = fhttp://<nautobot_url>/api/extras/job-results/
{response.json()['result']['id']}/
response_job_result = requests.request("GET", job_result_url,
headers=headers)
print(json.dumps(response_job_result.json(), indent=2))
```

You will then see the resulting data:

```
{
  "id": "f39faca0-5968-4824-9aef-8c8731722a13",
  "display": "b9e04d7f-dfae-4bb7-b33b-ee835e958ec3",
  "url": "http://<nautobot_url>/api/extras/job-results/f39faca0-5968-
4824-9aef-8c8731722a13/",
  "created": "2023-04-20T03:16:59.218829Z",
  "completed": "2023-04-20T03:16:59.297910Z",
  "name": "plugins/sandbox.jobs.demo_job/JobName",
  "job_model": {
    "display": "Example job",
    "id": "4b3013c0-d5c3-4405-b350-827a9cc6f042",
    "url": "http://<nautobot_url>/api/extras/jobs/4b3013c0-d5c3-4405-
b350-827a9cc6f042/",
    "source": "plugins",
    "module_name": "sandbox.jobs.demo_job",
    "job_class_name": "JobName",
```

```
      "grouping": "sandbox.jobs.demo_job",
      "name": "Example job",
      "slug": "plugins-sandbox-jobs-demo_job-jobname"
    },
    "obj_type": "extras.job",
    "status": {
      "value": "failed",
      "label": "Failed"
    },
    ... OUTPUT OMITTED FOR BREVITY ..."  "custom_fie"ds": {}
}
```

Note that you can see the status based on the result of the Job at the JSON path of `status`. You also get a `completed` timestamp for the Job's execution.

Job permissions

To execute a Job, a user must have the `run` permission.

This was covered in *Chapter 7*, in the *Additional actions* section.

Regarding Jobs and permissions, it is up to the Job author to ensure a user that does not have access to edit/view an object, such as a circuit, can run a Job; that is, if a user can't add or edit a circuit, do you want them to be able to add/edit a circuit from a Job? This may be the workflow you want – for example, run this Job to update a circuit instead of doing so manually so that standards are followed.

In short, you need to consider if you want to maintain or bypass permissions within a Job. It depends on why the Job exists and who the target users are for the Job. Again, you may very well want a project manager to execute Jobs but not make manual changes to objects in Nautobot.

Summary

In this chapter, we did a deep dive into Nautobot Jobs. We looked at how Nautobot Jobs are a great place to build out workflows for network automation, especially when you already have Python scripts you wish you could easily expose to your team and others within your organization. Beyond network automation tasks, we covered that Jobs can be used for verifying and validating data in Nautobot, with the option to create downloadable files. Beyond user and API-executed Jobs, we also looked at Job Hooks, Job buttons, and Job schedules and approvals, all of which provide enhanced functionality for teams looking to simplify the user experience for their network automation workflows.

In the next chapter, we will learn about the Nautobot app ecosystem and dive into all that Nautobot apps have to offer as you continue your network automation journey.

Data-Driven Network Automation Architecture

Over the last few chapters, we looked at an introduction to Nautobot, got Nautobot up and running, learned about its extensibility features, and delved into Nautobot platform administration. The focus has been on Nautobot as a **Source of Truth (SoT)** and overall data platform. However, Nautobot is more than an SoT—it is a network automation platform. All of the data that is stored in Nautobot can be used to power a data-driven network automation architecture. Over the next few chapters, we'll review the relationship between data and network automation, seeing the power of Nautobot as an automation platform. Before we get into Nautobot's role in driving network automation, we'll look at the evolution of managed networks and the need for a data-driven approach to network automation architecture.

The following are the main topics that will be covered in this chapter:

- Data-driven network automation architecture

- Evolution of managed networks

- SoT with Nautobot

- Automation and orchestration

- Modern network monitoring—telemetry and observability

Data-driven network automation architecture

Over the last few years, the industry has been taken by storm with engineers learning network automation through Python, Ansible, and a number of other open source and commercial products. This is undoubtedly great for the industry. After all, network automation is the future, but it leaves many organizations short on how to get there. Based on *Network to Code*'s extensive experience in this space, one of the largest skill gaps we see in the market today is an understanding of network automation architecture. Through these last few years working with customers from around the world, we've been able to build a network automation architecture that depicts and highlights the major components of network automation, the role they play, and how they can work together.

The following is a high-level visualization of this architecture:

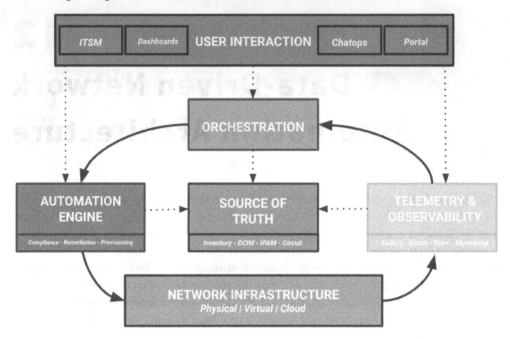

Figure 12.1 – Reference network automation architecture

Throughout this chapter, we are going to focus on four of the components depicted in the architecture: SoT, automation, orchestration, and telemetry and observability.

Given that Nautobot is an SoT and at the center of network automation that stores the intended state of the network, let's start with reviewing SoT, building on what was covered in *Chapter 1*.

Evolution of managed networks

First, to understand what the ideal state of network management is, let's take a brief overview of the current state of network operations before taking a deep look at operating networks with automation later in the chapter.

Manually managed networks

Most networks fall into the category of not being automated. That is, when a request or outage happens, a qualified network administrator reviews the situation, makes a determination on what to do, and addresses the situation. This leaves little room for documenting the changes, is error-prone, and generally leads to inconsistency in results such as configuration standards.

When a network administrator looks at a configuration that *"seems out of the ordinary,"* they are generally unable to understand when that configuration was added, who added that configuration, and, most importantly, why that configuration is there. This leads to following the status quo bias, often spoken as, *"If it isn't broken, don't fix it."* However, it is often those out-of-the-ordinary configurations that lead to outages or broken designs down the road.

Power tool automated networks

The first level of automation tends to be the power tool—that is, scripts, playbooks, jobs, and so on that are generally written in a very one-to-one manner. The scripts tend to be named sanely and describe exactly what action is going to happen. The following is a list of scripts:

- `add_vlan.py`
- `bounce_port.py`
- `create_firewall_rule.py`
- `get_vpn_stats.py`

Without even explaining these scripts, you likely can imagine exactly what is going to happen. You can also imagine that it will likely expect a few parameters or ask you a few questions, such as *"What device would you like to add the VLAN to?"* and *"What VLAN would you like to add?"* This is where the challenges stem from. While the actions become much more efficient and it is easy to get started, you still do not know the when, who, or why the configurations are there. The data is likely not stored anywhere, and if by chance it is stored, it is only stored transactionally; that is to say, there is no place to go to say, *"What are all the data points used to create the data that makes up all of the configurations I see today?"* If you have ever been asked to do an audit on your firewall rules, you will know the pain that your ITSM system has the changes, but it is impossible to reconstruct due to free-form notes in tickets, partial data, and not understanding the full state of the firewall at any given time.

Legacy and domain network management managed networks

Scripts aren't new. While Python is currently the de facto language for network automation, Perl (and other scripting) has been around for decades. When scripting wasn't enough, organizations and teams would buy network management tools. These network management tools usually got off to a good start when getting deployed. Over time, their limitations became more apparent. When you upgraded the network devices, your network management tool(s) didn't support them. When you changed vendors, your network management tool(s) didn't support them. When you needed an API, your tool(s) didn't support them. When your vendor added new features you wanted to automate, your tool(s) didn't support them!

In summary, these tools were either too narrowly focused or they weren't extensible. You got what you got, and you couldn't add functionality—you hoped the vendor gave you what you needed in the next release until they didn't, and the network management tool was no longer valuable.

Infrastructure as Code (IaC) automated networks

The next level of automation to consider is the IaC approach. This is often referred to as NetDevOps, being the intersection of networking and DevOps.

That is to say that your data is treated the same way as your code, namely in a **version control system** (**VCS**) such as GitHub or Bitbucket. IaC has many properties that are advantageous in your network automation configuration management strategy:

- **Consistency** – Since configurations are generated a single way, they are more consistent
- **Traceability** – The VCS provides an audit trail of who made the updates and a comment system to document "*why*" the change was made
- **Collaboration** – VCSs provide the ability to collaborate on changes to scripts, templates, and other software artifacts
- **Approvals** – Through collaboration, you have the ability to have peer review (and approve) changes
- **Documentation** – The data becomes a de facto documentation store

IaC is most popularly described in either YAML or **HashiCorp Configuration Language** (HCL) but can be described in any data or markup language, such as JSON, TOML, XML, and so on. Data structures can be built and directly consumed by the infrastructure through automation, often using REST APIs, DevOps, or automation tooling, and through a traditional network CLI. Let's look at an example of YAML data and a Jinja network configuration template:

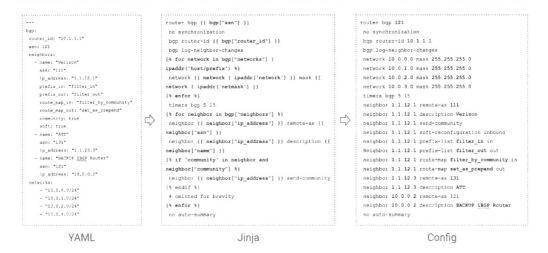

Figure 12.2 – Showing how YAML and Jinja can generate an intended configuration

From this example, you can see how data can be used to generate configuration. While at first glance, it may seem no better than maintaining the original configuration—after all, the configuration still needs to be generated, and now you need to maintain this data in a separate place.

This concept introduces one of the key tenets of IaC which is having an intended state, which was originally described in the *Understanding Source of Truth* section in *Chapter 1*. By separating the actual state (configuration running on the device) from the intended state (defined in your IaC SoT with data and templates), you can now see the power of the features described previously:

- **Consistency**: I know the configuration was generated from a YAML file and Jinja template; there is no room for other configurations to creep in

- **Traceability**: I can use the `git blame` feature to understand when, why, and how data was updated

- **Collaborative**: I can ask for help or review on my changes

- **Approvals**: I can wait for a peer or manager to approve my changes.

- **Documentation**: I can view the data centralized in the YAML files

---

Pro tip

`git blame` is used to annotate each line in a file to see who made changes and when the last change to that line was.

All of these features and outcomes reinforce the intended state paradigm that an SoT is a **Source of Intent (SoI)**. Had I "just trusted" the configuration on the device, there are an infinite amount of reasons why the configuration may not be as intended. I do not have to scour through years' worth of tickets, logs, emails, instant messages, and so on trying to find out why the configurations are the way they are.

While IaC has its benefits, it also has drawbacks when considering the following questions:

- Do people have the right skills to contribute?

- Should someone need to know Git to update data?

- How do other tools query the data stored in a Git repository?

- How is it possible to apply business logic to ensure data compliance and quality for data stored in Git?

- Which teams need access to view and/or update the data? Which interfaces work best for them? Is Git the right approach?

- When managing hundreds or thousands of lines in a file or files themselves, you have to ask, "*Is it the right approach?*" due to the amount of complexity, how tedious it is, and how apt to human error it is.

By reflecting on the answers to these questions, you'll see that IaC also requires a complementary approach for managing data and network automation. We'll talk about what that means next.

Nautobot automated networks

There is no denying that an IaC design has provided a cosmic leap forward for automation and has put the SoT at the front and center of automation. However, there are many other things to consider when managing SoT:

- **Scalability** – How to manage thousands or even millions of records

- **Data storage** – Where the data is stored and how it is exposed

- **Data quality** – The ability to enforce data quality to ensure organizational standards are always enforced and/or reported on

- **Consumption models** – All of the ways the data can be consumed, such as APIs, GraphQL, Django ORM, UIs, database queries, and so on

- **User experience** – How the data is consumed (viewed and updated) for different types of users, from non-technical to technical

- **Queriability** – The ability to query the data to ask intelligent questions

- **Performance** – How to manage the demands of a growing large-scale deployment

- **Concurrency** – The ability for many systems, applications, and people to access the data at the same time

- **Atomic Consistent Isolated Durable (ACID)** – A core concept for databases that ensures your data will not become corrupt

Nautobot continues to be developed and enhanced with these characteristics in mind to accommodate modern networks at scale. That comes not only in the feature set of exposing data in new and interesting ways but also in the consideration of integrations that truly enable network automation.

So, what exactly does network automation look like with Nautobot? Here's an overview:

- Has defined network models that can be leveraged while also tailoring them through features such as custom fields and computed fields

- Allows users to build custom data models using the robust Nautobot developer API

- Directly integrate YAML/JSON already stored in a Git repo and map it to a device in Nautobot

- Has the ability to perform and execute robust tasks through Nautobot Jobs

- Easily convert any Python script into a self-service Nautobot Job

- Access and aggregate data through REST APIs and GraphQL, drastically simplifying client tooling working with Nautobot programmatically

- Use the same REST and GraphQL APIs to integrate Nautobot with third-party automation and orchestration tools

- Dynamically trigger Webhooks and Job Hooks based on Nautobot events

- Build automation off of Nautobot Dynamic Groups that allow automation to easily query groups instead of building complex logic

- Use Nautobot Golden Config to perform backups, generate intended configuration, check for device configuration compliance, and execute config remediation plans

- Natively integrate Git repositories that include Jinja templates, backups, and intended configs directly with Nautobot

- Dynamically federate and synchronize other data sources with Nautobot using the Nautobot **Single SoT (SSoT)** application; for example, ServiceNow, Infoblox, and so on

- Take advantage of Nautobot ChatOps to chat with Nautobot or any other third-party system

- Ensure Enterprise standards are codified and "bad data" cannot be entered in Nautobot

- Build and view data center floor plans with Nautobot Floor Plan

With capabilities that include those just listed, Nautobot can play a role in various parts of a network automation architecture. We are going to go into each functional block of the network automation architecture in *Figure 12.1* and examine the role of Nautobot.

SoT with Nautobot

An SoT is central to any complete network automation architecture and central to the core of what the Nautobot platform and this book intend to showcase and highlight.

So far, we have considered SoT-driven network automation primarily under the prism of configuration management, but there are many other considerations and use cases that an SoT grounded with a solid network automation architecture aims to solve.

Let's take a look at how integrations and extensibility can bring a system to life and make it a critical part of your automation infrastructure.

Integrations and extensibility

While we have already briefly covered and alluded to a key configuration management integration with the Nautobot Golden Config App (`https://github.com/nautobot/nautobot-app-golden-config`), there are many more integrations to consider for Nautobot and SoT in general:

- Inventory integrations to any system that also needs an up-to-date inventory.
- Network management integrations to properly manage the infrastructure.
- Data expansion integrations to codify network designs. This enables you to populate data consistently in Nautobot with a minimal amount of inputs; for example, a self-service form in which you select you want to add a "Large" site with 10 edge devices. From there, all data is generated and populated into Nautobot.
- ITSM integrations for requested objects as well as **incident management** (**IM**) data.
- NetDevOps interfaces and tooling such as APIs, SDKs, Ansible, Terraform, and so on.

> **Note**
> Nautobot extensibility is covered in great detail in *Chapters 13*, *14*, and *15*.

A SoT by itself is helpful, but these integrations are what change an SoT from a document store or even a configuration management tool to a mission-critical part of your infrastructure.

SoT life cycle

Let's take a look at how Nautobot can be used in a full life cycle use case of data for a new campus site:

1. Data is populated with a minimal amount of inputs. Users provide only a few inputs:

    *   Site name

    *   Region

    *   Services provided at the site: data, voice, video servers, and so on

    *   Number of switches

    *   Serial numbers

2. Data is expanded to generate device names, subnets, IP addresses, VLANs, connections, and so on. This is made possible in Nautobot by codifying network designs using Nautobot Jobs and apps such as Design Builder (described more in *Chapter 13*).

3. The preceding network and IT data points are used to generate intended configurations using Nautobot Golden Config.

4. The intended configurations are reviewed and approved and finally deployed to your network with the Nautobot Golden Config application.

5. Based on a change in the device's Statuses and Roles, the following steps occur:

    *   Nautobot SSoT updates data in your ITSM tool

    *   Nautobot SSoT updates your network management tools

6. Day 2 requested items (for example, port changes) happen from your ITSM tool (or as a Nautobot Job):

    *   Data is updated in Nautobot

    *   Configurations are staged (and optionally deployed) from Nautobot

    *   APIs are consumed to further analyze and perform changes in Nautobot

7. Network operations leverage Ansible playbooks that use Nautobot Ansible Collection for common campus issues. The modules use the Nautobot REST or GraphQL APIs.

8. Monthly reports are generated to understand the security and end-of-life status leveraging the Nautobot Device Lifecycle App.

9. When a given device status is changed to **Decomm** or **Inactive**, the ITSM and network management tools are updated, again from the SSoT App.

Throughout this life cycle, you can see that there isn't that much time spent on configuration management. We call this out because, in network automation, configuration management is always the hot topic, but many steps to make that a reality start and stop with data.

This also means while your network infrastructure is changed or updated, as long as your data model stays consistent, it is far easier to manage and adapt if needed.

One of the largest changes with an SoT-first network automation strategy outside of configuration management is the integration with the telemetry stack. While we often think of devices and infrastructure being added to our network management stack in legacy monitoring solutions, it is helpful to start thinking about how an SoT can enrich the data and aid with the configuration of your telemetry stack. This can be thought of as the configuration management strategy for your telemetry stack. We will cover this concept in more detail later in the chapter.

Nautobot enablers for SoT

The following section includes many of the concepts and design considerations when deploying Nautobot as an SoT. It also covers specific features and integrations that make Nautobot a great SoT to power a network automation architecture.

Accuracy

The SoT is only as good as the data, so the data must be accurate. If the network operators do not trust the data, they will resort to just going to the network devices. Several strategies can be used, as well as associated Nautobot features, to help you with ensuring your data is accurate.

System of Record (SoR) is the idea that for a given data type (or attribute), data can only be *written* in one place. That is not to be confused with saying that data can only exist in one place, but all systems should agree that it is only written in one place and reflected in all other systems. Once you have two places the data should be written to, it is said you have a *source of lies*.

It is important to note the nuance of "for a given attribute." To take a real-life example, many companies' inventory must be stored in ServiceNow, so it is reasonable to designate the SoR for the device name, serial number, and hardware type as ServiceNow. However, configuration management attributes such as BGP ASNs, VLAN IDs on ports, and connections are rarely, if ever, modeled in ServiceNow. As such, you could designate Nautobot as the SoR for those attributes. Leveraging Nautobot's SSoT framework (`https://github.com/nautobot/nautobot-app-ssot`), it can be simple to keep the SoR model while also keeping data synchronized.

Here is an example of what that looks like in Nautobot, with SSoT aggregating data between Nautobot and ServiceNow (either direction is supported):

Figure 12.3 – Nautobot SSoT synchronization of data with ServiceNow

In *Figure 12.3*, Nautobot is acting as the SoR for the data types shown in the **Data Mappings** panel on the right in the **Nautobot** column. The corresponding field for each data type in ServiceNow is shown in the **ServiceNow** column.

In the next section, we talk about SoT data aggregation that helps to address the problem of distributed data by providing a read-only view of all the data, but for now, it is enough to understand the subtle differences between an SoR and an SoT. Again, our view is that an SoR is the authoritative source of information (where write operations exist) for a given data type, but SoRs can be aggregated into an SoT. An SoT is often an SoR for multiple pieces of data as well.

Dynamic network import dynamically discovers an existing network. It is often a great strategy when starting out to import data from the network. Some SoT purists will argue against this, on the premise that data should only be added to the SoT by humans or machines that understand the desired state. However, in real-world organizations, this is both unrealistic and unsympathetic to the plight of managing even moderate-sized networks. In addition, manually adding thousands or even millions of records is costly, timely, and error-prone. We will review alternatives to ensure the data ends up being as reliable and valid as if humans did accurately curate it.

Leveraging network import as an ongoing strategy should be avoided in most scenarios, as it reinforces bad practices such as trusting inaccurate data that could then be used for automation efforts, resulting in unforeseen outages.

When leveraging the network import strategy, it is important to think about what data you want to import. Configurations that are good to import are generally not easily compressed (meaning they are tied 1:1 from the device to the data in Nautobot), such as serial numbers and IP addresses on interfaces. Global configurations such as DNS servers and NTP servers should just be built on what they should be (the data in the SoT), not what configurations are there, since there is little variance on what should exist.

So, as you consider doing a dynamic network import into Nautobot, you should consider if you want this once to populate the initial data or to do it on an ongoing basis. If it is the latter, you may need to question if Nautobot is really the SoT because it would no longer be the SoI. However, if you're doing it even once, you should be building Nautobot Jobs, Data Validation and Compliance Rules, and Config Compliance Rules to ensure the data in Nautobot is of high quality.

Configuration compliance is a great tool to understand the accuracy of your data. The premise of the Nautobot Golden Config App is to take the intended configuration that is generated from your data with Jinja templates and compare those configurations to the actual device configurations (sourced from backups).

At Network to Code, we often call this the "*Christmas tree*" approach, iterate by comparing the intended versus actual configurations, and then view the compliance report, which is green and red, and make adjustments. This iterative approach tends to be far more efficient than trying to get the data and templates correct the first time. As an example, if you initially thought that all devices had the same three NTP servers configured on them, it is far easier to add the three IPs and have that rendered with a template that builds an intended configuration. This intended configuration is then used for basic compliance, in contrast to verifying manually by SSHing to dozens of devices or setting up multiple meetings to ask. Viewing the report and checking the first few non-compliant devices, it is generally easy to find any simple issues. This iteration is easy with Nautobot Golden Config because it shows which commands are "extra" or "missing" from your intended configuration, allowing you to incrementally update your templates and data to be correct.

Here is an example showing extra and missing configuration commands using Nautobot Golden Config:

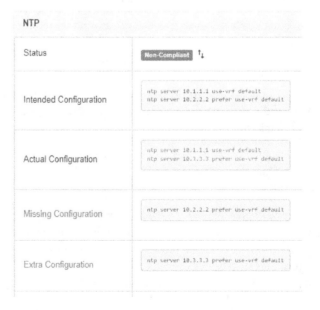

Figure 12.4 – Nautobot Golden Config compliance example

Data validation at creation or update time helps ensure bad data is never entered into the SoT. Leveraging the traditional IaC approach, you can use something such as JSON Schema to provide type enforcement, such as ensuring that an IP address provided is, in fact, a valid IP address and not `500.1.1.1`, or that speed is provided in integer format. But what about more advanced data validation? While possible, it usually adds complexity. Nautobot provides type enforcement out of the box. While type enforcement is a huge step forward, generally speaking, your data is unique to your organization. This is where Nautobot's data validation framework comes into place.

Let's say, for instance, that all devices should follow the format of `<site-code>-<two-char-function><two-digit-sequence-number>`, such as `GOT-RT01`. That enforcement must be done firm by firm. Luckily, the Nautobot Data Validation Engine App can provide enforcement for both simple and complex use cases. The Data Validation Engine App can provide the following validation via the UI:

- **Regular expression (RegEx)**
- Min/max value
- Required fields
- Unique values

Let's take a look at a few examples.

We'll start with adding **Required Rule**, which is used to ensure that a given field on a form is required when adding or updating an object. After installing the Data Validation Engine App, you'd see the following in the navbar. You would click the + sign to add a new Required Rule:

Figure 12.5 – Data Validation Engine App menu for rule creation

If we added a new rule that requires a name on a device and you tried adding a new device, you'd see the following. It has an error message that states, **Device Name is required when adding a new device:**

Add a new device

Device	
Name	Name
	• Device Name is required when adding a new device.
Role	leaf x ▾
	The function this device serves

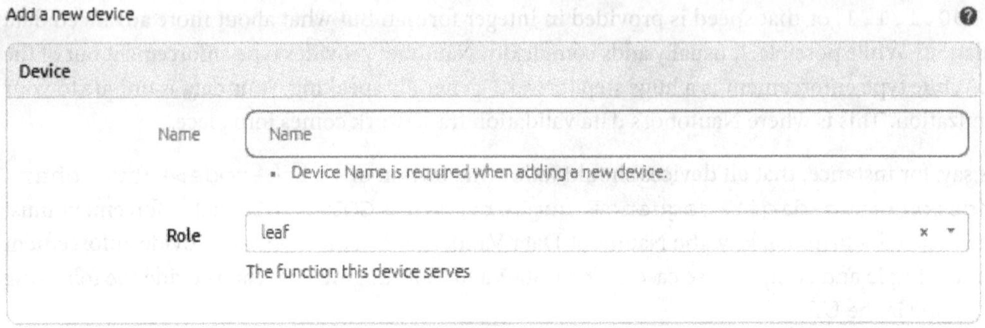

Figure 12.6 – Data Validation Required Rule violation

In order to add this rule and try it out, you'd add it with the following fields in the creation form:

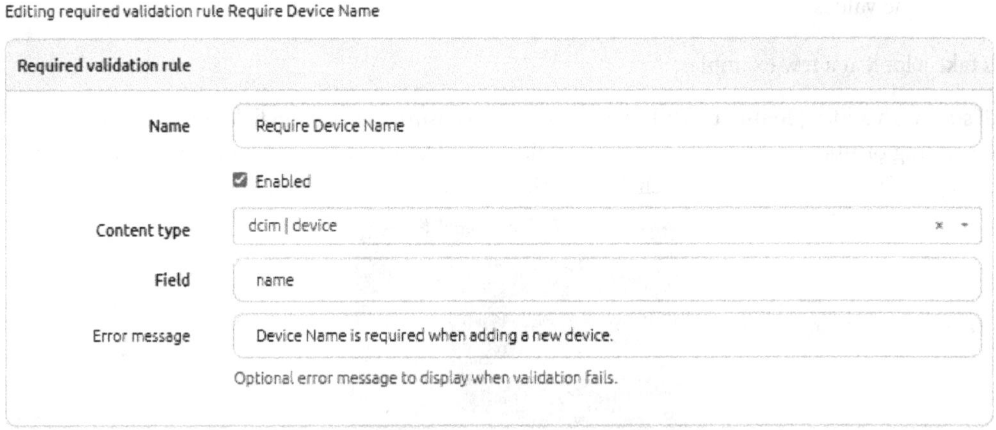

Editing required validation rule Require Device Name

Required validation rule	
Name	Require Device Name
	☑ Enabled
Content type	dcim	device x ▾
Field	name
Error message	Device Name is required when adding a new device.
	Optional error message to display when validation fails.

Figure 12.7 – Data Validation Required Rule creation form

Going one step further, you could use a RegEx rule, also shown in the navbar, so it is not just required but needs to conform to an actual standard. Here is an example of the error that could be displayed:

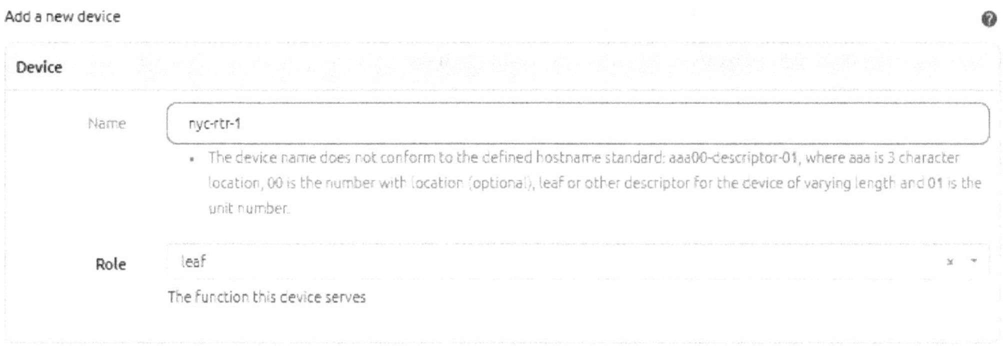

Add a new device   ❼

Device

| Name | nyc-rtr-1 |

• The device name does not conform to the defined hostname standard: aaa00-descriptor-01, where aaa is 3 character location, 00 is the number with location (optional), leaf or other descriptor for the device of varying length and 01 is the unit number.

| Role | leaf | x ▾ |

The function this device serves

Figure 12.8 – RegEx rule violation

And here is the configuration of that RegEx validation rule required to give that error message:

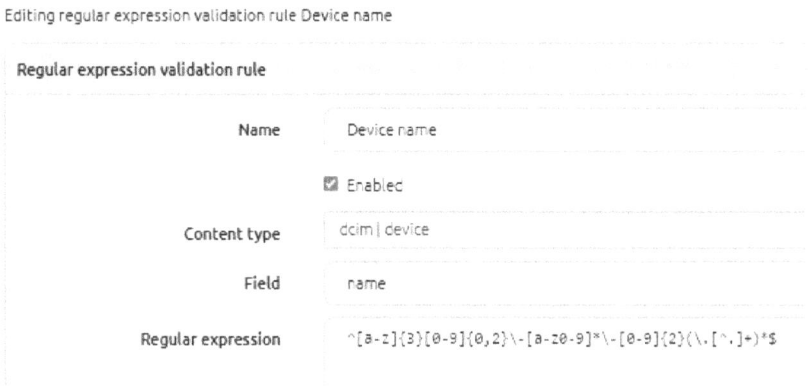

Editing regular expression validation rule Device name

Regular expression validation rule

| Name | Device name |

☑ Enabled

| Content type | dcim | device |

| Field | name |

| Regular expression | ^[a-z]{3}[0-9]{0,2}\-[a-z0-9]*\-[0-9]{2}(\.[^.]+)*$ |

Figure 12.9 – RegEx rule creation form

Let's look at one more example, this time using a Min/Max Rule:

Figure 12.10 – Min/Max Rules in the Data Validation Rules menu

We'll show how you can enforce that a VLAN must be between 11 and 500. Maybe your network has reserved VLANs or you just want to ensure that range is used before opening up the larger scope of VLAN IDs to be used.

The following is the creation form of the min/max validation rule that allows you to do this:

Editing min max validation rule Max VLAN ID

Min max validation rule	
Name	Max VLAN ID
	☑ Enabled
Content type	ipam | VLAN ✕ ▾
Field	vid
Min	11
	When set, apply a minimum value contraint to the value of the model field.
Max	500
	When set, apply a maximum value contraint to the value of the model field.
Error message	All VLANs must be between 11 and 500.
	Optional error message to display when validation fails.

Figure 12.11 – Min/Max validation rule creation form

And with that new rule, if you try to add a VLAN outside of the defined range in the rule, you'd see this error message added in the preceding form:

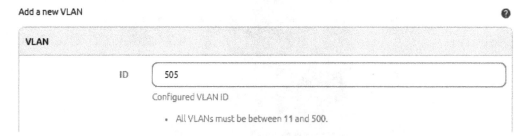

Add a new VLAN   ❓

VLAN	
ID	505
	Configured VLAN ID
	• All VLANs must be between 11 and 500.

Figure 12.12 – Min/Max validation rule violation

Beyond that, you can extend it to anything that can be described in Python, which is essentially any reasonable use case leveraging Nautobot's Custom Validator framework. This means you can also interact with third-party systems as data is managed in Nautobot. Imagine ensuring that no one in your organization can remove "*a critical interface or device*" or ensuring no one can update an interface if its data is not in sync with the real device. This is possible with the Custom Validator framework. Whenever you are clicking **Create** or **Update** on a form in Nautobot, this Python logic can run to enforce advanced Enterprise standards. One example may be you never want anyone to modify any backbone device. Here is an example of that:

```python
# custom_validators.py
from nautobot.apps.models import CustomValidator


class DeviceValidator(CustomValidator):
    """Custom validator that ensures Backbone devices can't be modified."""

    model = 'dcim.device'

    def clean(self):
        if self.context['object'].role == "Backbone":
            self.validation_error({
                "device": "It is not possible to modify any Backbone devices."
            })


custom_validators = [DeviceValidator]
```

Figure 12.13 – Custom Validator to enable Enterprise standards

This file would need to be distributed within a Nautobot App; this is covered in more detail in *Chapter 13*.

Fail fast, fail often is a good mantra to have when it comes to data. The power of simply not letting bad data enter avoids many future issues.

Data compliance is similar to both data validation and configuration compliance. Data compliance leverages the same framework that data validation uses through the Data Validation Engine App, but is not enforced on creation or update; instead, it only reports on data inconsistencies in a very similar fashion to configuration compliance. This offers you flexibility in that you want to prevent bad data, as described previously. However, in certain cultures or organizations, there are non-technical people managing data. Certain organizations prefer to get data added, verify the data is clean, and if it is not, the right team(s) clean the data before it's used to drive automation.

Nautobot's Data Validation Engine App is equipped to handle several data compliance use cases:

- Data exists and a new standard has been created; enforcing data validation would break any updates being made
- Many standards exist, and tooling is required to reasonably get a single standard enforced
- Data reporting allows you to understand the state of the data

You define data compliance similarly to custom validators shown previously. If you are interested in seeing additional details, check out the *Data Compliance* official docs (`https://docs.nautobot.com/projects/data-validation/en/latest/user/app_data_compliance/#example`).

A *feedback loop* provides a mechanism to understand if data is accurate. The premise is applied to both the configuration and data compliance methodologies. However, this concept extends beyond that. A common loop is to parse the current configuration and compare it to the current data.

Providing a concrete example, parsing a device name, and determining if the device role is the same as the device name indicates provides you with valuable information. If you parse `GOT-RT01` and determine that the device's role is "**Firewall**," that would be a clear indication of bad data. Note that this is a more general philosophy described and made possible with Nautobot with its data and the Nautobot Jobs framework (parsing configs and automatically comparing them against the data in the SoT is not an out-of-the-box feature).

SoT aggregation

The centralization of data is a key concept in the Nautobot philosophy and architecture. While there is no single feature that provides aggregation of data, Nautobot has a suite of features that make up Nautobot's data aggregation strategy, some of which have already been covered. Highlights that make up the solution are the following:

- Nautobot SSoT
- GraphQL
- Git as a Data Source

Nautobot SSoT App is an open source Nautobot App designed to enable you to use Nautobot as your network's SSoT, unifying data from any number of SoRs behind Nautobot's UI and APIs. This means once data is aggregated into Nautobot, you can view it or fetch it with the Nautobot UI and API, respectively. Additionally, this App makes it quick and easy to develop and integrate specialized Nautobot Jobs to synchronize data from other systems ("data sources") into Nautobot and/or from Nautobot to other systems ("data targets") as desired. In short, whatever system(s) you need to synchronize with Nautobot, you can support via Nautobot SSoT (as Nautobot Jobs). Nautobot SSoT is covered in great detail in *Appendix 2*.

GraphQL is a query language for APIs. It allows a user to programmatically request and receive data from a server. One of GraphQL's defining features is the capability to allow the user to specify the exact data needed, even if the data components are in different data models or domains. And, unlike REST API queries, a GraphQL query will return only the specific, requested data, allowing you to even change the keys that are used in the response with GraphQL aliasing. Additionally, the level of entry to get started with GraphQL is much easier than for REST APIs, given there is a single endpoint in Nautobot versus having to worry about different endpoints for different object and model types. With GraphQL, you have the ability to query data from across all of your SoRs with a single Nautobot

GraphQL query. You get to access all of the data relationships in a single API call. Remember—GraphQL is covered in *Chapter 7* in much more detail.

Git as a Data Source is well suited to store YAML/JSON data that is mapped to locations such as DNS servers, NTP servers, Syslog servers, and so on that don't often change. It can also be used to map data to platform types, vendors, and so on, so there is a lot of flexibility to map YAML/JSON data to the right devices in Nautobot.

Data from Git is synchronized with Nautobot via the *Git as a Data Source* feature. Data in Git (YAML or JSON files) is synchronized as Nautobot config contexts, which allows seamless integrations with the SoT Aggregation view. This means you can have YAML data for a given device (or type) stored in a Git repo, sync it to Nautobot, and then access that data directly from GraphQL and also use it as part of your SSoT strategy.

When using Git and YAML as the SoT for various network device properties, it is important to define what the expected model—for example, the format of the YAML files—is. Nautobot provides native integrations with JSON Schema with config contexts, as we looked at in *Chapter 5*. This ensures that even your structured data is schema-enforced. JSON Schema can validate YAML and JSON files and data.

Data governance

As data becomes more critical in the realm of network automation, it is a natural evolution to consider data provenance, which includes auditing data, examining change logs to look at how the data has changed over time, understanding who owns the data (which internal or external stakeholder), and if there is a classification required for the data.

In order to better support the governance of data, one high-level (and newer) table in Nautobot is the **Data Provenance** table, which is located on the **Advanced** tab of a detailed object view. It includes a summary of data that was already in Nautobot, namely when the object was first created, when it was last updated, who initially created the data, and who last updated it. The following is an example:

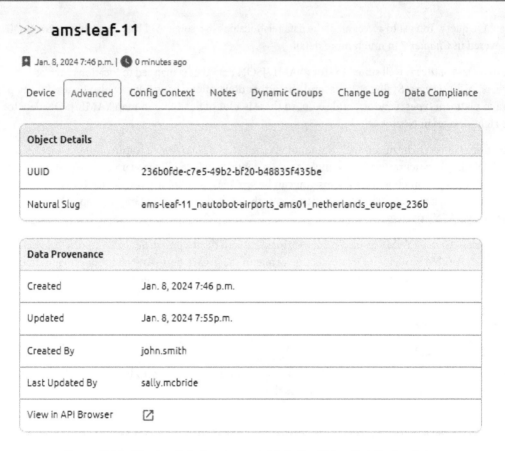

Figure 12.14 – Viewing Data Provenance table in the Object Details view (who created and last updated an object and when this was done)

The goal is to eventually have this type of data at an attribute (or field) level compared to an object level.

While it was already covered in *Chapter 4*, it is important to remember you can also access the Change Log from a detailed object view as well. This shows a summary of all changes made to a specific object, such as a device, as shown in the following screenshot:

Figure 12.15 – Viewing Change Log on a given object for an audit trail

Network model flexibility and extensibility

All network management tools, network operating systems, APIs, and so on have specific data model requirements. This is a programmatic truism and one of the reasons that static typing is so pervasive. That is to say, a tool must know what data you are to provide it and what data should be returned.

As an example, your average tool or network operating system will not consider how you get the data to its system, and simply leave it as a prerequisite for using the tool or system. Tools do not provide ETL capabilities or other convenience methods to easily orchestrate data among other tools or systems. That is to say, tools operate in a manner in which they are the SoT and SoR for all data, full stop. This is not an unreasonable manner in which to operate for a given vendor as they cannot consider all the ways that an operator may want to provide the data.

That being said, the largest challenge in a workflow is often getting valid data. This is often conflated as "*getting the configuration correct*" or "*getting the device into all of the systems*," but is more accurately a data problem. The Nautobot ecosystem has been developed with this challenge in mind, and it shapes the design and many of the features created.

Tools will occasionally allow for multiple types of directly translatable data formats such as YANG, XML, JSON, TOML, and so on, as well as potentially multiple protocols such as NETCONF, RESTCONF, REST API, gRPC, GNMI, and so on.

Network models often exhibit characteristics such as the following:

- Generally leverage data models such as YANG, XML, or JSON Schema definitions, or models built in an open source or custom framework
- Tightly integrated with the protocol (NETCONF, GNMI)
- Tightly coupled to the specific vendor(s)
- Tightly coupled to the model standard, such as OpenConfig or IETF

- Explicit in the model, and often not **DRY (Don't Repeat Yourself)**—a common principle in software development to ensure you are modularizing and creating the right abstractions so that you are not duplicating code and patterns

The advent of standard network data models such as the aforementioned OpenConfig or IETF is an attractive one. However, they have not gained mainstream adoption because there are several real challenges with such models:

- Agreement of the standard

- Vendors naturally want to work on their feature differentiators, which are by very nature not going to be in the open standard

- The models have to be complex to support all features

  - The community has largely been CLI-driven

  - Adoption is slow since learning modeling is challenging and learning complex models is much more challenging

- These models do not have nearly the same amount of documentation and training that is produced by vendors and training companies that the CLI has

- Even with the complexity of the models, there is still often a gap in having a specific feature you need

It is still attractive to have a single model that defines your entire network, but these complexities make it difficult to jump into such scenarios. The current state of the network industry seems to be in a wait-and-see approach, without major movement in the last 2 to 3 years.

The state of industry models is why so many features have been built in Nautobot to ease the burden of tailoring and customizing data models. Through the use of custom fields, computed fields, relationships, Git as a Data Source, GraphQL, SSoT, and purely custom data models, there are predefined patterns to tailor Nautobot for any given network.

In the meantime, standardizing your network model in Nautobot and building adapters, ETL tools, or drivers is the prevalent strategy. Swapping from multiple adapters to a single OpenConfig model, if adoption is picked up at some point in the future, should be one of the easier changes to be made.

Automation and orchestration

Automation and orchestration may or may not be individual building blocks within a network automation architecture. For the purposes of simplification (and how most of the industry views it), we have collapsed automation and orchestration into a section here. But functionally, they are different functional blocks, as shown in *Figure 12.1*. It may be obvious that automation and orchestration make up a large portion of any network automation design, given it is this block that coordinates tasks and workflows and executes all communication with the network infrastructure. In general, we look at automation as the execution of tasks and orchestration as the ability to stitch different tasks or automations together.

For the purpose of this section, we will concentrate on automation and orchestration interactions with Nautobot as an SoT and the role Nautobot can play as a network automation platform.

Before we dive into the tech, let's take a step back and look at workflows, what they are, and why they're important.

Understanding workflows

The first step to any automation should be to truly understand the workflows desired to be automated. That is to say, before you automate you must understand—this is a key phrase that is often overlooked. You must understand and document your workflows before automating them! On the journey, you will often find that data is just as important to the automation journey as any configuration management task. Generating all of the prefixes, IP addresses, VLANs, and so on can easily be the majority of your work. Even in configuration generation, many engineers will have easy ways to generate configurations, provided they have the correct data.

We define workflows as the steps and processes required to perform an end-to-end check operation, such as performing a change or an upgrade on the network. A workflow would consider every step from opening a ticket, updating the ticket, generating data and commands or API calls, performing a change, rollback options, and updating the ticket (as one example of a high-level workflow).

Workflows can ultimately be managed in your tool of choice such as Ansible, Rundeck, Python, and even Nautobot Jobs for certain types of workflows. Organizations need to evaluate their own requirements and skills to understand what type of workflow orchestration system satisfies their business and technical requirements.

Workflows are often cross-domain, meaning that they will integrate with a variety of systems and tools. From a network perspective, this means LAN, WAN, DC, WLAN, cloud, and so on. Oftentimes, an ITSM tool or the actual orchestration tool (such as Rundeck) will be used as the frontend. When using an ITSM tool, it is recommended that the tool be used to simply provide a form to gather the user's intent; that system can then call the orchestration system in which all of your logic is created and then integrate it with your SoT. This allows for a clean **Separation of Concerns** (**SoC**) and does not position all of the automation on the ITSM team's shoulders. This is especially true in organizations where there are large lead times and even longer release cycles to make updates to the ITSM tool. Each organization being empowered to run its own workflows on a centralized orchestration platform tends to strike the right balance between SoC and operational efficiency.

Let's look at a few of the most common network automation workflows to understand their interactions with an SoT and how Nautobot can be used as an automation engine.

Configuration management and compliance

Configuration management, including configuration generation and configuration deployment, will of course be the cornerstone of any network automation architecture. The ability to generate configuration from data should take an SoT-first approach. As discussed already, the combination of data with a templating engine (such as Jinja) is powerful. Going beyond the power tools approach and storing your data opens up opportunities to massively simplify your configuration deployment strategies. That is to say, rather than develop a configuration generation strategy for each configuration feature, device type, or controller, you develop a single configuration generation strategy and develop different ways to update the data.

This paradigm highlights the shift in focus needed from that of orchestration systems to SoT. The common frontends to automation in an SoT-first network automation strategy are to your ITSM tool, Nautobot, or your orchestration system. That power tool that was created called `add_vlan.py` may still exist and even take in the same parameters, but the purpose of that script is to *add the data into Nautobot* and then call the configuration deployment process (which, by the way, may very well be a Nautobot Job—remember that Jobs are covered in *Chapter 11*). This is compared to the original goal of the script that didn't have a focus on data and just "*made the change*" of adding a VLAN. It is a shift in mindset—add and update the data, then drive the configuration change from that data.

Going one step further and building on what is covered in *Chapter 11* with Jobs, you could use a Job Hook that would be automatically triggered as soon as the data was updated in Nautobot too.

Advanced and production-grade config compliance and remediation, configuration generation, and configuration deployment can happen directly within Nautobot using Nautobot Golden Config. This will be covered in *Appendix 3* in great detail.

Adding data, generating configuration, reviewing the remediation plan, and finally deploying the configuration are all core components that work naturally with the traditional network engineering workflow. Here is an example of a basic remediation plan in the Nautobot Golden Config App:

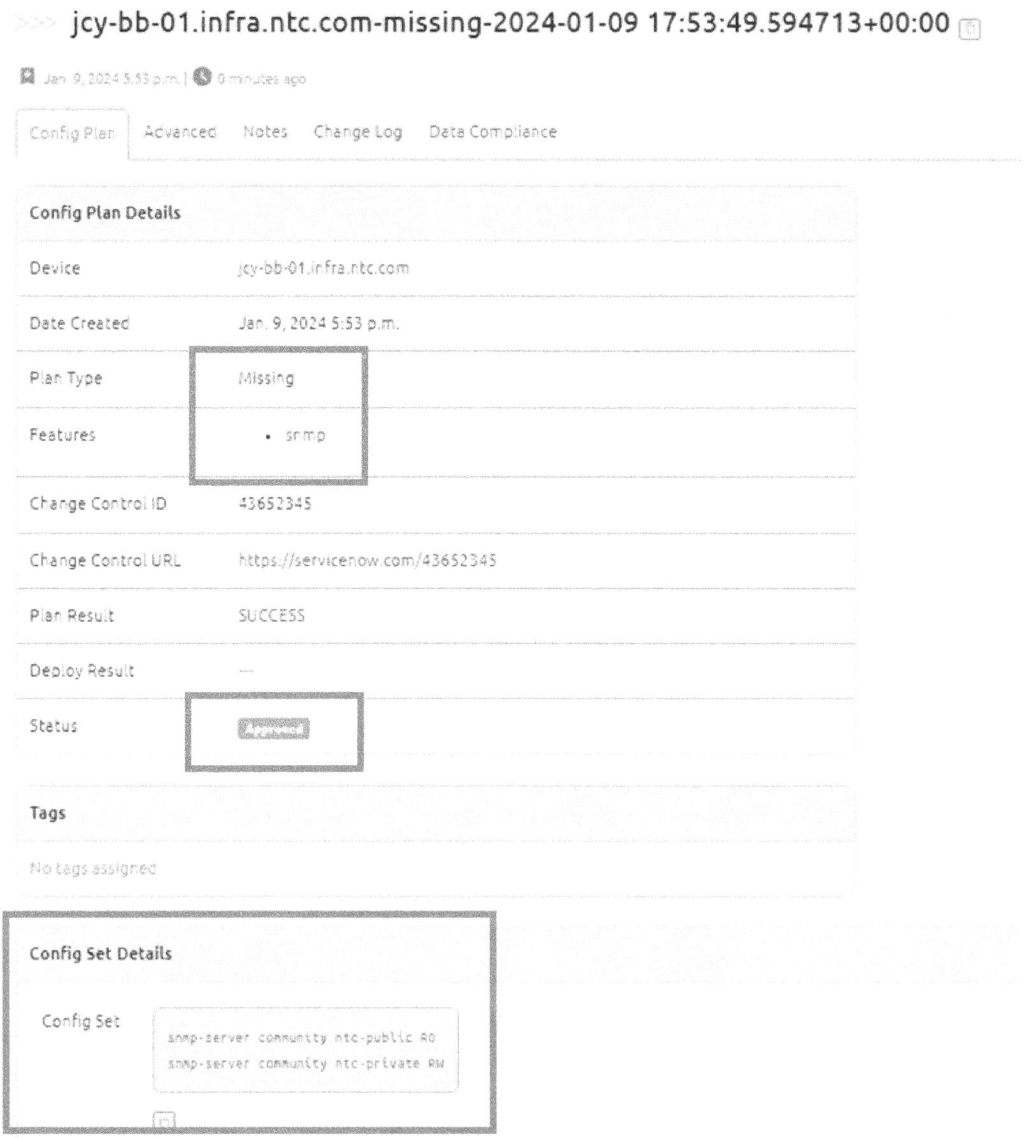

jcy-bb-01.infra.ntc.com-missing-2024-01-09 17:53:49.594713+00:00

Jan. 9, 2024 5:53 p.m. | 0 minutes ago

Config Plan   Advanced   Notes   Change Log   Data Compliance

Config Plan Details

Device	jcy-bb-01.infra.ntc.com
Date Created	Jan. 9, 2024 5:53 p.m.
Plan Type	Missing
Features	• snmp
Change Control ID	43652345
Change Control URL	https://servicenow.com/43652345
Plan Result	SUCCESS
Deploy Result	---
Status	Approved

Tags

No tags assigned

Config Set Details

Config Set	snmp-server community ntc-public RO
	snmp-server community ntc-private RW

Figure 12.16 – Viewing a Config Plan from Nautobot Golden Config

This is a Config Plan generated from Golden Config. It has the ability to track change control information and it showcases that this plan is going to send the **Missing** SNMP to one device to get the device back into a compliant state. You can also see this plan has been approved, which means the config is ready to send. Config Plans have a great amount of flexibility and are covered in *Appendix 3*.

Operational validation

Validating the state of the network pre- and post-change becomes a crucial part of the automation process. Automation is likely best not run at all if the configuration is not deemed healthy. The strategy to define *healthy* is often stored in the SoT, either directly or indirectly. An example of a direct definition would be to declare which network devices and interfaces you expect to see in LLDP. An indirect example would be implying that two network devices connected should have an LLDP neighbor attached. Note that right now, there are no default models to store this data in Nautobot. However, it has been widely done to date through custom apps or through config contexts in Nautobot. Be on the lookout for future enhancements here.

Validation can happen as part of your automation and orchestration system but is often included as part of a visibility or modern network telemetry stack too. This allows you to offload the collection of operational state to a system that is already capturing this data.

The key from a Nautobot and data perspective is to try to decouple the "rules" from the tool (any tool doing validation) and define the rules as data in Nautobot; for example, routing table size should be in a given range, the specific quantity of neighbors to be seen on a leaf or spine (and optionally, who the neighbors should be), the desired range for bandwidth utilization, and so on. The team in charge of the "rules" or desired state may not be the team to consume that data and configure the validation or monitoring tools.

OS upgrades

Building on the previous examples, it may be more clear on what we're referring to between the SoT interactions needed by automation and orchestration systems. Let's look at one more example, which is another prime network automation workflow in that of OS upgrades.

A few of the data points required for OS upgrade automation include the following:

- Approved OS images to be used
- Preferred OS images to be used
- OS images that should *not* be used
- Size of image
- Quantity of images to retain on a target device
- MD5 checksum of a given image
- Device types that can use a given OS image
- Filename(s) of images (OS file and kickstart, if applicable)
- OS image storage location
- Transport protocol of images to the device

With this list not even being exhaustive, when you start thinking about data (along with workflows), it starts to transform how you think about automation. So, instead of thinking about the commands needed to do an upgrade, think about the data required to perform upgrades. From there, examine processes, understand workflows, and then build automation around the data to execute the automation.

Nautobot enablers for automation and orchestration

The following section includes many of the options available to you to use Nautobot as an automation platform going beyond an SoT-only platform.

Each topic reviewed here is meant to be terse and only a short summary because each area has been covered in a specific section of the book already or is covered later in the book. We'll call that out as appropriate in each feature or function being described.

Model extensibility

We've mentioned the flexibility and extensibility of Nautobot models several times already. We talked about many of the options that are configurable in the UI, ranging from computed fields, custom fields, config contexts, and relationships features in *Chapter 6*. We also have a great write-up on how to create custom models in *Chapter 15*. The power of this functionality is that you can store the intent of *your* network, the desired operational state, and other use cases such as required data for OS upgrades that we just discussed, but also those other use cases that are purely custom that you need an extensible SoT to solve.

Nautobot Jobs

We spent a full chapter discussing Nautobot Jobs in *Chapter 11*, so they should need no introduction here. Based on the maturity or requirements of your organization, Jobs may be a great choice for a significant portion of automation. It may also make sense to trigger jobs from your enterprise workflow orchestrator. But in the same vein, based on certain Nautobot Apps, you may want to trigger your orchestrator from a Nautobot Job. There are so many options available. The great thing is you have the flexibility to choose what makes sense for your environment.

Remember, Jobs are written in Python and offer self-service, scheduling, approvals, and even Job Hooks. There is some lightweight orchestration in that you can call Job B if Job A passes; else, call Job C. Be sure to reread *Chapter 11* if there are any questions.

APIs – REST, GraphQL, and Webhooks

We also spent time in *Chapter 8* on these topics, but these programmatic interfaces are what enable Nautobot as part of a full network automation stack. Remember—REST APIs are everything you'd expect in a traditional API, but be sure to leverage the power of GraphQL (with aliasing) to specify exactly the data you want to get back in certain API calls. It'll be sure to clean up client-side code and automations that are fetching data from Nautobot. Finally, Webhooks continue to be a great way to update third-party systems upon data changing in Nautobot. Once Webhooks aren't good enough, the natural evolution is to use Job Hooks.

Golden Config

Configuration backups, generating intended configuration, performing compliance, and deploying remediation commands are arguably the most common workflows in network automation. They are all included (free and open source) in the Golden Config App, which you'll get a deep dive into in *Appendix 3*.

Platform extensibility

Beyond just *model extensibility*, which was mentioned earlier, remember that Nautobot has *full* platform extensibility. As you continue to hear more about the Golden Config App in *Appendix 3*, remember that Golden Config is a Nautobot App. This means it uses the developer API (covered in *Chapter 14*), and you can build any lightweight or full-fledged application using Nautobot as a platform. The value added here is taking advantage of the data already inside Nautobot; but more than that, you get to use its packaging, security, SSO, APIs, GraphQL, Jobs, logging, and any other feature inside an application you may want to build. There is a lot of one-time infrastructure and code offloaded when you build an App on top of Nautobot.

Modern network monitoring – telemetry and observability

Traditional network management tools relied heavily on SNMP and, generally speaking, required a lot of overhead to manage the data or configuration of the system. While traditional network management platforms have made some progress with pollers going beyond SNMP, by integrating with NetFlow, APIs, SQL, and more, the design is still primarily to poll data and generate graphs, for which it is configured. This is a sane tactic that keeps the design reasonable but overall limiting compared to modern telemetry stacks.

Modern telemetry stacks can poll data in nearly any fashion, store data with nearly any amount of metadata, and allow queries and dashboards to be built with that data at any time. This change in philosophy does not come without a design cost, one that ultimately requires much more configuration— configuration to *define the pollers and to enrich the data* when storing the data.

In this section, we're going to explore how Nautobot can be used to complement a modern telemetry and observability stack.

Data enrichment

Let's first understand the power that metadata provides with a modern telemetry stack. You can do interesting things with your monitoring because of this metadata, such as the following:

- Develop queries of data that you had not thought of before; for example, query all standby routers on the East Coast or query all devices connected to a certain PDU

- Build dashboards with that data

- Obtain granular-level data over a long period of time

- Poll data that is not just SNMP

This is all dependent on a proper data enrichment implementation. As an example, if you have a circuit that you are monitoring, you would be able to make certain queries, presuming you have *enriched* the data with other data, such as the interface with the circuit provider, circuit speed, site, region, cluster, and tenant. These queries could include the following:

- Total bandwidth for all tenants' circuits in America for a specific carrier; for example, Verizon or BT

- Capacity in percentage for Acme (fake company) ATT circuits in the world

- Remaining capacity for a given site

- All errors for a given customer or stakeholder over the last month

Once again, it is important to know that you can only accomplish this if you have data enrichment. Realistically, it is nearly impossible for even a moderate network to accomplish this without an SoT. Just consider being able to make observability queries based on any metadata already in Nautobot!

The general design using Nautobot is accomplished via the following:

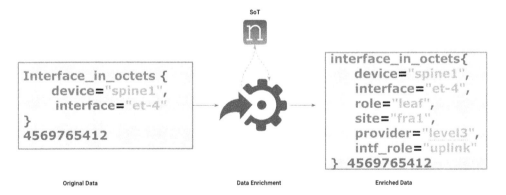

Figure 12.17 – Data enrichment with Nautobot

Generally, this is accomplished by regenerating your configuration for your telemetry stack daily and then deploying that configuration. That means, rather than going to your network management stack to make updates to the system, you simply go to Nautobot and update the data for any changes (for example, an upgraded circuit bandwidth), which is likely needed for configuration management anyway, and everything else "*just works.*"

The power to ask different questions of your data when you have a proper SoT-generated telemetry and observability stack is powerful. The examples included previously only scratch the surface. You can build any query imaginable, as long as you have the data, such as how many errors on AP connected ports or the average TCAM used for all L2 switches.

Data normalization

Each vendor and poller tends to return data in a different manner. Providing a mechanism to normalize the data so that regardless of vendor or polling type, data is stored the same way, allows you to keep all other aspects of the solutions the same. An entire query can be misaligned if the data is not normalized:

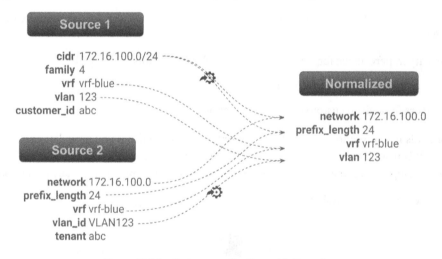

Figure 12.18 – Data normalization with Nautobot

Data collection

While SNMP polling is still the primary means by which data is collected, the landscape is changing. Streaming telemetry from the system versus systems that poll are becoming more and more popular. While architecturally it makes little difference in the push (streaming) versus pull (polling) model, there is a performance impact on the workload of the pollers.

With so many different options now for data collection such as SNMP, syslog, GNMI, NetFlow, APIs, customer pollers, and more, there is a lot more configuration that needs to be created. Once again, tight integration with the SoT is required, primarily based on the device role and physical hardware to generate the expected configuration.

Closed loop network automation

Closed-loop network automation is a strategy that ensures that one part of the automation stack informs the next until the situation is completed. This could be for configuration changes, outages, troubleshooting, and so on. This can also be thought of as intent-based networking.

High-level points of consideration when designing your closed-loop system include the following:

- Message distribution
- Data collection
- Decision process
- Verification
- Alerts
- Alarms
- Actions

To highlight the idea of closed-loop automation, specifically as related to an SoT-first automation stack, consider the following use cases:

- Querying the telemetry and SoT when an internet circuit goes down to determine what impact it will have on the network
- Developing a **Zero-Touch Provisioning** (**ZTP**) process by observing a device being added to the network, checking the SoT for its validity, configuring the device, checking the state (such as interface status), and notifying the network team of completion
- Observing an AP going down, ensuring that all clients that are associated with it re-register to other APs, and opening a ticket with local hands if it hasn't come back up after 10 minutes

The importance of SoT is to enable the understanding and relationships of the metadata associated with each piece of data for any decision tree that can be built. From there, you can understand services, applications, communication paths, and overall impact rooted in any element of the network:

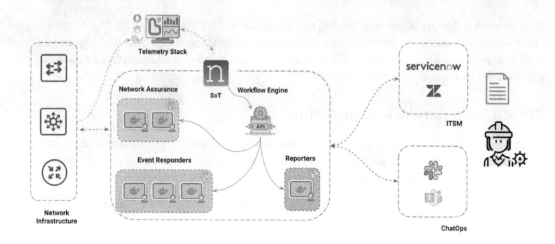

Figure 12.19 – Closed-loop network automation

As you can see, a closed-loop automation takes in multiple parts of the network automation architecture, combining the SoT, orchestration, and telemetry into a cohesive system in which each step informs the other.

User interactions

A network automation stack is absolutely nothing if not used. If it is not used, we are in no better place than yesterday. So, you must consider how users are going to interact with your network automation stack. This section covers user interactions, a key piece of the network automation architecture shown in *Figure 12.1*.

The following are a few common considerations with and without Nautobot to think about as you embark on a user interaction strategy:

- Make sure to explore ChatOps and some of the Nautobot ChatOps integrations that already exist. Within minutes, you can read and write data to any IT system with an API. If your service desk can engage with an MS Teams or Slack bot instead of using ServiceNow, would they prefer it? You should know that answer. Again, the goal is not to "force one solution" on all consumers of network automation.

- If you are a larger organization, you probably want to have self-service with ServiceNow or one of the bigger ITSM players. It makes sense, and it'll make you look good. If this is the case, you are likely going to explore data aggregation between the ITSM and Nautobot, and since you're using Nautobot, you should explore ITSM Self-Service forms that trigger Nautobot Jobs

- Using Nautobot Jobs as a way to enable team-based network automation. If your team has Netmiko and NAPALM scripts already, you can likely convert them to Nautobot Jobs for self-service and have that self-service accessing the data in Nautobot.

- Git is still going to be used to version certain data and artifacts. For whatever YAML or JSON data you're storing in Git, be sure to take advantage of the Git as a Data Source feature in Nautobot. You do not want to manage JSON/YAML via the UI in Nautobot.

- Nautobot also exposes Prometheus metrics as HTTP endpoints. Any app can also create its own metrics, thus allowing you to collect this data, store it in a time-series database, and later graph it in the tools of your choice such as Grafana, as one example. You can collect objects in Nautobot, execute chat commands, and use many other metrics so that you can visualize these and trend them over time.

Summary

It should be clear that there is a lot involved in devising an end-to-end data-driven network automation architecture. This chapter highlighted the key architectural building blocks for network automation, including SoT, automation, orchestration, modern network monitoring with telemetry and observability, and user interactions. You also learned how Nautobot can play an integral role in each block. Now, it is up to you to see where and how it makes sense to integrate Nautobot into your network automation architecture.

In the next chapter, we will shift gears from network automation architectures and dive into the world of Nautobot Apps, learning how to create and build custom data models and Apps.

Part 4:
Nautobot Apps

This part brings to life how you stay in control of your operational models and reduce tool sprawl with Nautobot. While Nautobot has rich core features and built-in extensibility options, there are always reasons to have tailored solutions, from enterprise network requirements and data classification to integrations with custom in-house tools that have been around for years. From building new network data models and custom workflows to full-blown packaged solutions, the last part of the book provides a deep dive into Nautobot Apps. You will gain an understanding of the Nautobot ecosystem and its developer APIs, and how to get started creating your own Nautobot Apps.

This part consists of the following chapters:

- *Chapter 13, Learning about the Nautobot App Ecosystem*
- *Chapter 14, Intro to Nautobot App Development*
- *Chapter 15, Building Nautobot Data Models*
- *Chapter 16, Automating with Nautobot Apps*

13

Learning about the Nautobot App Ecosystem

As you have seen from the prior chapters, Nautobot is a powerful Network Source of Truth. It has great APIs, flexibility, and a number of integrations, such as `pynautobot` and the Nautobot Ansible collection. Throughout this book, you have also seen countless references to various Nautobot Apps. In *Part 4* of this book and starting with this chapter, we are going to demystify what Nautobot Apps are all about, showcase several open source Nautobot Apps, and then dive into how you can create your own in *Chapters 14* to *16*.

In this chapter, we will cover the following:

- Nautobot Apps overview
- Why Nautobot Apps?
- Nautobot Apps ecosystem
- What's possible with Nautobot Apps?
- Nautobot Apps Administration

Nautobot Apps overview

Nautobot Apps are to Nautobot as iPhone Apps are to the iPhone. Imagine a world that has the internet, but your smartphone has no Apps. Your phone would still be a great communication device. It would still be a platform that you can use to make calls and browse the internet. Similarly, Nautobot without apps can still power network automation by serving as the **Network Source of Truth** of your automation stack or even carrying out automation while using Nautobot Jobs. Adding Nautobot Apps to Nautobot takes Nautobot to the next level and allows you to tailor Nautobot to do anything imaginable. While that sounds cliche, it is just as cliche to create an iPhone that can do anything! Similarly, consider platforms such as ServiceNow. They have Apps that allow you to tailor data and automation; Nautobot is no different.

The truth is, the more data that are stored in Nautobot, the more workflows that can be automated from Nautobot. In Nautobot, with this data, you may want to build Nautobot Apps to really take advantage of automating networks and performing device audits.

Nautobot Apps allow you to customize Nautobot by updating the UI and adding data models and even APIs. One of the major benefits of this approach is minimizing tool sprawl across a given team. We'll talk more about this later.

For now, a great way to think about Nautobot Apps is adding new functionality to the Nautobot. The most user-friendly way to understand this is by looking at the Nautobot navbar. Take note of **DEVICE LIFECYCLE, SECURITY**, and **GOLDEN CONFIG** in the next figure (this appears in the public sandbox instance `https://demo.nautobot.com`). They were all added by Nautobot Apps:

Figure 13.1 – Nautobot Apps were used to create the top-level navbar items

Your app may not need a top-level item in the navbar, but maybe you just want to add a menu item underneath a current dropdown. This is possible, too, with a little effort put into Nautobot Apps development. If an app does not create new menu items on new or existing top-level items, the app menus are injected under the **Plugins** navbar item. Note that in Nautobot v2.2, the **Plugins** menu item has changed to **Apps**:

Figure 13.2 – Nautobot Apps-created menu items

This is shown by looking at the **Circuit Maintenances** app. It has been used to inject menu items within the Nautobot-provided **Circuits** dropdown.

Why Nautobot Apps?

Nautobot Apps can significantly transform how an organization goes about network automation. Like any decision, it is not one that should be taken lightly. In this section, we are going to highlight the common reasons many organizations and customers have been using and creating their own Nautobot Apps.

Flexibility

One of the primary reasons there have been hundreds of Nautobot Apps (including those made by customers of Network to Code) built by users is because of the flexibility. As you'll see in the *What's possible with Nautobot Apps?* section, you can do just about anything inside Nautobot. When you think about new or unforeseen requirements, such as the need to track new types of data (anything from network features to financial cost modeling of the network), the ability to view data in Nautobot in a particular way, or the need for a solution that automates common workflows in your environment, it is possible to create a Nautobot app for them.

Nautobot's developer API and its flexible architecture allow developers and automation engineers to keep creating Nautobot Apps. Anything you can do in Nautobot can be offered from a Nautobot app. For example, you can add a new data model to Nautobot in an app. Your new data model can then leverage Custom Fields or GraphQL. You can also distribute Jobs within an app. We'll cover all of this and more in this and the next three chapters.

Because these apps are primarily written in Python, the power of an app is up to the developer. Therefore, while anything is possible, the primary use cases for these apps are as follows:

- Adding data models to Nautobot
- Adding APIs to Nautobot
- Adding new pages (views) to Nautobot
- Augmenting existing views and APIs
- Distributing jobs
- Creating full solutions that incorporate all of the above

We'll showcase these through numerous screenshots in the *What's Possible with Nautobot Apps?* section.

Access to SoT data

As you've seen throughout the book, there can be quite a bit of data stored in Nautobot. If you're building an application or even a set of scripts that need access to the desired state of the network, building a Nautobot app simplifies the process.

Consider what you learned in *Chapter 11* when building Nautobot Jobs. Building jobs that have dynamic dropdowns and that are tied to the desired intent of the network provides a great user experience for network operators who are looking to configure or troubleshoot their network. While it is always possible to perform or build apps and tools outside of Nautobot (and there may be valid reasons to do so), keep in mind that you're getting everything else Nautobot has to offer, such as logging, security, APIs, and SSO just to name a few, all in a platform that will already be deployed in your environment.

Accelerated development

Over the years and through customer engagements at Network to Code, we've seen countless homegrown tools written in PHP, Perl, Flask, and Django. They usually start with some splash page with common automation such as IP/DNS requests, executing backups, looking up where an IP or MAC is on the network, resolving FQDNs, and a plethora of other common operational tasks. In the vast majority of these cases, there is no authentication, a lack of logging, no predefined ways to contribute, and usually between one and three people who built the tooling (often no longer employed at the organization or on the team they were once a part of when they built the tooling).

Using Nautobot Apps accelerates the development of any new app because Nautobot has a predefined, standardized developer API platform to build on top of. Going beyond having a Python API that can be used to build apps, you get the non-functional requirements that we just mentioned out of the box. This includes authentication, fine-grained authorization, and RBAC, APIs, and GraphQL, as well as custom fields for data models, using Git within an app, logging, task workers, the ability to use or distribute jobs with direct access to the Django ORM, and so on. To build all of that equates to literally dozens of years of development time that would be saved. Why would you need to build a logging system or worry about SSO? It doesn't make sense for the vast majority of use cases.

To emphasize this point even more, in theory, you can even choose to not use Nautobot for its core inventory and Source of Truth use cases and use it as an app platform just because of the development framework it provides. If you do go down this route, you can "*hide*" access to anything you don't want your user to see with the advanced RBAC provided by Nautobot.

Reduced tool sprawl

One of the greatest benefits in the long term is reduced tool sprawl, which goes beyond accelerated development. Network teams have many tools deployed, and the main reason is usually that the original tool was good for a while, new requirements emerged, and a new tool was needed to solve a new challenge or use case. Team members leave, and there is no longer expertise in using that tool, so it sits stale. The longer it sits, the more risk there is of it going down because no one knows the what, when, and why of its deployment in the first place.

It's important to note that Nautobot Apps don't need to be advanced or robust. There may be a new requirement to track MAC addresses or IP addresses or maybe to request IP addresses. These are small requirements that can be added to Nautobot as individual apps. The value is really gained by building around Nautobot as a platform. There are predictable and defined ways to do things. There is a common approach to grant permissions. Nautobot as a platform minimizes the need to deploy a new tool for every new use case, drastically reducing tool sprawl and potentially saving millions of dollars in tools and the associated operational costs related to those tools; remember this when you're building the justification of Nautobot for your organization.

Nautobot Apps ecosystem

The most popular Nautobot Apps today are developed by Network to Code. They can be found in the Nautobot GitHub organization (`https://github.com/nautobot`). You can also browse the Nautobot docs page (`https://docs.nautobot.com/`) and explore all of the Network to Code Apps from there too. If you recall way back in *Chapter 1*, we provided a quick list of the Nautobot Apps that are available. In this section, we are going to go a little deeper to better understand key use cases of the prebuilt open source apps.

Golden Config

The most popular Nautobot app is **Golden Config**. Golden Config solves several of the most common problems in the network industry today and does it in a very flexible and NetDevOps-type manner.

> **Important note**
> We have a Golden Config deep dive in *Appendix 3*.

The main problems Golden Config solves are the following:

- Organizations don't have a backup solution

- Organizations don't have an architecture or solution to generate intended configurations

- Organizations don't have a way to perform configuration compliance (and do it offline)

- Organizations don't have a way to conduct configuration remediation integrated with compliance

Golden Config addresses these problems, which many organizations have been facing for decades. Golden Config allows you to do the following:

- Perform network backups, or you can easily point to a Git repo that already contains your backups

- Generate intended configurations rendered with the data in Nautobot using Jinja templates, or you can easily point to a Git repo that already contains your intended configurations

- Perform config compliance (per feature and per device type) based on those backups and intended configurations, meaning it is possible to use Nautobot for compliance and use other tools to do backups or generate configurations

- Perform configuration remediation based on the compliance results

The underpinnings of Golden Config use the Python Nornir network automation library, making it extremely powerful based on other open source technology.

Here is a visual that showcases how Golden Config allows you to easily see the missing and extra configuration on a per-feature and per-device basis:

Figure 13.3 – Viewing Nautobot Config Compliance for a single feature

You could then automate the remediation of these devices too. As stated earlier, there is a deep dive into Golden Config in the *Appendix 3*, so we are only scratching the surface here.

As a reminder, the best way to navigate the docs for each app is to start at the main docs page at https://docs.nautobot.com/.

Nornir

The **Nornir** Nautobot app is an app built to enable Nautobot app developers that are conducting automation using Nornir. The Nornir app can also be used to enable Nautobot Jobs too. It streamlines working with the Nautobot inventory and credentials for apps and jobs. It is a must if you're trying to automate your network with Nornir from within Nautobot.

Device Onboarding app

The **Device Onboarding** app is going through a bit of a change, so if you haven't used it or upgraded in a while, you should check out the latest version. It supports a few different workflows while it continues to evolve:

- When provided with an IP address and credentials, it will connect to a device, log in, and retrieve a device's Platform, Manufacturer, Management Interface, Management IP Address, and Serial Number. It will create any of these objects in the event they don't already exist in order to onboard the device into Nautobot. For example, if a given Manufacturer or Platform is not in Nautobot already, a specific instance (Manufacturer) will be created to onboard the device. This can be run or rerun to ensure the correct data are in Nautobot.

- When provided with a device's backed-up configuration, it can parse data with ntc-templates and/or Batfish to populate data in Nautobot.

The Device Onboarding app is extremely useful when you have an inventory with which to get started but need to onboard key information such as interfaces and IPs from your network devices.

Device Lifecycle Management (DLM)

Nautobot **DLM** adds key data models to Nautobot, enabling key workflows such as the following:

- Tracking end-of-sale/end-of-life devices
- Tracking support contract cost and which contract is associated with which devices
- Tracking vulnerabilities and CVEs
- Tracking which software images are approved

The following screenshot shows a view of software images from DLM:

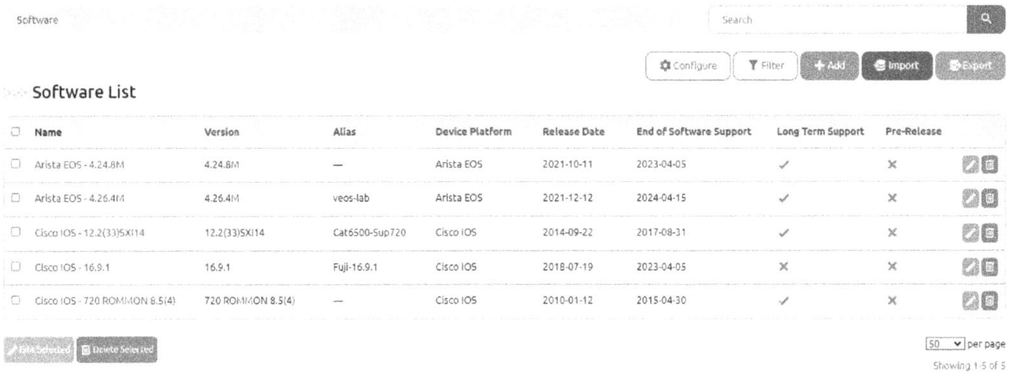

Name	Version	Alias	Device Platform	Release Date	End of Software Support	Long Term Support	Pre-Release	
Arista EOS - 4.24.8M	4.24.8M	—	Arista EOS	2021-10-11	2023-04-05	✓	✗	
Arista EOS - 4.26.4M	4.26.4M	veos-lab	Arista EOS	2021-12-12	2024-04-15	✓	✗	
Cisco IOS - 12.2(33)SXI14	12.2(33)SXI14	Cat6500-Sup720	Cisco IOS	2014-09-22	2017-08-31	✓	✗	
Cisco IOS - 16.9.1	16.9.1	Fuji-16.9.1	Cisco IOS	2018-07-19	2023-04-05	✗	✗	
Cisco IOS - 720 ROMMON 8.5(4)	720 ROMMON 8.5(4)	—	Cisco IOS	2010-01-12	2015-04-30	✓	✗	

Figure 13.4 – Viewing software images using Device Lifecycle Management

> **Important note**
>
> Software and software image models are a part of Nautobot Core in v2.2 and will be removed from DLM in the future.

Data Validation Engine

Nautobot is only as good as the data it has in it. If the data are of dubious quality or do not follow standards, they may as well not be in Nautobot. The **Data Validation Engine** app allows you to create rules that either ensure no bad data can be added to Nautobot or allow you add the data to Nautobot and report on it after the fact. Based on the culture of your organization, you have the flexibility to decide which will work best for you.

Some common examples of using Data Validation Engine include the following:

- To create min/max rules, such as for VLAN IDs, e.g., do not allow the creation of a VLAN ID of less than 10 or do not allow an FW policy ID of greater than 100.
- To create regular expression (RegEx) rules for any data being added to Nautobot, e.g., ensure the location and device names adhere to a RegEx expression/pattern.
- To ensure that certain fields (or attributes) are always required even though they are optional in Nautobot's default data model, for example, when always requiring an interface description, serial number, or asset tag on an interface or device.
- You can also use custom validators (Python code) and view the results within the UI as well. This is the approach to take if you want to analyze data already in Nautobot. It also offers great flexibility because you can use Python for any validation you'd like. You'd build a Nautobot Custom Validator in Python that can have basic or advanced logic and then view the rules that are violated in the UI.

Single Source of Truth (SSoT)

The **Single Source of Truth** (**SSoT**) app serves the primary use case of orchestrating data flows in your environment, ensuring that a user can make a change in a single place and have that change propagate to the other systems that need that data. This allows Nautobot to aggregate data from (and to) other data sources.

Consider the cases of doing network automation and needing IP addresses, details on circuits and circuit terminations, or information on security policies or groups. This type of data may already exist in other databases. If it does, you don't need to replace those databases; you can use SSoT to aggregate data from those systems into Nautobot.

You may also want to sync data to other systems. For example, if you have an existing IPAM, would it be valuable to push interface or device information as an attribute of an IP address to your IPAM? This is possible with SSoT.

Without diving into the details, it is good to note that SSoT is widely used across large enterprises. It is primarily used as a framework that offers a pattern for building data integrations into Nautobot, but it also comes with prebuilt integrations such as Infoblox, ServiceNow, Cisco ACI, Device42, Arista CloudVision, and IP Fabric.

Here is a visual of using SSoT to sync data from Nautobot to ServiceNow:

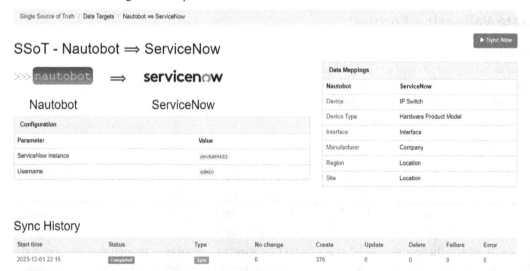

Figure 13.5 – SSoT integration between Nautobot and ServiceNow

> **Important note**
>
> The SSoT app is covered in detail in *Appendix 2*.

Network data models

There are three data model apps to highlight: **Firewall Models**, **BGP models**, and **WLAN Models**. Note that the WLAN models are in the `nautobot-app-network-models` GitHub repository. All of these apps are good to highlight because their primary focus is adding data models to Nautobot. They don't carry out automation. They allow you to store more data in Nautobot to either document your network or power your network automation.

Here is a subset of the models for the **Firewall Models** app:

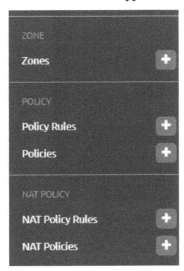

Figure 13.6 – Viewing models from the Firewall Models app

Here are the models offered by the BGP Models app:

Figure 13.7 – Viewing models from the BGP Models app

Keep this in mind when you start considering building an app. Your app doesn't need to carry out automation. You may just want to add more data models to Nautobot.

Design Builder app

Managing designs and ensuring they are adhered to has been no easy feat for the last few decades in the world of networking. Designs are built and deployed but are then usually modified with little to no visibility of which devices are running which design until the point where there is no standard design for the devices. Nautobot and network automation help because they force you to think about data first and then build automation around that data, helping to solve the problem. However, in larger organizations, it is still a tedious task to manually plan a site, add and manage the data, and then add those data into Nautobot.

Design Builder is an app that allows users to generate a YAML spec that documents network designs in a predefined and predictable manner. For example, the spec may have t-shirt size designs for small, medium, and large branch sites. The spec can include every piece of data needed to onboard new sites to Nautobot, from devices and interfaces to IP addressing, VLANs, and circuits.

Here is a snippet and example of a Design Builder spec:

```
---
sites:
  - name: "LWM1"
    status__name: "Staging"
    prefixes:
      - prefix: "10.37.27.0/24"
        status__name: "Reserved"
    devices:
      - name: "LWM1-LR1"
        status__name: "Planned"
        device_type__model: "C8300-1N1S-6T"
        device_role__name: "Edge Router"
        interfaces:
          - name: "GigabitEthernet0/0"
            type: "1000base-t"
            description: "Uplink to backbone"
            status__name: "Planned"
```

> **Important note**
>
> For a great introduction to Design Builder, check out the blog from *Network to Code* at `https://blog.networktocode.com/post/design-builder/`.

After the Design Builder spec is built, you can run a Nautobot job to add a new design and its data to Nautobot. As a user, you'd enter the minimum required inputs in the form of a job, and the rest of the data would then be generated and added to Nautobot. From there, you can use your automation tooling as you normally would for deployments and compliance.

Design Builder also allows you to see which objects are part of which design and even allows you to restrict editing objects that were deployed as part of a Design Builder design.

An enterprise looking to standardize and codify network designs would do well to check out Design Builder.

Circuit Maintenance app

As you learned already, Nautobot has data models for circuits, allowing you to track circuits and providers. However, a common requirement in enterprises is the ability to track maintenance notifications from providers (or carriers). It is estimated that there are two to four maintenance notifications per year for any given circuit. The provider typically emails an operations team when the maintenance is going to take place, and then the ops team must plan around it, ensuring that change windows are managed. Sometimes, they need to direct traffic to a standby device during the actual maintenance window.

The **Circuit Maintenance** app adds the Circuit Maintenance model to Nautobot, allowing you to see all maintenance notifications on a single screen. The nice thing is that each notification has a relationship back to the actual circuit that is also stored in Nautobot. From there, you can understand which device and interface terminate the circuit, having all data in a single place.

The following is an example of a detailed Maintenance object view:

Figure 13.8 – Detailed Circuit Maintenance object view

Going beyond just being an app that adds data models, Circuit Maintenance also includes automation that reads emails from the providers, parses them, and then properly adds Circuit Maintenance notifications to Nautobot.

Secrets Providers app

Nautobot does not store any secrets in the database. Out of the box, you can have Nautobot reference secrets that are stored as environment variables or in files on the local file system. If you do this, you'll likely have some other automation system, such as a CI/CD pipeline or config management solution with which to manage those env vars or files.

Many larger organizations prefer to integrate with an enterprise secrets management platform such as HashiCorp Vault or AWS Secrets Manager. When using the **Secrets Provider** app, you can natively integrate Nautobot with the secrets provider of your choice (many come with the app by default), and you can add your own if it isn't already supported.

The following figure shows the **Provider** dropdown when adding a new secret after installing the Secrets Provider app:

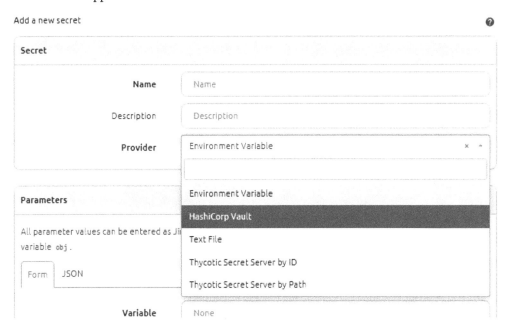

Figure 13.9 – Viewing integrations in the Secrets Provider app

By using the Secrets Provider app, Nautobot dynamically fetches secrets from your secrets management platform, improving the overall security of your infrastructure.

Floor Plan app

In *Chapter 5*, we saw how Nautobot allows you to populate racks specifying the exact unit position of a device. The **Floor Plan** app allows you to visualize the floor plan for your given locations, from colocations to data centers and to the room of your choice. You can place racks onto the floor plan and highlight the hot and cold aisles or just show reserved tiles for future growth.

The following is a good example of what a Nautobot Floor Plan looks like:

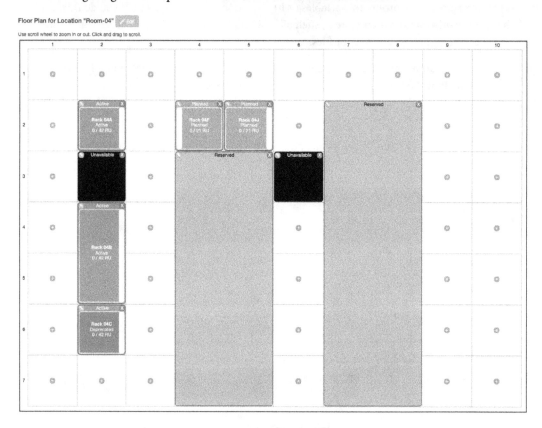

Figure 13.10 – Example of a Nautobot floor plan

As you can see, you can visualize racks and also reserve larger parts of the room for expansion or call out those larger CRAC units that are in large data centers.

ChatOps

The Nautobot **ChatOps** app is both a framework and solution that comes with existing chat integrations for Nautobot Cisco ACI & Meraki, Ansible AWX/Tower, Arista CloudVision, Grafana, IP Fabric, and Panorama. When you deploy the Nautobot ChatOps app, it creates a chat server API that is exposed to Nautobot and then communicates with chat applications such as Microsoft Teams, Slack, Webex Teams, and Mattermost. All four of these chat platforms are supported.

In this model, when you issue a chat command in Slack, it will get routed to Nautobot and Nautobot will respond and broker the communication with the integrated system.

The ChatOps framework is the foundation that enables each integration to be built in a predictable manner. The framework is what allows new chat commands to be built in minutes once the initial integration is created. When deploying the Nautobot ChatOps app, you get a few commands built-in that allow you to interact with Nautobot directly. These commands allow you to query data from Nautobot:

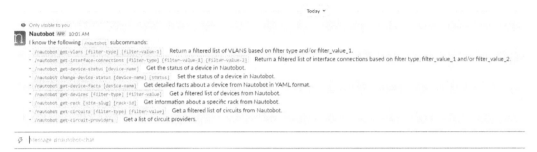

Figure 13.11 – Built-in Nautobot commands in Nautobot ChatOps

The following is an example of Nautobot ChatOps using the Meraki integration:

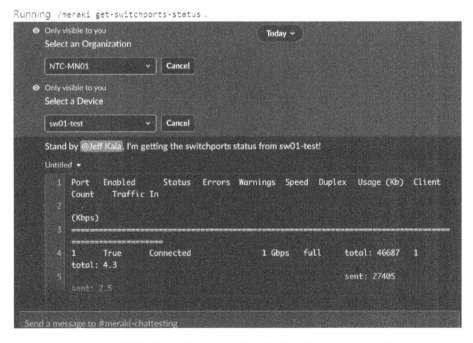

Figure 13.12 – Executing a Nautobot ChatOps Meraki command

The Nautobot ChatOps app can be used to integrate into any IT system with an API. The most common use case is enabling internal stakeholders to gain access to their data in Nautobot or other tools. However, the opportunities are there should you want to use ChatOps for making basic, repeatable changes or perform ops tasks such as bouncing a switchport or assigning a VLAN to an interface.

Welcome Wizard app

As you saw in *Chapter 4*, before you can add your first device, there are a few required objects to add to Nautobot such as Location Types, Locations, Platforms, Device Types, and Roles. The Welcome Wizard app is a nice app for those getting started, and it streamlines adding new objects to Nautobot with wizard-like banners to ensure you know what is needed in order to add certain object types.

What's possible with Nautobot Apps?

At this point, you've seen firsthand what a few existing Nautobot Apps do, so you are probably gaining an understanding of what is possible with Nautobot Apps. In this section, we are going to take it one step deeper by showing screenshots of what is possible to ensure there is no ambiguity because each picture is worth a thousand words, right?

Creating data models

When you extend Nautobot by creating Nautobot Apps, one of the most common requirements and needs is to create a new data model. There may be custom device types in your environment (such as domain- or vertical-specific devices), from IoT devices to healthcare, as well as manufacturer-networked systems on the plant floor. Once Nautobot is adopted and used for automation, more and more devices are added because the data can help make automation decisions. However, it can also help you understand applications and impact, e.g., which applications or end-user services are impacted when a device goes down.

However, you may want to model network features that Nautobot doesn't have models for, or maybe your deployment is unique and warrants its own model for carrier or backbone routing.

The key to remember is that once data are stored in Nautobot, you get the benefit of tracking the data, running audits, having the data programmatically accessible, and, of course, being able to use the data to power network automation.

The following shows two figures that you get out-of-the-box when creating models in Nautobot: the object list view and the object detailed view. There is no frontend work required to add these to Nautobot:

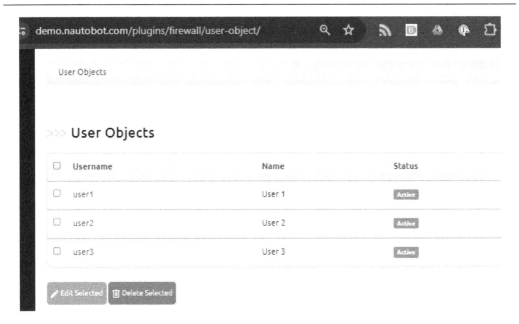

Figure 13.13 – User Objects list view from the Firewall Models app

Here is the user object detailed view:

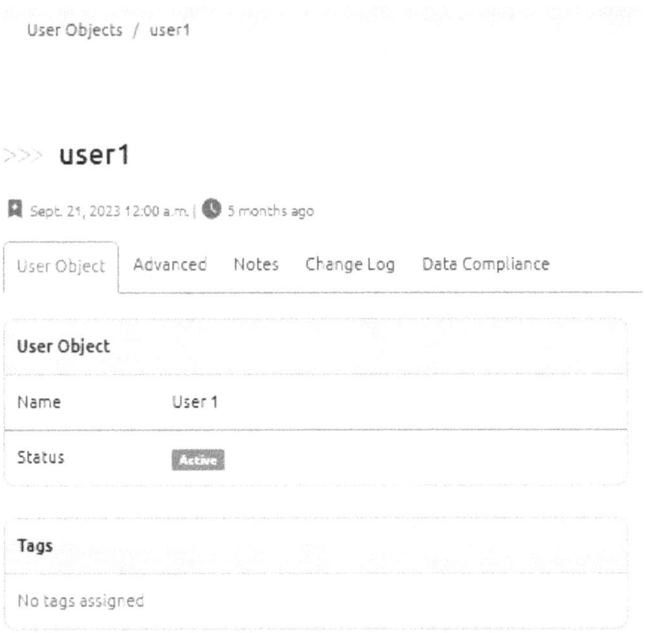

Figure 13.14 – User Object detailed view from the Firewall Models app

Additionally, when you create new models, you can reference and/or access data that are already in Nautobot. You can see this in the Device Lifecycle Management app, which references the Device Type in the second column in the following image:

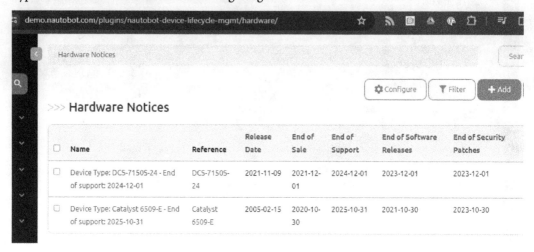

Figure 13.15 – Hardware Notices referencing device type in the Device Lifecycle Management app

Remember that no frontend work is required to create lists and detailed views for your models. For more advanced use cases, you can also customize or override the HTML views as well.

Creating APIs

If you want to move beyond creating models, it is only natural that you will want or need the associated RESTful APIs. The following is an example showing APIs generated for the models for the same user object of the model shown previously in *Figures 13.13* and *13.14* in the Firewall Models app:

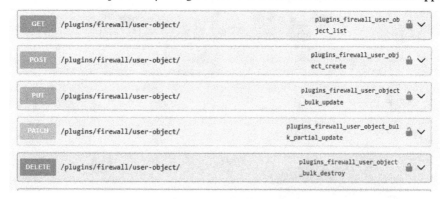

Figure 13.16 – Viewing API specs for Nautobot Apps

When APIs are created, they are also added to the API docs (interactive API documentation) that you first started to use in *Chapter 8*.

> **Important note**
>
> It is also possible to create custom APIs that do not map to a single or new model in Nautobot. While GraphQL helps this use case for retrieving data from Nautobot, you can also create a wrapper or facade API that is simply a proxy to other systems.

Creating UI elements to enhance the user experience

This section shows how you can enhance the user experience by updating a Nautobot object's detailed view.

We previously showed that each object you create gets a detailed view that shows its attributes. This is the same as all core models in Nautobot.

However, when you create a Nautobot app, you also have the ability to inject content into a core object's detailed view. For example, in the following figure, we can see how the Device Lifecycle Management app has injected **Hardware Notice** information into a device type's detailed view:

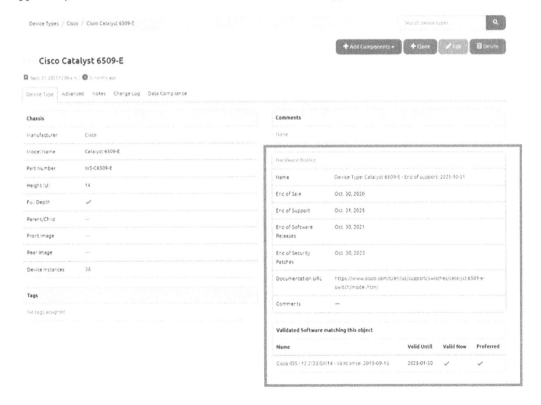

Figure 13.17 – Device Lifecycle Management app used to inject content into Device Type (detailed view)

Maybe you don't want to display the data using the detailed view. Maybe you'd prefer to add data into a tab on an existing core object's detailed view. That is possible, too.

The following figure shows how the Golden Config app created a Configuration Compliance tab for a device's detailed view:

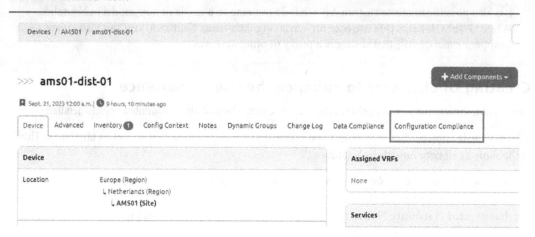

Figure 13.18 – Golden Config app used to create a custom tab for each device's detailed view

Your custom Nautobot app can also inject content into the homepage. Here is an example of the data added to the homepage using the Firewall Models app:

Security	
Security Policies Firewall Policies	3
NAT Policies NAT Policies	3
Capirca Policies Firewall Policies	0
Security Rules Firewall Policies	5
NAT Rules NAT Policies	3

Figure 13.19 – Homepage content added by the Firewall Models app

Next, let's look at how Nautobot Apps can distribute jobs.

Distributing jobs

In *Chapter 11*, you learned about Nautobot Jobs. We primarily focused on managing jobs using `$JOBS_ROOT`, but we alluded to other ways that jobs can be distributed, either by using Git as a data source or as an app. If your job has several Python dependencies and/or you'd like to publish one or more jobs to a Python package index internally (within your own organization) or externally (publicly available (e.g., PyPI)), then you may want to create an app that is basically an encapsulation or distribution mechanism for jobs.

In fact, if you log in to the Nautobot public demo instance at `https://demo.nautobot.com`, every job was added as part of an app. The **Capirca Jobs** were added from the Firewall Models app; the **Circuit Maintenance** jobs were added from the Circuit Maintenance app; the **CVE Tracking** and **Device/Software Lifecycle Reporting** jobs were added from the Device Lifecycle Management app, and finally, the **Golden Configuration** jobs were added using the Golden Config app:

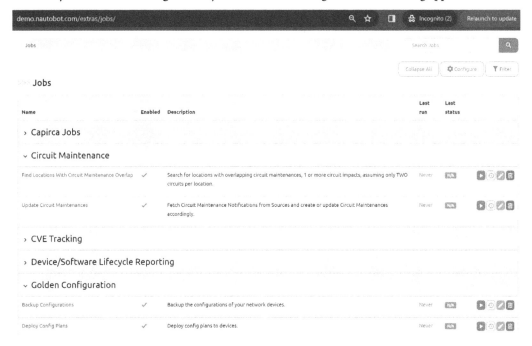

Figure 13.20 – Jobs added to Nautobot from Nautobot Apps

Nautobot Jobs form the foundation to perform network automation tasks directly from Nautobot. Remember, you can use the Nautobot Nornir app to accelerate job development if you are looking to automate devices using Nornir. If you need or prefer to use other libraries, of course, you can do that, as you can control what goes inside a job and app.

Creating network automation solutions

When you start to consider all that is possible with Nautobot (including the fact that you can have custom HTML and JavaScript as part of an app), from adding data models, APIs, and new pages (or views) to even distributing jobs as part of a Nautobot app, you shouldn't forget that apps can be as lightweight or robust as needed.

You may have ideas that will end up yielding quick-win Nautobot Apps. Those are great places to start. However, when you consider more advanced requirements, you can also build robust Nautobot Apps that meet the exact needs of your data and users and that are fully accessible from the UI and API with role-based access controls.

The best example of this is the Golden Config app. You may have a complex workflow for load balancers or firewalls. Instead of trying to build tools or utilities from scratch, it may be advantageous to build one or more Nautobot Apps.

The following are more examples of pages created by the Golden Config app. The first one also shows that you can add custom buttons to pages, too:

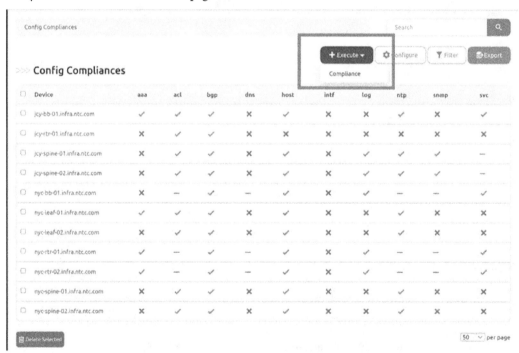

Figure 13.21 – Custom page created using the Golden Config app (with button)

Since you control your own apps, you can get as creative as you'd like by using any dependency required, such as those to create charts and graphs. The following is an example of charts added by using the Golden Config app:

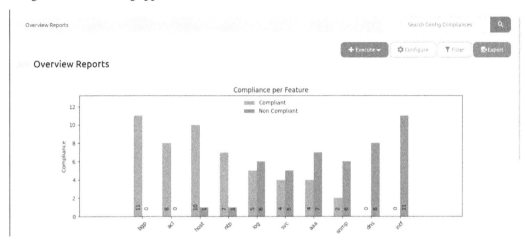

Figure 13.22 – Bar graphs reports added by using the Golden Config app

This means that if you'd like to add a library that'll make charts even better looking, you can do that, too.

Nautobot Apps administration

Now that we have provided an overview of the common Nautobot Apps and shown what is possible, let's review how you'd manage the installation of the apps.

We are going to show how to install the BGP Models, Firewall Models, and the Data Validation Engine apps.

Installing Nautobot Apps

First, let's ensure we have a pip requirements file (local_requirements.txt) in the $NAUTOBOT_ROOT (while logged in as the nautobot user):

```
root@nautobot-dev:~# sudo -iu nautobot
nautobot@nautobot-dev:~$
```

To add an Nautobot app that can then be installed, open the local_requirements.txt file and add it to the file. After you add all three of the package names of the apps we're installing and cat the file, you'll see the following:

```
nautobot@nautobot-dev:~$ cat $NAUTOBOT_ROOT/local_requirements.txt
nautobot-firewall-models
```

```
nautobot-bgp-models
nautobot-data-validation-engine
```

> **Important note**
>
> The app-specific docs are the best place to reference in order to know the Python package name of any given app. You can also search PyPI for Nautobot at https://pypi.org/search/?q=nautobot.

It is recommended to pass the --upgrade flag when installing new Nautobot apps, as this ensures that the compatible versions of Nautobot and the apps are installed together:

```
nautobot@nautobot-dev:~$ pip3 install --upgrade -r $NAUTOBOT_ROOT/
local_requirements.txt
```

At the bottom of the output, you'll see that the apps have been installed:

```
Downloading nautobot_firewall_models-2.0.3-py3-none-any.whl (114 kB)
                            115.0/115.0 kB 2.7 MB/s eta 0:00:00
Downloading nautobot_bgp_models-0.20.1-py3-none-any.whl (68 kB)
                            68.5/68.5 kB 5.0 MB/s eta 0:00:00
Downloading nautobot_data_validation_engine-3.1.0-py3-none-any.whl (1.8 MB)
                            1.8/1.8 MB 40.6 MB/s eta 0:00:00
Downloading capirca-2.0.9-py3-none-any.whl (232 kB)
                            232.3/232.3 kB 9.8 MB/s eta 0:00:00
Downloading abs1_py-2.1.0-py3-none-any.whl (133 kB)
                            133.7/133.7 kB 5.2 MB/s eta 0:00:00
Downloading mock-5.1.0-py3-none-any.whl (30 kB)
Installing collected packages: ply, mock, abs1-py, capirca, nautobot-firewall-models, nautobot-data-validation-engine, nautobot-bgp-models
Successfully installed abs1-py-2.1.0 capirca-2.0.9 mock-5.1.0 nautobot-bgp-models-0.20.1 nautobot-data-validation-engine-3.1.0 nautobot-firewall-models-2.0.3 ply-3.11
nautobot@nautobot-dev:~$
```

Figure 13.23 – Verifying that Nautobot app Python packages have been installed

You can also add the version specifier to the end of any app to ensure no major upgrades are performed for apps either but note that this will increase the likelihood of needing to review this file every time.

We also need to update nautobot_config.py to add the newly installed app. We'll be updating the PLUGINS and PLUGINS_CONFIG variables.

> **Important note**
>
> Plugins are synonymous with apps.

We are primarily using the defaults from the app's docs. There is no default configuration required for the Data Validation Engine:

```
# nautobot_config.py

PLUGINS = ["nautobot_firewall_models", "nautobot_bgp_models",
"nautobot_data_validation_engine"]
```

```
PLUGINS_CONFIG = {
    "nautobot_firewall_models": {
        "default_status": "Active",
        "allowed_status": [],
        "capirca_remark_pass": True,
        "capirca_os_map": {
            "cisco_ios": "cisco",
            "arista_eos": "arista",
        },
    },
    "nautobot_bgp_models": {
        "default_statuses": {
            "AutonomousSystem": ["Active", "Available", "Planned"],
            "Peering": ["Active", "Decommissioned", "Deprovisioning",
"Offline", "Planned", "Provisioning"],
        }
    }
}
```

Be sure to view the docs for each app to understand the required and optional parameters. The right place to start is always here: `https://docs.nautobot.com`.

When all is said and done, we'll see the following in `nautobot_config.py`:

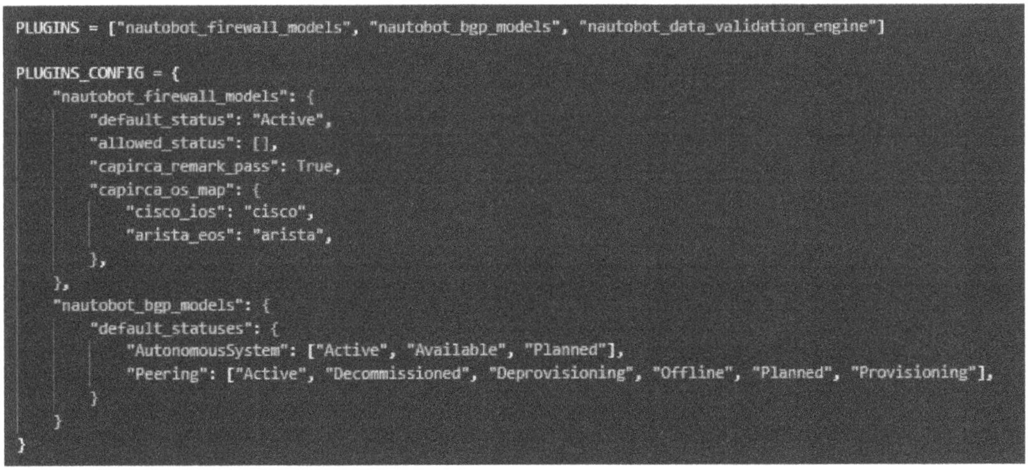

```
PLUGINS = ["nautobot_firewall_models", "nautobot_bgp_models", "nautobot_data_validation_engine"]

PLUGINS_CONFIG = {
    "nautobot_firewall_models": {
        "default_status": "Active",
        "allowed_status": [],
        "capirca_remark_pass": True,
        "capirca_os_map": {
            "cisco_ios": "cisco",
            "arista_eos": "arista",
        },
    },
    "nautobot_bgp_models": {
        "default_statuses": {
            "AutonomousSystem": ["Active", "Available", "Planned"],
            "Peering": ["Active", "Decommissioned", "Deprovisioning", "Offline", "Planned", "Provisioning"],
        }
    }
}
```

Figure 13.24 – Viewing PLUGINS and PLUGINS_CONFIG in nautobot_config.py

Once you save `nautobot_config.py`, perform a post upgrade and restart services (this is good practice for app installations, but in reality, restarting services may not be needed for a given app based on what it is providing). Let's walk through this:

```
nautobot@nautobot-dev:~$ nautobot-server post_upgrade
# output omitted for brevity - this may take a few mins
Exit the Nautobot user's session.
nautobot@nautobot-dev:~$ exit
logout
```

Then, restart Nautobot services:

```
root@nautobot-dev:~# systemctl restart nautobot nautobot-worker
nautobot-scheduler
root@nautobot-dev:~#
```

At this point, go to the Nautobot UI. Click **PLUGINS | Installed Plugins**:

Figure 13.25 – Viewing installed apps in Nautobot

You should then see all three apps that were just installed:

>>> Installed Plugins

Name	Description
Firewall & Security Models	Nautobot App to model firewall and security objects. Allows users to model policies in a vendor-neutral manner and use that data to drive network security automation.
BGP Models	Nautobot BGP Models Plugin.
Data Validation Engine	Provides UI to build custom data validation rules for data in Nautobot.

Figure 13.26 – Viewing the list of installed apps (in the Installed Plugins list)

Let's go back to the navbar and see each of the menu items that were all just added by each app.

The **ROUTING** menu item was created by the BGP Models app, and the **SECURITY** menu item was created by the Firewall Models App:

Figure 13.27 – Viewing navbar items created using the BGP and Firewall Models apps

The Data Validation Engine app does not have its own top-level navbar item; rather, it creates menu items underneath the **EXTENSIBILITY** menu item:

Figure 13.28 – Viewing menu items added from the Data Validation Engine app

At this point, you can follow the same process to add more Nautobot Apps.

Uninstalling Nautobot Apps

In this section, we are going to remove the Data Validation Engine app we just installed. Generally speaking, you should get in the habit of looking at the instructions in the app's docs:

Figure 13.29 – Viewing instructions for installing and uninstalling an app

You'll always look in the **Administrator Guide** section for install and uninstall instructions.

Let's walk through uninstalling the Data Validation Engine app.

Perform the data migration required:

```
root@nautobot-dev:~# sudo -iu nautobot
nautobot@nautobot-dev:~$
nautobot@nautobot-dev:~$ nautobot-server migrate nautobot_data_
validation_engine zero
```

```
nautobot@nautobot-dev:~$ nautobot-server migrate nautobot_data_validation_engine zero
Operations to perform:
  Unapply all migrations: nautobot_data_validation_engine
Running migrations:
  Rendering model states... DONE
  Unapplying nautobot_data_validation_engine.0006_add_field_defaults... OK
  Unapplying nautobot_data_validation_engine.0005_remove_slugs_alter_tags... OK
  Unapplying nautobot_data_validation_engine.0004_created_datetime... OK
  Unapplying nautobot_data_validation_engine.0003_datacompliance... OK
  Unapplying nautobot_data_validation_engine.0002_required_unique_types_regex_context... OK
  Unapplying nautobot_data_validation_engine.0001_initial... OK
16:29:08.400 INFO    nautobot.extras.utils :
  Refreshed Job "System Jobs: Export Object List" from <ExportObjectList>
16:29:08.407 INFO    nautobot.extras.utils :
  Refreshed Job "System Jobs: Git Repository: Sync" from <GitRepositorySync>
16:29:08.413 INFO    nautobot.extras.utils :
  Refreshed Job "System Jobs: Git Repository: Dry-Run" from <GitRepositoryDryRun>
16:29:08.419 INFO    nautobot.extras.utils :
  Refreshed Job "Capirca Jobs: Generate FW Config via Capirca." from <RunCapircaJob>
16:29:08.425 INFO    nautobot.extras.utils :
  Refreshed Job "nautobot_data_validation_engine.jobs: Run Registered Data Compliance Rules" from <RunRegisteredDataComplianceRules>
16:29:08.430 INFO    nautobot.extras.utils :
  Refreshed Job "nautobot_data_validation_engine.jobs: Delete Orphaned Data Compliance Data" from <DeleteOrphanedDataComplianceData>
nautobot@nautobot-dev:~$
```

Figure 13.30 – Performing an uninstall of the Data Validation Engine app

Remove the `nautobot_data_validation_engine` element from the `PLUGINS` list in `nautobot_config.py`:

```
PLUGINS = ["nautobot_firewall_models", "nautobot_bgp_models"]
```

Uninstall the Python package:

```
nautobot@nautobot-dev:~$ pip3 uninstall nautobot-data-validation-
engine
```

Remove `nautobot-data-validation-engine` from the `local_requirements.txt` file:

```
nautobot@nautobot-dev:~$ cat local_requirements.txt
nautobot-firewall-models
nautobot-bgp-models
nautobot@nautobot-dev:~$
nautobot@nautobot-dev:~$ exit
root@nautobot-dev:~#
```

Restart Nautobot services (for good measure):

```
root@nautobot-dev:~# systemctl restart nautobot nautobot-worker
nautobot-scheduler
root@nautobot-dev:~#
```

At this point, all of the app's data have been deleted, and the app has now been uninstalled from the environment.

Summary

This chapter was the first of four that explore Nautobot Apps. We covered an overview of Nautobot Apps, reviewed why you should consider building Nautobot Apps, provided an overview of many of the common open source Nautobot Apps, spent time showing what is possible firsthand, serving as a teaser for the next three chapters, and finally, we showed how you install and uninstall apps.

In the next chapter, we move on from theory and show how you can start building your own Nautobot app by exploring the Nautobot app developer API.

14

Intro to Nautobot App Development

In this chapter, you will be introduced to the Nautobot Developer API that is used to extend Nautobot and create Nautobot Apps. By the end of this chapter, you will know how to set up a basic Python development environment for working with Nautobot, you will know the internals and structure of a Nautobot App, and you will have a good idea of the types of features and extensions you can enable in your App.

This chapter consists of two primary sections. In the first section, we will walk through the steps involved in creating and running a new basic Nautobot App using the Nautobot App Cookiecutter project template. The second section of this chapter will provide an overview of the many different features of the Nautobot Developer API that a Nautobot App can make use of, including concrete examples of existing Apps that use many of these features.

In this chapter, we will cover the following topics:

- Setting up your system for Nautobot App development
- Starting a Nautobot App with Cookiecutter
- Exploring the Nautobot Developer API
- Conclusion

Setting up your system for Nautobot App development

The steps described in this section assume you are working on a Linux system, specifically running Ubuntu 20.04 LTS. While it's quite possible to develop for Nautobot on a different Linux distribution, or even on a Mac or a Windows system (the latter via the Windows Subsystem for Linux), these alternative approaches are out of scope for this chapter.

One relatively straightforward option for setting up a suitable system is Digital Ocean (`https://www.digitalocean.com/`), which gives you the ability to create a "droplet" running an OS of your choice for a few dollars per month. The remainder of this chapter was validated on a droplet with 4 CPUs and 8 GB of RAM running Ubuntu 20.04 LTS.

Let's get started.

Installing Docker

The standard Nautobot App development workflow uses Docker (`https://docs.docker.com/get-started/overview/`), which lets you run applications in one or more isolated environments called containers. This chapter will not make you an expert at using Docker, but we'll walk you through the basics of what you need to get started.

You'll need to install Docker, as it isn't included in most systems by default. On Ubuntu 20.04, the easiest way is to use its built-in `apt` software management application.

> **Tip**
>
> All of the commands in this chapter assume you are logged in as the root user (which is the default in Digital Ocean). In addition, this chapter assumes a fresh Ubuntu install for this chapter.

First, you'll install a few basic pieces of software: `curl`, which is a command line utility for downloading from the internet; `software-properties-common`, which lets you reconfigure `apt` itself; and `tree`, which lets you view the contents of a directory structure as a tree:

```
root@nautobot-app-dev:~# apt install curl software-properties-common
tree
Reading package lists... Done
Building dependency tree
Reading state information... Done
…
Setting up software-properties-common (0.99.9.12) ...
Processing triggers for man-db (2.9.1-1) ...
Processing triggers for dbus (1.12.16-2ubuntu2.2) …
root@nautobot-app-dev:~#
```

Next, you'll configure `apt` to be able to download and install Docker:

```
root@nautobot-app-dev:~# curl -fsSL https://download.docker.com/linux/
ubuntu/gpg | apt-key add -
OK
root@nautobot-app-dev:~# add-apt-repository "deb [arch=amd64] https://
download.docker.com/linux/ubuntu focal stable"
```

```
...
Fetched 91.8 kB in 1s (86.4 kB/s)
Reading package lists... Done
root@nautobot-app-dev:~#
```

Next, install Docker:

```
root@nautobot-app-dev:~# apt install -y docker-ce
Reading package lists... Done
Building dependency tree
Reading state information... Done
...
Setting up docker-ce (5:24.0.7-1~ubuntu.20.04~focal) ...
Created symlink /etc/systemd/system/multi-user.target.wants/docker.
service → /lib/systemd/system/docker.service.
Created symlink /etc/systemd/system/sockets.target.wants/docker.socket
→ /lib/systemd/system/docker.socket.
Processing triggers for man-db (2.9.1-1) ...
Processing triggers for systemd (245.4-4ubuntu3.17) ...
root@nautobot-app-dev:~#
```

You can now check to confirm that Docker is installed and running:

```
root@nautobot-app-dev:~# systemctl status docker
● docker.service - Docker Application Container Engine
     Loaded: loaded (/lib/systemd/system/docker.service; enabled;
vendor preset: enabled)
     Active: active (running) since Mon 2023-12-11 14:59:14 UTC; 1min
27s ago
TriggeredBy: ● docker.socket
       Docs: https://docs.docker.com
   Main PID: 11775 (dockerd)
      Tasks: 10
     Memory: 27.7M
     CGroup: /system.slice/docker.service
             └─11775 /usr/bin/dockerd -H fd:// --containerd=/run/
containerd/containerd.sock
root@nautobot-app-dev:~#
```

Now that you have Docker installed, you're all set to be able to run a Docker-based development environment for a Nautobot app. The next step will be to install Python and several Python packages that will help you in creating that environment.

Installing Python 3, Pip, Cookiecutter, and Poetry

In this section, you'll install Python 3 and its package manager `pip`, then install the Python software packages `cookiecutter` and `poetry`. `cookiecutter` will be used to create the skeleton of a Nautobot app, and `poetry` will be used to manage the Python dependencies of that App as you develop it.

> **Tip**
>
> `tip` and `poetry` are Python package managers, similar to `apt` but specialized for Python software. Generally speaking, you should prefer to use these tools instead of `apt` when installing Python packages, except when (as below) using `apt` to install Python and `pip` itself.

You'll use `apt` to do the initial installation of `pip`:

```
root@nautobot-app-dev:~# apt install -y python3-pip
Reading package lists... Done
Building dependency tree
Reading state information... Done
...
Setting up python3-dev (3.8.2-0ubuntu2) ...
Processing triggers for libc-bin (2.31-0ubuntu9.9) ...
Processing triggers for man-db (2.9.1-1) ...
Processing triggers for mime-support (3.64ubuntu1) ...
root@nautobot-app-dev:~#
```

Now that `pip` is installed, you'll use it to install the Python `cookiecutter` software (https://www.cookiecutter.io/). This is a program designed to take a set of template files (a "Cookiecutter template") and combine that with some user-provided information to produce a new set of files (a "baked cookie").

Nautobot provides a Cookiecutter template specifically for starting to develop a new Nautobot App, which is much more convenient than trying to bootstrap an App from nothing.

Telling `pip` to install `cookiecutter` will install it and any related software that it depends on:

> **Tip**
>
> One advantage of developing on a dedicated server such as a Digital Ocean droplet, and running as the root user, is that you can freely make changes to the system-wide Python installation without concern for its impact to other users or programs. If you were working in a multi-user system, or one running other software, you wouldn't generally want to install `cookiecutter` globally as we're doing here, but would instead use an isolated Python environment called a `virtualenv`. Setting up a `virtualenv` isn't difficult but it's out of scope for this chapter.

```
root@nautobot-app-dev:~# pip install cookiecutter
Collecting cookiecutter
  Downloading cookiecutter-2.5.0-py3-none-any.whl (39 kB)
...
Successfully installed arrow-1.3.0 binaryornot-0.4.4 charset-
normalizer-3.3.2 cookiecutter-2.5.0 markdown-it-py-3.0.0 mdurl-0.1.2
pygments-2.17.2 python-dateutil-2.8.2 python-slugify-8.0.1
requests-2.31.0 rich-13.7.0 text-unidecode-1.3 types-python-
dateutil-2.8.19.14 typing-extensions-4.9.0
~#
```

The next piece of Python software we'll install is `poetry` (`https://python-poetry.org/`), which is much like `pip`, but is designed a bit more specifically to manage dependencies within a Python software development project as it's being worked on.

You could use `pip` to install `poetry`, but `poetry` provides its own installation script that it recommends using, so we'll use that. (This may seem contrary to our earlier recommendation to just use `pip` across the board, but if you opened and read the installation script, you'd see that it actually *is* using `pip` behind the scenes, just wrapped in some additional setup and validation logic specific to `poetry`.)

You can download and run the script in one step, using `curl` to download it and then pass it into `python3` to execute (since it's a Python-based script):

```
root@nautobot-app-dev:~# curl -sSL https://install.python-poetry.org |
python3 -
Retrieving Poetry metadata
# Welcome to Poetry!
This will download and install the latest version of Poetry, a
dependency and package manager for Python.
```

It will add the `poetry` command to Poetry's bin directory, located at the following path:

```
/root/.local/bin
```

You can uninstall at any time by executing this script with the `--uninstall` option, and these changes will be reverted.

```
Installing Poetry (1.7.1): Done
Poetry (1.7.1) is installed now. Great!
```

To get started you need Poetry's bin directory (`/root/.local/bin`) in your PATH environment variable.

Add `export PATH="/root/.local/bin:$PATH"` to your shell configuration file.

Alternatively, you can call Poetry explicitly with `/root/.local/bin/poetry`.

You can test that everything is set up by executing the following:

```
`poetry --version`
root@nautobot-app-dev:~#
```

As the output says, you'll need to configure your PATH environment variable so that it can see where Poetry is installed, so let's do that:

```
root@nautobot-app-dev:~# export PATH="/root/.local/bin:$PATH"
root@nautobot-app-dev:~#
```

Now that you have these packages installed, the next step is to use them to create and configure the development environment for your new Nautobot App.

Starting a Nautobot App with Cookiecutter

The `cookiecutter` software can use any one of a vast number of templates to create a new project, but we want to use the one provided by Nautobot at `https://github.com/nautobot/cookiecutter-nautobot-app`.

`cookiecutter` will download the template, prompt you for information, and use that information to generate a new project.

```
root@nautobot-app-dev:~# cookiecutter https://github.com/nautobot/
cookiecutter-nautobot-app.git --directory=nautobot-app
  [1/18] codeowner_github_usernames (): @glennmatthews
```

You can fill in your GitHub account username if you want to, or just hit *Return* to accept the default. This is just used to fill in some blanks in the project template; it doesn't send data out to GitHub or anything like that.

```
  [2/18] full_name (Network to Code, LLC): Glenn Matthews
```

Enter your name or your company's name in the preceding line.

```
  [3/18] email (info@networktocode.com):
```

Enter your email address. Again, this is just to configure the template; you're not signing up for a mailing list or anything like that.

```
  [4/18] github_org (nautobot):
```

The GitHub organization that this project would be published to, if you decide to publish it later.

```
  [5/18] app_name (my_app): nautobot_ip_acls
```

Here we recommend that for now, you enter `nautobot_ip_acls` as the App name, as we'll be building this template into a full-featured IP **Access Control List** (**ACL**) App in the next chapter. This will be intelligently used to customize the defaults for the rest of the App template, as seen below:

```
[6/18] verbose_name (Nautobot Ip Acls): Nautobot IP ACLs
[7/18] app_slug (nautobot-ip-acls):
[8/18] project_slug (nautobot-app-nautobot-ip-acls): nautobot-app-
ip-acls
[9/18] repo_url (https://github.com/nautobot/nautobot-app-ip-acls):
[10/18] base_url (nautobot-ip-acls): ip-acls
```

The above are used to generate some basic code structure for your App. The defaults are a little bit off from best practices (as of this writing) so you'll want to override them as in the preceding snippet.

```
[11/18] min_nautobot_version (2.0.0): 2.1.0
[12/18] max_nautobot_version (2.9999):
```

The preceding code determines the range of Nautobot versions that will be supported by this App. Unless you need to support an older Nautobot version (such as your existing Nautobot deployment that for some reason isn't yet running the latest-and-greatest), we'd recommend that you specify the version you want to develop against as `min_nautobot_version`.

```
[13/18] camel_name (NautobotIpAcls): NautobotIPACLs
[14/18] project_short_description (Nautobot IP ACLs):
```

In the preceding, we can see more code and documentation generation parameters.

```
[15/18] model_class_name (None):
```

If you want the App to be prepopulated with an example of a data model class, you can enter its name here, but for now, we'll skip that.

```
[16/18] Select open_source_license
    1 - Apache-2.0
    2 - Not open source
    Choose from [1/2] (1): 1
```

The project template defaults to declaring your App as licensed under Apache-2.0 (the same as Nautobot itself) but you can change this if you're writing an App that would be private to your company.

```
[17/18] docs_base_url (https://docs.nautobot.com):
[18/18] docs_app_url (https://docs.nautobot.com/projects/nautobot-
ip-acls/en/latest):
```

The preceding code determines where the documentation for this App will eventually live. You can leave these as default for now.

> **Tip**
>
> Don't worry yet about the following instructions printed by Cookiecutter about specific commands to run; we'll deal with those shortly.

```
Congratulations! Your cookie has now been baked. It is located at /
root/nautobot-app-ip-acls.
⚠️ Before you start using your cookie you must run the following
commands inside your cookie:
* poetry lock
* cp development/creds.example.env development/creds.env
* poetry install
* poetry shell
* invoke makemigrations
* black . # this will ensure all python files are formatted correctly,
may require `sudo chown -R <my local username> ./` as migrations may
be owned by root
The file `creds.env` will be ignored by git and can be used to
override default environment variables.
root@nautobot-app-dev:~#
```

And just like that, you now have the framework for a new project called "nautobot-app-ip-acls", which contains the Python package (more specifically, the Nautobot App) nautobot_ip_acls.

Exploring the App structure

Now that you have created your App framework, let's go inside it and take a look at everything that's already included. Change directory into the newly created project directory:

```
root@nautobot-app-dev:~# cd nautobot-app-ip-acls
root@nautobot-app-dev:~/nautobot-app-ip-acls#
```

Let's take a quick look around the structure that's been provided for you. You can use tree for this purpose.

> **Tip**
>
> Don't worry if your specific project structure looks slightly different than the following, as the cookiecutter template may evolve over time.

```
root@nautobot-app-dev:~/nautobot-app-ip-acls# tree .
.
├── LICENSE
```

The LICENSE file specifies the software license (defaulting to Apache-2.0) for this project.

```
├── README.md
```

This is a generic README document template. You'll want to fill this in eventually with more information.

```
├── changes
```

As you develop this app over time, you will want to document your changes. Cookiecutter uses towncrier (https://towncrier.readthedocs.io/en/stable/) as a way to manage your change history using text files in this directory, though we won't go into that in this chapter.

```
├── development
│   ├── Dockerfile
│   ├── creds.example.env
│   ├── development.env
│   ├── development_mysql.env
│   ├── docker-compose.base.yml
│   ├── docker-compose.dev.yml
│   ├── docker-compose.mysql.yml
│   ├── docker-compose.postgres.yml
│   ├── docker-compose.redis.yml
│   └── nautobot_config.py
│   └── towncrier_template.j2
```

The files in the development directory define a Docker Compose-based development environment for this application, including Nautobot's dependencies and the option to run either PostgreSQL or MySQL as the backend database. You generally shouldn't need to change the files here.

This nautobot_config.py provides a development-specific set of Nautobot configuration; it shouldn't be used for a production deployment, but it's ideal for development.

```
├── docs
│   ├── admin
```

The files in docs provide a template for documenting your App. The admin subdirectory contains boilerplate for documenting the installation and administration of this App. If you open these Markdown files in a text editor you'll see many sections prompting you to fill in appropriate information. These are not needed at the moment, but are useful to complete if you plan to distribute this app eventually.

```
│       ├── assets
```

By default, the cookiecutter assumes this will be an "official" Nautobot App maintained by *Network to Code*, and uses some documentation formatting based on this assumption. You may eventually want to replace any or all of the files in this subdirectory if that's not the case.

```
│       ├── dev
```

The `docs/dev` subdirectory provides boilerplate for documenting how to develop this App. As with the admin documentation, you'll eventually want to flesh this out with your own specifics.

```
|   ├── images
|   |   └── icon-ip-acls.png
```

You're recommended to provide a custom icon for your App, but this default file will work for the moment. You can replace it later.

```
|   ├── index.md
```

This is the landing page for your App's documentation. Like the other files, it's a template that you're encouraged to eventually fill out with more details.

```
|   ├── requirements.txt
```

The `docs/requirements.txt` file defines Python packaging requirements for rendering the documentation from Markdown format to HTML. This is useful if, for example, you wish to publish your documentation on `https://readthedocs.org`.

```
|   └── user
```

The final `docs` subdirectory, `user`, contains user-facing documentation templates. It's highly recommended to fill these out if you plan to distribute this App.

```
├── invoke.example.yml
├── invoke.mysql.yml
```

We'll talk about Invoke shortly, but these two files are just examples of how you might create an `invoke.yml` file to change Invoke's default behavior in this project.

```
├── mkdocs.yml
```

The `mkdocs.yml` file is the configuration for the `mkdocs` software that can be used to render your project's documentation from Markdown to HTML.

```
├── nautobot_ip_acls
|   ├── __init__.py
```

The preceding directory contains the actual Python source code for your App. As you can see, there's not a lot here to begin with. We'll take a look inside shortly. This is where we will spend a lot of time in future chapters.

```
|   └── tests
|       ├── __init__.py
|       ├── test_api.py
|       └── test_basic.py
```

The `tests` subdirectory contains example Python unit tests for your App. Developing and running test automation is always a good idea for any software project.

```
├── pyproject.toml
```

One of the most important files in the project, `pyproject.toml` is what actually defines this App as a Python software package that can be packaged, distributed, and installed elsewhere. We'll take a closer look shortly.

```
└── tasks.py
```

This last `tasks.py` file defines "tasks", command-line commands that can be run by Invoke (more on this tool shortly). The Cookiecutter template has conveniently provided you with a predefined set of tasks to handle most of the common things you might want to do while developing this project.

Exploring pyproject.toml

As mentioned above, `pyproject.toml` is the latest standard way to define a Python package. It replaces older Python packaging files such as `setup.py`, `setup.cfg`, or `requirements.txt`. It also serves as a shared configuration file for various Python development tools such as Poetry when working on this project.

If you want to learn more about the history of and justification for this file, you can refer to **Python Enhancement Proposal** (**PEP**) 518 at `https://peps.python.org/pep-0518/`, but that's up to you.

You can manually edit this file in most cases, but in a few scenarios, it's preferable to use a tool such as Poetry to do so automatically. Let's take a look inside—it's a long file, but not a difficult one to understand:

```
root@nautobot-app-dev:~/nautobot-app-ip-acls# cat pyproject.toml
[tool.poetry]
name = "nautobot-ip-acls"
version = "0.1.0"
description = "Nautobot IP ACLs"
authors = ["Glenn Matthews <info@networktocode.com>"]
license = "Apache-2.0"
readme = "README.md"
homepage = "https://github.com/nautobot/nautobot-app-ip-acls"
repository = "https://github.com/nautobot/nautobot-app-ip-acls"
keywords = ["nautobot", "nautobot-app", "nautobot-plugin"]
classifiers = [
   ...
]
```

All of the above is informational metadata about this app, mostly populated from the information you provided when running Cookiecutter.

```
packages = [
    { include = "nautobot_ip_acls" },
]
```

The `packages` list defines which Python modules are included in this package.

```
[tool.poetry.dependencies]
python = ">=3.8,<3.12"
# Used for local development
nautobot = "^2.1.0"
```

The `tool.poetry.dependencies` section defines the Python dependencies for installing/running this App. By default, as you can see here, it's just a recent version of Python plus Nautobot itself, but you could add additional dependencies as needed (such as an SDK for some other system) via the `poetry add <dependency>` command later. (This is one of those cases where you shouldn't edit this part of the file by hand—always use `poetry add`, `poetry remove`, or `poetry update` to make changes here.)

```
[tool.poetry.group.dev.dependencies]
```

This section specifies the *development* dependencies for this App. These are Python packages that are useful when developing the App itself, but are not needed when actually installing the App into a production Nautobot environment. The below may change as the Cookiecutter template evolves over time, but as of the time of this writing, here's what you get:

```
bandit = "*"
```

Bandit (`https://bandit.readthedocs.io/`) is a tool used to find potential security vulnerabilities in your code.

> **Tip**
>
> The `"*"` means that any version of this tool is permissible; as such, when you run `poetry lock` later in this chapter, you'll be sure to get the then-latest version of this tool and those that follow.

```
black = "*"
```

Black (`https://black.readthedocs.io/`) is a tool used to automatically format your code in a self-consistent style.

```
coverage = "*"
```

Coverage (`https://coverage.readthedocs.io/`) enhances the built-in unit test automation framework to measure and report on how well the tests cover all of your project's code.

```
django-debug-toolbar = "*"
```

Django Debug Toolbar (`https://django-debug-toolbar.readthedocs.io/`) makes it easier to measure and troubleshoot various aspects of how your app is using the underlying Django software that Nautobot is built on top of.

```
flake8 = "*"
```

Flake8 (`https://flake8.pycqa.org/`) is a fast code analysis tool to check for and flag various bugs and incorrect code in your project.

```
invoke = "*"
```

Invoke (`https://www.pyinvoke.org/`) is a tool for automating various tasks. We'll take a closer look shortly.

```
ipython = "*"
```

IPython (`https://ipython.org/`) enhances Python's built-in command shell.

```
pylint = "*"
pylint-django = "*"
pylint-nautobot = "*"
```

Pylint (`https://pylint.readthedocs.io/en/stable/`) is an in-depth code analysis tool. It's slower than Flake8 but a lot more comprehensive. The App Cookiecutter template includes a couple of Pylint extensions that improve its usefulness to developing Django apps and specifically Nautobot Apps:

```
ruff = "*"
```

Ruff (`https://docs.astral.sh/ruff/`) is a relatively newer, very fast Python code analysis tool. It's not as comprehensive as Pylint (yet), but it can replace Flake8 and Black for most purposes; and indeed by the time you read this, we may have already removed those two from the Cookiecutter template.

```
yamllint = "*"
```

Yamllint (`https://yamllint.readthedocs.io/en/stable/`) enforces a consistent style when writing YAML files.

```
toml = "*"
```

The `toml` library is a helper library for parsing TOML-formatted files, including `pyproject.toml` itself.

```
Markdown = "*"
...
mkdocstrings-python = "1.5.2"
```

The Markdown and MkDocstrings packages and their plugins all work together when rendering the App documentation from Markdown format into rendered HTML.

```
[tool.black]
line-length = 120
target-version = ['py38', 'py39', 'py310', 'py311']
include = '\.pyi?$'
exclude = '''
  ...
'''
```

The `tool.black` section defines the configuration of the Black tool to match Nautobot's conventions. This will likely be replaced soon with the equivalent Ruff configuration.

```
[tool.pylint.master]
...
[tool.pylint.basic]
...
[tool.pylint.messages_control]
...
[tool.pylint.miscellaneous]
...
[tool.pylint-nautobot]
...
```

The preceding `tool.pylint` subsections define the configuration of Pylint.

```
[tool.ruff.lint]
select = [
  ...
]
ignore = [
  ...
]
```

The preceding `tool.ruff` subsection(s) define the configuration of Ruff.

```
[build-system]
requires = ["poetry_core>=1.0.0"]
build-backend = "poetry.core.masonry.api"
```

The preceding `build-system` section is the configuration of the Python package itself to use Poetry.

Finally, `pyproject.toml` concludes with configuration for the Towncrier tool for managing your App's changelog and development history:

```
[tool.towncrier]
package = "nautobot-ip-acls"
directory = "changes"
filename = "docs/admin/release_notes/version_X.Y.md"
template = "development/towncrier_template.j2"
start_string = "<!-- towncrier release notes start -->"
issue_format = "[#{issue}](https://github.com/nautobot/nautobot-app-
ip-acls/issues/{issue})"
...
```

All of this may seem like a lot, but this is where Cookiecutter really shines—all of the preceding configuration is already provided for you, and most of it you'll never need to change in the future. Just knowing that it's there is good enough for most purposes.

Post-Cookiecutter tasks and Poetry

As you have seen so far, there are a lot of useful Python packages that this project can make use of. You could `pip install` each of them, but now you've got a project set up, so you can instead let the project's configuration be used with Poetry to handle the installation for you.

First, you'll want to run `poetry lock`. This detects the latest allowed versions of all of your project's dependencies and dev-dependencies and creates a `poetry.lock` file that caches that information for you:

```
root@nautobot-app-dev:~/nautobot-app-ip-acls# poetry lock
Updating dependencies
Resolving dependencies... (36.0s)
Writing lock file
root@nautobot-app-dev:~/nautobot-app-ip-acls#
```

That's actually all you need to do manually with Poetry at the moment, but for future reference, here are a few other useful Poetry subcommands:

- `poetry init` – Generates a new `pyproject.toml` for a new project (here, we used Cookiecutter for this purpose instead).

- `poetry install` – Creates a new Python virtual environment and installs any Python package dependencies into this environment. This is useful if you're doing pure-Python development locally, but because Nautobot includes some non-Python dependencies, like a database, we're going to use a Docker setup (which does include `poetry install` as a step in building the Docker containers for this project) instead of running this command ourselves.

- `poetry shell` – After running `poetry install`, this command "activates" the resulting virtual environment so that you may execute commands within it, including all of the various Python packages that were installed thus.

Introducing Invoke

GNU Make is commonly used in open source projects for providing common tasks for managing or developing the project. Nautobot uses Invoke as a replacement for Make for common operations such as building Docker containers or running tests. Invoke was chosen because it is less arcane than Make and allows tasks to be written in Python instead of relying on shell commands. Instead of a Makefile, Invoke reads tasks from the `tasks.py` file in the root of your project.

Once more, we'll use `pip` to install this tool:

```
root@nautobot-app-dev:~/nautobot-app-ip-acls# pip install invoke
Collecting invoke
  Downloading invoke-2.2.0-py3-none-any.whl (160 kB)
    |█████████████████████████████████████| 160 kB 19.1 MB/s
Installing collected packages: invoke
Successfully installed invoke-2.2.0
root@nautobot-app-dev:~/nautobot-app-ip-acls#
```

Now that it's installed, Invoke can list the available commands provided by the `tasks.py` file found in your project root:

```
root@nautobot-app-dev:~/nautobot-app-ip-acls# invoke --list
Available tasks:
  autoformat (a)          Run code autoformatting.
  backup-db               Dump database into `output_file` file from
`db` container.
  bandit                  Run bandit to validate basic static code
security analysis.
  black                   Check Python code style with Black.
  build                   Build Nautobot docker image.
…
  unittest                Run Nautobot unit tests.
  unittest-coverage       Report on code test coverage as measured by
'invoke unittest'.
  vscode                  Launch Visual Studio Code with the appropriate
Environment variables to run in a container.
  yamllint                Run yamllint to validate formatting adheres to
NTC defined YAML standards.
root@nautobot-app-dev:~/nautobot-app-ip-acls#
```

The exact list of tasks may vary as the Cookiecutter template evolves over time, and we won't cover every one of them in this chapter. For now, the important task to start with is `build`, which, as the name suggests, will build the Nautobot Docker image that we'll rely on to develop and test this App.

Building the Docker image

The Nautobot App Cookiecutter template uses Docker (specifically Docker Compose) to assemble multiple virtualized containers into a fully functional Nautobot system. Some of these containers use "off-the-shelf" Docker images, such as the standard images for PostgreSQL and Redis, but we need to build our own image containing our new App to run as a part of the system. To do this, we use the `invoke build` command:

```
root@nautobot-app-dev:~/nautobot-app-ip-acls# invoke build
Building Nautobot with Python 3.11...
Running docker compose command "build"
#0 building with "default" instance using docker driver
#1 [nautobot internal] load build definition from Dockerfile
#1 transferring dockerfile: 3.95kB done
#1 DONE 0.0s
#2 [nautobot internal] load metadata for ghcr.io/nautobot/nautobot-
dev:2.1.0-py3.11
#2 DONE 0.4s
```

There will be a lot more output here, which we'll omit to save space, but in the end you should see something like the following, indicating that it successfully finished building the new Docker image containing your App:

```
#16 [nautobot] exporting to image
#16 exporting layers
#16 exporting layers 2.4s done
#16 writing image
sha256:2e23cc00553fe362008bb8c06cb37fff390a773456b7fb43610d6896c1aec
318 done
#16 naming to docker.io/nautobot-ip-acls/nautobot:2.1.0-py3.11 done
#16 DONE 2.4s
root@nautobot-app-dev:~/nautobot-app-ip-acls#
```

As simply as that, you now have a newly built Docker image, ready to run as a part of a Nautobot system.

Defining credentials

Now that the image is built, you can use either the `debug` or `start` tasks to run Nautobot with your app; the difference between these two tasks is that `invoke debug` will continue to run and will display all output from Docker to your terminal, while `invoke start` will start the Docker containers in the background and return control to you when they're ready to run. For now, let's use `invoke debug` so we can see what is happening in more detail:

```
root@nautobot-app-dev:~/nautobot-app-ip-acls# invoke debug
Starting  in debug mode...
Running docker compose command "up"
env file /root/nautobot-app-ip-acls/development/creds.env not found:
stat /root/nautobot-app-ip-acls/development/creds.env: no such file or
directory
root@nautobot-app-dev:~/nautobot-app-ip-acls#
```

Oops! We forgot an important step, one which Cookiecutter even warned us about when we created the project—we don't have a `development/creds.env` file, which defines the environment variables to use for credentials within and among our Docker environment. Fortunately, there's a template that we can copy from:

```
root@nautobot-app-dev:~/nautobot-app-ip-acls# cp development/creds.
example.env development/creds.env
root@nautobot-app-dev:~/nautobot-app-ip-acls#
```

You should then open the `development/creds.env` file in your text editor of choice and fill in appropriate non-default values for each variable.

> **Tip**
>
> Most importantly, given that you're developing on a system that's exposed to the internet, you should change the values of NAUTOBOT_SECRET_KEY, NAUTOBOT_SUPERUSER_NAME, NAUTOBOT_SUPERUSER_PASSWORD, and NAUTOBOT_SUPERUSER_API_TOKEN, as all of these are relevant to you or anyone else being able to log in to the Nautobot instance you're about to build and run. Note also that in current versions of Nautobot, the NAUTOBOT_SUPERUSER_API_TOKEN must be exactly 40 alphanumeric characters long, and you may get an error at startup if it isn't.

The other credentials are mostly used for inter-service authentication within the Nautobot instance, rather than external access, and so are less critical to change, but it's a good practice to update them as well regardless.

That done, let's try again.

Running Nautobot

Now that you have an appropriate credentials file defined, you can run `invoke debug` again, this time successfully.

```
root@nautobot-app-dev:~/nautobot-app-ip-acls# invoke debug
Starting  in debug mode...
```

```
Running docker compose command "up"
 redis Pulling
 db Pulling
```

The first output you'll see, as shown in the preceding snippet, is related to Docker downloading some additional container images from the internet—specifically, the standard images for Redis and PostgreSQL, as those are an important part of the overall Nautobot system. This is a one-time operation typically, as Docker will cache these images locally for later reuse.

Once that's done, you'll see Docker creating all of the various containers needed for the Nautobot system to run:

```
Network nautobot-ip-acls_default  Creating
Network nautobot-ip-acls_default  Created
Volume "nautobot-ip-acls_postgres_data"  Creating
Volume "nautobot-ip-acls_postgres_data"  Created
Container nautobot-ip-acls-docs-1  Creating
Container nautobot-ip-acls-redis-1  Creating
Container nautobot-ip-acls-db-1  Creating
Container nautobot-ip-acls-nautobot-1  Creating
Container nautobot-ip-acls-worker-1  Creating
Container nautobot-ip-acls-beat-1  Creating
```

Next, you'll see the logs from each of these containers—there will be a lot here as Nautobot and its related services all start up:

```
Attaching to nautobot-ip-acls-beat-1, nautobot-ip-acls-db-1, nautobot-
ip-acls-docs-1, nautobot-ip-acls-nautobot-1, nautobot-ip-acls-redis-1,
nautobot-ip-acls-worker-1
nautobot-ip-acls-redis-1   | 1:C 13 Dec 2023 22:33:45.911 #
oO0OoO00oO00o Redis is starting oO0OoO00oO00o
nautobot-ip-acls-redis-1   | 1:C 13 Dec 2023 22:33:45.911 # Redis
version=6.2.14, bits=64, commit=00000000, modified=0, pid=1, just
started
nautobot-ip-acls-redis-1   | 1:C 13 Dec 2023 22:33:45.911 #
Configuration loaded
nautobot-ip-acls-redis-1   | 1:M 13 Dec 2023 22:33:45.912 * monotonic
clock: POSIX clock_gettime
nautobot-ip-acls-db-1      | The files belonging to this database
system will be owned by user "postgres".
nautobot-ip-acls-db-1      | This user must also own the server
process.
```

You'll see Nautobot preparing the database by running database migrations:

```
nautobot-ip-acls-nautobot-1   |   Applying ipam.0032_ipam__namespaces_
finish... OK
nautobot-ip-acls-nautobot-1   |   Applying ipam.0033_fixup_null_
statuses... OK
nautobot-ip-acls-nautobot-1   |   Applying ipam.0034_status_
nonnullable... OK
nautobot-ip-acls-nautobot-1   |   Applying ipam.0035_ensure_all_
services_fit_uniqueness_constraint... OK
nautobot-ip-acls-nautobot-1   |   Applying ipam.0036_add_uniqueness_
constraints_to_service... OK
nautobot-ip-acls-nautobot-1   |   Applying ipam.0037_data_migration_
vlan_group_name_uniqueness... OK
nautobot-ip-acls-nautobot-1   |   Applying ipam.0038_vlan_group_name_
unique_remove_slug... OK
```

Finally, if all goes well, you should see the following:

```
nautobot-ip-acls-nautobot-1   |  Starting development server at
http://0.0.0.0:8080/
nautobot-ip-acls-nautobot-1   |  Quit the server with CONTROL-C.
nautobot-ip-acls-nautobot-1   |  22:38:25.555 INFO nautobot
init__.py                          setup() :
nautobot-ip-acls-nautobot-1   |    Nautobot initialized!
```

There will continue to be more output on your terminal as the services continue running, but at this point you should be able to open a web browser and connect to your Nautobot development instance at http://<server IP address>:8080.

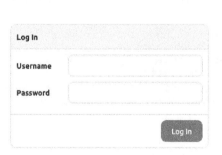

Figure 14.1 – Nautobot Login Page

Click the **Log In** button and log in with the administrator credentials that you defined in `development/creds.env` earlier, and you'll see the Nautobot home page.

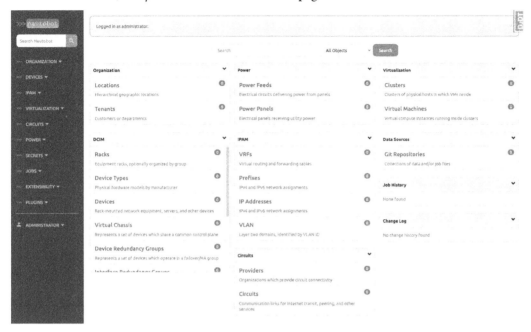

Figure 14.2 – Nautobot home page

Next, you can select the **Plugins** menu and the **Installed Plugins** menu item (which will be renamed to **Apps** and **Installed Apps** in the next minor release of Nautobot), and you can see that this Nautobot instance is running your new App:

Figure 14.3 – Nautobot Installed Plugins view

You can click on the App's name in the table to see more information about it, just as you've defined earlier in this chapter:

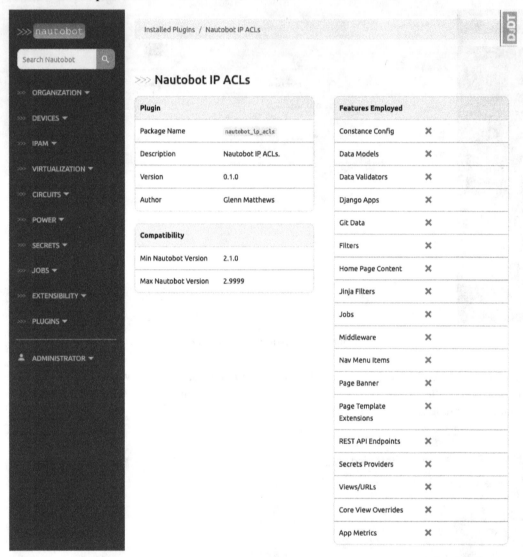

Figure 14.4 – Details about your App

You can also see on the right side of the page that there are many features that a Nautobot App can make use of, and currently this App isn't using any of them. Never fear! In the remainder of this book we'll walk you through developing an App that uses many of these features.

Exploring the Nautobot Developer API

The remainder of this chapter is focused on providing an overview of the Nautobot developer API, or in other words, all of the extension points that Nautobot provides for Apps to make use of to extend and customize Nautobot's operation and functionality. For those API features that are simple to use, we'll show you how to experiment with adding them to your new App, while more complex features will wait until the more focused development work of the following chapters. We'll also provide examples of specific existing Apps that use the different parts of the API.

You'll need the ability to make changes to your App's code to explore these features here; I recommend that you leave `invoke debug` running in the terminal and just open a new terminal session to do the development work. (You can hit *Ctrl + C* to stop Nautobot when you're done.) This way if your code accidentally has syntax errors or the like (it happens to all of us) you'll be able to see them immediately appear in the debug output.

Configuring a Nautobot App

Nautobot provides a special configuration class called `NautobotAppConfig`, which is a Nautobot-specific wrapper around Django's built-in `AppConfig` class. It is used to declare Nautobot App functionality within a Python package. Each App must provide its own subclass, defining its name, metadata, and default and required configuration parameters.

To load the configuration for your App, Nautobot looks for the `config` variable within your App's `__init__.py` found at the root of your App's package directory. Let's look at ours so far:

```
root@nautobot-app-dev:~/nautobot-app-ip-acls# cat nautobot_ip_acls/__
init__.py
"""Plugin declaration for nautobot_ip_acls."""
# Metadata is inherited from Nautobot. If not including Nautobot in
the environment, this should be added
from importlib import metadata
from nautobot.apps import NautobotAppConfig
__version__ = metadata.version(__name__)
```

The preceding little bit of code defines `nautobot_ip_acls.__version__` to be the same version number defined in your `pyproject.toml`, ensuring that it always stays in sync between your packaging and your Python code.

```
class NautobotIPACLsConfig(NautobotAppConfig):
    """App configuration for the nautobot_ip_acls app."""
    name = "nautobot_ip_acls"
    verbose_name = "Nautobot IP ACLs"
    version = __version__
    author = "Glenn Matthews"
    description = "Nautobot IP ACLs."
```

```
        base_url = "ip-acls"
        required_settings = []
        min_version = "2.1.0"
        max_version = "2.9999"
        default_settings = {}
        caching_config = {}
    config = NautobotIPACLsConfig  # pylint:disable=invalid-name
```

As you can see, Cookiecutter filled in a lot of these values for us. Every `NautobotAppConfig` subclass *must* define its name attribute, and the following additional attributes can be defined as well:

Name	Default	Description
author	""	Author's name
author_email	""	Author's email address
base_url	Same as name	Base path to use for App URLs (/plugins/<base_url>/...)
config_view_name	None	Named URL for a "configuration" view defined by this App, if any
default_settings	{}	A dictionary of configuration parameters that can be defined in settings.PLUGINS_CONFIG and their default values
description	""	Brief description of the App's purpose
docs_view_name	None	Named URL for a "documentation" view defined by this App, if any
home_view_name	None	Named URL for a "home" or "dashboard" view defined by this App, if any
installed_apps	[]	A list of additional Django application dependencies to automatically enable when the App is activated (you must still make sure these underlying dependent libraries are installed)
label	Same as name	Shorter alternative name for the App
max_version	None	Maximum version (string) of Nautobot with which the App is compatible
middleware	[]	A list of additional Django middleware classes to enable in addition to Nautobot's built-in middleware

Name	Default	Description
`min_version`	None	Minimum version (string) of Nautobot with which the App is compatible
`required_settings`	[]	A list of any configuration parameters that must be defined in `settings.PLUGINS_CONFIG` before the App can be used
`verbose_name`	label.title()	Human-friendly name for the App
`version`	""	Version number (semantic versioning – `https://semver.org/` – is highly encouraged)

If you like, you can experiment with editing the `author` or adding an `author_email`. After you do so and save changes to `__init__.py`, you'll see the Nautobot server process automatically restart to pick up these changes:

```
nautobot-1  | 22:07:36.779 INFO  django.utils.autoreload :
nautobot-1  |   /source/nautobot_ip_acls/__init__.py changed,
reloading.
...
nautobot-1  | 22:07:43.147 INFO  nautobot          __init__.
py                             setup() :
nautobot-1  |   Nautobot initialized!
```

Then you can refresh your browser window to see your change has taken effect:

Nautobot IP ACLs

Plugin

Package Name	nautobot_ip_acls
Description	Nautobot IP ACLs.
Version	0.1.0
Author	Joe User (joe@example.com)

Figure 14.5 – Updated author information

After you're done experimenting with this feature, I'd recommend reverting your changes to __init__.
py—in fact, I'd encourage you to do this after each of the experimental code changes you'll be making
in this chapter, as *Chapter 15* will assume you're starting from a "clean" freshly baked cookie for the
Nautobot IP ACLs app. Alternatively, you can leave your changes in place, and plan to delete or move
the nautobot-app-ip-acls directory at the end of this chapter and re-bake a fresh cookie at
the start of the next chapter—it's up to you.

Extending the existing Nautobot UI

Nautobot Apps allow you to extend the default Nautobot UI in various ways.

Adding a banner

An App can optionally provide a function that renders a custom banner on any or all of Nautobot's
various views.

Nautobot looks inside each installed App for a banner.py file, and specifically for a function inside
that file named banner(). If it finds such a function, it will call it when any page view is rendered,
and if the function returns a Banner object instance, that will be displayed as part of the page.

This function currently receives a single argument, context, which is the Django rendering context in
which the current page is being rendered. You can access various information about the request being
rendered, such as context.request (the Django request that triggered this view), context.
request.user (the Nautobot User instance that's calling this request), and so on.

You can try this feature easily enough yourself—create nautobot_ip_acls/banner.py and
populate it as follows:

```
from django.utils.html import format_html
from nautobot.apps.ui import Banner, BannerClassChoices


def banner(context, *args, **kwargs):
    # Banner content greeting the user
    content = format_html(
        "<div>"Hello, <strong>{}</strong>!" 🖐</div>",
        context.request.user,
    )
    return Banner(
        content=content,
        banner_class=BannerClassChoices.CLASS_SUCCESS
    )
```

In the preceding code, we're using Django's format_html API to construct an HTML fragment to
include in the banner, and we include a reference to the user who's viewing the page (derived from
the context argument).

> **Tip**
>
> Because this is an entirely new file in your App, Django won't automatically detect it and restart the server to include it—so instead you'll need to hit *Ctrl+C* in the terminal where your `invoke debug` command is running to stop the Nautobot environment, then rerun `invoke debug`, at which point the new file will be detected as expected. You'll need to do this any time you add an entirely new file to your App.

Figure 14.6 – A custom banner

More complex banner code can include conditional logic to determine whether a banner is displayed at all (your `banner()` function can return `None` if no banner is needed), provide different banners in different contexts, and so forth. As an example, the **Nautobot Welcome Wizard** App only shows a **Getting started** banner when a user is on the Nautobot home page, the App is configured to enable its banner, and the user has permission to make use of the App's features.

Adding home page content

Apps can optionally add content (panels and subpanels) to the Nautobot home page.

Nautobot looks inside each installed App for a `homepage.py` file that defines a layout list containing one or more `HomePagePanel` object instances. Each `HomePagePanel` has a name and a weight that defines where (in what order) the panel appears relative to others on the home page by default (in Nautobot 2.1 and later, users can reorder these panels if desired by dragging and dropping them). A `HomePagePanel` can in turn contain `HomePageItem` and/or `HomePageGroup` instances that define individual rows within the standard Nautobot home page panel appearance, or can provide an entirely custom HTML template as an alternative. As this is a fairly flexible feature, more specifics can be found in the documentation at `https://docs.nautobot.com/projects/core/en/v2.1.7/development/core/homepage/`. But for now, let's do a simple example. Let's say you're interested in keeping an eye on the number of Status records in Nautobot and want to see this on the home page – information that Nautobot doesn't show there by default. Create `nautobot_ip_acls/homepage.py` as follows:

```python
from nautobot.apps.ui import HomePageItem, HomePagePanel
from nautobot.extras.models import Status

layout = (
    HomePagePanel(
        name="Organization",
        items=(
            HomePageItem(
                name="Statuses",
                model=Status,
                weight=300,
                link="extras:status_list",
                description="Customizable object statuses",
                permissions=["extras.view_status"],
            ),
        ),
    ),
)
```

Note that we're extending the existing **Organization** panel in this case, and the `weight=300` is there to ensure that this item appears after the existing **Locations** and **Tenants** items in that panel (UI item weights in Nautobot are typically multiples of 100 by default).

Once again, stop and restart the Nautobot `invoke debug` task, and when it comes back up, you should be able to see your new subpanel on the Nautobot home page:

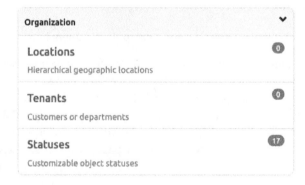

Figure 14.7 – A custom home page item, adding a "Statuses" counter and link

You can experiment with changing the preceding code to create a new panel rather than augmenting the existing one (hint: just change the `name` of the `HomePagePanel` you're using), or adjusting the weight parameter to reposition this item within the existing panel.

Here's a real-world example: the **Security** panel added to the home page by the Nautobot Firewall Models App to display information about that App's data, specifically the number of records of various Firewall Models data:

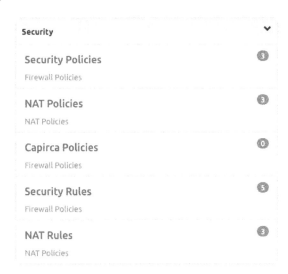

Figure 14.8 – Home page panel added by the Nautobot Firewall Models App

Through judicious use of this feature, your Nautobot deployment's home page can be fine-tuned to specifically reflect your particular Apps and use cases, distinct from any other Nautobot instance in existence. It's a pretty powerful feature!

Adding navigation menu items

Apps can optionally extend the existing Nautobot navigation bar layout by adding new top-level menus and/or adding menu items to existing menus. This allows you to add whatever navigational shortcuts your app wants to make available to users.

Nautobot looks inside each app for a `navigation.py` file that contains a `menu_items` list, which in turn contains instances of `NavMenuTab`, `NavMenuGroup`, `NavMenuItem`, and `NavMenuButton` objects. Much like the home page panel definitions, each nav menu tab, group, and item has a `name` and a `weight` that specifies how the menus are constructed.

Let's add a simple menu item. Create `nautobot_ip_acls/navigation.py` as follows:

```python
from nautobot.apps.ui import NavMenuGroup, NavMenuItem, NavMenuTab

menu_items = (
    NavMenuTab(
        name="IPAM",
        groups=(
            NavMenuGroup(
                name="IP ACLs",
                weight=250,
                items=(
                    NavMenuItem(
                        link="ipam:ipaddress_list",
                        name="Standard IP ACLs",
                        permissions=["ipam:view_ipaddress"],
                        buttons=(),
                    ),
                ),
            ),
        ),
    ),
)
```

We're cheating a bit here since we don't actually have an IP ACL model and its corresponding links and permissions yet (that will come in the next chapter), so for now our **Standard IP ACLs** menu will actually just link to the **IP Addresses** list view. But this still should demonstrate the idea once you restart the Nautobot server and refresh your browser:

Figure 14.9 – A custom navigation menu item

Just as with the home page panels, you can add to existing `NavMenuTabs` and `NavMenuGroups` by reusing their names in your `navigation.py`, or if you provide an entirely new name (as we did with our **IP ACLs** `NavMenuGroup`) you can create new menu groups or even new top-level menu tabs.

As a real-world example, here's the Nautobot core **Circuits** menu with some extra menu items injected by the Nautobot Circuit Maintenance App in order to work with circuit maintenance notifications sent by a service provider:

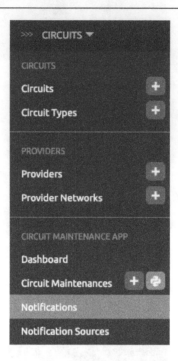

Figure 14.10 – Navigation menu extensions from the Nautobot Circuit Maintenance App

Just as with home page customization, customizing the navigation menus in Nautobot is key to the usability and discoverability of your App's features and functionality. Almost every App will want to extend the navigation menus in some way, except maybe for Apps that are purely focused on other areas (such as perhaps a ChatOps extension or a provider of custom Prometheus metrics).

Extending object detail views

Apps can inject custom content into certain areas of the "detail" views of any Nautobot data model. This can be done to show additional context, related information provided by your App, and so forth.

Nautobot looks in each installed App for a `template_content.py` file containing a `template_extensions` list consisting of `TemplateExtension` subclasses. Each `TemplateExtension` applies to a specific Nautobot data model's detail views, and can implement any of the following methods as desired:

- `left_page()` – Returns HTML content to display on the left half of the page

- `right_page()` – Returns HTML content to display on the right half of the page

- `full_width_page()` – Returns HTML content to display across the entire bottom of the page

- `buttons()` – Returns HTML content to display alongside the action buttons in the top-right corner of the page

- `detail_tabs()` – Returns a list of dicts describing additional "tabs" to add to the page.

Each of these methods receives no parameters directly when called, but has access to a `self.context` variable that contains information including the following:

- `object` – The object being viewed

- `request` – The current request

- `settings` – Global Django/Nautobot settings object

- `config` – App-specific configuration parameters

For more technical details, please refer to `https://docs.nautobot.com/projects/core/en/v2.1.7/development/apps/api/ui-extensions/object-detail-views/`.

Let's try a simple example, adding a panel to the Status detail view that just renders some basic HTML. Create a `nautobot_ip_acls/template_content.py` file as follows:

```python
from nautobot.apps.ui import TemplateExtension


class StatusContent(TemplateExtension):
    model = "extras.status"

    def right_page(self):
        return """
        <div class="panel panel-default">
            <div class="panel-heading">
                <strong>Status Information</strong>
            </div>
            <div class="panel-body">
                Hello!
            </div>
        </div>
        """


template_extensions = [StatusContent]  # Important to include!
```

Restart the Nautobot instance again, then navigate to the list view for statuses (under **Organization >
Statuses** in the navigation menu) and click on a row in the table to view the detail view. Your content
should appear here now:

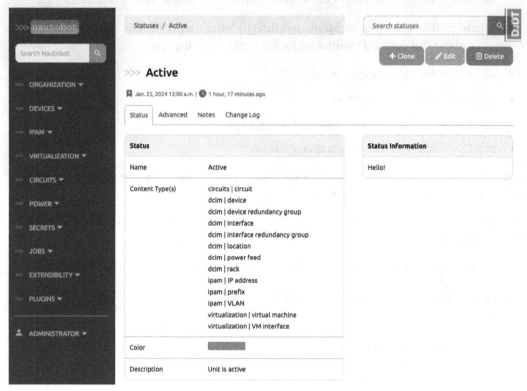

Figure 14.11 – Status detail view showing custom right_page content

A very common use case for Apps is to extend the Device detail view.

Here are a couple of screenshots from `https://demo.nautobot.com` showing extra tabs (**Data
Compliance** and **Configuration Compliance**) injected by the Nautobot Data Validation App and
Nautobot Golden Config App respectively, as well as right-hand content (**BGP Routing Instances**,
BGP Peerings, **Device Onboarding**, etc.) injected by the Nautobot BGP Models App, the Nautobot
Device Lifecycle App, and the Nautobot Device Onboarding App. Whew!

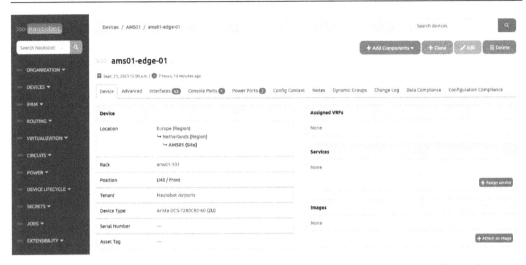

Figure 14.12 – The upper portion of the Device detail view with additional App-defined tabs

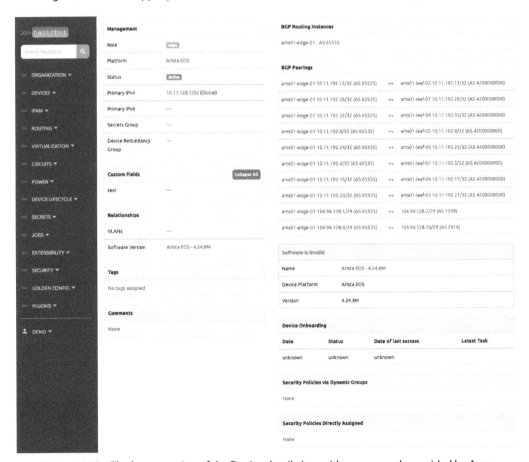

Figure 14.13 – The lower portion of the Device detail view with many panels provided by Apps

As you can see, this is a tremendous amount of additional information added to the Device detail view just by installing several Nautobot Apps. Depending on what Apps you use or write, other object detail views may similarly grow to show a wealth of information—perhaps in your use case you need to add to the Circuit detail view, or the Virtual Machine detail view, or any of the many other Nautobot object types.

Replacing views altogether

> **Tip**
> This is advanced functionality—use with caution!

You may outright replace any existing Nautobot views by providing an `override_views` dict in an App's `views.py` file. The keys of this dict are the Nautobot URL pattern names to replace, and the values are the new views that will take over the response to those URLs.

A simple example to override the device detail view at `/dcim/devices/<id>/`, which is registered in the `dcim` namespace and simply named `device`:

```python
# views.py
from django.shortcuts import HttpResponse
from django.views import generic


class DeviceViewOverride(generic.View):
    def get(self, request, *args, **kwargs):
        return HttpResponse(("Hello world! I'm a view which "
                             "overrides the device object detail "
view."))


override_views = {
    "dcim:device": DeviceViewOverride.as_view(),
}
```

A word of caution here—overriding an existing view is a very powerful feature, but you must keep in mind that doing so makes it *impossible* for users to access the original view. Pay close attention to what features or functionality the original view may provide, and make sure that you're not blocking users from doing something that they may have no other way to accomplish.

Extending core functionality

In addition to extending the UI, Apps can extend a number of Nautobot features through code, above and beyond the platform's built-in extensibility features such as custom fields and relationships.

Adding Jinja2 filters

An App can define custom Jinja2 filters that will be available when rendering templated content such as computed fields and custom links, as discussed in *Chapter 6*.

Nautobot looks in each installed app for a `jinja_filters.py` file. In this file, any function decorated with the `library.filter` decorator will be recognized as a Jinja2 filter.

Let's try this out. Create `nautobot_ip_acls/jinja_filters.py` as follows:

```
from django_jinja import library

@library.filter
def my_uppercase(text):
    return text.upper()
```

After you restart the Nautobot process, you can define a computed field (**Extensibility** > **Computed Fields**) as follows:

Figure 14.14 – Defining a computed field that uses your custom Jinja2 filter

Next, you can navigate to the detail view of any Status object, and you'll see this field is rendered using your filter:

Figure 14.15 – Computed field rendered with custom Jinja2 filter

Obviously, this specific example is rather contrived, but don't take that as meaning that this isn't a useful feature to be aware of. A half-dozen lines of Python code in a custom filter to solve a common text-construction challenge in your use cases can save you dozens of lines of repeated Jinja2 templating throughout Nautobot.

Implementing custom data validators

Apps can register custom data validation classes to enforce business rules for any Nautobot data model. For example, you might implement a Device validator that enforces your organization's naming conventions for devices, enforces that every Device must have an asset tag recorded in Nautobot, and so forth.

The Nautobot Data Validation Engine App builds on this framework to provide a low-code way for users to define certain types of basic data validation rules, but you will probably find it more useful to understand the underlying data validation framework and be able to implement more complex rules via code.

Nautobot looks in each installed App for a `custom_validators.py` file that defines a `custom_validators` list of `CustomValidator` subclasses, each of which defines a rule or rules for a specific Nautobot data model. Each `CustomValidator` subclass should implement a `clean()` method that receives no parameters when called, but can access the object to validate via `self.context["object"]`. If the `clean()` method raises a `ValidationError` (the helper API `self.validation_error()` can be used here), the contents of this error will be displayed in the UI or returned via the REST API, providing the user with relevant guidance as to the problem and its solution.

Let's try it. Create `nautobot_ip_acls/custom_validators.py` as follows:

```
from nautobot.apps.models import CustomValidator

class RoleLowerCaseNameValidator(CustomValidator):
    """Company policy is that Role names must be lower-case."""

    model = "extras.role"

    def clean(self):
        role = self.context["object"]
        if role.name.lower() != role.name:
            self.validation_error({"name": "Names must be lower case
only."})

custom_validators = [RoleLowerCaseNameValidator]  # Important!
```

Restart the Nautobot process, and navigate to **Organization** > **Roles**. Try to create or edit a Role whose name contains capital letters, which is perfectly valid in baseline Nautobot, but note that your new custom validator now prevents you from saving your changes in this case:

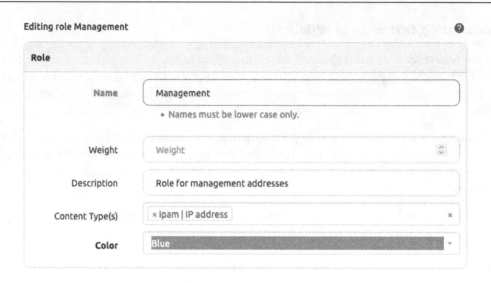

Figure 14.16 – Custom validator in action

As a real-world example, the Nautobot Circuit Maintenance App implements a custom validator for the circuit Provider model, which is used to enforce global uniqueness of entries in the emails_ circuit_maintenances custom field, since it would be incorrect for the given email address, from which circuit maintenance notifications are received, to correspond to more than one distinct Provider.

Figure 14.17 – Custom validator in the Nautobot Circuit Maintenance App

Custom validators are a tremendously powerful way to customize Nautobot for your specific business needs and processes. Anywhere you have problems with inconsistently formatted text or missing or incomplete data can potentially have corrections enforced through the appropriate definition of a custom validator in an App.

Implementing secrets providers

An App can define and register additional providers (sources) for secret data, allowing Nautobot to retrieve secret values from additional systems or data sources. The Nautobot Secrets Providers App provides reference implementations for many secrets providers, but perhaps your organization has a bespoke secrets system that requires Nautobot integration.

Nautobot looks in each installed App for a `secrets.py` file containing a `secrets_providers` list of `SecretsProvider` subclasses. To define a new `SecretsProvider` subclass, you must specify the following:

- A unique `slug` string identifying this provider
- A human-readable `name` string (optional; the `slug` will be used if this is not specified)
- A Django form for entering the parameters required by this provider, as an inner class named `ParametersForm`
- An implementation of the `get_value_for_secret()` API to actually retrieve the value of a given secret

Most secrets provider APIs are complex enough that a full example here would be overly long to include, but as a simple (**highly insecure**!) example, you could define a "constant-value" provider that simply stores a constant value in Nautobot itself and returns this value on demand.

> **Warning**
> This is an intentionally simplistic example and should not be used in practice! Sensitive secret data should never be stored directly in Nautobot's database itself.

You can create `nautobot_ip_acls/secrets.py` as follows:

```
from django import forms
from nautobot.apps.forms import BootstrapMixin
from nautobot.apps.secrets import SecretsProvider

class ConstantValueSecretsProvider(SecretsProvider):
    """
    Example - this just returns a user-specified constant value.

    Obviously this is insecure and not something you'd want to
actually use!
    """
```

```
        slug = "constant-value"
        name = "Constant Value"

        class ParametersForm(BootstrapMixin, forms.Form):
            """
            User-friendly form for specifying required parameters.
            """
            constant = forms.CharField(
                required=True,
                help_text="Constant secret value. DO NOT USE FOR REAL
DATA"
            )

        @classmethod
        def get_value_for_secret(cls, secret, obj=None, **kwargs):
            """
            Return the value defined in the Secret.parameters "constant"
key.

            A more realistic example would make calls to external APIs,
etc.,
            to retrieve a secret from another system as desired.

            Args:
                secret (Secret): The secret whose value should be
retrieved.
                obj (object): object (Django model or similar) providing
context
                    for the secret's parameters.
            """
            return secret.rendered_parameters(obj=obj).get("constant")

secrets_providers = [ConstantValueSecretsProvider]
```

After restarting Nautobot, you should now be able to navigate to **Secrets** > **Secrets** and create a new secret, at which point "constant-value" should now be available as a new secrets provider to use, and when selected, Nautobot will present you with the **Parameters** form to enter the constant value to use for this secret:

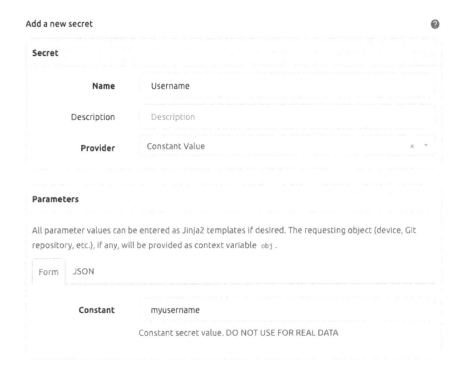

Figure 14.18 – Your custom SecretsProvider in action

> **Tip**
> Be sure to remove this code after you're done experimenting with it, as again, this specific
> `SecretsProvider` implementation is highly insecure.

Extending object filtering

Apps can extend any model-based `FilterSet` and `FilterForm` classes that are provided by
the Nautobot core, allowing for new data filters to be used in the UI and REST API to facilitate your
organization-specific use cases for viewing and manipulating Nautobot data.

Nautobot looks in each installed App for a `filter_extensions.py` file containing a `filter_`
`extensions` list of `FilterExtension` subclasses, each of which corresponds to a specific Nautobot
data model's filtering logic. Each `FilterExtension` can define a `filterset_fields` dict
and/or a `filterform_fields` dict, corresponding to the `FilterSet` and/or `FilterForm`.

This one's a bit more complex to understand than some of the other extension points, at least until you have an understanding of QuerySets and FilterSets (which you'll get some hands-on experience within the next chapter), so we won't provide a hands-on example here. But as a real-world example, the Nautobot Floor Plans App defines a `FilterExtension` for the Rack data model to allow filtering the Rack list to select only those Racks that are associated with a specific Floor Plan record via the Location model. This clearly wouldn't make sense to build into the Nautobot core since Floor Plans are specific to this App, but by using the `FilterExtension` API this extension exists right alongside the core Rack filters:

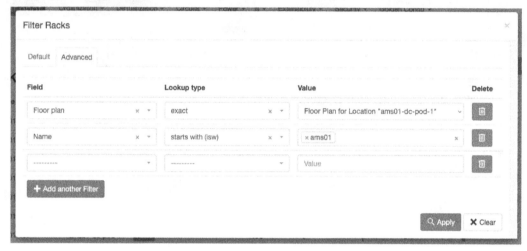

Figure 14.19 – Filter extension provided by the Nautobot Floor Plans App

Filter extensions are a useful feature to be aware of, because although they aren't as often used as many other App features, when they're the right tool for the job, they can solve challenges that are all but impossible to address in any other way.

Loading additional data types from a Git repository

An App can register additional types of data that a Git repository can be configured to provide to Nautobot. When an appropriately configured Git repository is refreshed in Nautobot, the App will automatically be called to process the data discoverable in this repository.

Nautobot looks in each installed App for a `datasources.py` file containing a `datasource_contents` list of `DatasourceContent` records describing a data type and the callback function to invoke when a Git repository is refreshed containing that data type. The callback function operates much like part of a Nautobot Job, in that it has access to both the Git repository record and an associated JobResult, which it can use to log information about its progress and outcomes.

This too is a bit complex to cover as a hands-on example here, but as a real-world example, the Nautobot Golden Config App makes heavy use of this feature for storing and retrieving device configuration backups and similar data in Git. As you can see in the following screenshot, these additional data types appear in the Nautobot UI directly alongside the four Git data types that are currently built into Nautobot itself:

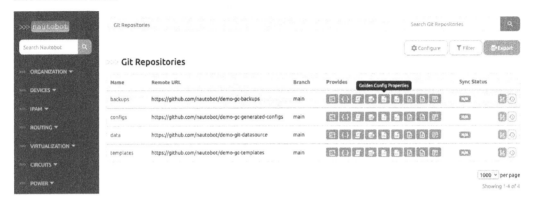

Figure 14.20 – Additional Git repository data types added by the Nautobot Golden Config App

Through implementation of this feature, you can enable various GitOps-style workflows that integrate seamlessly with Nautobot.

Publishing jobs

In *Chapter 11*, you learned about jobs; in addition to storing jobs in a Git repository or a local filesystem, jobs can belong to a Nautobot App.

Nautobot will look in each installed App for a `jobs.py` file that when imported calls the `register_jobs()` API function, much like an equivalent file in a Git repository. There's not much difference in terms of the actual authoring and running of such jobs, but there is at least one thing that an App-packaged job can do that other jobs generally can't. Because Apps can contain Django HTML templates, an App-provided job can define and make use of a custom job submission form, potentially with advanced features such as custom JavaScript logic to show and hide various form fields as the user inputs their data, and so forth.

To do this, you would author the Django template like any other in the app (in `nautobot_ip_acls/templates/nautobot_ip_acls/mycustomjobform.html` or equivalent), then in the `Job`. `Meta` class, declare a `template_name` string referencing this template (`nautobot_ip_acls/mycustomjobform.html`) and it will be used when presenting the job to the user. I'm not presently aware of an open source Nautobot App that makes use of this feature to point you to as an example, but I do know that it's been used successfully in any number of private Apps.

Adding entirely new functionality

Because a Nautobot App is also a Django App, your App can take full advantage of all of the features of Django's own APIs and capabilities, though in many cases Nautobot also provides specific extensions, derivations, or customizations of the basic Django feature set. This section won't attempt to reproduce Django's own documentation, but will call out a few Nautobot-specific idioms to be aware of. If you're not already at least somewhat familiar with Django, don't worry if you don't fully understand the nuances of this section; this is intended more as a best-practices quick reference than anything else. These features will also be explored more fully in a hands-on way in *Chapter 15*.

Database model base classes

Instead of Django's basic `django.db.models.Model` base class, you'll almost certainly want to use one of Nautobot's customized and enhanced subclasses as the basis of your new data model classes.

The main classes to be aware of are `BaseModel`, `OrganizationalModel`, and `PrimaryModel`, all of which are available for apps to consume as imports from the `nautobot.apps.models` module. The inherent capabilities provided by inheriting from these various parent models differ as follows, and are compared against Django's base Model class for context:

Feature	Model	BaseModel	OrganizationalModel	PrimaryModel
UUID primary key	✗	✓	✓	✓
Object permissions	✗	✓	✓	✓
validated_save()	✗	✓	✓	✓
Change logging	✗	✗	✓	✓
Custom fields	✗	✗	✓	✓
Relationships	✗	✗	✓	✓
Notes	✗	✗	✓	✓
Tags	✗	✗	✗	✓

Table 14.1 – Data model and Django base model comparison table

As a general rule of thumb, most new data models should use `PrimaryModel`, with the exception of models that are intentionally bare-bones and do not need to support the full feature set, which should use `BaseModel`. Examples in Nautobot itself of `BaseModel` classes are `CablePath`, `ObjectChange`, and `JobLogEntry`.

The extras_features model decorator

A number of Nautobot's extensibility features must be "opted into" when defining a new data model in an App. Some are done by inheriting from an appropriate base class or mixin, but others are flagged through the @extras_features decorator applied to a model. For example, for a model to support the implementation of custom validators against it, and also be included in Nautobot's GraphQL API schema, you would need to mark it with @extras_features("custom_validators", "graphql"). The full list of possible extras features are defined in nautobot.extras. constants.EXTRAS_FEATURES; at the time of this writing they are as follows:

```
EXTRAS_FEATURES = [
    "cable_terminations",
    "config_context_owners",
    "custom_links",
    "custom_validators",
    "dynamic_groups",
    "export_template_owners",
    "export_templates",
    "graphql",
    "job_results",
    "locations",
    "statuses",
    "webhooks",
]
```

Implementing views with NautobotUIViewSet

While you can (and in some cases may need to) implement individual Django views for special purposes, in many cases you'll find yourself wanting to create a standard suite of views for each data model you define—"list", "detail", "create/edit", "delete", "bulk import", "bulk edit", and "bulk delete" are the common cases. To make this easier, Nautobot provides a NautobotUIViewSet class that you can use to greatly simplify the implementation of this suite of model views. *Chapter 15* will provide worked examples of how to define a NautobotUIViewSet subclass and its prerequisites for custom data models.

Extending view templates

The standard set of object views in Nautobot also corresponds to a standard set of generic templates provided by Nautobot. In many cases, you won't need to write custom templates at all, but Django's template system also makes it very straightforward to "subclass" a generic template to customize it to your needs.

Template naming is very intuitive in `NautobotUIViewSet` for cases where you do need to write a custom template. In your `<app_name>/templates/<app_name>` folder, name your templates following the convention `{model_name}_{action}.html`, where the action can be one of the following:

- `list`
- `retrieve`
- `create`
- `update`
- `destroy`
- `bulk_create`
- `bulk_update`
- `bulk_destroy`

We'll show worked examples of such templates in *Chapter 15*.

Since in many cases the `create` and `update` templates for a model will be identical, you are not required to create both. If you provide an `<app_name>/<model_name>_create.html` file but not an `<app_name>/<model_name>_update.html` file, then when you update an object, Nautobot will fall back to discovering and using the `<app_name>/<model_name>_create.html` template, and vice versa.

Adding REST API endpoints

Apps can declare custom endpoints on Nautobot's REST API to retrieve or manipulate models or other data. These behave very similarly to views, except that instead of rendering arbitrary HTML content using a template, data is returned in JSON format using a serializer.

Nautobot uses the Django REST Framework (`https://www.django-rest-framework.org/`), which makes writing API serializers and views very simple. As usual, though, Nautobot extends and subclasses Django REST framework's basic features to customize them for Nautobot's specific needs. Thus, for example, you'll want to inherit from `NautobotModelSerializer` instead of `ModelSerializer`, `NautobotModelViewSet` instead of `ModelViewSet`, and so forth.

Again, *Chapter 15* will provide several practical examples as we revisit this topic in more detail.

Summary

In the first half of this chapter, we walked you through using the Nautobot App Cookiecutter to set up the skeleton of a new App called `nautobot_ip_acls`. In the second half, we gave a broad but necessarily brief overview of the myriad features available for an App to implement or extend, with quick hands-on examples of many of these features.

In the next chapter, we'll bring these two concepts together more fully as we walk through the implementation of a Nautobot IP ACLs App that starts with your App skeleton and implements a fully functional App.

15

Building Nautobot Data Models

As you learned from reading all prior chapters, Nautobot provides a comprehensive set of data models as a baseline but, inevitably, there will be some use cases for your organization that are not fully covered by the built-in data models. This chapter covers two approaches to solving this problem by developing a custom Nautobot App using the developer API that was introduced in *Chapter 14*.

The first approach involves using Nautobot's built-in extensibility features in combination with developing an App that provides a custom UI and REST API built atop those features. The second approach takes things a step further by developing an App that provides entirely new models (database tables) as well.

In this chapter, we will cover:

- A real-world use case for custom Apps
- Data model design
- Building an App around existing data models
- Building an App with custom data models
- Exercises or next steps

This chapter will be easier to follow if you have previous experience with Django and Python, especially if it becomes necessary to troubleshoot any errors introduced by typos and the like, but even without that experience, it should be possible to follow along and perform the exercises with careful reading.

A real-world use case for custom Apps

The example use case that we'll use for this chapter (as well as the following chapter) is an organization that wants to have a source of truth for their standard IP ACLs that are configured on various devices in their network. This is a data type that Nautobot (currently, at least) does not provide built-in models for, but it's also a fairly simple and straightforward data type to model. By the end of this chapter, you'll understand how to implement a custom Nautobot App to solve this requirement in two different ways and should be able to envision how this App could be extended to support more complex future requirements.

Data model design

Before you begin implementing a Nautobot App, you should always spend some time making sure you understand the structure of the data that the App will require and use, and that's where we'll start.

Gathering representative data and requirements

In a typical "brownfield" environment where you have an existing network that you want to represent and document, you can logically start from the actual configuration of actual devices, whether this is represented as configuration CLI, JSON, NETCONF XML, or some other format. Conversely, if you're starting a new "greenfield" network, you should have architectural or other design documents spelling out how the newly deployed devices will be configured, and you can use those documents as a starting point.

In either case, you should review the available information to identify what configuration parameters (or other data) are relevant to your organization's needs, as well as how those parameters will be common or different across various devices or deployments. The goal here is not to comprehensively model all possible configuration knobs of all possible features, but instead, to develop the simplest possible data model that covers the spectrum of likely configuration within your organization.

Data for this exercise

For this example of IP ACLs, let's say you can gather ACL configuration from a representative sample of devices across our existing network.

A Cisco device might have a configuration like this:

```
!
ip access-list standard 1
 10 permit ip 10.0.0.0 0.255.255.255
 20 deny ip any
!
interface GigabitEthernet1
  ip access-group 1 in
```

An Arista device might have a configuration as follows:

```
ip access-list standard numberone
 10 deny 10.1.1.1/32
 20 permit 10.1.0.0/16
 30 deny any
interface ethernet 5
 ip access-group numberone out
```

Understanding the data requirements

While the syntax differs between platforms, the *data* being modeled is consistent:

- On each *device*, there are zero or more *access lists*, each with an *identifier* (a number on Cisco, and a string on Arista) that is unique within that device

- For each *access list* on a device, there are a series of *entries*, each with a unique *sequence number*, an *action* (permit or deny), and a *network* (which may be an entire subnet or a single host, or it may be any)

- For each *interface* on a device, an access list can optionally be applied *inbound* and/or *outbound*

We happen to know that the `any` string in both configurations is just shorthand for the network `0.0.0.0/0`, so in the data model, you could choose either of these two representations, depending on which makes the model simpler to populate and consume.

In YAML form, data for a single device might look something like this:

```
access_lists:
  "1":
    - action: permit
      network: 10.0.0.0/8
      seq: 10
    - action: deny
      network: any
      seq: 20
  "2":
    - action: deny
      network: 10.1.1.1/32
      seq: 10
    - action: permit
      network: 10.1.0.0/16
      seq: 20
    - action: deny
      network: any
      seq: 30
interfaces:
  "GigabitEthernet1":
    acl_in: 1
  "GigabitEthernet2":
    acl_out: 2
  "GigabitEthernet3":
    acl_in: 1
    acl_out: 2
```

A note on size and scoping

For now, let's keep the scope of this project very small and limited—only "standard" IP ACLs are being modeled, not the more complex and more powerful "extended" ACLs that most platforms support. Better to start with something small but immediately achievable and useful than to try to take on the entire problem space at once!

Considering composability, reusability, and deduplication of data

When developing a Nautobot App, remember that your data model does not have to start from nothing. Nautobot already models many network components, and by this point in the book, your Nautobot deployment should already have some data populated for many of these models. Furthermore, the modularity and relational aspect of Nautobot's data models are one of its key strengths to take advantage of. When developing any new data model, you should always keep this in mind and think about how your model can be *combined* with the existing data models, how these existing models can be *reused* to reduce the amount of new data you have to manage, and how your data model can *deduplicate* information that recurs in many places across your network.

In the case of our IP ACL App, we know that Nautobot already has data and data models for `Device`, `Interface`, `Prefix`, and `IPAddress`, so we don't need to reinvent data models for any of those. Our data models for ACLs should reference these existing models rather than duplicate this information within the ACL data. We know that while individual devices may have specific access list requirements, in many cases, there will be common ACLs that get reused across many distinct devices that share a common role within the network, so if possible, we should model the ACLs in a way that supports and reflects this reuse.

Considering built-in Nautobot extensibility features

Before developing an entirely new data model, you should also consider whether any of Nautobot's built-in extensibility features could provide a simpler and sufficient solution to your data modeling needs. Let's take a look at tags, custom fields, computed fields, relationships, and config contexts. Remember, these were all initially covered in *Chapter 6*. Make sure to reference that chapter if needed:

- *Tags* – These are probably not suitable to represent either ACL definitions or the association of ACLs to a device or an Interface since they're globally defined and we'd have to potentially have a different tag for each ACL.

- *Custom fields* – We could potentially use a pair of custom fields on the Interface to represent the name of the inbound and/or outbound ACLs assigned to that Interface.

- *Computed fields* – These generally rely on remixing data that's already stored in Nautobot in one form or another, and we don't currently have anywhere available to store the ACL definition, so these probably won't work here.

- *Relationships* – These provide a way to define new interconnections between existing models. But a device and its Interfaces are already intrinsically linked, and we don't have an existing model for ACLs to link to either of those.

- *Config contexts* – Now, this has potential for our project. Config contexts are well-suited for deduplication of data since a single config context can define data that applies to a large group of related devices. Additionally, since the config context can contain arbitrary structured data (typically represented as YAML or JSON), optionally enforced by a schema (using JSON Schema), we certainly could store a dictionary of ACL definitions as part of a config context. The main drawback here would be that config contexts can't easily reference other data model objects (such as Interfaces or prefixes) except as strings within the JSON data, so we would miss out on some data-integrity assurances (such as guaranteeing that the interface that an ACL is assigned to actually exists). But it's worth considering.

Config contexts, custom fields, and relationships are all incredibly powerful and flexible ways to extend Nautobot's functionality without writing a single line of Python code, but they have limitations as well. Most notably, none of them provide a straightforward way to define an entirely new data model that exists independently of the built-in data models. If there's a self-contained "thing" that you need to represent, then you'll probably need a new data model.

When the data model suggests you should build an App

Depending on the outcome of your data modeling exercise, you may or may not end up actually needing to build an App. It may turn out that the existing Nautobot models and extensibility features are perfectly adequate to your needs. On the other hand, it's also quite possible that either the existing extensibility features are suitable for your *data modeling* needs but not for your user interface (including the REST API) requirements, or you may have determined that you do in fact need a new set of data models. We'll address both of these possibilities in the following sections of this chapter.

Building an App around existing data models

For this section, let's say that you've decided that, for now, modeling your ACL configuration via config contexts is the way to go and, therefore, you don't need to develop a new set of data models at this time. However, you find the existing generic UI and API presented by config contexts to be somewhat lacking and want some ACL-specific presentation of the data. This is a case where it makes sense to develop your own Nautobot App to extend the UI and REST API of Nautobot. We'll call this app `nautobot_ip_acls`.

This chapter does not cover in detail all of the prerequisite steps of setting up a Nautobot App development environment and creating a minimal template of your App—for that information, please refer back to *Chapter 14*—but to recap, here are the basic steps on a freshly initialized DigitalOcean Droplet:

```
root@nautobot-app-dev:~# sudo apt install curl software-properties-
common
```

```
root@nautobot-app-dev:~# curl -fsSL https://download.docker.com/linux/
ubuntu/gpg | sudo gpg --dearmor -o /usr/share/keyrings/docker-archive-
keyring.gpg
root@nautobot-app-dev:~# echo "deb [arch=$(dpkg --print-architecture)
signed-by=/usr/share/keyrings/docker-archive-keyring.gpg] https://
download.docker.com/linux/ubuntu $(lsb_release -cs) stable" | sudo tee
/etc/apt/sources.list.d/docker.list > /dev/null
root@nautobot-app-dev:~# apt update
root@nautobot-app-dev:~# apt install docker-ce
root@nautobot-app-dev:~# apt install -y python3-pip
root@nautobot-app-dev:~# pip install cookiecutter
root@nautobot-app-dev:~# curl -sSL https://install.python-poetry.org |
python3 -
root@nautobot-app-dev:~# export PATH="/root/.local/bin:$PATH"
root@nautobot-app-dev:~# cookiecutter https://github.com/nautobot/
cookiecutter-nautobot-app
  --directory nautobot-app
  [1/18] codeowner_github_usernames (): glennmatthews
  [2/18] full_name (Network to Code, LLC):
  [3/18] email (info@networktocode.com):
  [4/18] github_org (nautobot):
  [5/18] app_name (my_app): nautobot_ip_acls
  [6/18] verbose_name (Nautobot Ip Acls): Nautobot IP ACLs
  [7/18] app_slug (nautobot-ip-acls):
  [8/18] project_slug (nautobot-app-nautobot-ip-acls): nautobot-app-
ip-acls
  [9/18] repo_url (https://github.com/nautobot/nautobot-app-ip-acls):
  [10/18] base_url (nautobot-ip-acls): ip-acls
  [11/18] min_nautobot_version (2.0.0): 2.1.0
  [12/18] max_nautobot_version (2.9999):
  [13/18] camel_name (NautobotIpAcls): NautobotIPACLs
  [14/18] project_short_description (Nautobot IP ACLs):
  [15/18] model_class_name (None):
  [16/18] Select open_source_license
    1 - Apache-2.0
    2 - Not open source
    Choose from [1/2] (1): 1
  [17/18] docs_base_url (https://docs.nautobot.com):
  [18/18] docs_app_url (https://docs.nautobot.com/projects/nautobot-
ip-acls/en/latest):

Congratulations! Your cookie has now been baked. It is located at /
root/nautobot-app-ip-acls.

root@nautobot-app-dev:~# cd nautobot-app-ip-acls
root@nautobot-app-dev:~/nautobot-app-ip-acls# poetry lock
root@nautobot-app-dev:~/nautobot-app-ip-acls# cp development/creds.
example.env development/creds.env
```

```
root@nautobot-app-dev:~/nautobot-app-ip-acls# vi development/creds.env
root@nautobot-app-dev:~/nautobot-app-ip-acls# invoke build
root@nautobot-app-dev:~/nautobot-app-ip-acls# invoke debug
```

After completing all of the preceding steps, you should once again have a running Nautobot system with a skeleton of a Nautobot App installed.

Data model based on extensibility features

Let's say you've decided to model the ACL definitions for a set of devices into a config context using the following data structure:

```
{
    "access_lists": {
        "1": [
            {"action": "permit", "network": "10.0.0.0/8", "seq": 10},
            {"action": "deny", "network": "any", "seq": 20}
        ],
        "2": [
            {"action": "deny", "network": "10.1.1.1/32", "seq": 10},
            {"action": "permit", "network": "10.1.0.0/16", "seq": 20},
            {"action": "deny", "network": "any", "seq": 30}
        ]
    }
}
```

> **Note**
>
> In the Nautobot UI, you can view config context data as either JSON or YAML, but inputting the value via the UI can currently only be done in JSON.

Let's populate some data into our Nautobot App development environment to include a config context following this structure. This should be straightforward to do if you've worked through *Chapters 4, 5, and 6*, and in fact, you could certainly reproduce the same data in the development environment as you entered into your Nautobot deployment in those chapters. We'd recommend not using a "production" device in this chapter and the next one though! Either define a lab or development device or (for now) create an entirely fictitious one instead:

- Create a location type and specify that it can be used with devices
- Create a location of this location type
- Create a manufacturer
- Create a device type using this manufacturer
- Create a role and specify that it can be used with devices

- Create a device using this device type, role, and location
- Create a config context, specify that it applies to all devices in your location, and populate its data with the data structure described previously

All of the preceding should be straightforward for you to accomplish if you've made it this far in this book. When you're done, you should be able to see the config context as shown at the bottom of the following figure:

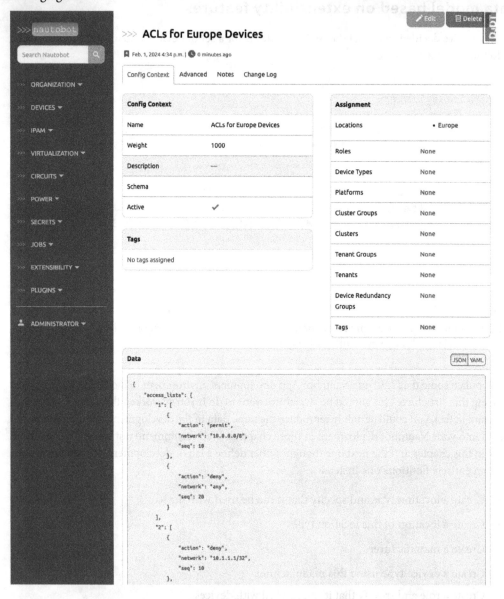

Figure 15.1 – Config context defining ACL data

Let's also say that you've decided to model the ACL-to-Interface assignment as a pair of custom fields on the Interface model, `acl_in` and `acl_out`, each optionally storing the name/identifier of the ACL that applies.

Create these two custom field definitions and make sure they are specified to apply to the `dcim |` `interface` content type.

When you're done, these custom fields should be present in the UI accordingly:

Figure 15.2 – Custom fields defining ACL association to Interfaces

> **Note**
>
> It's a good practice to use keys in the config context data (and keys for custom fields) that consist only of letters, numbers, and underscores, as it makes it easier to access this data in various programming contexts.

Go ahead and edit your device to add a few Interfaces to it (hint: use the **Add Components** button from the **Device** detail view), including populating the custom fields with references to ACLs **1** and **2** as described in your config context:

csr1

Feb. 1, 2024 4:31 p.m. | 12 minutes ago

Device   Advanced   Interfaces 3   Status   LLDP Neighbors   Configuration   Config Context   Notes   Dynamic Groups   Change Log

Interfaces   Filter   Configure

Name	Status	Label	Enabled	Type	Mode	Description	IP Addresses	ACL In	ACL Out	
GigabitEthernet1	Active	—	✓	Virtual	—	—		1	2	
GigabitEthernet2	Active	—	✓	Virtual	—	—		2	—	
GigabitEthernet3	Active	—	✓	Virtual	—	—		—	1	

Rename   Edit   Disconnect   Delete   + Add Interfaces

Figure 15.3 – Interfaces added to the device with populated ACL custom fields

Adding an ACL overview to the Device detail view

In this section, you'll use the `TemplateExtension` feature of the App API, as introduced in *Chapter 14*, to add some summary information about ACLs to the detail view of any given device.

Defining a template extension

Create `nautobot_ip_acls/template_content.py` and populate it as follows:

```python
from nautobot.apps.ui import TemplateExtension

class IPACLDeviceTemplateExtension(TemplateExtension):
    """Add IP ACL information to the Nautobot Device detail view."""
    model = "dcim.device"

    def right_page(self):
        """Add content on the right side of the view."""
        return """<strong>Hello from the IP ACL App!</strong>"""

template_extensions = [IPACLDeviceTemplateExtension]
```

At this point, your file structure should look like this:

```
root@nautobot-app-dev:~/nautobot-app-ip-acls# tree -I __pycache__
nautobot_ip_acls/
nautobot_ip_acls/
├── __init__.py
├── template_content.py   # underline added to highlight newly created
file
└── tests
    ├── __init__.py
    ├── test_api.py
    └── test_basic.py
```

As we covered in *Chapter 14*, the Nautobot development server that we're running will automatically restart when it detects changes to a file that it already knows about, but it isn't smart enough to automatically detect entirely new files. Stop the running `invoke  debug` session with *Ctrl+C* and then rerun it to start the server again so that it can detect the new file. Once the server is up and running again, you can then navigate to your **Device** detail view and you should see **Hello from the IP ACL App!** in bold on the right side:

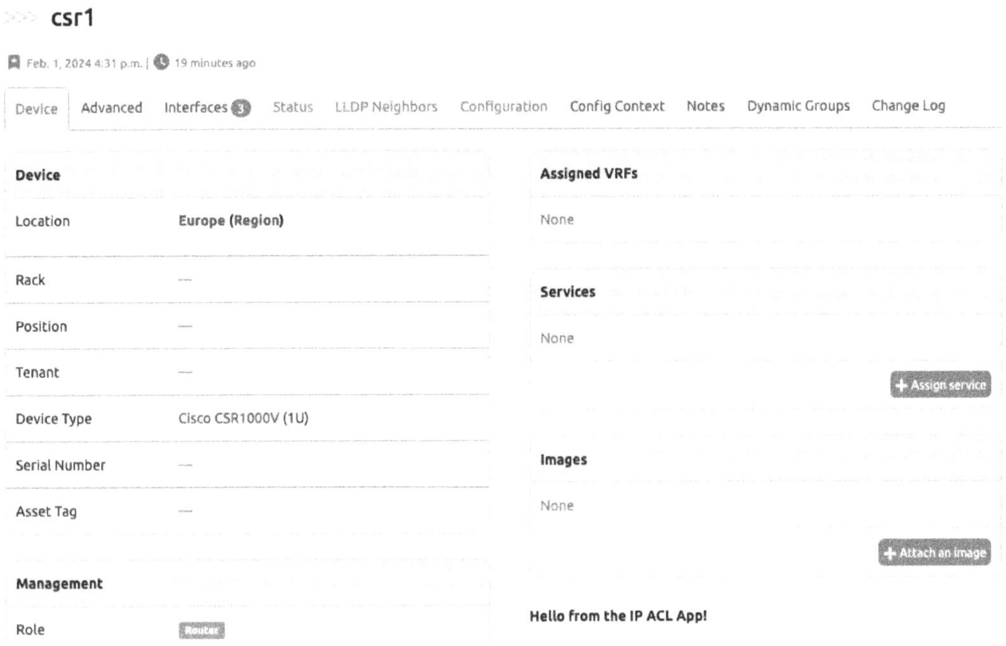

>>> **csr1**

🔖 Feb. 1, 2024 4:31 p.m. | 🕐 19 minutes ago

| Device | Advanced | Interfaces ❸ | Status | LLDP Neighbors | Configuration | Config Context | Notes | Dynamic Groups | Change Log |

Device

Location	**Europe (Region)**
Rack	—
Position	—
Tenant	—
Device Type	Cisco CSR1000V (1U)
Serial Number	—
Asset Tag	—

Management

| Role | Router |

Assigned VRFs

None

Services

None

➕ Assign service

Images

None

➕ Attach an image

Hello from the IP ACL App!

Figure 15.4 – Hello from the IP ACL App! is added to the Device detail view

There's a bit of magic involved here that we introduced briefly in the previous chapter, but it's worth revisiting in more detail now.

When Nautobot starts up, it looks in every installed App for a Python submodule called `template_content`, and in that module, it looks for a list called `template_extensions`. It then associates each `TemplateExtension` in that list with the model that it declares (here, `dcim.device` maps to the `device` model in the `Nautobot dcim` app). When the detail view for that model is rendered, Nautobot iterates over each `TemplateExtension` that applies to that model and calls its `left_page`, `right_page`, and `full_width_page` methods (if any; see also the overview of these methods in *Chapter 14*) and inserts the HTML that they return into the rendered page template.

Populating the template extension with useful information

Now, let's enhance the template extension so that it actually renders something more useful. Edit it as follows:

```
def right_page(self):
    """Add content on the right side of the view."""
    # Get the object that was provided as template context;
    # in this case, the Device object itself.
    device = self.context["object"]
```

```python
# Render the config context of this Device as a Python dict:
config_context_data = device.get_config_context()
# Get the contents of the "access_lists" key in the dict:
acls = config_context_data.get("access_lists", {})
# Construct the HTML to contain this data
# Start a panel containing an unordered list:
output = """
    <div class="panel panel-default">
    <div class="panel-heading"><strong>IP ACLs</strong></div>
    <div class="panel-body">
    <ul>
"""
# Add list entries based on the available data:
for acl_id, acl_entries in acls.items():
    output += (
        f"<li>ACL <code>{acl_id}</code> "
        f"with {len(acl_entries)} entries</li>"
    )
# End the list and the panel
output += "</ul></div></div>"
return output
```

Let's walk through the preceding code. First, you get the object (in this case, a device) that was provided as `context` to the template extension, and call its `get_config_context` method to render the overall config context into a Python dictionary. You retrieve the contents of the `access_lists` key within this dictionary into an `acls` variable and start constructing an HTML panel (styled via CSS classes to match similar panels already on the page). Next, you iterate over the items in `acls` to populate a list within the panel, and finally, you close the HTML panel and return the constructed HTML fragment.

Now, you're rendering a panel that embeds information about the ACLs defined in the aggregate config context for this device, but in a much more concise form than the config context data itself. Since this was an update to an existing file in your App, the Nautobot development server will have automatically restarted when it detected your changes, so you can simply refresh your web browser and should now see your new template content (assuming, of course, that this device has any ACLs in its config context).

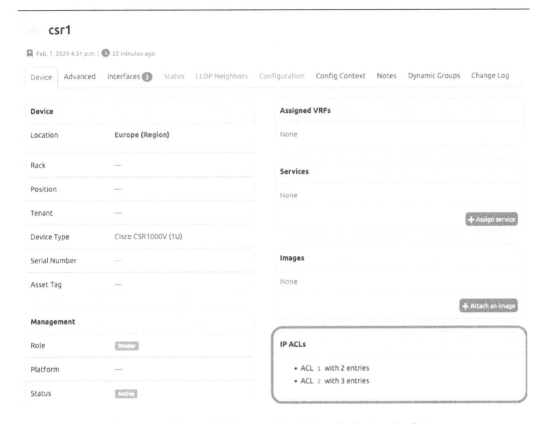

Figure 15.5 – Custom IP ACLs panel is added to the Device detail view

You could, of course, extend this function to include more information, such as a **No ACLs present** text if acls is empty, and so forth. You also don't have to build raw HTML directly inside the function but can instead use Django's HTML templating capabilities to render something more complex, if desired. But what we have here is useful enough for the moment and serves to demonstrate this feature.

Adding ACL details as a Device tab

Next, let's say you want to render the full ACL definitions for a device into the Nautobot UI, but you may be concerned that this would take up too much space on the main **Device** detail view. Fortunately, Apps can also add entire new dedicated tabs to model detail views, so let's do that next, adding an **IP ACLs** tab to the **Device** detail view. Here's what we want it to look like:

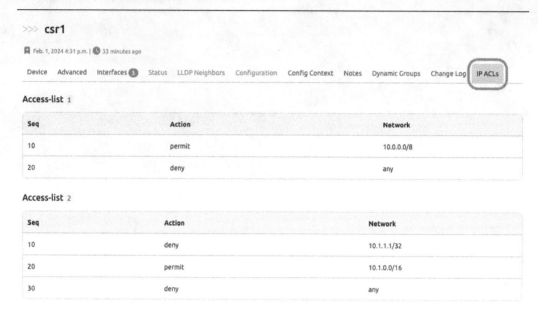

Figure 15.6 – Target appearance for a custom IP ACLs tab in the Device detail view

This may look dauntingly complex to implement but it actually isn't all that complicated. Let's proceed.

Adding a Detail tab to the template extension

Extend your IPACLDeviceTemplateExtension class to include a detail_tabs function as well:

```python
from django.urls import reverse

from nautobot.apps.ui import TemplateExtension

class IPACLDeviceTemplateExtension(TemplateExtension):
    model = "dcim.device"

    def detail_tabs(self):
        return [
            {
                "title": "IP ACLs",
                "url": reverse(
                    "plugins:nautobot_ip_acls:device_ip_acls",
                    kwargs={"pk": self.context["object"].pk}
                ),
            },
```

```
    ]

    def right_page(self):
        # unchanged from before, hence omitted here for brevity
```

This is pointing to a separate URL for a new Django view that we're going to write. The URL is in the `plugins:nautobot_ip_acls` namespace (in other words, it's a URL defined by our App), and the specific URL pattern to look up is the one named `device_ip_acls`, which takes a single `pk` argument, corresponding to the primary key (ID) of the device being viewed.

Implementing the view for the Detail tab

Next, we need to write the Django view that will be displayed in this new ,tab and define the URL that will be used by the user to access this tab. Create `nautobot_ip_acls/views.py`:

```
from nautobot.apps import views
from nautobot.dcim.models import Device


class DeviceIPACLsView(views.ObjectView):
    queryset = Device.objects.all()
    template_name = "nautobot_ip_acls/device_ip_acls.html"

    def get_extra_context(self, request, instance):
        return {"config_context": instance.get_config_context()}
```

Here, we're using Nautobot's built-in `views.ObjectView`, for which we only need to provide a QuerySet (the set of possible objects it can display) and a template name, which is a Django template file to render with the selected object. In this specific use case, we also need to implement the `get_extra_context` function and have it provide the device's rendered config context as additional information that the template file can access since that's where our IP ACL information is currently stored.

Next, create `nautobot_ip_acls/urls.py` as follows:

```
from django.urls import path

from nautobot_ip_acls import views

urlpatterns = [
    path(
        "devices/<uuid:pk>/ip-acls/",
        views.DeviceIPACLsView.as_view(),
        name="device_ip_acls",
    ),
]
```

Then, create nautobot_ip_acls/templates/nautobot_ip_acls/device_ip_acls. html (first creating the directory if needed with mkdir -p nautobot_ip_acls/templates/ nautobot_ip_acls/):

```
{% extends 'dcim/device.html' %}
{% block content %}
{% for acl_id, acl_entries in config_context.access_lists.items %}
<h4>Access-list <code>{{ acl_id }}</code></h4>
<div class="panel panel-default">
<table class="table table-hover table-headings">
  <thead><tr><th>Seq</th><th>Action</th><th>Network</th></tr></thead>
  {% for entry in acl_entries %}
    <tr><td>{{ entry.seq }}</td><td>{{ entry.action }}</td><td>{{
entry.network }}</td>
  {% endfor %}
</table>
</div>
{% endfor %}
{% endblock content %}
```

This is the most complex file we've created yet, and the first Django HTML template, so let's spend some time looking at it.

First, it's an extension of the built-in dcim/device.html template, which is the base template for every tab in the **Device** detail view. In this template, we only override the block called content, which defines the HTML content of the tab itself. (We don't want to override the entire template, as that includes things such as the page header and footer, the **Add Components** button, and many other pieces of baseline functionality that don't need to be changed for our purposes.)

Within the content block, we iterate over the items under the access_lists key in the config_ context object (which, you'll recall, our view class provided via its get_extra_context function). We define a heading for the ACL, and then a table of ACL entries, which we populate by iterating over the entries in the config context data.

Now that all of the preceding has been done, your file structure should look like this:

```
root@nautobot-app-dev:~/nautobot-app-ip-acls# tree -I __pycache__
nautobot_ip_acls/
nautobot_ip_acls/
├── __init__.py
├── template_content.py
├── templates
│   └── nautobot_ip_acls
│       └── device_ip_acls.html
├── tests
```

```
|   ├── __init__.py
|   ├── test_api.py
|   └── test_basic.py
├── urls.py
└── views.py
```

Since you created new files that weren't previously known to Nautobot, you'll again need to press *Ctrl + C* for the `invoke debug` process and restart it. After waiting for the restart and then refreshing your browser, you should see a new **IP ACLs** tab in the **Device** detail view, and clicking on it should send you to the new view containing your rendered content:

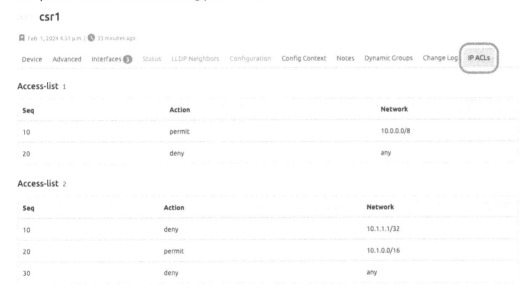

Figure 15.7 – Device detail view with custom IP ACLs tab

Looks great!

Adding a new Devices/ACLs view

Next, you're going to add an "auditing" view, providing, at a glance, the ability to see which devices don't yet have their ACL information recorded into a config context. Note the **ACL Audit Status** column in the following figure:

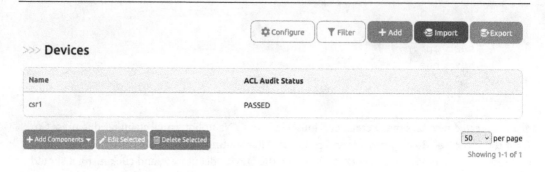

Figure 15.8 – Target appearance for an ACL auditing view

To do this, we'll write another new view and URL pattern, make this view use a new data table, and then make this new view discoverable in the Nautobot navigation bar.

Implementing the data table

This time, we'll start with the data table. Create `nautobot_ip_acls/tables.py`:

```python
import django_tables2

from nautobot.apps.tables import BaseTable
from nautobot.dcim.models import Device

class DeviceAuditTable(BaseTable):
    name = django_tables2.Column(linkify=True)
    acl_audit_status = django_tables2.Column(
        verbose_name="ACL Audit Status",
        empty_values=(),
        orderable=False
    )

    def render_acl_audit_status(self, value, record):
        config_context = record.get_config_context()
        if (
            "access_lists" in config_context
            and len(config_context["access_lists"]) > 0
        ):
            return "PASSED"
        return "FAILED"

    class Meta(BaseTable.Meta):
        model = Device
        fields = ["name", "acl_audit_status"]
```

This is using the same `django_tables2` library that Nautobot uses for its own data tables. It defines a single table of `Device` objects with two columns, one showing the device name (and link to the detail view for that object) and one custom column that renders a **PASSED/FAILED** status based on the contents of each device's config context.

Implementing a List ViewSet

For the associated view to display this table, we'll use the `ObjectListViewMixin` class that Nautobot provides. In `nautobot_ip_acls/views.py`, add this new view:

```python
from nautobot.apps import views
from nautobot.dcim.models import Device
from nautobot.dcim.filters import DeviceFilterSet
from nautobot_ip_acls.tables import DeviceAuditTable


class DeviceIPACLsAuditViewSet(views.ObjectListViewMixin):
    queryset = Device.objects.all()
    filterset_class = DeviceFilterSet
    table_class = DeviceAuditTable


class DeviceIPACLsView(views.ObjectView):
    # unchanged from before, omitted for brevity
```

Then, add this view into `nautobot_ip_acls/urls.py` (since we're using a `ViewSet` class this time rather than a standalone view, the pattern is a bit different):

```python
from django.urls import path
from nautobot.apps.urls import NautobotUIViewSetRouter
from nautobot_ip_acls import views

router = NautobotUIViewSetRouter()
router.register("devices", views.DeviceIPACLsAuditViewSet)
urlpatterns = [
    path(
        "devices/<uuid:pk>/ip-acls/",
        views.DeviceIPACLsView.as_view(),
        name="device_ip_acls",
    ),
]
urlpatterns += router.urls
```

At this point, your file structure should look like this:

```
root@nautobot-app-dev:~/nautobot-app-ip-acls# tree -I __pycache__
nautobot_ip_acls/
nautobot_ip_acls/
├── __init__.py
├── tables.py
├── template_content.py
├── templates
│   └── nautobot_ip_acls
│       └── device_ip_acls.html
├── tests
│   ├── __init__.py
│   ├── test_api.py
│   └── test_basic.py
├── urls.py
└── views.py
```

You should now be able to point your browser to `/plugins/ip-acls/devices/` and have it display this new table:

Figure 15.9 – Implemented ACL Audit Status custom view

> **Note**
> If you have a large number of devices and/or a large number of config contexts, this view may take quite some time to render, as it has to calculate the rendered config context for each device in the table before it can determine its audit status. This is a clear downside to this particular data-modeling approach, which is why we'll present and implement an alternative approach in the last part of this chapter.

Implementing a navigation menu item

Before we move on, let's add this view to the Nautobot navigation bar so that users don't have to guess the correct URL to reach this view.

Create `nautobot_ip_acls/navigation.py` as follows:

```python
from nautobot.apps.ui import NavMenuTab, NavMenuGroup, NavMenuItem

menu_items = (
    NavMenuTab(
        name="Devices",
        groups=(
            NavMenuGroup(
                name="Devices",
                items=(
                    NavMenuItem(
                        link="plugins:nautobot_ip_acls:device_list",
                        name="IP ACL Audit",
                        permissions=["dcim.view_device"],
                    ),
                ),
            ),
        ),
    ),
)
```

Rather than defining a new menu for our menu item, we're adding the **IP ACL Audit** menu item, linking to our new view, under the **Devices** section of the **Devices** menu that already exists in Nautobot. Also note that since we're extending the device data model rather than creating a new data model of our own at this point, the appropriate permission to require to access this menu item is `dcim.view_device`.

> **Note**
>
> If you did want to create a new menu, you'd use a unique name for `NavMenuTab` and also specify a `weight` attribute to influence where this new menu appears relative to the existing menus.

Stop and restart your `invoke debug` process, then refresh your browser and you should see the new link in the menu:

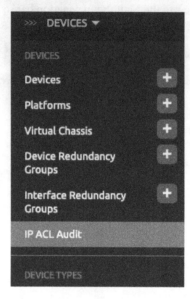

Figure 15.10 – IP ACL Audit menu item added to the navigation menu

Adding ACL details as a REST API endpoint

It's generally a good practice to make sure that any information accessible through the UI is also consumable by software, usually via the REST or GraphQL APIs. In this particular approach, since we're not adding a new distinct data model, there's not much we can do to add to Nautobot's GraphQL interface, but we *can* add a new REST API.

For a new data model, the easiest approach would be to use the `ModelSerializer` classes provided by `django-rest-framework` and extended by Nautobot, but in this case, we don't have a separate data model to serialize, so you'll instead implement the JSON encoding of the data directly. Fortunately, this is very straightforward in this case since the config context data is already structured JSON.

Start by setting up `nautobot_ip_acls/api/views.py` (you'll also need to create the `nautobot_ip_acls/api/` directory and an empty file at `nautobot_ip_acls/api/__init__.py`) as follows:

```
from django.shortcuts import import get_object_or_404
from rest_framework.permissions import IsAuthenticated
from rest_framework.response import Response
from rest_framework.views import APIView
```

```python
from nautobot.dcim.models import Device

class DeviceIPACLsView(APIView):
    permission_classes = [IsAuthenticated]

    def get(self, request, pk=None, format=None):
        device = get_object_or_404(
            Device.objects.restrict(self.request.user, "view"), pk=pk
        )
        config_context = device.get_config_context()
        return Response(
            {"access_lists": config_context.get("access_lists", {})}
        )
```

Next, we need to register this view to a URL pattern. Create `nautobot_ip_acls/api/urls.py`:

```python
from django.urls import path

from nautobot_ip_acls.api.views import DeviceIPACLsView

urlpatterns = [
    path(
        "devices/<uuid:pk>/ip-acls/",
        DeviceIPACLsView.as_view(),
        name="device_ip_acls"
    ),
]
```

Following Nautobot's usual REST API patterns, we're identifying the device of interest by its **primary key (PK)** in this implementation. You could, of course, use a different identifier such as the device's name, but there's a lot to be said for self-consistency. You can obtain the PK of any given object in Nautobot by going to its detail view and selecting the **Advanced** tab, where the PK is identified as the object's UUID.

>>> **csr1**

📑 Feb. 1, 2024 4:31 p.m. | 🕐 50 minutes ago

| Device | Advanced | Interfaces ③ | Status | LLDP Neighbors | Configuration | Config Context |

Object Details

UUID	feb8474b-8f49-4e12-97e2-cd10a9fc7793
Natural Slug	csr1__europe_feb8

Figure 15.11 – Device detail view Advanced tab showing its UUID

At this point, your file structure should appear as follows:

```
root@nautobot-app-dev:~/nautobot-app-ip-acls# tree -I __pycache__
nautobot_ip_acls/
nautobot_ip_acls/
├── __init__.py
├── api
│   ├── __init__.py
│   ├── urls.py
│   └── views.py
├── navigation.py
├── tables.py
├── template_content.py
├── templates
│   └── nautobot_ip_acls
│       └── device_ip_acls.html
├── tests
│   ├── __init__.py
│   ├── test_api.py
│   └── test_basic.py
├── urls.py
└── views.py
```

To verify that this REST API endpoint is available and working, after restarting `invoke debug` once again, you can use Nautobot's built-in OpenAPI (Swagger) UI. Navigate to `http://<your server ip>:8080/api/docs/` and scroll down to the **plugins** heading, and you should see your new REST API endpoint listed:

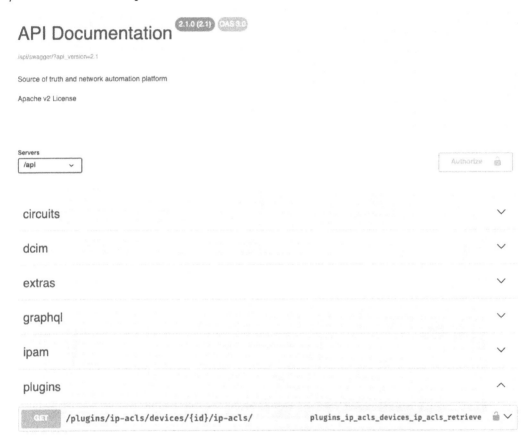

Figure 15.12 – Nautobot Swagger UI

Click on this endpoint to expand it, then click the **Try it out** button. Enter the previously obtained device PK into the **id** field and click **Execute**. You should see the expected response containing the list of IP ACLs defined for this device:

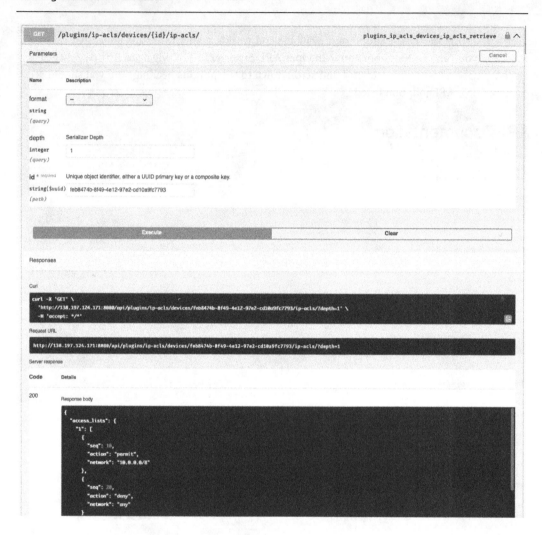

Figure 15.13 – Your App's REST API is fully operational

Review

In this section, you have successfully learned how to develop a Nautobot App that can do the following:

- Add content to an existing object detail view, both on the main view as well as on a separate dedicated tab

- Add an entirely new list view displaying a table of objects with specific information about each object

- Add links to the Nautobot navigation menus

- Add and verify a new Nautobot REST API endpoint that presents specific data

Nicely done!

Building an App with custom data models

As you have seen, it's quite possible to implement a useful Nautobot App without adding any custom data models at all. However, you've probably also noticed some of the limitations of the approach we took in the previous sections of this chapter:

- The config context can't store direct references to other database objects, only information about them. That is to say, the ACL entries you entered defined networks as strings, and could *not* directly reference the corresponding Nautobot `Prefix` or `IPAddress` objects. One consequence of this is potential issues with data integrity—for example, deleting a `Prefix` or `IPAddress` object from Nautobot does not result in any automatic change to the ACL definitions that were using that network.

- Because a device's config context data is derived on the fly rather than stored in the database, performance can suffer when inspecting a large number of devices at the same time.

- While it's possible to extend the Nautobot REST API with custom views to serve a specific purpose, it's not generally possible to extend the GraphQL API similarly, although the inherent flexibility of GraphQL can sometimes make up for this.

For these and other reasons, you may find that your specific use case is better served by implementing a custom data model or models as part of your App, and that is what we will explore in the remainder of this chapter.

Designing the ACL data models

Going back to the data model requirements we determined earlier in the chapter, we can see that we probably need two distinct data models for this App, which we will call `ACL` and `ACLEntry`:

```
access_lists:
  "1":  # an ACL
    # an ACLEntry
    - action: permit
      network: 10.0.0.0/8
      seq: 10
    # another ACLEntry
    - action: deny
      network: any
      seq: 20
```

```
    "2":  # another ACL
      - action: deny
        network: 10.1.1.1/32
        seq: 10
      - action: permit
        network: 10.1.0.0/16
        seq: 20
      - action: deny
        network: any
        seq: 30
interfaces:  # already modeled by Nautobot
  "GigabitEthernet1":
    acl_in: "1"  # new relationship
  "GigabitEthernet2":
    acl_out: "2"  # another relationship
  "GigabitEthernet3":
    acl_in: "1"
    acl_out: "2"
```

An ACL:

- Has a unique identifier, which, depending on the target platform, might be a number or text—so we'll model this as a string.

  To keep this example simple, we'll assume that the network never uses the same identifier to define different ACLs on different devices; any time a given ACL identifier appears in your network, it refers to an identically configured access list. Of course, this may not be true in your actual network (but maybe it should!).

- Is associated with any number of Interfaces as an inbound_acl or outbound_acl (or both).

An ACLEntry:

- Belongs to a specific ACL.

  To keep this example simple, we won't attempt to deduplicate data even further by allowing a single ACLEntry to be reused across multiple different access lists; these will be separate ACLEntry records in the database even if they represent the same rule.

- Has an integer sequence_number, which is unique within its containing ACL.

- References a specific Nautobot Prefix object or none at all (representing the any case).

  To keep this example simple, we won't account for the case where an ACLEntry might instead reference a single specific IPAddress object rather than a Prefix object; by the time you're done with this exercise, you should be reasonably comfortable with extending the App to account for that case.

- Can either permit or deny the prefix in question.

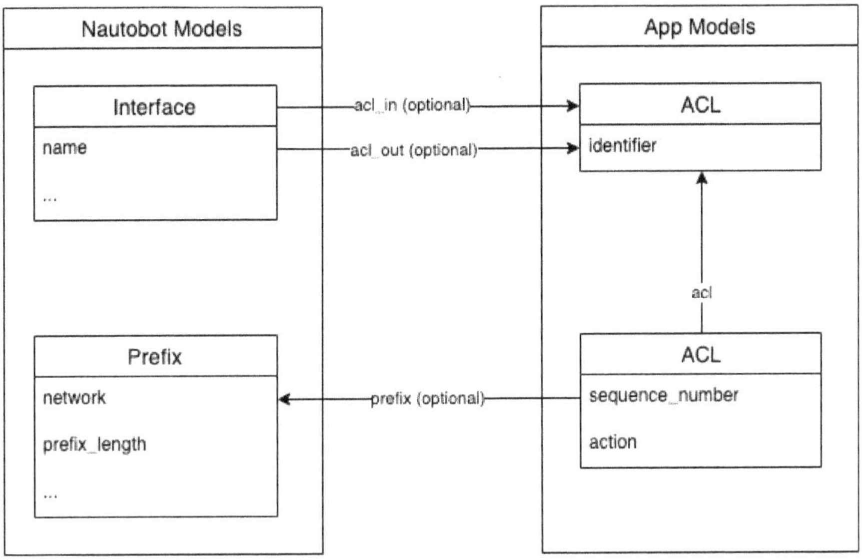

Figure 15.14 – Data model for Nautobot IP ACLs App

Implementing the ACL data models

We'll translate these models into Django classes as follows.

Implementing the ACL model

First, we'll create the ACL class in `nautobot_ip_acls/models.py`:

```
from django.db import models
from django.urls import reverse
from nautobot.apps.models import PrimaryModel


class ACL(PrimaryModel):
    identifier = models.CharField(max_length=255, unique=True)

    def __str__(self):
        return self.identifier

    class Meta:
        verbose_name = "ACL"
```

Let's break this down a little:

- `PrimaryModel` is the class used in Nautobot for full-featured data models that support most or all of Nautobot's extensions to base Django functionality. It's generally the class you'll want to use for any data model that represents a distinct "thing" in the network.

- We're using a `CharField` for the identifier since we know that identifiers are integers on some platforms but text strings on other platforms, and we want to support both.

- Note that we're using `unique=True` on the `identifier` field to specify that this must be globally unique within Nautobot's database.

Next, we need to ensure that this new model gets implemented in Nautobot's database.

First, run `invoke makemigrations`; this will generate a file along the lines of `nautobot_ip_acls/migrations/0001_initial.py`, which basically is a set of instructions to Django telling it how to translate your Python code to SQL. In more advanced App development use cases, you might need to edit or even manually construct a `migrations` file, but for a simple model like this, Django can easily handle it for you:

```
root@nautobot-app-dev:~/nautobot-app-ip-acls# invoke makemigrations
Running docker compose command "ps --services --filter status=running"
Running docker compose command "exec nautobot nautobot-server
makemigrations nautobot_ip_acls"
Migrations for 'nautobot_ip_acls':
  nautobot_ip_acls/migrations/0001_initial.py
    - Create model ACL
```

Next, run `invoke migrate`. This takes the generated `migrations` script and actually applies it to your Nautobot database. You should see output similar to this:

```
root@nautobot-app-dev:~/nautobot-app-ip-acls# invoke migrate
Running docker compose command "ps --services --filter status=running"
Running docker compose command "exec nautobot nautobot-server migrate"
Operations to perform:
  Apply all migrations: admin, auth, circuits, contenttypes, database,
dcim, django_celery_beat, django_rq, extras, ipam, nautobot_ip_acls,
sessions, social_django, taggit, tenancy, users, virtualization
Running migrations:
  Applying nautobot_ip_acls.0001_initial... OK
```

You can verify that this model is now defined by using the `invoke nbshell` command as follows, though of course, there are no specific ACL records actually present in your database yet:

```
root@nautobot-app-dev:~/nautobot-app-ip-acls# invoke nbshell
Running docker compose command "ps --services --filter status=running"
Running docker compose command "exec nautobot nautobot-server nbshell"
# Django version 3.2.23
# Nautobot version 2.1.0
# Nautobot IP ACLs version 0.1.0
Python 3.11.7 (main, Dec 19 2023, 03:30:20) [GCC 12.2.0]
Type 'copyright', 'credits' or 'license' for more information
IPython 8.12.3 -- An enhanced Interactive Python. Type '?' for help.
In [1]: from nautobot_ip_acls.models import ACL
In [2]: ACL.objects.all()
Out[2]: <RestrictedQuerySet []>
```

One-to-many relationships and our data model

You may notice that we're missing something for this model—it doesn't define any reference to Nautobot's Interface model yet. This is due to the nature of how relational databases such as PostgreSQL and MySQL work, which is reflected in how Django handles data relationships. It's a common mental hurdle that must be overcome during initial data modeling work:

- For a *many-to-one* relationship from model A to model B, model A can declare a `ForeignKey` referencing model B.

- For a *one-to-one* relationship from model A to model B, model A can declare a `OneToOneField` referencing model B, or model B can define a `OneToOneField` referencing model A.

- For a *many-to-many* relationship from model A to model B, model A can declare a `ManyToManyField` referencing model B, or model B can declare a `ManyToManyField` referencing model A.

- For a *one-to-many* relationship from model A to model B (which is what we need here—one ACL to many Interfaces), this is not representable in a database as a field on model A—it must be defined as a field on model B as a `ForeignKey` pointing to model A. Django's representation of the models as Python objects will provide a "reverse relationship" representing the "many" side of the relationship (from model A back to model B), but the actual data in the database must be an attribute of model B.

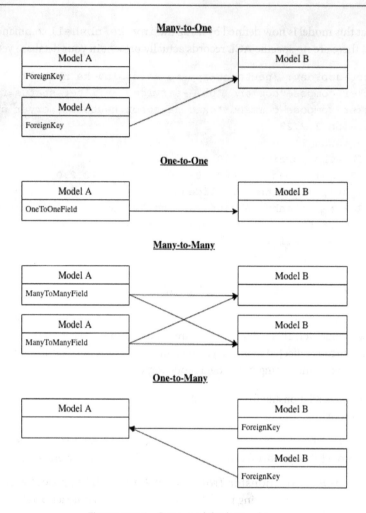

Figure 15.15 – Data model relationships

Django (and Nautobot) do not provide a mechanism for Apps to change the field definitions on a core Nautobot model, and the `Interface` model is defined by Nautobot itself, so we don't have the ability in our App to add a `ForeignKey` from the core `Interface` model to our App's ACL mode, as we cannot directly edit the `Interface` class definition to include foreign keys to our ACL class. Are we out of luck then? Can our data model not be implemented in an App? Fortunately, the answer is no—this is one of the use cases that Nautobot's **Relationships** feature is designed to account for.

For the moment, we'll just manually define the possibility of "in" and "out" relationships between `Interface` and ACL via the Nautobot UI. (It's possible to write code in your App that will automatically define these relationships when the App is installed into Nautobot, but that's a bit more advanced than we need to cover at this time.) Navigate to **Extensibility** > **Relationships** in the Nautobot UI, and click the **Add** button. Fill in the form as follows, then click **Create**:

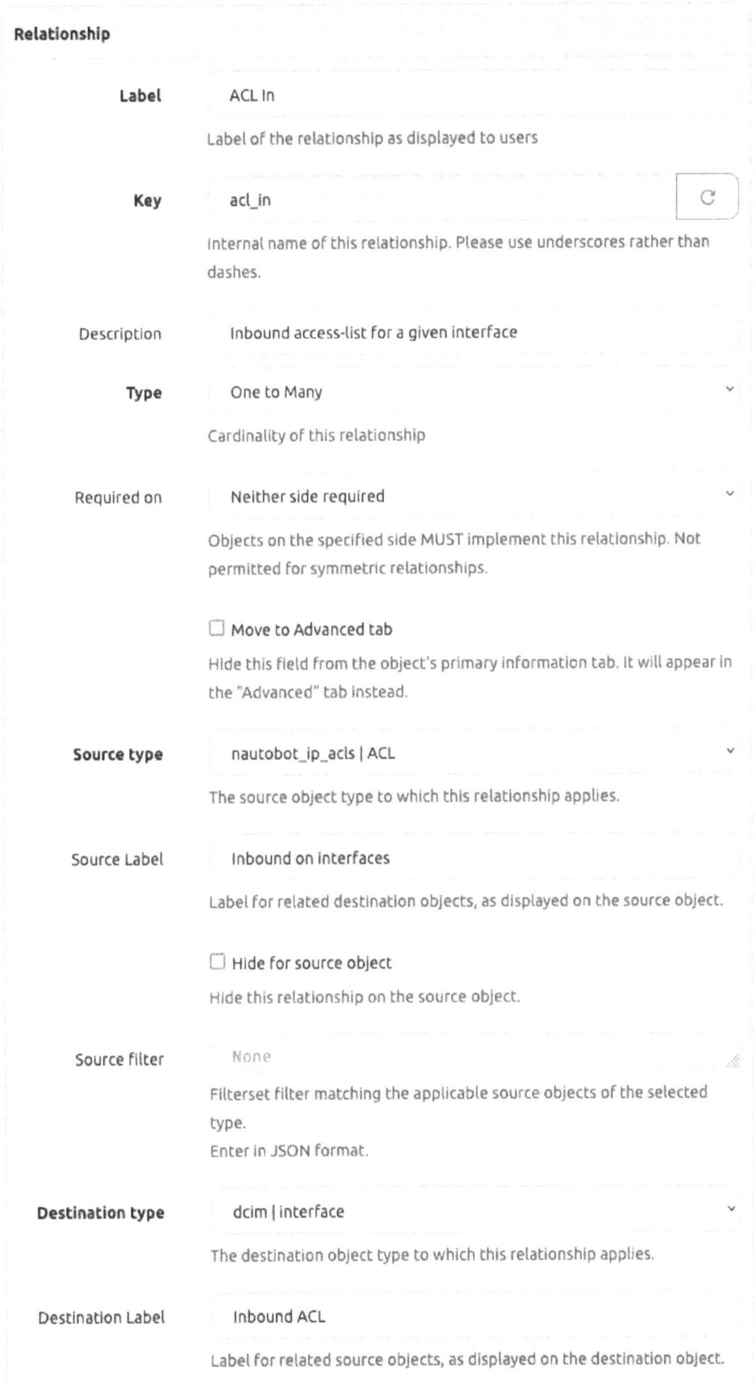

Relationship

Label   ACL In
Label of the relationship as displayed to users

Key   acl_in   ⟳
Internal name of this relationship. Please use underscores rather than dashes.

Description   Inbound access-list for a given interface

Type   One to Many   ⌄
Cardinality of this relationship

Required on   Neither side required   ⌄
Objects on the specified side MUST implement this relationship. Not permitted for symmetric relationships.

☐ Move to Advanced tab
Hide this field from the object's primary information tab. It will appear in the "Advanced" tab instead.

Source type   nautobot_ip_acls | ACL   ⌄
The source object type to which this relationship applies.

Source Label   Inbound on interfaces
Label for related destination objects, as displayed on the source object.

☐ Hide for source object
Hide this relationship on the source object.

Source filter   None
Filterset filter matching the applicable source objects of the selected type.
Enter in JSON format.

Destination type   dcim | interface   ⌄
The destination object type to which this relationship applies.

Destination Label   Inbound ACL
Label for related source objects, as displayed on the destination object.

Figure 15.16 – Creating an ACL In relationship

Then, repeat the process for a second relationship:

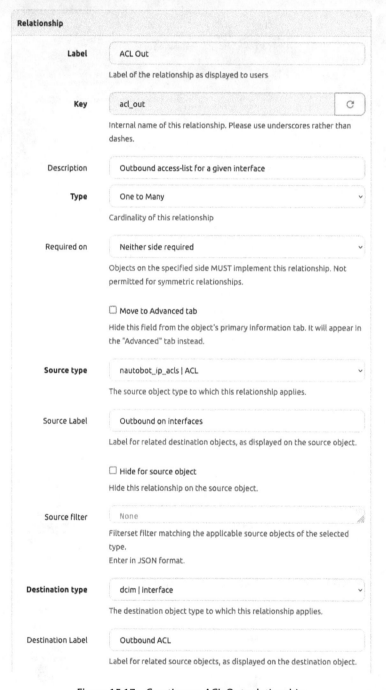

Figure 15.17 – Creating an ACL Out relationship

You can now navigate from the **Device** detail view to the detail view for a specific Interface and see that the two new relationships are present, though not populated yet:

Relationships

Inbound ACL	—
Outbound ACL	—

Figure 15.18 – Custom relationships in the Interface detail view

> **Note**
>
> On navigating to some views, you may get an error message saying **NoReverseMatch: Reverse for 'acl-list' not found**—this is due to the fact that we haven't yet implemented the UI views for the ACL model. Don't worry, we'll fix this soon!

Implementing the ACLEntry data model

Next, add the ACLEntry data model to nautobot_ip_acls/models.py:

```python
from django.db import models
from django.urls import reverse
from nautobot.apps.models import BaseModel, ChangeLoggedModel,
PrimaryModel

class ACL(PrimaryModel):
    # unchanged, omitted here

class ACLEntry(BaseModel, ChangeLoggedModel):
    acl = models.ForeignKey(ACL, on_delete=models.CASCADE)
    sequence_number = models.PositiveSmallIntegerField()
    prefix = models.ForeignKey(

        "ipam.Prefix", blank=True, null=True, on_delete=models.CASCADE
    )
    action = models.CharField(
        max_length=10,
        choices=[("permit", "Permit"), ("deny", "Deny")]
    )
```

```python
    def __str__(self):
        return (
            f"ACL {self.acl}: {self.sequence_number} {self.action} "
            f"{self.prefix or 'any'}"
        )

    def get_absolute_url(self):
        return reverse(
            "plugins:nautobot_ip_acls:aclentry",
            kwargs={"pk": self.pk}
        )

    class Meta:
        ordering = ("acl__identifier", "sequence_number")
        unique_together = [("acl", "sequence_number")]
        verbose_name = "ACL entry"
        verbose_name_plural = "ACL entries"
```

Let's break this down a little:

- Because we don't need to support custom fields, relationships, and other features on individual `ACLEntry` records, we're not using `PrimaryModel` for this model but are instead building it from several lower-level Nautobot mixin classes.

- The foreign keys here use `on_delete=models.CASCADE`, which ensures that when the related objects (`ACL` or `Prefix`) are deleted, any `ACLEntry` referencing those objects will be automatically deleted as well. If we wanted the existence of an `ACLEntry` to *prevent* related objects from being deleted, we could use `models.PROTECT` instead, but that can be a source of user frustration in some cases.

- The `choices` parameter to the `action` field is a list of (database value, human-readable equivalent) tuples.

- The `unique_together` constraint specifies that we cannot reuse a given sequence number within a given ACL.

Once again, run `invoke makemigrations` and `invoke migrate` to generate and then apply the migrations corresponding to this new data model:

```
root@nautobot-app-dev:~/nautobot-app-ip-acls# invoke makemigrations
Running docker compose command "ps --services --filter status=running"
Running docker compose command "exec nautobot nautobot-server
makemigrations nautobot_ip_acls"
Migrations for 'nautobot_ip_acls':
  nautobot_ip_acls/migrations/0002_aclentry.py
    - Create model ACLEntry
```

```
root@nautobot-app-dev:~/nautobot-app-ip-acls# invoke migrate
Running docker compose command "ps --services --filter status=running"
Running docker compose command "exec nautobot nautobot-server migrate"
Operations to perform:
  Apply all migrations: admin, auth, circuits, contenttypes, database,
dcim, django_celery_beat, django_rq, extras, ipam, nautobot_ip_acls,
sessions, social_django, taggit, tenancy, users, virtualization
Running migrations:
  Applying nautobot_ip_acls.0002_aclentry... OK
```

Implementing the REST API

Since our ACL and ACLEntry models are full-fledged Nautobot data models, we can use a lot of Nautobot's built-in functionality to define the REST API for working with them, and doing this first will actually make it easier for us to implement a UI next.

Implementing the serializers and nested serializers

Serializers are how django-rest-framework can easily translate a Django model into its JSON representation, and vice versa. Nautobot takes this a step further by defining its own Serializer base classes that provide Nautobot-specific functionality; we'll use these as the basis for our ACL and ACLEntry serializer logic.

Create nautobot_ip_acls/api/serializers.py as follows:

```python
from rest_framework import serializers
from nautobot.apps.api import (
    NautobotModelSerializer, TaggedModelSerializerMixin
)
from nautobot_ip_acls.models import ACL, ACLEntry


class ACLSerializer(
    NautobotModelSerializer, TaggedModelSerializerMixin
):
    class Meta:
        model = ACL
        fields = ["__all__"]


class ACLEntrySerializer(NautobotModelSerializer):
    class Meta:
        model = ACLEntry
        fields = ["__all__"]
```

This almost seems too good to be true, but for most basic data models, that's all that's needed to create REST API serializers in Nautobot:

- Declare which model the serializer applies to. This is needed for each model.

- Declare `fields = ["__all__"]`, which means "serialize all fields defined by this model." You would only need to change this if you wanted to serialize extra data beyond the database fields of the model, or if you wanted to explicitly exclude certain fields, such as those that might contain sensitive data.

Implementing the API ViewSets

Next, we add the following code to `nautobot_ip_acls/api/views.py` to define ViewSets that use these serializers. We also add placeholders for the filtering capabilities that we'll implement a bit later on:

```python
from django.shortcuts import get_object_or_404
from rest_framework.permissions import IsAuthenticated
from rest_framework.response import Response
from rest_framework.views import APIView

from nautobot.apps.api import NautobotModelViewSet
from nautobot.dcim.models import Device

# from nautobot_ip_acls import filters
from nautobot_ip_acls import models
from nautobot_ip_acls.api import serializers

class DeviceIPACLsView(APIView):
    # unchanged, omitted here

class ACLViewSet(NautobotModelViewSet):
    queryset = models.ACL.objects.all()
    serializer_class = serializers.ACLSerializer
    # filterset_class = filters.ACLFilterSet

class ACLEntryViewSet(NautobotModelViewSet):
    queryset = models.ACLEntry.objects.all()
    serializer_class = serializers.ACLEntrySerializer
    # filterset_class = filters.ACLEntryFilterSet
```

There's not a lot to explain here—each ViewSet simply defines the dataset that it references, the serializer it uses to represent that data, and optionally, the FilterSet that it will use to allow users to filter the data.

Finally, we need to define the URLs for these views in `nautobot_ip_acls/api/urls.py`:

```
from django.urls import include, path
from nautobot.apps.api import OrderedDefaultRouter
from nautobot_ip_acls.api.views import (
    ACLViewSet, ACLEntryViewSet, DeviceIPACLsView
)


router = OrderedDefaultRouter()
router.register("acls", ACLViewSet)
router.register("acl-entries", ACLEntryViewSet)

urlpatterns = [
    path(
        "devices/<uuid:pk>/ip-acls/",
        DeviceIPACLsView.as_view(),
        name="device_ip_acls"
        ),
    path("", include(router.urls)),
]
```

Your file structure should now appear as follows:

```
root@nautobot-app-dev:~/nautobot-app-ip-acls# tree -I __pycache__
nautobot_ip_acls/
nautobot_ip_acls/
├── __init__.py
├── api
│   ├── __init__.py
│   ├── serializers.py
│   ├── urls.py
│   └── views.py
├── migrations
│   ├── 0001_initial.py
│   ├── 0002_aclentry.py
│   └── __init__.py
├── models.py
├── navigation.py
├── tables.py
├── template_content.py
├── templates
│   └── nautobot_ip_acls
│       └── device_ip_acls.html
├── tests
```

```
|     ├── __init__.py
|     ├── test_api.py
|     └── test_basic.py
├── urls.py
└── views.py
```

Your App is growing quite nicely at this point. Let's see what's next.

Validating the REST API

Now that you've implemented the REST API for your models, you can point your browser to `http://<your server IP>:8080/api/docs/` and scroll down to the **plugins** section to see that your new REST APIs are automatically discovered and documented:

Method	Endpoint	Name
GET	/plugins/ip-acls/acl-entries/	plugins_ip_acls_acl_entries_list
POST	/plugins/ip-acls/acl-entries/	plugins_ip_acls_acl_entries_create
PUT	/plugins/ip-acls/acl-entries/	plugins_ip_acls_acl_entries_bulk_update
PATCH	/plugins/ip-acls/acl-entries/	plugins_ip_acls_acl_entries_bulk_partial_update
DELETE	/plugins/ip-acls/acl-entries/	plugins_ip_acls_acl_entries_bulk_destroy
GET	/plugins/ip-acls/acl-entries/{id}/	plugins_ip_acls_acl_entries_retrieve
PUT	/plugins/ip-acls/acl-entries/{id}/	plugins_ip_acls_acl_entries_update
PATCH	/plugins/ip-acls/acl-entries/{id}/	plugins_ip_acls_acl_entries_partial_update
DELETE	/plugins/ip-acls/acl-entries/{id}/	plugins_ip_acls_acl_entries_destroy
GET	/plugins/ip-acls/acl-entries/{id}/notes/	plugins_ip_acls_acl_entries_notes_list
POST	/plugins/ip-acls/acl-entries/{id}/notes/	plugins_ip_acls_acl_entries_notes_create
GET	/plugins/ip-acls/acls/	plugins_ip_acls_acls_list

Figure 15.19 – REST API endpoints for ACL and ACLEntry data

If you like, you can experiment with the various API endpoints to populate some data for these models, but it may be easier to wait until you have a UI to do so. Onward to the next section!

Implementing the UI

Since our `ACL` and `ACLEntry` models are full-fledged Nautobot data models, we can use a lot of Nautobot's built-in functionality to define the UI for working with them. To get the full functionality, in addition to the REST API serializers we wrote in the previous section, you'll need to write code for a data table, code for filtering that data, forms for editing that data, templates for displaying the data, and views for displaying all of these.

Implementing the tables

In `nautobot_ip_acls/tables.py`, we'll add a table for each of our two models:

```python
import django_tables2
from nautobot.apps.tables import (
    BaseTable, ButtonsColumn, TagColumn, ToggleColumn
)
from nautobot.dcim.models import Device
from nautobot_ip_acls.models import ACL, ACLEntry

class DeviceAuditTable(BaseTable):
    # unchanged, omitted here

class ACLTable(BaseTable):
    pk = ToggleColumn()
    identifier = django_tables2.Column(linkify=True)
    tags = TagColumn(url_name="plugins:nautobot_ip_acls:acl_list")
    actions = ButtonsColumn(ACL)

    class Meta(BaseTable.Meta):
        model = ACL
        fields = ("pk", "identifier", "tags", "actions")
```

Let's break this down a little:

- The `pk` column is used to provide a checkbox at the left side of the table that can be used to select multiple records for operations such as bulk editing and bulk deletion; generally, you'll want to include this in all of your model tables.

- Note the use of `linkify=True` on the `identifier` column; this makes it so the text in this column will automatically link to the detail view for the corresponding object.

- Note that we didn't define a `tags` field on the `ACL` model; this comes in automatically to the model from our use of `PrimaryModel` as a base class, but we do need to explicitly include it in the table.

- The `actions` column provides quick **Edit** and **Delete** buttons for individual records in the table.

- The `Meta` class defines which `model` the table applies to, and the `fields` attribute defines the order in which the columns will be rendered by default. In general, the pk column should always be first and the `actions` column should be last.

Continue in this same file as follows:

```
class ACLEntryTable(BaseTable):
    pk = ToggleColumn()
    acl = django_tables2.Column(linkify=True)
    sequence_number = django_tables2.Column(linkify=True)
    prefix = django_tables2.Column(linkify=True)
    actions = ButtonsColumn(ACLEntry)

    class Meta(BaseTable.Meta):
        model = ACLEntry
        fields = ("pk", "acl", "sequence_number", "prefix", "action",
"actions")
```

Here are some notes about the preceding code:

- Using `linkify=True` on the `acl` and `prefix` columns allows for automatic cross-linking to the detail views for those related objects

- Since the `action` column (not to be confused with `actions`) is just plain text, we don't need to explicitly declare a column for it—just including it in `fields` is sufficient

Implementing the FilterSets

Next, we'll define the FilterSets for both models; these can be used to control the filtering of the data in both the UI and the API. Let's start with `nautobot_ip_acls/filters.py`:

```
from nautobot.apps.filters import (
    BaseFilterSet,
    NaturalKeyOrPKMultipleChoiceFilter,
    NautobotFilterSet,
    SearchFilter,
    TagFilter,
)
from nautobot_ip_acls.models import ACL, ACLEntry


class ACLFilterSet(NautobotFilterSet):
    q = SearchFilter(filter_predicates={"identifier": "icontains"})
```

```
    tags = TagFilter()

    class Meta:
        model = ACL
        fields = ["identifier"]
```

By using `NautobotFilterSet`, the filtering of the ACL table will automatically support filtering by features such as any relationships and custom fields that exist for ACLs. Additionally, we're defining how free-text search of the ACL table will work (case-insensitive search of the `identifier` field only), and declaring the ability to filter by assigned tags as well.

Continue this file as follows:

```
class ACLEntryFilterSet(BaseFilterSet):
    q = SearchFilter(
        filter_predicates={"acl__identifier": "icontains"}
    )
    acl = NaturalKeyOrPKMultipleChoiceFilter(
        field_name="identifier", queryset=ACL.objects.all()
    )

    class Meta:
        model = ACLEntry
        fields = ["acl", "sequence_number", "action"]
```

Because the `ACLEntry` model doesn't support relationships and custom fields, we use the `BaseFilterSet` class here instead of `NautobotFilterSet`. We use `NaturalKeyOrPKMultipleChoiceFilter` to allow filtering by the related `acl` either by its UUID primary key or by its identifier all in a single filter field.

At this point, you can, if you wish, go back to nautobot_ip_acls/api/views.py and uncomment the references to these FilterSets, which will make it possible for the REST API to use them as well.

Here is the file structure at this point:

```
root@nautobot-app-dev:~/nautobot-app-ip-acls# tree -I __pycache__
nautobot_ip_acls/
nautobot_ip_acls/
├── __init__.py
├── api
│   ├── __init__.py
│   ├── serializers.py
│   ├── urls.py
│   └── views.py
├── filters.py
```

```
├── migrations
│   ├── 0001_initial.py
│   ├── 0002_aclentry.py
│   └── __init__.py
├── models.py
├── navigation.py
├── tables.py
├── template_content.py
├── templates
│   └── nautobot_ip_acls
│       └── device_ip_acls.html
├── tests
│   ├── __init__.py
│   ├── test_api.py
│   └── test_basic.py
├── urls.py
└── views.py
```

It is coming along nicely!

Implementing the forms

We'll just implement the "create/edit" forms for our two models at this time. As a future exercise, you might want to implement the full set of Nautobot model UI functionality including forms for bulk editing and CSV importing. Write the following in `nautobot_ip_acls/forms.py`:

```python
from django import forms
from nautobot.apps.forms import (
    BootstrapMixin, NautobotModelForm, DynamicModelChoiceField
)
from nautobot.ipam.models import Prefix
from nautobot_ip_acls.models import ACL, ACLEntry


class ACLForm(NautobotModelForm):
    class Meta:
        model = ACL
        fields = ["identifier"]


class ACLEntryForm(BootstrapMixin, forms.ModelForm):
    acl = DynamicModelChoiceField(
        queryset=ACL.objects.all(), label="ACL"
    )
    prefix = DynamicModelChoiceField(
        queryset=Prefix.objects.all(), required=False
```

```
    )

    class Meta:
        model = ACLEntry
        fields = ["acl", "sequence_number", "prefix", "action"]
```

Here again, we use a full-featured class (NautobotModelForm) for the full-featured ACL model, and a combination of more basic classes (BootstrapMixin and forms.ModelForm) for the less feature-rich ACLEntry model. We use DynamicModelChoiceField for the related ACL and Prefix models; this class allows Nautobot to populate options for these fields on the fly, rather than pre-populating all available options when initially rendering the page, allowing the create/edit views to load much more quickly.

Implementing the templates

For most of the basic views for our ACL and ACLEntry models, we actually don't need to write a template at all—Nautobot's built-in generic templates will do the job quite well. The only exception to this is the detail view, which will need a template for each model to handle rendering its distinctive fields.

Create nautobot_ip_acls/templates/nautobot_ip_acls/acl_retrieve.html:

```
{% extends 'generic/object_retrieve.html' %}

{% block content_left_page %}
<div class="panel panel-default">
    <div class="panel-heading">
        <strong>ACL</strong>
    </div>
    <table class="table table-hover panel-body attr-table">
        <tr>
            <td>Identifier</td>
            <td>{{ object.identifier }}</td>
        </tr>
    </table>
</div>
{% endblock content_left_page %}
```

Also create nautobot_ip_acls/templates/nautobot_ip_acls/aclentry_retrieve.html:

```
{% extends 'generic/object_retrieve.html' %}
{% load helpers %}

{% block content_left_page %}
<div class="panel panel-default">
```

```html
<div class="panel-heading">
    <strong>ACL Entry</strong>
</div>
<table class="table table-hover panel-body attr-table">
    <tr>
        <td>ACL</td>
        <td>{{ object.acl|hyperlinked_object }}</td>
    </tr>
    <tr>
        <td>Sequence Number</td>
        <td>{{ object.sequence_number }}</td>
    </tr>
    <tr>
        <td>Action</td>
        <td>{{ object.action }}</td>
    </tr>
    <tr>
        <td>Prefix</td>
        <td>
            {% if object.prefix %}
                {{ object.prefix|hyperlinked_object }}
            {% else %}Any
            {% endif %}
        </td>
    </tr>
</table>
</div>
{% endblock content_left_page %}
```

Note here that we're extending the generic `generic/object_retrieve.html` template and loading the set of additional template tags provided by Nautobot's `helpers` module, which includes the `hyperlinked_object` tag that we're using to automatically hyperlink to the detail views of the related `ACL` and `Prefix` objects. Just as we did earlier in this chapter, we're not overriding the entire template but just the `content_left_page` block, which defines the left side of the rendered template.

This is the file structure now:

```
root@nautobot-app-dev:~/nautobot-app-ip-acls# tree -I __pycache__
nautobot_ip_acls/
nautobot_ip_acls/
├── __init__.py
├── api
│   ├── __init__.py
```

```
|    ├── serializers.py
|    ├── urls.py
|    └── views.py
├── filters.py
├── forms.py
├── migrations
|    ├── 0001_initial.py
|    ├── 0002_aclentry.py
|    └── __init__.py
├── models.py
├── navigation.py
├── tables.py
├── template_content.py
├── templates
|    └── nautobot_ip_acls
|      ├── acl_retrieve.html
|      ├── aclentry_retrieve.html
|      └── device_ip_acls.html
├── tests
|    ├── __init__.py
|    ├── test_api.py
|    └── test_basic.py
├── urls.py
└── views.py
```

One more piece of the puzzle remains—the ViewSets to make use of the FilterSets, forms, serializers, and tables that we've created so far.

Implementing the views

Now, we can finally implement the views to tie all of these components together! In `nautobot_ip_acls/views.py`, add the following:

```python
from nautobot.apps import views
from nautobot.dcim.models import Device
from nautobot.dcim.filters import DeviceFilterSet
from nautobot_ip_acls import forms, filters, models, tables
from nautobot_ip_acls.api import serializers
from nautobot_ip_acls.tables import DeviceAuditTable


class DeviceIPACLsAuditViewSet(views.ObjectListViewMixin):
    # unchanged, omitted here
```

```
class DeviceIPACLsView(views.ObjectView):
    # unchanged, omitted here

class ACLUIViewSet(views.NautobotUIViewSet):
    filterset_class = filters.ACLFilterSet
    form_class = forms.ACLForm
    lookup_field = "pk"
    queryset = models.ACL.objects.all()
    serializer_class = serializers.ACLSerializer
    table_class = tables.ACLTable

class ACLEntryUIViewSet(views.NautobotUIViewSet):
    filterset_class = filters.ACLEntryFilterSet
    form_class = forms.ACLEntryForm
    lookup_field = "pk"
    queryset = models.ACLEntry.objects.all()
    serializer_class = serializers.ACLEntrySerializer
    table_class = tables.ACLEntryTable
```

Next, add these views to `nautobot_ip_acls/urls.py`:

```
from django.urls import path
from nautobot.apps.urls import NautobotUIViewSetRouter
from nautobot_ip_acls import views

app_name = "nautobot_ip_acls"
router = NautobotUIViewSetRouter()
router.register("acls", views.ACLUIViewSet)
router.register("acl-entries", views.ACLEntryUIViewSet)
router.register("devices", views.DeviceIPACLsAuditViewSet)

urlpatterns = [
    path(
        "devices/<uuid:pk>/ip-acls/",
        views.DeviceIPACLsView.as_view(),
        name="device_ip_acls",
    ),
]
urlpatterns += router.urls
```

We've written a lot of code here, so let's check that it all works. Point your browser to `http://<your server IP>/plugins/ip-acls/acls/` and it should load, showing an (empty, so far) list of ACLs:

Figure 15.20 – An empty ACL table

Click the **Add** button to go to the form you've defined and fill in some data:

Add a new ACL

ACL

Identifier   1

Relationships

Inbound on   × GigabitEthernet1   × GigabitEthernet3   ×
interfaces   Inbound access-list for a given interface

Outbound on   × GigabitEthernet2   × GigabitEthernet3 | ×
interfaces   GigabitEthernet1

                    GigabitEthernet2

                    GigabitEthernet3

Notes

Figure 15.21 – Creating an ACL record

Note that the ACL edit form automatically includes related features such as the **Inbound on Interfaces** and **Outbound on Interfaces** relationships that you defined earlier as well as the option for adding notes to the ACL.

The user experience here is a bit suboptimal still as we can only select the Interfaces by name, lacking important context such as the names of the devices each Interface belongs to. This is a limitation of the generic **Relationships** feature in Nautobot itself at present. It's not so bad in this simple example since we only have one device anyway for now, but it's something you might think about addressing as a future exercise in order to make this App more fully usable in a real deployment. (For now, you can at least link the relationship "in reverse" by editing an Interface and defining its associated ACLs, rather than editing an ACL and trying to define its associated Interfaces—the Relationship UI in Nautobot will work either way.)

> **Hint**
>
> Your steps would probably be to implement a custom form HTML template in `nautobot_ip_acls/templates/nautobot_ip_acls/aclentry_create.html`, add custom **Device** dropdowns to the form above each of these **Relationship** dropdowns, and use Nautobot's built-in support for related form fields acting as filters so that selecting a device restricts the set of Interfaces presented to only include those Interfaces that belong to the device. Not a simple exercise, but very much possible and within the scope of the kinds of problems that an App can solve with a bit of work!

Hit **Create** and you should be redirected to your detail view for the ACL:

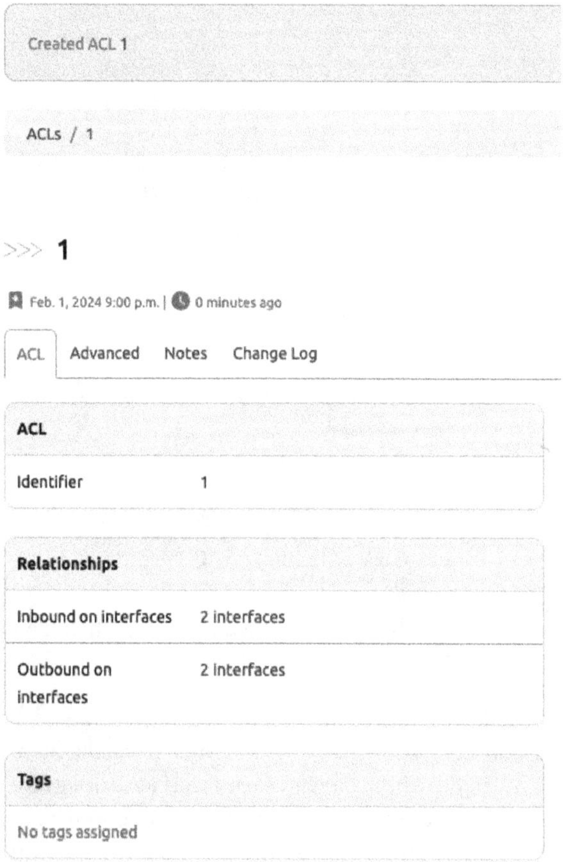

Created ACL 1

ACLs / 1

>>> **1**

Feb. 1, 2024 9:00 p.m. | 0 minutes ago

| ACL | Advanced | Notes | Change Log |

ACL

| Identifier | 1 |

Relationships

| Inbound on interfaces | 2 interfaces |
| Outbound on interfaces | 2 interfaces |

Tags

No tags assigned

Figure 15.22 – An ACL record

Note here as well that Nautobot automatically adds the related relationships and tags for this record to the view, as well as the tabs for viewing related notes and change log entries.

Next, before creating some ACLEntry records, make sure that you've created a few Prefix records so that you have something for an ACLEntry to refer to. Once that's done, you can add an ACL entry by navigating to /plugins/ip-acls/acl-entries/add/:

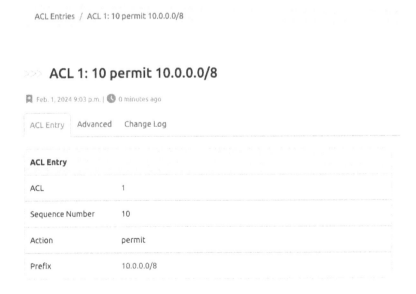

Figure 15.23 – Form for creating a new ACL entry

Create the entry and you should see its detail view:

ACL Entries / ACL 1: 10 permit 10.0.0.0/8

ACL 1: 10 permit 10.0.0.0/8

Feb. 1, 2024 9:03 p.m. | 0 minutes ago

ACL Entry   Advanced   Change Log

ACL Entry

ACL	1
Sequence Number	10
Action	permit
Prefix	10.0.0.0/8

Figure 15.24 – The created ACL Entry record

At this point, your plugin is usable if you know the URLs, but as before, let's make these views reachable through Nautobot's navigation menu instead.

Implementing navigation

This time, we're going to add an entire new menu just for our App because we're just that proud of it. In nautobot_ip_acls/navigation.py, overwrite the old contents with the following:

```python
from nautobot.apps.ui import NavMenuGroup, NavMenuTab, NavMenuItem

menu_items = (
    NavMenuTab(
        name="IP ACLs",
        weight=1000,
        groups=(
            NavMenuGroup(
                name="ACLs",
                items=(
                    NavMenuItem(
                        link="plugins:nautobot_ip_acls:acl_list",
                        name="ACLs",
                        permissions=["nautobot_ip_acls.view_acl"],
                    ),
                    NavMenuItem(
                        link="plugins:nautobot_ip_acls:aclentry_list",
                        name="ACL Entries",
                        permissions=[
                            "nautobot_ip_acls.view_aclentry"
                        ],
                    ),
                ),
            ),
        ),
    ),
)
```

Success

At this point, if you refresh Nautobot in your browser, you should see the new menu:

Figure 15.25 – The new navigation menu for IP ACLs

From here, you can reach the list views for `ACL` and `ACLEntry`, showing off the tables that you defined and the data you've populated so far. (Note that unlike most of the menu items provided by core Nautobot, your new menus don't have the + buttons yet as a shortcut to adding new ACLs or ACL entries. Those are easy enough to add but it would have made the preceding code excerpt a bit unwieldy in length. Addressing this is one of the suggested exercises presented at the end of the chapter.)

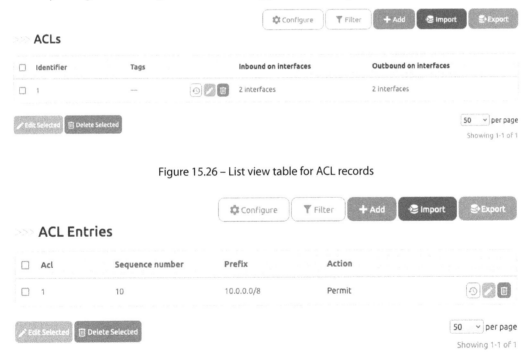

Figure 15.26 – List view table for ACL records

Figure 15.27 – List view table for ACL Entries records

At this point, you have a fully functional data model-based Nautobot App—congratulations!

Exercises or next steps

There's always room for bug fixes and enhancements! Consider exploring on your own any or all of the following possibilities:

- The **ACL Entries** table has a column header of **Acl**, which should really say **ACL**. This is easy to fix by adding a `verbose_name` attribute on the `Column` declaration in your `tables.py` file.

- The create/edit form for ACL records doesn't have a field to let you specify tags to attach to the ACL at present. Can you see what's missing in the `ACLForm` definition to fix this?

- It would be nice if the detail view for a given ACL included a listing of all associated `ACLEntry` records in order. How might you go about extending the `acl_retrieve.html` template to do that?

- The **IP ACLs** menu is currently missing the **add** buttons that most Nautobot models provide in their menus. Refer back to *Chapter 14* for examples of how you might add these.

- Currently, the **Edit Selected** and **Delete Selected** buttons on the table/list views, which should enable the bulk editing and bulk deletion of a set of selected objects, do not actually work yet. You'll need to implement a subclass of Nautobot's `NautobotBulkEditForm` class and set it as `bulk_update_form_class` on your ViewSet; this should only take a few lines of code to implement.

- Currently, the dropdowns for selecting Interfaces to attach to an ACL are not very usable when data in Nautobot grows, as we saw earlier in the chapter. To address this, you might implement a custom form HTML template in `nautobot_ip_acls/templates/nautobot_ip_acls/aclentry_create.html` to include additional form fields and some custom JavaScript, making it easier to look up specific devices and filter to only show the Interfaces attached to those devices.

- How might you alter or augment this App's data models to support `ACLEntry` records that reference a specific IP address as an alternative to a prefix? What alterations to the tables, forms, filters, views, and templates would be required to support this?

- How might you alter or augment this App's data model to support "extended" IP ACLs?

- What other use cases for ACLs could you support, besides assigning them to Interfaces?

Summary

In this chapter, we learned about data modeling within Nautobot and explored two different approaches to implementing a sample Nautobot App for dealing with IP ACL data.

In the following chapter, we'll build on this App by implementing basic configuration rendering based on the data stored in Nautobot and the App, then implementing a Job for pushing the rendered configuration to a device.

16

Automating with Nautobot Apps

After developing a custom Nautobot app to store data relevant to your particular organization's needs in *Chapter 15*, you can logically proceed to the next step: automating the application of this data to your actual day-to-day network operations. This chapter will cover a number of ways that you can achieve this goal through Nautobot by further extending your app and through the use of jobs, which were initially covered in *Chapter 11*.

The chapter will cover the following main topics:

- A real-world use case for network automation in a Nautobot app

- Building an App for network automation

- Next steps on your journey

A real-world use case for network automation in a Nautobot app

The app developed in *Chapter 15* lets you enter device IP ACL configuration into Nautobot; the goal now is to take it one step further in the automation workflow and let Nautobot's IP ACL information drive device configuration. This continues to highlight the value and relationship of network data and network automation we first talked about in *Chapter 1*. Having more data in Nautobot offers more value in driving even more automation directly from Nautobot.

Design requirements

There are three basic requirements for this solution:

- Must be able to record IP ACL definitions in Nautobot (already accomplished in *Chapter 15*)
- Nautobot must be able to render the full ACL config for a given device on demand
- Nautobot must be able to push the rendered config to that device on demand

As noted, this will build heavily on the IP ACLs data model that we developed an App for in *Chapter 15*.

Building an App for network automation

We'll pick up where we left off in the previous chapter with our Nautobot IP ACLs App, extending it first to render the IP ACL configuration CLI based on the data stored in Nautobot, then to use Netmiko to push this rendered configuration directly to network devices.

> **Note**
>
> Much as in the previous chapter, this is something of a contrived example to help you get your hands dirty with App development and network automation using Apps. In your real network and real deployment of Nautobot, you'd probably want to use the Nautobot Golden Config App and/or Nautobot Firewall Models App to manage and implement this sort of device configuration, rather than writing your own app specifically for this purpose.

Rendering IP ACL config using Jinja2

We're going to use Jinja2 as our templating engine for the rendering of the IP ACL configuration CLI. As you may recall, Nautobot uses this library natively as part of its Export Templates feature, among other areas; more generally, Jinja2 can be used to generate any plaintext output format you need, from CSV to JSON to XML to YAML, or in the case we'll implement here, Cisco IOS CLI configuration.

Resuming from where we left off in the previous chapter, we currently have the following app file structure:

```
root@nautobot-app-dev:~/nautobot-app-ip-acls# tree -I __pycache__
nautobot_ip_acls/
nautobot_ip_acls/
├── __init__.py
├── api
│   ├── __init__.py
│   ├── serializers.py
│   ├── urls.py
│   └── views.py
├── filters.py
```

```
├── forms.py
├── migrations
│    ├── 0001_initial.py
│    ├── 0002_aclentry.py
│    └── __init__.py
├── models.py
├── navigation.py
├── tables.py
├── template_content.py
├── templates
│    └── nautobot_ip_acls
│        ├── acl_retrieve.html
│        ├── aclentry_retrieve.html
│        └── device_ip_acls.html
├── tests
│    ├── __init__.py
│    ├── test_api.py
│    └── test_basic.py
├── urls.py
└── views.py
```

Your next step is to add a Jinja2 template for the Cisco IOS IP ACL configuration. Thinking toward the future, you might eventually want to have templates for configuration of other platforms or operating systems, so let's create a subdirectory for such templates:

```
root@nautobot-app-dev:~/nautobot-app-ip-acls# mkdir nautobot_ip_acls/
templates/nautobot_ip_acls/config
```

By convention, Jinja2 template filenames end in .j2, so let's create a nautobot_ip_acls/templates/nautobot_ip_acls/config/cisco_ios.j2 template file, as follows:

```
{%- for acl in acls %}
ip access-list standard {{ acl.identifier }}
{%- for entry in acl.aclentry_set.all() %}
{%- if entry.prefix %}
  {{ entry.sequence_number }} {{ entry.action }} {{ entry.prefix.
network }} {{ entry.prefix.prefix.hostmask }}
{%- else %}
  {{ entry.sequence_number }} {{ entry.action }} any
{%- endif %}
{%- endfor %}
!
{%- endfor %}
```

In Jinja2 template syntax, {%...%} is a statement (effectively logic to execute), and {{...}} is an expression (data to insert into the rendered file). So, what we're doing is iterating over all ACLs provided to the template and generating an `ip access-list standard` line for each one. Underneath that, we iterate over all entries associated with that ACL.

> **Note**
>
> `ACL.aclentry_set` is a default Django *reverse lookup* property that was autogenerated when we declared `ACLEntry.acl` as a foreign key to an ACL record; using this syntax lets us walk "backward" from the ACL to the ACL entries linked to it.

For each ACL entry, the template will generate the corresponding configuration statement, which varies slightly depending on whether it is associated with a prefix or whether it isn't (implying "any," per our data model design).

As a quick test, we can use Nautobot's `nbshell` Python shell to render this template:

```
root@nautobot-app-dev:~/nautobot-app-ip-acls# invoke nbshell
...
# Django version 3.2.23
# Nautobot version 2.1.0
# Nautobot IP ACLs version 0.1.0
Python 3.11.7 (main, Dec 19 2023, 03:30:20) [GCC 12.2.0]
Type 'copyright', 'credits' or 'license' for more information
IPython 8.12.3 -- An enhanced Interactive Python. Type '?' for help.
In [1]: from nautobot.apps.utils import render_jinja2
In [2]: from nautobot_ip_acls.models import ACL
In [3]: with open("nautobot_ip_acls/templates/nautobot_ip_acls/config/
cisco_ios.j2", "rt") as handle:
    ...:     template = handle.read()
    ...:
In [4]: print(render_jinja2(template, {"acls": ACL.objects.all()}))
ip access-list standard 1
 10 permit 10.0.0.0 0.255.255.255
!
```

Looking promising!

Next, you'll want to extend the template to also attach the access lists to interfaces. This will be a bit more complex, due to the necessity of using Nautobot Relationships to implement the access-list-to-interface associations rather than a simple foreign key on the interface model, but we can make it work. First, add the following to the end of your `nautobot_ip_acls/templates/nautobot_ip_acls/config/cisco_ios.j2` file:

```
{%- for interface, acls in interface_acls.items() %}
interface {{ interface }}
```

```
{% if acls["in"] %}ip access-group {{ acls["in"] }} in{% endif %}
{% if acls["out"] %}ip access-group {{ acls["out"] }} out{% endif %}
!
{%- endfor %}
```

Here, too, we can test this in nbshell. In the following example, I'm using a csr device, created as test data in the previous chapter, which has ACLs assigned to GigabitEthernet1 through GigabitEthernet3 interfaces via Nautobot Relationships. Feel free to replicate this configuration, or try it with whatever other device(s) and interface(s) you've got in your own dataset:

```
root@nautobot-app-dev:~/nautobot-app-ip-acls# invoke nbshell
...
# Django version 3.2.23
# Nautobot version 2.1.0
# Nautobot IP ACLs version 0.1.0
Python 3.11.7 (main, Dec 19 2023, 03:30:20) [GCC 12.2.0]
Type 'copyright', 'credits' or 'license' for more information
IPython 8.12.3 -- An enhanced Interactive Python. Type '?' for help.

In [1]: interface_acls = {}
In [2]: for assoc in RelationshipAssociation.objects.
filter(relationship__key="acl_in"):
    ...:     if assoc.destination.device.name == "csr1":
    ...:         interface_acls.setdefault(assoc.destination.name, {"in":
None, "out": None})
    ...:         interface_acls[assoc.destination.name]["in"] = assoc.
source.identifier
    ...:
```

The preceding code iterates over all associations related to the acl_in relationship that we created in *Chapter 15* and uses them to construct a dictionary of dictionaries, where the first-level dictionary keys are the interface names on the csr1 device, the second-level key is the word in, and the leaf values are the ACL identifiers. Next, we do something very similar to iterate over the acl_out relationship's associations and set an out key on the same dictionary with any outbound ACLs:

```
In [3]: for assoc in RelationshipAssociation.objects.
filter(relationship__key="acl_out"):
    ...:     if assoc.destination.device.name == "csr1":
    ...:             interface_acls.setdefault(assoc.destination.name,
{"in": None, "out": None})
    ...:             interface_acls[assoc.destination.name]["out"] = assoc.
source.identifier
    ...:
In [4]: interface_acls
Out[4]:
```

```
{'GigabitEthernet1': {'in': '1', 'out': None},
 'GigabitEthernet3': {'in': '1', 'out': '1'},
 'GigabitEthernet2': {'in': None, 'out': '1'}}
```

Now that we have this data structure, we can use it with our config template to render the target device's ACL configuration:

```
In [8]: from nautobot.apps.utils import render_jinja2
In [9]: from nautobot_ip_acls.models import ACL
In [10]: with open("nautobot_ip_acls/templates/nautobot_ip_acls/
config/cisco_ios.j2", "rt") as handle:
    ...:            template = handle.read()
    ...:
In [11]: print(render_jinja2(template, {"acls": ACL.objects.all(),
"interface_acls": interface_acls}))
ip access-list standard 1
 10 permit 10.0.0.0 0.255.255.255
!

interface GigabitEthernet1
 ip access-group 1 in

!
interface GigabitEthernet3
 ip access-group 1 in
 ip access-group 1 out
!
interface GigabitEthernet2

 ip access-group 1 out
!
```

Success! Now, we just need to be able to send this configuration to an actual device.

Writing a job to push config to a device using Netmiko

For the purposes of this example, we'll use the Netmiko library, which Nautobot includes as a dependency, as our means of sending configuration to a device. In other cases, you might use a different library such as NAPALM or Nornir, but here, we'll use Netmiko for its relative simplicity.

Create a `nautobot_ip_acls/jobs.py` file and start it as follows:

```
import os.path
from netmiko import ConnectHandler
from nautobot.apps.jobs import Job, ObjectVar, register_jobs
from nautobot.apps.utils import render_jinja2
```

```python
from nautobot.dcim.models import Device
from nautobot.extras.choices import SecretsGroupAccessTypeChoices,
SecretsGroupSecretTypeChoices
from nautobot.extras.models import RelationshipAssociation


class ConfigureIPACLs(Job):
    device = ObjectVar(model=Device)

    def run(self, device):


        with open(os.path.join(
            os.path.dirname(__file__),
            "templates",
            "nautobot_ip_acls",
            "config",
            "cisco_ios.j2"
        )) as handle:
            template = handle.read()

        relevant_acls = set()
        interface_acl_ids = {}

        for assoc in RelationshipAssociation.objects.filter(
            relationship__key="acl_in"
        ):
            if assoc.destination.device == device:
                interface_acl_ids.setdefault(
                    assoc.destination.name, {"in": None, "out": None}
                )
                interface_acl_ids[assoc.destination.name]["in"] = (
                    assoc.source.identifier
                )
                relevant_acls.add(assoc.source)

        for assoc in RelationshipAssociation.objects.filter(
            relationship__key="acl_out"
        ):
            if assoc.destination.device == device:
                interface_acl_ids.setdefault(
                    assoc.destination.name, {"in": None, "out": None}
                )
                interface_acl_ids[assoc.destination.name]["out"] = (
                    assoc.source.identifier
```

```
        )
        relevant_acls.add(assoc.source)

config = render_jinja2(
    template,
    {
        "acls": relevant_acls,
        "interface_acls": interface_acl_ids
    }
)
self.logger.info(f"```\n{config}\n```")
```

The preceding code should look pretty similar to what we already did manually in `nbshell` earlier. We're finding all of the ACLs that are relevant to the device specified by the user when submitting this Job, and also figuring out which interfaces these ACLs need to apply to. We then render the configuration template based on this information and log the rendered configuration as a message to the user. (As Job logs support Markdown formatting, we wrap the config in triple backticks so that it will be rendered as a preformatted code block in the UI.)

Next, it's time to continue to actually configure the device. We'll use Nautobot's native Secrets functionality that was introduced in *Chapter 4* to look up the connection credentials needed (we haven't actually configured these credentials yet in Nautobot, but we'll get to that before we run this Job). Continue writing the Job as follows:

```
username = device.secrets_group.get_secret_value(
    SecretsGroupAccessTypeChoices.TYPE_GENERIC,
    SecretsGroupSecretTypeChoices.TYPE_USERNAME,
    obj=device,
)
password = device.secrets_group.get_secret_value(
    SecretsGroupAccessTypeChoices.TYPE_GENERIC,
    SecretsGroupSecretTypeChoices.TYPE_PASSWORD,
    obj=device,
)
```

And next, use Netmiko to connect to and configure the device, and log the output returned:

```
with ConnectHandler(
    device_type="cisco_ios",
    host=str(device.primary_ip.address.ip),
    username=username,
    password=password,
) as net_connect:
    output = net_connect.send_config_set(
```

```
            config.split("\n")
        )
    self.logger.info(f"```\n{output}\n```")
```

Here, we're hardcoding the device-type string to use with Netmiko; in practice, you'd normally want to use something dynamic, such as `device.platform.network_driver_mappings["netmiko"]`, as a way to make this Job more reusable.

Finally, we just need to tell Nautobot that this Job is one that it should discover and register, with the following line at the end of the file, outside the Job definition block:

```
class ConfigureIPACLs(Job):
    device = ObjectVar(model=Device)

    def run(self, device):
        # same as before, omitted for brevity


register_jobs(ConfigureIPACLs)
```

Since this is a new Job, you'll need to run `nautobot-server post_upgrade` to ensure the discovery of the Job, using the `invoke post-upgrade` command:

```
root@nautobot-app-dev:~/nautobot-app-ip-acls# invoke post-upgrade
...
18:06:43.125 DEBUG   nautobot.core.celery __init__.
py                   register_jobs() :
  Registering job nautobot_ip_acls.jobs.ConfigureIPACLs
```

If you see the preceding two lines, or ones similar to them, that indicates that Nautobot correctly saw your new Job. There will be additional output from this command, but that's the part we're interested in here.

If you see an error message here instead, most likely you have a syntax error in your Job code, so double-check that it matches what we've provided previously.

If you see no mention of your Job at all, most likely you're missing the `register_jobs(ConfigureIPACLs)` line at the end of your file, or else it's present but incorrectly indented to fall under the Job class definition instead of being a top-level (unindented) line of code.

Next, as a security feature, Nautobot defaults any newly discovered Jobs to be disabled from being run, so you'll need to go into the Nautobot UI and enable this job. To do that, navigate to **Jobs | Jobs** from the menu, then click the *pencil* icon to the right of the `ConfigureIPACLs` job entry. Check the **Enabled** checkbox in the resulting form and then click the **Update** button:

696 Automating with Nautobot Apps

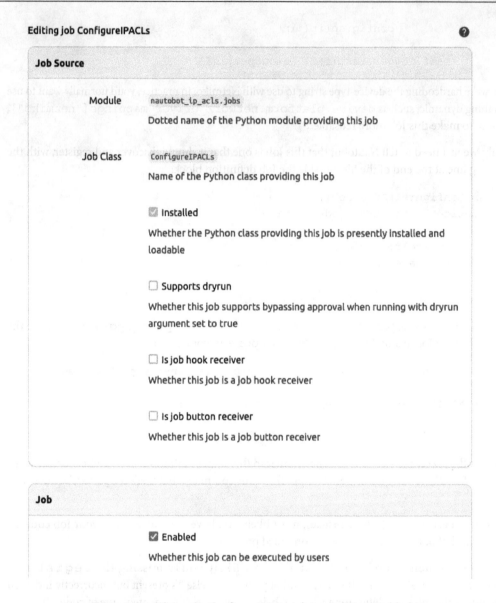

Figure 16.1 – Editing the ConfigureIPACLs job to mark it as Enabled

Next, before we can test out the Job, you'll need to set up a "real" device in Nautobot to run the Job against.

Preparing the device and related data in Nautobot

We can't easily provide you with a device to test this Job against, but hopefully, you have a lab that you can try it out with. If not, you may have some luck using one of the sandbox devices that Cisco Systems currently maintains through `https://devnetsandbox.cisco.com/`. It's definitely easier if you're using a device that you have full control of, however.

You'll need to enter your device into Nautobot, following steps similar to these:

1. Create a manufacturer if you don't have one already.
2. Create a device type belonging to the manufacturer.
3. Create a role and configure it to apply to the `dcim.device` content type.
4. Create a location type and configure it to permit `dcim.device` content types.
5. Create a location using this location type.
6. Create a device using the aforementioned location, role, and device type, and set its **Status** option to **Active**.
7. Use the **Add Components** button on the **Device** detail view to add interfaces to this device, named to correspond to the interfaces present on your device.
8. Add the device's IP address to whichever interface is the management interface for your device and mark it as the primary IP for the device.

Next, you'll want to create several IP ACL records through your App's UI and add ACL entries to them. Edit the non-management interfaces in Nautobot and assign inbound and/or outbound ACLs to them. (Do not assign ACLs to the management interface, as a misconfigured ACL on this interface could block you from accessing the device!)

In the following screenshot, I've used the **Configure** button to customize which columns are displayed in the **Device Interfaces** table, hiding a number of columns that are irrelevant to this example and adding **Inbound ACL** and **Outbound ACL** columns:

Figure 16.2 – Device with interfaces, IPs, and ACLs defined in Nautobot

We'll use text-file-based Secrets to provide the device username and password for this example, though in a real-world deployment, you'd likely want to use whatever enterprise password storage solution you have access to, such as HashiCorp Vault. Shell into the Nautobot container environment using `invoke cli`, and create these files:

```
root@nautobot-app-dev:~/nautobot-app-ip-acls# invoke cli
Running docker compose command "ps --services --filter status=running"
Running docker compose command "exec nautobot bash"
root@5c17b87cf69a:/source# echo -e "myusername" > csr1_username.txt
root@5c17b87cf69a:/source# echo -e "mypassword" > csr1_password.txt
```

(Of course, use the actual credentials provided for your device in place of `myusername` and `mypassword`.)

Returning to the Nautobot UI, create secrets referencing these two files by navigating to **Secrets | Secrets** and creating two records as follows:

Figure 16.3 – Two secret records created

Next, create a secrets group (**Secrets | Secret Groups**) and add these two secrets to this group with the appropriate **Access Type** and **Secret Type** values:

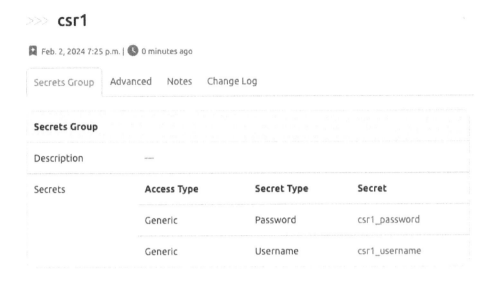

Figure 16.4 – Secrets group record containing the two secrets

Edit the device that you previously created and assign this secrets group to the device:

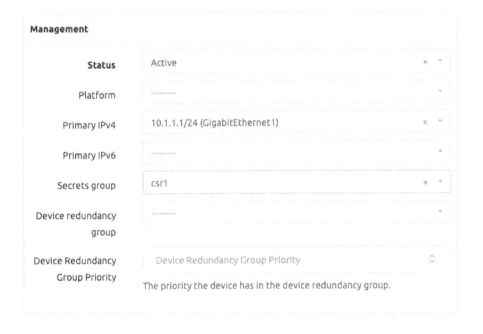

Figure 16.5 – Adding the secrets group to the device

You now have enough information entered into Nautobot for your Job to be able to successfully interact with the device. The preceding activity may seem like a lot of steps, but it's the sort of thing you should only really need to do once per device, and it's definitely something you can automate in and of itself if you have a large number of devices with similar profiles to enter.

Running the job

After all of the aforementioned steps, actually running the Job may seem a bit anticlimactic. Navigate to **Jobs | Jobs** and click the *run* (right-facing triangle) button next to your **ConfigureIPACLs** job. Select your device from the drop-down menu:

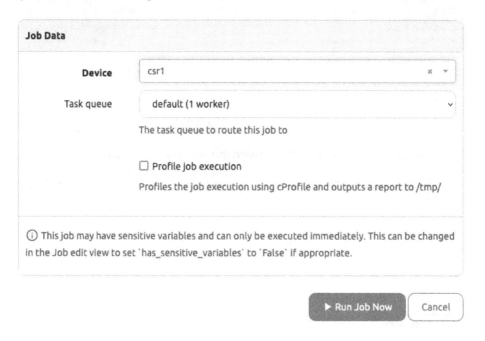

Figure 16.6 – Preparing to run the job

Next, click **Run Job Now**. You should be redirected to the **Job Result** view for the just-scheduled job, and shortly, the job output should appear. If all is well, you should see something like the following, showing the rendered configuration followed by the response received from the device via Netmiko. (In the following screenshot, the Job failed because I didn't actually have my lab device up and running at this time, but close enough for the purpose of demonstration!)

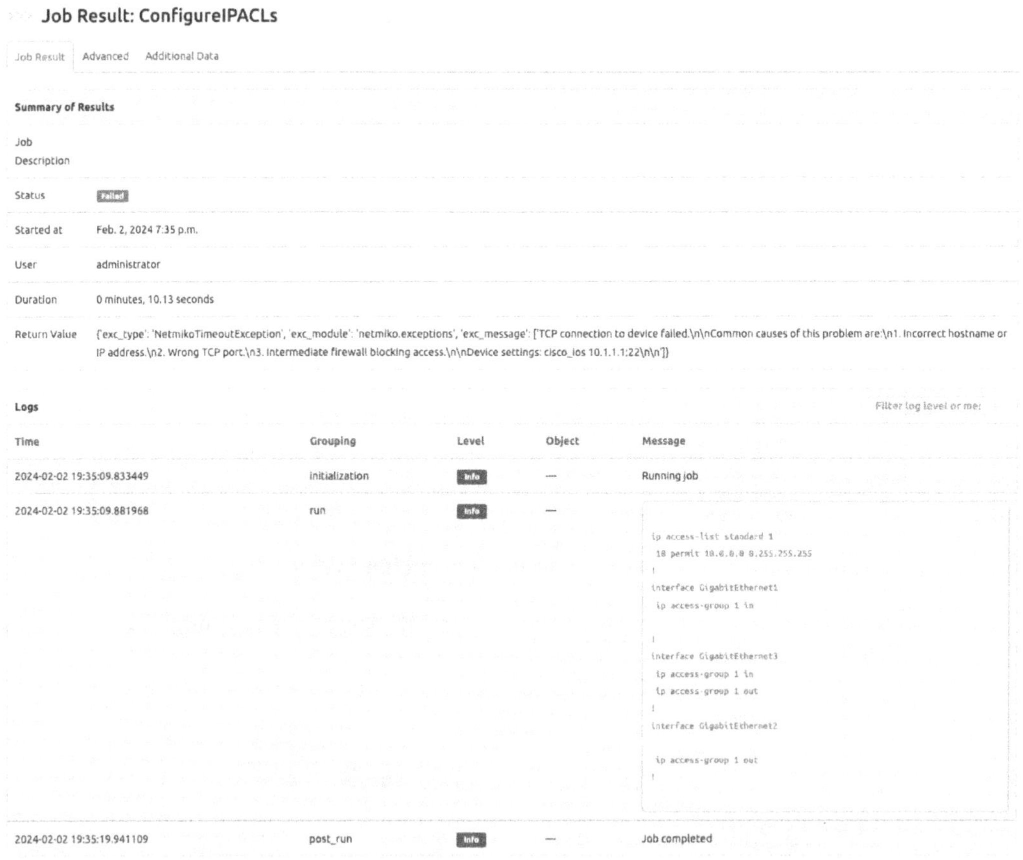

Job Result: ConfigureIPACLs

Job Result   Advanced   Additional Data

Summary of Results

Job Description	
Status	Failed
Started at	Feb. 2, 2024 7.35 p.m.
User	administrator
Duration	0 minutes, 10.13 seconds
Return Value	{'exc_type': 'NetmikoTimeoutException', 'exc_module': 'netmiko.exceptions', 'exc_message': ['TCP connection to device failed.\n\nCommon causes of this problem are:\n1. Incorrect hostname or IP address.\n2. Wrong TCP port.\n3. Intermediate firewall blocking access.\n\nDevice settings: cisco_ios 10.1.1.1:22\n\n']}

Logs   Filter log level or mes

Time	Grouping	Level	Object	Message
2024-02-02 19:35:09.833449	initialization	Info	---	Running job
2024-02-02 19:35:09.881968	run	Info	---	ip access-list standard 1 10 permit 10.0.0.0 0.255.255.255 ! interface GigabitEthernet1 ip access-group 1 in ! interface GigabitEthernet3 ip access-group 1 in ip access-group 1 out ! interface GigabitEthernet2 ip access-group 1 out !
2024-02-02 19:35:19.941109	post_run	Info	---	Job completed

Figure 16.7 – Job output showing rendered configuration and an attempt to configure the device

Hopefully, in your case, you have an actual device to test your app and job against, but as you can see in *Figure 16.7*, even without a "real" device to test against, you can at least demonstrate that your job runs.

Next, let's explore a way to streamline the job execution process a bit further.

Adding a job button to enable one-click configuration

Job buttons are a way to connect a job (specifically, a job subclass called `JobButtonReceiver`) to a model detail page so that the Job can easily be run via a single button click. Let's add a **Push IP ACLs to Device** button to **Device** detail views.

In `nautobot_ip_acls/jobs.py`, add the following and change the `register_jobs(ConfigureIPACLs)` line at the end of the file as follows:

```
from nautobot.apps.jobs import JobButtonReceiver
class ConfigureIPACLsButton(JobButtonReceiver):
    def receive_job_button(self, obj):
        job = ConfigureIPACLs()
        job.job_result = self.job_result
        job.run(obj)


register_jobs(ConfigureIPACLs, ConfigureIPACLsButton)
```

Instead of a run method, `JobButtonReceiver` needs to implement a `receive_job_button` method, which takes a single argument: the object that the button is being clicked from. Here, we are just having the `ConfigureIPACLsButton` job do some clever redirection back to the `ConfigureIPACLs` job to save ourselves from needing to write a bunch of duplicated code.

Next, we'll again run `invoke post-upgrade` to get Nautobot to detect the new Job:

```
root@nautobot-app-dev:~/nautobot-app-ip-acls# invoke post-upgrade
Running docker compose command "ps --services --filter status=running"
Running docker compose command "exec nautobot nautobot-server post_
upgrade"
...
19:41:21.420 DEBUG    nautobot.core.celery __init__.
py                    register_jobs() :
  Registering job nautobot_ip_acls.jobs.ConfigureIPACLsButton
...
19:41:24.036 INFO     nautobot.extras.utils utils.py         refresh_
job_model_from_job_class() :
  Created Job "nautobot_ip_acls.jobs: ConfigureIPACLsButton" from
<ConfigureIPACLsButton>
```

Just as before, this job will be disabled by default when first discovered, so we need to enable it. Once again, navigate to **Jobs** | **Jobs**, click the *pencil* icon to its right to edit it, and enable it for running, as before:

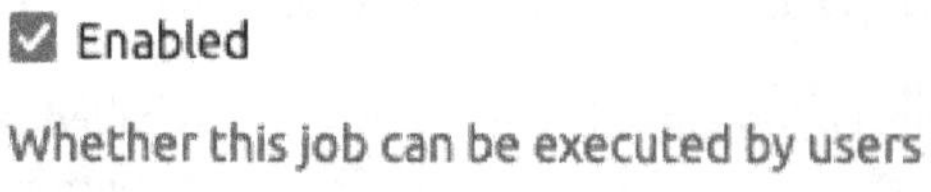

☑ Enabled

Whether this job can be executed by users

Figure 16.8 – Marking JobButtonReceiver as Enabled for running

Next, you'll create a job button database record to enable this particular `JobButtonReceiver` subclass to be included in **Device** detail views. Navigate to **Jobs | Job Buttons** and click the **Add** button, then fill in the form as follows and submit it:

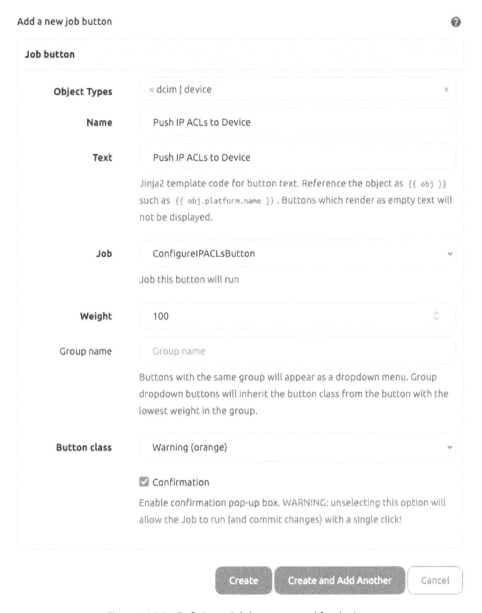

Figure 16.09 – Defining a Job button record for devices

Now, you can navigate to **Devices | Devices** and select your device. In its detail view, you should now see the new job button in the top right of the tab bar:

Figure 16.10 – Device detail view showing the new Job button

Click this button, and (assuming you checked **Confirmation** when creating the job button), click again to confirm the popup that appears:

Figure 16.11 – Job button confirmation prompt

Here, instead of being automatically redirected to the **Job Result** view, you'll remain on the **Device** detail view, but a banner should appear with a link to the job result:

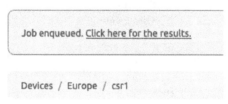

Figure 16.12 – Banner with link to job result

Click the provided link, and you should see the now familiar **Job Result** view:

Figure 16.13 – Job result from Job button

And just like that, you now have the ability to push IP ACLs to a device with a single click!

Next steps on your journey

Consider exploring any of the following possibilities:

- Could you write an "onboarding" or "importer" Job that would connect to a device, issue a `show running-config` or equivalent, parse the output (possibly using a library such as https://github.com/networktocode/ntc-templates), and use it to construct ACL and ACL entry database records in Nautobot for you?

- Could you add extra Jinja2 templates to render configuration for platforms other than Cisco IOS?

- Could you extend the `ConfigureIPACLs` Job to inspect the device's `platform.network_driver` attribute in order to automatically select between multiple Jinja2 templates and configure Netmiko appropriately to connect to that device?

- Could you write a `JobHookReceiver` subclass, similar to `JobButtonReceiver`, that would automatically regenerate and re-push the configuration to a device when an ACL, ACL entry, and/or interface is edited in Nautobot?

- Can you enhance the Nautobot Job based on what you learned in *Chapter 11*?

- If you're interested in exploring ACL modeling in Nautobot in more depth, take a look at `https://docs.nautobot.com/projects/firewall-models/en/latest/` for a more full-featured existing Nautobot App.

Summary

In this chapter, we have explored the fundamental capabilities of using an App and Jobs to render and implement device configuration. The sky really is the limit when it comes to network automation using Nautobot, Apps, and Jobs together. Hopefully, your mind is already awash with the possibilities for your own networks!

Appendix 1
Nautobot Architecture

In *Chapter 3*, we learned how to install the fundamental elements of Nautobot—databases, Redis servers, Celery workers, and schedulers. In this appendix, we'll look at how these services work, along with more advanced use cases.

Nautobot components and services

We'll be taking a look at the following components that comprise a Nautobot installation:

- **Database**: This is where the "source of truth" lives. The database is where Nautobot stores all of its information about your data models and their relationships to each other.

- **Redis**: This is an open source in-memory database that is used for fast storage and retrieval of key/value pairs. It can be installed as a system service or as a separate container through Docker, Podman, or Kubernetes. This is used in Nautobot for low-latency caching and as a task queue for Nautobot jobs.

- **Redis Sentinel**: Sentinel is an alternate deployment mode in Redis 2.8+ for running Redis with high availability. It provides replication of your Redis database to multiple servers and handles failovers, monitoring, and notifications. In short, Redis Sentinel is the preferred method for deploying a robust distributed Redis database for Nautobot.

- **Celery Beat scheduler**: This is the process that manages the monitoring and dispatching of scheduled jobs in Nautobot. It is started using the `nautobot-server celery beat` command.

- **Celery Worker**: This is the process that performs the execution of Nautobot jobs. It is started using the `nautobot-server celery worker` command.

- **uWSGI**: This is the web server that allows the Nautobot Python code to communicate with web clients on the HTTP/HTTPS protocols. It is required to host your web UI, Rest API, and GraphQL API interfaces and is the core service in the Nautobot architecture. The uWSGI server is started when running the `nautobot-server start` command.

These services can be seen in the following figure:

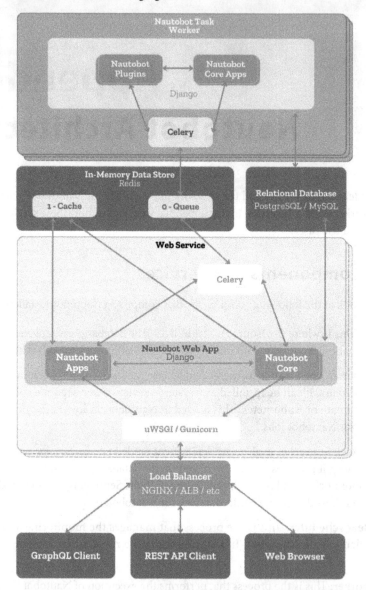

Figure A1.1 – Nautobot services

We'll start by looking at the Nautobot database.

Database: PostgreSQL or MySQL

Like most modern web applications, the database is where Nautobot's data is stored. All of the data model information, discussed in *Chapter 2*, is stored here. The database should be one of the first components that you consider when deploying Nautobot. There are two databases that are supported: PostgreSQL and MySQL. It's up to you which database to use as they both work well with Nautobot. Ideally, you'll want to install the database on a dedicated virtual machine or container or use a dedicated database server. For testing or development purposes, most modern machines with at least 16GB of RAM will handle running Nautobot and the database simultaneously.

In-memory data store: Redis

In *Chapter 3*, we learned about the task queue, where Nautobot workers listen for new tasks from the Nautobot web app. Nautobot uses an in-memory data store service called Redis for its task queue. This task queue is then utilized by Celery, the framework for running Jobs in Nautobot. Jobs power most of the advanced automation features of Nautobot, such as loading data from Git repositories or pushing configuration changes to devices. Celery tasks are messages that contain all of the details to run a job, such as the job name and any arguments supplied for the job. The tasks are published to Redis, and then consumed by one or more Celery workers, which we refer to as Nautobot workers. Here's a step-by-step explanation to better understand the flow of events:

1. A user requests a job execution for a specific job in the Nautobot web UI or REST API. This request may also include data required to run the job, such as a list of devices to run the job against.

2. Nautobot gathers all of the required information from the web request and packages it into a task request. This process is known as serialization.

3. The serialized request is then sent to Redis.

4. Redis places the message into a queue.

5. Celery workers, which are separate processes running in the background, continuously monitor the queue for new messages.

6. When a worker detects a new message in the queue, it retrieves (or "consumes") the message.

7. The worker then deserializes the message back into a task.

8. Finally, the worker executes the task.

The following is a figure showing how this process works at a high level:

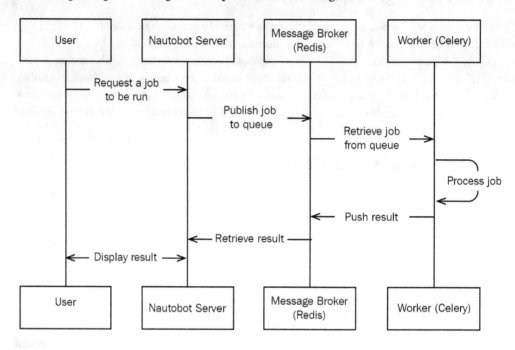

Figure A1.2 – Conceptual job workflow

Using Redis allows for tasks to be distributed across multiple workers, potentially on different machines, which can help to scale the application and improve performance. The next figure shows how multiple workers interact with Redis and the database to share the work from multiple jobs. You may notice that the worker does not store the result on the message broker but instead saves it directly to the database. Before Nautobot version 2.0, the worker would save a result to the message broker and a similar but not identical result to the database. These were merged into a single database entry in version 2.0:

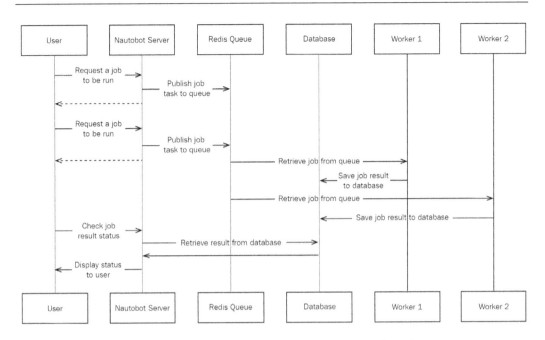

Figure A1.3 – Job workflow in Nautobot 2.x with multiple workers

In-Memory Data Store High-Availability: Redis Sentinel

Redis Sentinel is a system designed for managing Redis server instances. It provides several key features:

- **Monitoring**: Sentinel constantly checks your Redis servers to ensure they are functioning properly.

- **Notifications**: Sentinel can notify you about issues with your Redis servers.

- **Automatic failover**: If a master Redis server is not functioning, Sentinel can automatically promote a slave server to be the new master.

- **Configuration provider**: Sentinel can provide other Redis clients with the information about configuration and the current master–slave setup.

You would want to use Redis Sentinel for the following reasons:

- **High availability**: Sentinel helps to ensure that your system remains operational even in the case of a failure. It does this by automatically switching to a replica server if the main server fails.

- **Disaster recovery**: Sentinel can help to minimize downtime and data loss in the event of a failure.

- **Scalability**: By managing the configuration of your Redis servers, Sentinel makes it easier to add or remove servers as your needs change.

Redis is required for Nautobot to run, so you may want to consider using Redis Sentinel if you need to perform maintenance on your Redis server without impacting Nautobot. Sentinel is one of two methods for deploying Redis across multiple servers and it operates in an active/standby configuration. If you would like to distribute your Redis store across multiple active servers, Redis Cluster may be a better option. A typical Nautobot installation does not put enough load on Redis to require a Redis Cluster installation, so it won't be discussed in detail here.

Job execution: Celery Worker(s)

A Celery worker runs jobs in Nautobot. You can run as many workers as you need but they must all have access to the Nautobot database and Redis servers. By default, the Celery worker will run at most n concurrent tasks and reserve an additional $4n$ tasks, where n is the number of CPU cores on the system. You can change this behavior by sending the following arguments to the `nautobot-server celery worker` command:

```
nautobot-server celery worker --concurrency <n>
    Number of concurrent tasks to process
nautobot-server celery worker --prefetch-multiplier <n>
    Number of tasks to prefetch for each concurrent process (default
4)
```

Since Nautobot jobs spend a lot of their time waiting for responses from network devices or remote APIs, it's usually safe to increase the number of concurrent processes as long as your server has the memory capacity. Nautobot uses fork-based multiprocessing, which makes a copy of the entire process including its memory so each forked process will use 100–200 MBs of memory. If you notice your workers are reaching 100% CPU utilization, you may need to add more workers to your deployment.

Celery makes it easy to scale your workloads horizontally. This means that if you need more capacity, you add more servers running the Celery worker process. Adding more workers may not be necessary, depending on what kind of capacity issues you're experiencing. If your workers' CPUs and memory capacities are not being fully utilized but you still have jobs stacking up in the queue, you may need to try separating your jobs into distinct queues.

Job queues: Celery task queues

If you're running a combination of long and short running jobs, you may find that your queue is becoming full and jobs are taking a long time to start. This problem can usually be resolved by separating your long and short jobs into distinct Celery queues so that long running jobs don't block short jobs. New queues can be used in Nautobot by starting a worker process listening on that queue. You can select any name you would like for your queue, but Nautobot's default queue is named `default`. Here's an example of how you can start two Celery workers on the same server listening to different queues:

```
nautobot-server celery worker \
    --loglevel INFO \
```

```
    --concurrency 10 \
    --queues slow \
    --hostname 'worker1@%%h'

nautobot-server celery worker \
    --loglevel INFO \
    --concurrency 4 \
    --queues fast \
    --hostname 'worker2@%%h'
```

In this example, we had to specify the hostname of each worker using the --hostname argument, otherwise Celery would default to the server's hostname and there would be a name conflict. If your long running jobs are blocked by IO and not CPU, it should be safe to increase the concurrency value on this queue; but you should always monitor your worker's load and adjust your number of concurrent processes according to your use case.

It may seem like a good idea to allow the slow queue to pick up jobs from the fast queue if there's an idle process, but remember that by default Celery reserves up to five total tasks per process, one active task and four queued. This means that a "fast" job picked up by the slow worker could still end up waiting for four slow jobs to finish before it starts.

The task queues that a user can select when running a job are defined on each individual job. The list of queues may be defined by the job developer in the Python code, but they can also be overridden within Nautobot by editing a given job from the UI:

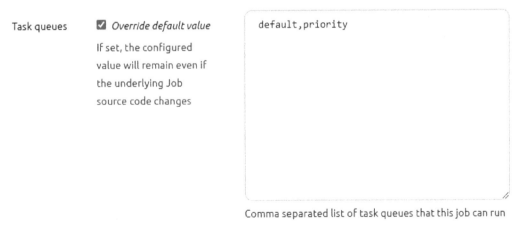

Figure A1.4 – Setting specific task queues for a given job

When a user runs a job, they will be presented with a list of queues to choose from, shown in the **Task queue** dropdown in *Figure A1.5*, which shows how many workers are listening on each queue. If a job is sent to a queue that no workers are listening to, that job will remain in pending status until the queue is manually purged with the `nautobot-server celery purge -Q queue_name` command or a worker starts listening to the queue and processes the job. It's important to ensure that the queue names on the workers and jobs match:

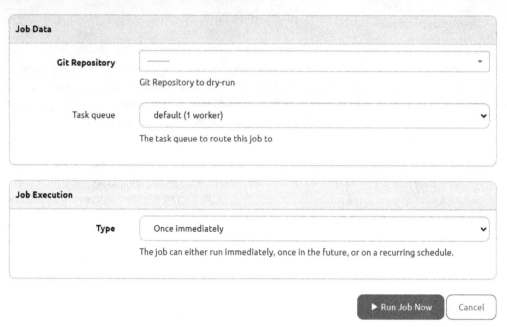

Figure A1.5 – Viewing the Task queue on the Job Execution screen

Workers can also listen on more than one queue simultaneously by passing multiple queue names to the queues argument when starting a Nautobot worker process:

```
nautobot-server celery worker --queues default,queue2,queue3
```

Job scheduler: Celery Beat

Nautobot jobs can be scheduled to run at regular intervals and optionally require approval to run. To accomplish this, Nautobot uses Celery Beat, the scheduling component of Celery. Scheduled jobs are stored in the database and a separate process, called the beat process, monitors the database for changes to the schedule. When a scheduled job is due to run, the beat process places the job in the Celery queue to be picked up by a worker. You have to ensure that only one beat scheduler is running at a time or you will end up with duplicated tasks.

Web server: uWSGI

Nautobot uses two separate components to serve its web requests. Django is the web framework where all of Nautobot's application-specific functions are executed, and uWSGI is the web server that handles the http and https requests from the user.

Django includes a development server to serve http requests, but it is not designed to scale for production use and isn't guaranteed to be free of security vulnerabilities. The Django development server is started if you run the `nautobot-server runserver` command as you have in prior chapters.

To serve web requests in production, Nautobot uses a purpose-built web server called uWSGI. uWSGI uses a standard Python interface called the **Web Server Gateway Interface** (**WSGI**, defined by PEP 333) to communicate with Django. The uWSGI server is started when you run the `nautobot-server start` command. A typical web request flow in Nautobot looks like this:

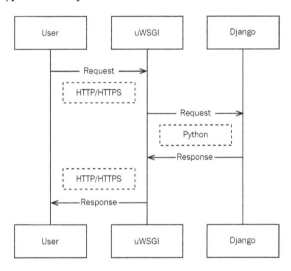

Figure A1.6 – Nautobot HTTP/web request workflow

The uWSGI server requires a configuration file in `ini` format that must be passed to the `nautobot-server start --ini <file>` command.

Here's an example configuration file:

```
[uwsgi]
; The IP address (typically localhost) and port that the WSGI process
should listen on
http = 127.0.0.1:8080

; Fail to start if any parameter in the configuration file isn't
explicitly understood by uWSGI
```

```
strict = true

; Enable master process to gracefully re-spawn and pre-fork workers
master = true

; Allow Python app-generated threads to run
enable-threads = true

;Try to remove all of the generated file/sockets during shutdown
vacuum = true

; Do not use multiple interpreters, allowing only Nautobot to run
single-interpreter = true

; Shutdown when receiving SIGTERM (default is respawn)
die-on-term = true

; Prevents uWSGI from starting if it is unable to load Nautobot
(usually due to errors)
need-app = true

; By default, uWSGI has rather verbose logging that can be noisy
disable-logging = true

; Assert that critical 4xx and 5xx errors are still logged
log-4xx = true
log-5xx = true

; Enable HTTP 1.1 keepalive support
http-keepalive = 1

;
; Advanced settings (disabled by default)
; Customize these for your environment if and only if you need them.
; Ref: https://uwsgi-docs.readthedocs.io/en/latest/Options.html
;

; Number of uWSGI workers to spawn. This should typically be 2n+1,
where n is the number of CPU cores present.
; processes = 5

; If using subdirectory hosting e.g. example.com/nautobot, you must
uncomment this line. Otherwise you'll get double paths e.g. example.
com/nautobot/nautobot/.
```

```
; Ref: https://uwsgi-docs.readthedocs.io/en/latest/Changelog-2.0.11.
html#fixpathinfo-routing-action
; route-run = fixpathinfo:

; If hosted behind a load balancer, uncomment these lines, the
harakiri timeout should be greater than your load balancer timeout.
; Ref: https://uwsgi-docs.readthedocs.io/en/latest/HTTP.
html?highlight=keepalive#http-keep-alive
; harakiri = 65
; add-header = Connection: Keep-Alive
; http-keepalive = 1
```

This appendix provided a comprehensive overview of the Nautobot architecture. We talked about supported databases, PostgreSQL and MySQL, the in-memory database Redis, and the Celery tools that are used to drive automation through Nautobot. To get the most out of your Nautobot deployment, you should run at least one Celery worker, a Celery Beat scheduler, and the uWSGI web server. But if you just want to dive in quickly and try out Nautobot, all that's required is a database, Redis, and Nautobot itself, as seen starting in *Chapter 3*.

Appendix 2
Integrating Distributed Data Sources of Truth with Nautobot

Through this book, you have discovered that Nautobot is capable of modeling your network infrastructure out of the box, and also of being extended to adapt to many use cases not covered by default. However, even with all this potential, it is pretty common that many environments already have other data sources that contain relevant network intent data, and that data can't be owned by Nautobot – such as for organizational boundaries or established processes. For example, you may already have a circuits database owned by the telco team or maybe an IPAM system owned by another network department, and you may need to keep them in use without changing their processes.

In these cases, we have to find ways to interact with all the required data sources to obtain the data that defines the network intent. This is the goal of this chapter.

In this chapter, we cover the following:

- Understanding distributed data sources
- Exploring the Nautobot SSoT framework
- Building your own SSoT integration

We start by defining the problem statement and the different approaches you might follow to solve it.

Understanding distributed data sources

As we already mentioned, it's pretty common that when establishing your Nautobot **Source of Truth** (**SoT**), there could be other data sources that contain relevant data. You may be asking yourself, *"What is relevant data?"* Well, everything you need to know in order to define the intended state of your network.

This can take many forms, as you have already seen from locations, racks, cables, devices, users, cloud services, IP addresses, circuits, and so on. The list is very long. There is a lot of data that you may require to create a device configuration. And, as we said, the data can already be found in some systems that are used for other purposes.

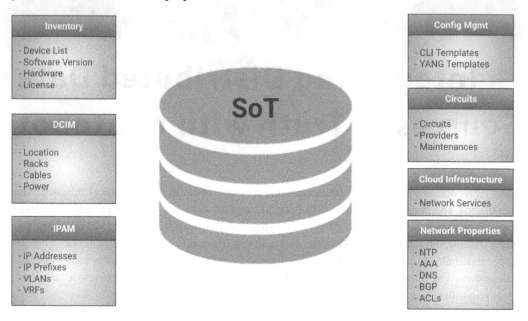

Figure A2.1 – Data types

A common example of this is DDI (DNS, DHCP, and IPAM) solutions, such as Infoblox or Bluecat, which maintain the data for these functionalities (closely connected) while delivering these services. Nautobot is capable of managing the IPAM, but it is not a DNS or a DHCP server. Thus, it is a very reasonable decision to maintain a DDI solution in parallel to Nautobot.

However, the data from Infoblox could find applications if reused in Nautobot (for example, to assign the IP addresses to the VLANs or interfaces, so the desired configurations can be generated), or the other way around. We may need to get all the IPAM information into Nautobot (i.e., sync from Infoblox to Nautobot) and then start incrementally updating Infoblox when Nautobot data changes (i.e., sync from Nautobot to Infoblox). Both serve valid use cases depending on the workflow desired. However, this highlights a key concept in distributed data sources, the **System of Record (SoR)**, already introduced in *Chapter 1*. This ensures you define one system that is the owner of a specific set of data. The other systems become consumers of that data.

The SoR is the owner of a specific dataset. It's usually defined by the type of data (e.g., inventory, IPAM, etc.) and organizational boundaries, but it could be much more granular. For instance, we could have an SoR for the prefix `10.0.0.0/8` and another one for the prefix `172.16.0.0/12`. Defining which are the SoRs of your datasets is crucial to keeping data in good shape.

Thus, we have to embrace that, in most cases, the data will be distributed across several data sources and it comes with several challenges to address.

Challenges of distributed data

Distributed data implicitly introduces some challenges in obtaining good data quality to feed your network automation solutions:

- **Data validation** is a key aspect of the source of truth (i.e., your automation will only be as good as the data sustaining it). For every data source, you need to enforce data validation rules consistently across all of them.

- Having data in multiple places can lead to **inconsistent data** because of data conflicts that require solving the ambiguity about which data source is right regarding a specific data item.

- With several data sources and destinations, **data may be out of sync** due to the difference between the time of synchronization and the time of consumption.

- Data could belong to multiple domains, so understanding and enforcing **data ownership** is necessary to get the truth.

- Keeping track of all systems, maintaining them, and enforcing common standards.

All these challenges can be mitigated by aggregating data.

Benefits of aggregating data

Finally, even though we may have distributed data sources, a critical requirement of an automation strategy is to have a consistent view of the data to enforce a clear direction of how the network should look. Let's look at a few reasons why aggregating and centralizing data makes sense for network automation:

- The data validation challenge is reduced when there is only one data source to validate. All the necessary information is in one place (sometimes the validation has to cross-check different data types), and only one set of validation rules has to be defined.

- Defining a **Single Source of Truth** (**SSoT**) implicitly enforces consistent data by using normalization and creating relationships between the data, and, ideally, enriching the aggregated data.

- Once the data is aggregated, at that moment, the data is in sync. This has to be considered carefully to understand that the synchronization expires and what it means for your automation.

- During data aggregation, the roles of the data sources are honored so the data ownership is warranted and tracked.

Now that we understand the pros and cons of distributed data management, we next explain the general approach to dealing with this type of environment.

Approaches to distributed data management

To overcome these challenges, you may choose different strategies, each one with its own pros and cons. In general terms, the following three represent the main options:

- **Establish integrations with every authoritative SoR**: This is a decentralized approach. The different network automation components (e.g., a Python script, an Ansible Playbook, a custom tool, a commercial tool, etc.) need to integrate with each one of the data sources. For example, having a script that generates a configuration for a device interface may require the retrieval of information from a DCIM system (where the interfaces are defined) and an IPAM system (where the IP addresses are defined), and finally, understand how both data elements are connected. We've seen dozens of data sources that need to be accessed to generate a full network configuration.

  Each integration requires the mechanics to retrieve the data, but most importantly, it requires establishing a process to normalize the data (to integrate with other data coming from the other data sources) and to create logic to connect the different data together for meaningful outcomes. As you can see, the complexity of this approach grows exponentially and expects client-side scripts to connect to each data store.

- **Stateful data aggregation**: Here, the strategy is to consolidate the data from different sources into a single source, so the consumers of the data only need to establish one integration and can delegate the normalization and interrelation logic to the single source of truth. Following the previous configuration example, the Python script only needs one integration to get the interface and its IP address.

  This allows simpler client implementations and also makes it easier to maintain consistency across the data, creating enriched relationships between the data that can be leveraged by the automation applications. Being stateful, we have to consider the potential out of sync between when data was aggregated and used.

- **Stateless data aggregation**: This also offers a single pane of glass-like stateful data aggregation, but the data aggregation is generated on the fly in this case. Every time the consumer of the data requests the data, an aggregator (you can think of it like an API gateway aggregating many API calls) requests the data from the different data sources and applies the normalization and transformation required with the latest data available. In this case, the downside is the extra time required to respond and the limitations to enriching data that's connected from various sources.

Nautobot supports both data aggregation strategies. A Nautobot App can act as an API gateway for data aggregation or be a data source for an external aggregator, but this is out of the scope of this book. On the other hand, there is a Nautobot SSoT App that serves as a framework to implement stateful data aggregation with easy-to-implement (and reuse) integrations leveraging any type of external integrations (e.g., REST & GraphQL APIs, event bus, etc.). This is what we will cover in this chapter.

After this short introduction to managing distributed data sources, we'll dive into the Nautobot SSoT framework that helps aggregate and centralize data, which usually provides more long-term success in network automation projects.

Exploring the Nautobot SSoT framework

Nautobot is used as the primary network source of truth in many network environments. But in most of them, other data sources are in use simultaneously. These data sources may be systems of record or systems that consume Nautobot's data within well-established business operational workflows, and replacing them may not be a top priority. For these cases, Nautobot has a SSoT framework (`https://docs.nautobot.com/projects/ssot/en/latest/`) that allows the development of integration with external systems to retrieve data, compare it with existing data, and finally perform the necessary operations: create, update, or delete the data to establish Nautobot as the aggregated source of truth. This doesn't dictate that Nautobot has to be the SoR of all the data, just the place where the data converges.

> **Pro tip**
>
> If Nautobot needs to import data from a source system but doesn't require continuous synchronization, it doesn't make sense to build an SSoT solution. You can use a simple import job that does a one-time import.

The Nautobot SSoT framework is a Nautobot App that facilitates developing integrations with other systems, providing the following features out of the box:

- A dashboard UI with all the registered integrations and a summary of the synchronization history
- The outcome of the data synchronization job execution for auditing
- Logging of the difference between the initial and final state of the data
- Predefined engineering patterns to synchronize data between any systems

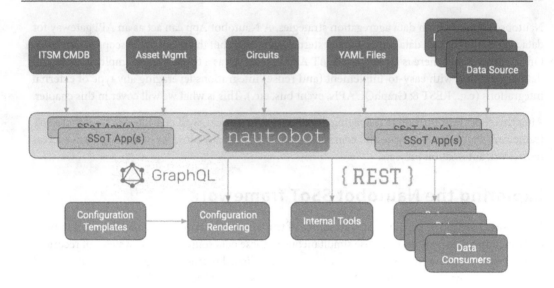

Figure A2.2 – Nautobot SSoT framework

At their core, the Nautobot SSoT integrations are the combination of Nautobot fobs and the DiffSync (https://github.com/networktocode/diffsync) Python library (also an open source project by Network to Code). The DiffSync library allows defining custom logic to compare data models from different systems and synchronize the resulting differences to bring both systems to the same state.

Figure A2.3 – Nautobot SSoT job components

> **Note**
>
> The DiffSync library is not dependent on Nautobot. It can be used for any kind of Python project where there is a need for data synchronization and aggregation.

Using the SSoT framework facilitates developing new integrations on your own. It saves countless hours compared to developing custom integrations due to its framework approach. The integrations you can build with SSoT can be any data source you can imagine (and you have access to) from simple YAML files (we will use it as an example in the last section of this appendix) to sophisticated APIs (as we show next when introducing some of the already available open source integrations).

Another important feature of the Nautobot SSoT framework is that Nautobot can be both a source or a destination of an integration. Indeed, Nautobot can consolidate data from many data sources, but can also propagate data to other systems. This is helpful to enrich existing IPAM and monitoring tools just to name a few.

But, before getting into more details, let's have a look at the Nautobot SSoT framework.

Getting started with the Nautobot SSoT framework

Let's get started!

Installing the Nautobot SSoT framework

The plugin is available as a Python package in PyPI and can be installed atop an existing Nautobot installation using pip:

```
nautobot@nautobot-dev:~$ pip install nautobot-ssot
Collecting nautobot-ssot
  Downloading nautobot_ssot-2.4.0-py3-none-any.whl.metadata (10 kB)

... omitted packages installation ...

Successfully installed colorama-0.4.6 diffsync-1.10.0 nautobot-
ssot-2.4.0 pydantic-1.10.14 structlog-22.3.0
nautobot@nautobot-dev:~$
```

Now, following the same process we did in previous chapters, the application has to be enabled in your nautobot_config.py file (use the editor of your preference):

```
nautobot@nautobot-dev:~$ vi /opt/nautobot/nautobot_config.py
```

Next, let's add the plugin to the PLUGINS list and uncomment the line:

```
# Enable installed plugins. Add the name of each plugin to the list.
#
PLUGINS = ["nautobot_ssot"]
```

To complete the installation, you need to apply the database migrations and installation steps that the new application requires:

```
nautobot@nautobot-dev:~$ nautobot-server post_upgrade
Performing database migrations...
Operations to perform:
  Apply all migrations: admin, auth, circuits, contenttypes, database,
dcim, django_celery_beat, django_celery_results, extras, ipam,
nautobot_ssot, sessions, silk, social_django, taggit, tenancy, users,
virtualization
```

```
Running migrations:
  Applying nautobot_ssot.0001_initial... OK
  Applying nautobot_ssot.0002_performance_metrics... OK
  Applying nautobot_ssot.0003_alter_synclogentry_textfields... OK
  Applying nautobot_ssot.0004_sync_summary... OK

... omitted some messages ...
```

Then, as root, you need to restart the Nautobot app and worker(s) to be aware of the new application:

```
root@nautobot-dev:~# systemctl restart nautobot nautobot-worker
nautobot-scheduler
```

If you navigate to **Plugins** > **Installed Plugins**, you will notice the SSoT.

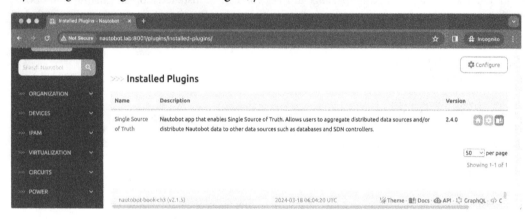

Figure A2.4 – Nautobot Installed Plugins

In the same **Plugins** tab, under the **Single Source of Truth** section, you have three options: **Dashboards**, **History**, and **Logs**. Let's explore them.

Using the Nautobot SSoT application

In the **Plugins** > **Single Source of Truth** > **Dashboard** section, you can see two lists, **Data Sources** and **Data Targets**.

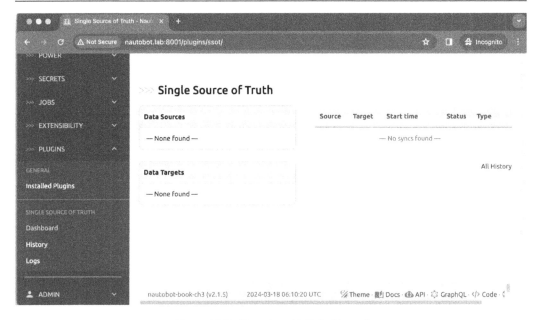

Figure A2.5 – Nautobot SSoT dashboard

In our case, we don't see anything. Why? Well, the reason is that we have not enabled any integrations, and we haven't activated the default Nautobot integrations (to sync to and from other Nautobot instances).

In PLUGINS_CONFIG, within nautobot_config.py, we have to activate the hidden example jobs:

```
PLUGINS_CONFIG = {
    "nautobot_ssot": {
        "hide_example_jobs": False,
    }
}
```

Then, restart the Nautobot services to take the new configuration and run post_upgrade to import the jobs into Nautobot. After this, we should have the two Nautobot integrations:

```
root@nautobot-dev:~# systemctl restart nautobot nautobot-worker
nautobot-scheduler
root@nautobot-dev:~# sudo -iu nautobot
nautobot@nautobot-dev:~$ nautobot-server post_upgrade
... omitted output ...
11:39:15.075 INFO    nautobot.extras.utils :
  Refreshed Job "SSoT Examples: Example Data Source" from
<ExampleDataSource>
11:39:15.082 INFO    nautobot.extras.utils :
  Refreshed Job "SSoT Examples: Example Data Target" from
<ExampleDataTarget>
... omitted output ...
```

Figure A2.6 – Default Nautobot SSoT jobs

Upon entering **Example Data Source**, we get a summary of this integration:

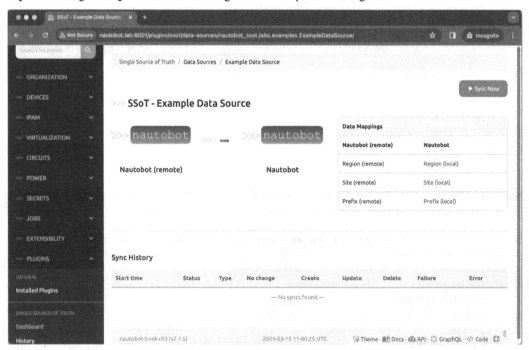

Figure A2.7 – Nautobot SSoT data mappings

You can observe that this job is taking data from an external Nautobot instance (over API) into the local Nautobot (with direct access to the database). On the right side, you see the three different data models to be converted. In this case, being Nautobot to Nautobot, the models are the same but in other integrations, they may differ. Finally, under **Sync History**, we can see that no execution has been performed yet. Let's change this.

If you click the top-right **Sync Now** button, you will get into the job execution mode. Jobs are not enabled to run by default, so you must enable it in the **Jobs -> Jobs** configuration panel.

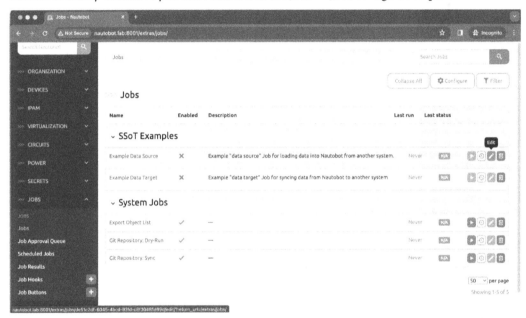

Figure A2.8 – Enable and tun Nautobot SSoT jobs

Now, in the **Example Data Source** sync job, you are ready to try the first data synchronization. By default, it takes the URL and token from the public Nautobot demo instance that we will use in this example. Also, notice that, by default, the **Dryrun** option is checked, meaning that the result of the data comparison is not going to be saved in our local database. This is very useful when we want to understand what would happen before actually committing to it.

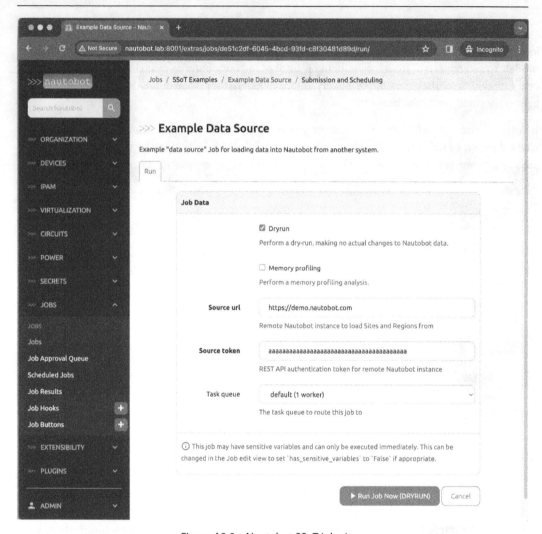

Figure A2.9 – Nautobot SSoT job view

Let's click on the **Run Job Now** button to get the first dryrun synchronization (it could take a few seconds). The next result may differ depending on the actual data you have in your local Nautobot instance and the data that the demo instance has (people can change it), so take it as a reference.

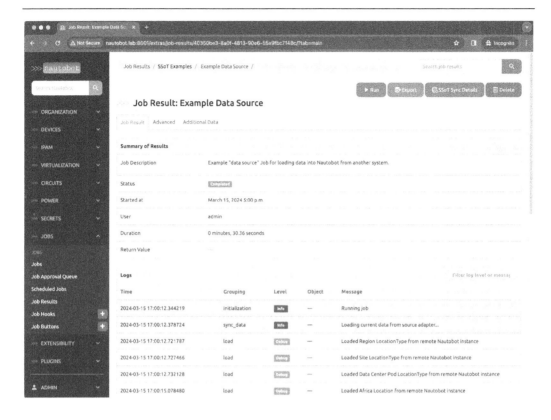

Figure A2.10 – Nautobot SSoT job result

In the **Job Result** page, you can see in debug mode (because we are running the instance in debug mode) all the steps are done. First, all the data from the source is loaded, then the data from the destination, and the difference is calculated.

If you click on the **SSoT Sync Details**, you can explore in detail the data changes:

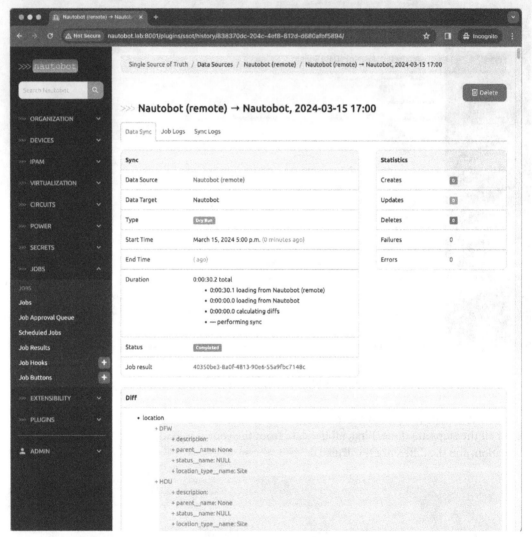

Figure A2.11 – Nautobot SSoT sync details with dryrun

You can see in green the data that will be created/updated, and in red, what will be removed. However, nothing has actually happened in our local Nautobot environment because, as you can also see in the previous screenshot, we ran a dryrun execution.

Now, let's be brave and run it with write access to see how our data in Nautobot changes.

> **Note**
>
> Before doing an actual synchronization, you may need to create a Status with the name NULL that is referenced by the data from the Nautobot demo instance. This is because the example job does not synchronize statuses and the dataset uses a custom one.

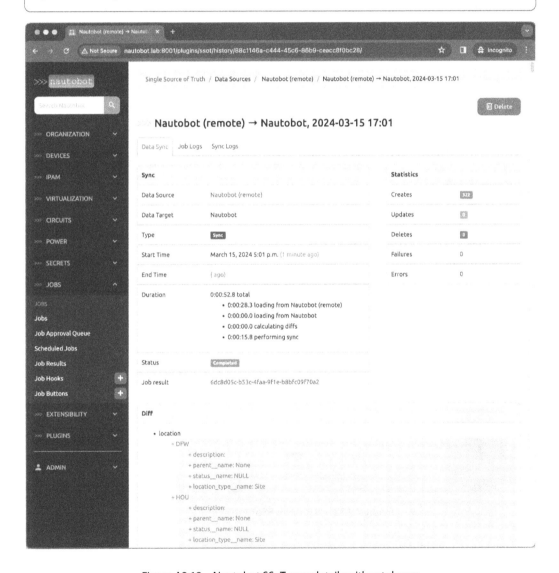

Figure A2.12 – Nautobot SSoT sync details without dryrun

This time, without data changing in between, the difference is the same, but in the **Statistics** section, we can see how many objects have been created, updated, and deleted. So, if the synchronization was successful, you should be able to go to the Sites and validate that new DFW and HOU sites have been created.

It's also important to notice potential failures or errors when some data is not able to be synchronized.

> **Tip**
> SSoT jobs are usually scheduled to provide continuous synchronization.

In the SSoT **Sync History** panel, under **Plugins** > **Single Source of Truth**, you can track the execution of all the SSoT jobs you may have. In our case, we can only see the two executions:

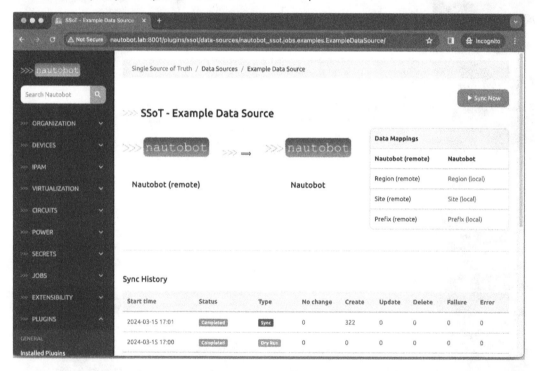

Figure A2.13 – Nautobot SSoT jobs history

This example is a built-in simple integration for demonstration purposes, but in the next section, we will use real integrations that are incorporated into the Nautobot SSoT framework.

Existing SSoT integrations

Similar to the example Nautobot integrations, the Nautobot SSoT framework allows the creation of any custom integration you may be interested in. Network to Code has created many integrations, and when those are reusable and ready to be shared with the community, these are consolidated into the Nautobot SSoT framework repository as package extensions.

Currently, there are six embedded integrations (and many more in the works):

- Cisco ACI

- Arista CloudVision

- Infoblox

- IP Fabric

- ServiceNow

- Device42

Some of these integrations are bidirectional (e.g., Infoblox and ServiceNow), and the others only take data into Nautobot.

Because we need an actual usable service to demonstrate these integrations, we have selected Cisco ACI from the always-on instance in DevNet Sandbox (`https://developer.cisco.com/site/sandbox/`). You could leverage your own Cisco ACI APIC controller or try other integrations.

The first step is to install the extra software needed by this integration (you may want to add it to your `local-requirements.txt` to reproduce it):

```
nautobot@nautobot-dev:~$ pip install nautobot-ssot[aci]
```

> **Note**
>
> In this case, no extra packages are needed, but other integrations may need to install particular SDKs and their dependencies to talk with the data systems.

Every integration will have its own configuration requirements. For Cisco ACI you can find them here (`https://docs.nautobot.com/projects/ssot/en/latest/admin/integrations/aci_setup/`). These should go into the `nautobot_config.py` configuration file (secrets and other config could be taken as environmental variables as recommended in the docs):

```
PLUGINS_CONFIG = {
    "nautobot_ssot": {
    "hide_example_jobs": False,
        "enable_aci": True,
```

```
        "aci_apics": {
        "NAUTOBOT_APIC_BASE_URI_DEVNET": "https://sandboxapicdc.cisco.
com",
        "NAUTOBOT_APIC_USERNAME_DEVNET": "admin",
        "NAUTOBOT_APIC_PASSWORD_DEVNET": "!v3G@!4@Y",
        "NAUTOBOT_APIC_VERIFY_DEVNET": False,
        "NAUTOBOT_APIC_SITE_DEVNET": "DevNet Sandbox",
    },
        # Tag which will be created and applied to all synchronized
objects.
        "aci_tag": "ACI",
        "aci_tag_color": "0047AB",
        # Tags indicating state applied to synchronized interfaces.
        "aci_tag_up": "UP",
        "aci_tag_up_color": "008000",
        "aci_tag_down": "DOWN",
        "aci_tag_down_color": "FF3333",
        # Manufacturer name. Specify existing, or a new one with this
name will be created.
        "aci_manufacturer_name": "Cisco",
        # Exclude any tenants you would not like to bring over from
ACI.
        "aci_ignore_tenants": ["common", "mgmt", "infra"],
        # The below value will appear in the Comments field on objects
created in Nautobot
        "aci_comments": "Created by ACI SSoT Integration",
    }
}
```

Now, with enable_aci configured to True, we run post_upgrade and restart Nautobot to get the Cisco ACI Data Source job:

```
nautobot@nautobot-dev:~$ nautobot-server post_upgrade

... Notice how after the Example Jobs you get some Cisco ACI data ...

17:12:14.139 INFO    nautobot.ssot.aci :
  Creating tags for ACI, interface status and Sites
17:12:14.155 INFO    nautobot.ssot.aci :
  Creating manufacturer: Cisco
17:12:14.183 INFO    nautobot.ssot.aci :
  Creating Site: DevNet Sandbox
17:12:14.196 INFO    nautobot.ssot.aci :
```

```
   Creating Device extra fields for PodID and NodeID
17:12:14.217 INFO     nautobot.ssot.aci :
   Creating Interface extra fields for Optics
... ommitted output ...
```

Now you have available the SSoT for Cisco ACI on the **Single Source of Truth** page:

Figure A2.14 – Nautobot SSoT Cisco ACI

> **Note**
> Remember that to run a job, you need to enable it (as we did previously with the example jobs).

If we run the Cisco ACI data source job against the API devnet (coming from the config), in dryrun mode, we will get something like the following execution (remember that the data output may differ).

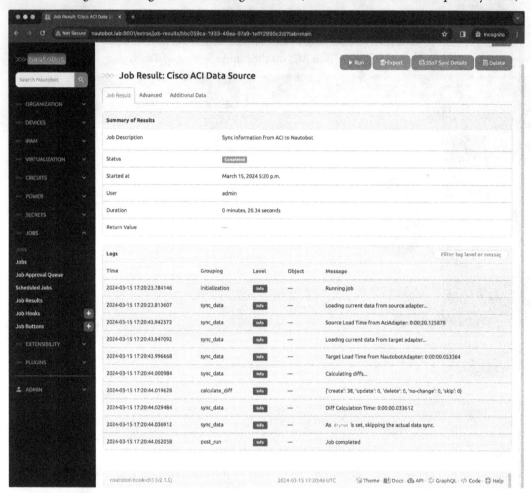

Figure A2.15 – Nautobot SSoT Cisco ACI results

In the preceding screenshot, you can see that the potential synchronization will create around 38 elements in Nautobot (it depends on the data coming from the DevNet instance).

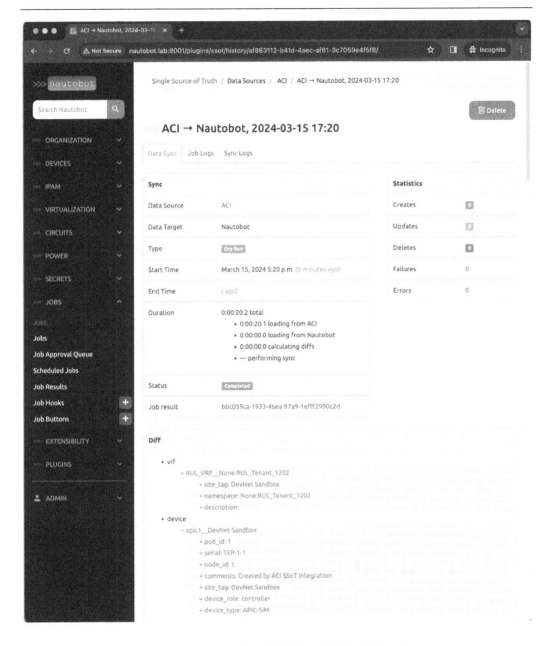

Figure A2.16 – Nautobot SSoT Cisco ACI detailed results

Having integrations available is great to get started, but most of the time you will need to create (or update) integrations with your own logic because data is not universal (it is rare for large enterprises to view and consume data the same exact way), and it usually requires customization resulting in data transformations. The next section covers tailoring an existing SSoT integration or creating a new one.

The SSoT framework allows synchronizing data from and to Nautobot. Synchronizing to Nautobot allows Nautobot to keep track of what exists in other systems such as data source or network controllers, and the other way around, allows Nautobot to synchronize data to other systems, including network changes if the system or controller takes the data to manage the network.

When paired with Nautobot's Data Validation Engine (`https://docs.nautobot.com/ projects/data-validation/en/latest/`) application, the SSoT application enables a *clean data ecosystem* whereby Nautobot can aggregate data from different sources, validate the data, and then distribute the validated data to other systems in the environment that require data. This capability to aggregate, validate, and distribute clean data produces more reliable automation because it allows users to make a change in a single location (the SoR), have that change validated, and then distribute that change throughout the environment.

Building your own SSoT integration

Building an integration on top of the Nautobot SSoT framework is not hard. At its core, every integration is a Nautobot job on steroids that can be set up as an independent job either in a stand-alone application or as an extra job, as in our instance.

The hot point is on the data modeling translation. The Nautobot SSoT framework leverages the DiffSync library that provides some helper methods to assist with this.

In this section, we will use a simple example that takes as data source a YAML file, and it loads the data into Nautobot. The complete code example can be found at `https://github.com/ PacktPublishing/Network-Automation-with-Nautobot/tree/main/appendix-02/ jobs`. You can get to use it by adding it to your Nautobot installation in the `jobs` folder:

```
nautobot@nautobot-dev:~$ tree /opt/nautobot/jobs/
/opt/nautobot/jobs/
├── __init__.py
├── example.yaml
└── example_ssot.py
```

Defining the data model mappings

Let's get started!

Understanding the source data

In this example, we use a simple YAML file (`example.yaml`) with the following data structure:

```
---
virtual_routing_instances:
  - name: "Orange"
    route_distinguisher: "61100:10"
```

```
          comments: "VPN Orange"
        - name: "Blue"
          route_distinguisher: "61100:20"
          comments: "VPN Blue"
        - name: "Green"
          route_distinguisher: "61100:30"
          comments: "VPN Green"

  prefixes:
    - prefix: "192.0.2.0/25"
      virtual_routing_instance: "Orange"
    - prefix: "192.0.2.128/25"
      virtual_routing_instance: "Orange"
    - prefix: "198.51.100.0/24"
      virtual_routing_instance: "Blue"
    - prefix: "203.0.113.0/24"
      virtual_routing_instance: "Green"
```

> **Note**
>
> In a real use case, you could integrate with any type of API and data structure. This is just a simple data structure for illustrating the idea.

This source data contains a group of VRFs and associated prefixes. As you may remember, the Nautobot data models that are similar to these are not exactly the same, so we will need to find common ground and enforce the necessary normalization.

Creating a data sync job

The Nautobot SSoT framework comes with several utilities to facilitate how to define models and easily integrate with Nautobot models (which is the most common use case). We will contain all our logic in the `example_ssot.py` file to keep it concise.

Establishing the DiffSync model

The first task is to find a common model that works well, especially toward the target. This approach simplifies the CRUD operations to synchronize the data. In this case, note how there is a reference to the Nautobot model (`_model`), and how we identify (`_identifiers`) the objects and the rest of the data to compare (`_attributes`):

```
... omitted library imports ...

class PrefixModel(NautobotModel):
```

```python
    """Prefix Model for DiffSync."""

    _model = Prefix
    _modelname = "prefix"
    _identifiers = ("network", "prefix_length", "namespace__name")
    _attributes = ("status__name", "type")

    network: str
    type: str = "network"
    namespace__name: str = "Global"
    prefix_length: int
    status__name: str = "Active"
    vrf_name: Optional[str] = None


class PrefixDict(TypedDict):
    """This typed dict is 100% decoupled from the `NautobotPrefix`
class defined above, and used to be referenced in a Many-to-many
relationship."""

    network: str
    prefix_length: int


class VRFModel(NautobotModel):
    """VRF Model for DiffSync."""

    _model = VRF
    _modelname = "vrf"
    _identifiers = ("name", "namespace__name")
    _attributes = ("rd", "description", "prefixes")

    name: str
    namespace__name: str = "Global"
    rd: Optional[str]
    description: Optional[str]
    prefixes: List[PrefixDict] = []
```

> **Note**
>
> It's interesting to notice that DiffSync is based on Pydantic, which enforces consistent typing of the data and provides implicit validation.

Loading data from the endpoints into DiffSync

Then, to allow DiffSync to calculate the difference between the source and target, we have to define how the data is loaded.

The Nautobot side is the easiest, as it only requires matching the models to the Nautobot models in a specific order (we need the prefixes first to be associated with the vrfs later):

```python
class ExampleNautobotAdapter(NautobotAdapter):
    """DiffSync adapter for Nautobot."""

    vrf = VRFModel
    prefix = PrefixModel
    top_level = (
        "prefix",
        "vrf",
    )
```

For the source, it's a bit more laborious, as you have to understand the data and convert it into the DiffSync models that will allow comparison with Nautobot. In real implementations, this method would take the data from an external source such as an API or a database:

```python
class ExampleRemoteAdapter(DiffSync):
    """DiffSync adapter for remote system."""

    vrf = VRFModel
    prefix = PrefixModel
    top_level = (
        "prefix",
        "vrf",
    )

    def load(self):
        dir_path = os.path.dirname(os.path.realpath(__file__))
        with open(os.path.join(dir_path, "example.yaml")) as yaml_
content:
            data = yaml.safe_load(yaml_content)

        for prefix in data["prefixes"]:
            network, prefix_length = prefix["prefix"].split("/")
            loaded_prefix = self.prefix(
                network=network,
                prefix_length=int(prefix_length),
                vrf_name=prefix["virtual_routing_instance"],
            )
```

```
            self.add(loaded_prefix)

    for vrf in data["virtual_routing_instances"]:
        prefixes = []
        for prefix in self.get_all("prefix"):
            if prefix.vrf_name == vrf["name"]:
                prefixes.append(
                    {
                        "network": prefix.network,
                        "prefix_length": prefix.prefix_length,
                    }
                )
        loaded_vrf = self.vrf(
            name=vrf["name"],
            rd=vrf["route_distinguisher"],
            description=vrf["comments"],
            prefixes=prefixes,
        )
        self.add(loaded_vrf)
```

> **Note**
>
> In this example, we haven't had to define the CRUD operations in the target side (Nautobot), because the `NautobotModel` class comes with an automatic method to infer them.

Defining a custom SSoT job

Finally, we defined a Nautobot job with the extra features of DataSource, offering the methods to calculate the difference of data and finally synchronize toward the target:

```
class ExampleYAMLDataSource(DataSource):
    """SSoT Job class."""

    def __init__(self):
        """Initialize ExampleYAMLDataSource."""
        super().__init__()
        self.diffsync_flags = self.diffsync_flags | DiffSyncFlags.
SKIP_UNMATCHED_DST

    class Meta:
        name = "Example YAML Data Source"
        description = "SSoT job example to get data from YAML"
        data_target = "Nautobot (remote)"
```

```
    def run(
        self, dryrun, memory_profiling, *args, **kwargs
    ):  # pylint:disable=arguments-differ
        """Run sync."""
        self.dryrun = dryrun
        self.memory_profiling = memory_profiling
        super().run(dryrun, memory_profiling, *args, **kwargs)

    def load_source_adapter(self):
        self.source_adapter = ExampleRemoteAdapter()
        self.source_adapter.load()

    def load_target_adapter(self):
        self.target_adapter = ExampleNautobotAdapter(job=self,
sync=self.sync)
        self.target_adapter.load()


jobs = [ExampleYAMLDataSource]
```

In the previous code snippet, note how you can leverage diffsync flags to customize the behavior of the comparison (globally or per model). In this case, SKIP_UNMATCHED_DST allows not comparing objects that don't exist in the target adapter.

> **Note**
> The jobs variable is used by Nautobot to detect available jobs.

Finally, we have to run a post_upgrade and restart the Nautobot services to be able to use the job just registered:

```
nautobot@nautobot-dev:~$ nautobot-server post_upgrade

... omitted output ...

nautobot@nautobot-dev:~$ exit

root@nautobot-dev:~# systemctl restart nautobot nautobot-worker
nautobot-scheduler
```

Using the custom SSoT job

Now, in the SSoT dashboard, you have a new SSoT job: **Example YAML Data Source**.

> **Note**
> Remember to enable the job to run it!

If you run it without dryrun mode, you will get the output like this.

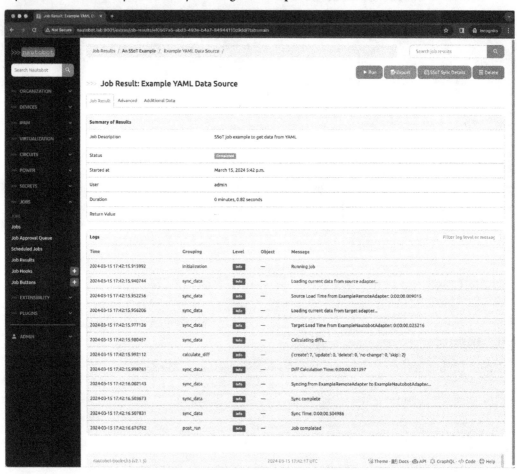

Figure A2.17 – Custom Nautobot SSoT results

And if you go into the SSoT sync details, you will see the data that has been created, the VRFs, and the Prefixes.

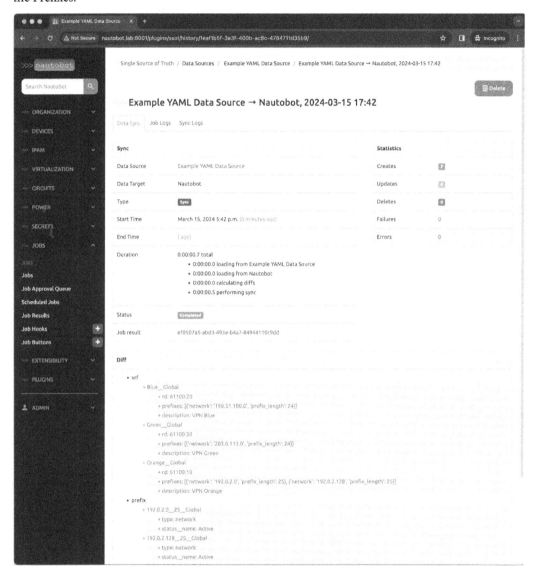

Figure A2.18 – Custom Nautobot SSoT detailed results

If you go into the objects views, for example, **VRFs**, you can see the three VRFs objects that were created.

Figure A2.19 – Custom Nautobot SSoT VRFs created

That was a very short introduction to SSoT job development. Hopefully, you have been able to grasp the idea of how you can leverage this framework to get started with your distributed SoT aggregation strategy. And, to get started, you have available a Cookiecutter template with the basic application structure to build a new SSoT plugin at `https://github.com/nautobot/cookiecutter-nautobot-app`.

Appendix 3
Performing Config Compliance and Remediation with Nautobot

If you've made it this far, you've seen countless references to Nautobot Golden Config throughout the book. That's because it is the most widely used app in the Nautobot ecosystem. The name Golden Config doesn't even do it justice because it does more than just define what a golden config is—Nautobot Golden Config enables backups, generates configurations, deploys configurations, performs compliance, and even performs remediation. It does all this in open source, providing maximum flexibility using popular Python libraries such as Netmiko, NAPALM, and Nornir.

This appendix is a short overview of Golden Config so that you can understand where and how it can be used in your infrastructure.

Please note that while this appendix explains how to do certain things and shows screenshots, it is not meant to be followed as a step-by-step guide. Golden Config is a product on its own and we'd practically need a whole book dedicated to showing every possible configuration and feature. Be on the lookout for more community material on getting up and running with Golden Config later in 2024.

In this appendix, we will cover the following:

- Why Golden Config
- Golden Config design
- Golden Config use cases
- Best practices and tips

Why Golden Config

Golden Config continues to gain popularity for network automation. Here are a few reasons why:

- It allows you to generate intended configurations from your Source of Truth, eliminating the need to have disparate tools.

- It embraces the Source of Truth, literally creating the intended configurations that would get sent, removing the need to have "rules" that only provide compliance. Golden Config goes beyond compliance providing built-in remediation.

- It is built with network engineers and network automation engineers in mind. Other solutions highlight only data models, but Golden Config allows you to see the configuration in the syntax the engineer has been trained in, the CLI, while providing flexibility and control for the power users, e.g., network automation engineers.

- It supports CLI, JSON, and even XML, which provides you with the ability to support nearly any device, controller, or system.

- It is built on the concept of Infrastructure as Code which brings all of the benefits that come with it while providing tangible results immediately for your network, thus providing a way to change how your network is managed.

- It allows you to crawl, walk, or run with your approach, separating the functions of backing up configurations, generating configurations, performing compliance, and even configuring and remediating devices.

Golden Config design

Golden Config's design may not follow the mental model that you are expecting, so it is important to understand the philosophy before diving in head first. The workflow it uses is to generate your configuration, back it up, parse the configuration into user-definable features (or config stanzas), and then compare those features *exactly*, as shown:

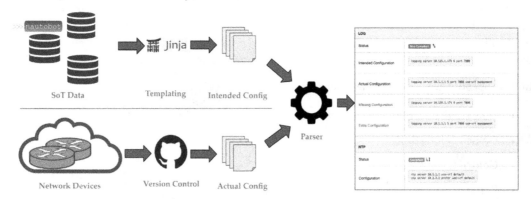

Figure A3.1 – Golden Config design

The premise of comparing those configurations exactly can cause confusion to a network engineer because we intrinsically know that `interface GigabitEthernet1` and `int Gig1` represent the same thing. To Golden Config, they are not an exact match and will never be compliant. This is true for many other reasons that may not affect a network engineer's pasting in config, such as whitespace, the usage of tabs, and literally anything that is not that exact match. For this reason, it is critical to be meticulous in matching your configurations by adjusting your Source of Truth data and configuration templates.

Generating the intended configurations requires rendering data—which is one of Nautobot's core features—with Jinja Templates, so it is a well-positioned solution for config generation. Nautobot already has access to your inventory and credentials (with Nautobot Secrets), so it is also well-positioned to be part of your backup strategy. For both config generation and backups, you can leverage Golden Config as a complete solution, but you can also use existing solutions if you have them deployed already. If you opt to use existing solutions, all you need to do is ensure the generated (intended) configurations and backups are stored in a Git repository. This highlights the NetDevOps philosophy within Golden Config of giving the user choice and flexibility. For those just getting started though, it is recommended to use Nautobot for backups and config generation.

From a lower-level design perspective, Golden Config uses popular open source network automation libraries, including Netmiko, NAPALM, and Nornir, for device connectivity. It also uses netutils for many internal functions. Beyond just using these projects, Golden Config embraces the notion that each user may need some form of customization for compliance and remediation, so you can build compliance through the Nautobot UI but also add custom Python functions if needed for your requirements. Again, this highlights the idea and design philosophy of being open, flexible, and extensible.

Finally, the remediation builds on open source and NetDevOps and is made possible by using a library called `hier_config`(`https://github.com/netdevops/hier_config`), which provides the ability to not only "ensure this NTP server exists" but to also "ensure only these NTP servers exist" (which means it is smart enough to know some configuration needs to be removed in order to make a configuration compliant.

Golden Config use cases

Let's start by covering the main use cases for Nautobot Golden Config, keeping in mind that the end goal is to perform compliance or remediate (and configure) devices from Nautobot.

When getting started with Golden Config, one of the first things you need to consider and ask yourself is "What devices are in scope for Golden Config?" meaning which devices do you want to perform compliance on? You'll see this list when you click **Golden Config** > **Configuration Overview** in the navbar:

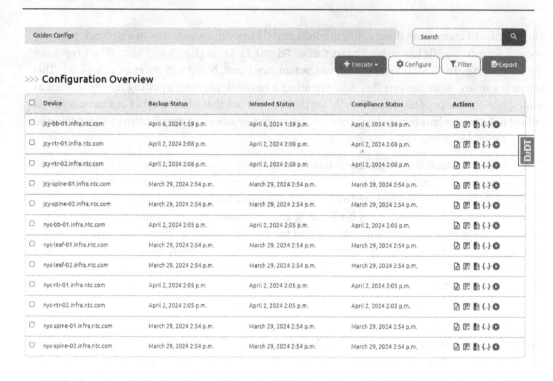

Figure A3.2 – High-level config overview for Golden Config

This table shows the status of the last time a Golden Config operation was performed for each device. It also has quick links to view the backup, intended config, compliance details, and data used to generate configs and execute Golden Config Jobs.

In order to define the scope of devices for Golden Config, you create a dynamic group and associate that group to Golden Config in **Golden Config** > **Settings**, as shown here:

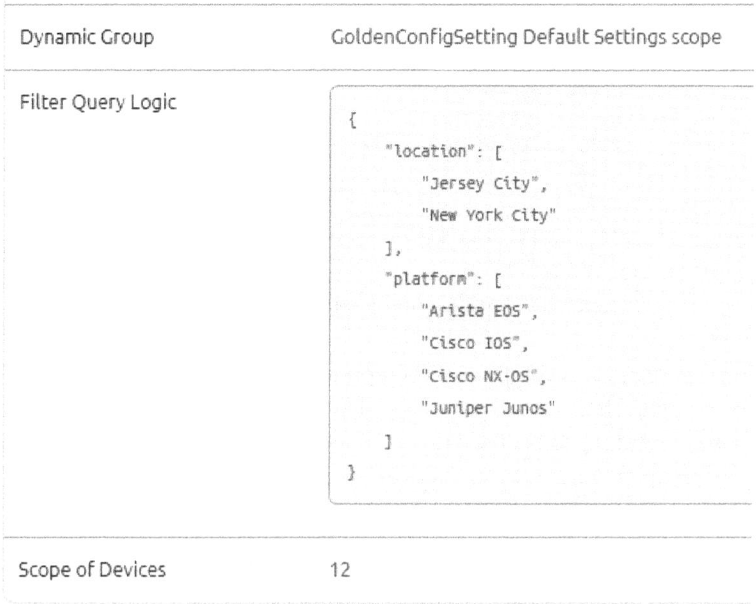

Dynamic Group	GoldenConfigSetting Default Settings scope
| Filter Query Logic | ```
{
    "location": [
        "Jersey City",
        "New York City"
    ],
    "platform": [
        "Arista EOS",
        "Cisco IOS",
        "Cisco NX-OS",
        "Juniper Junos"
    ]
}
``` |
| Scope of Devices | 12 |

Figure A3.3 – Dynamic group that returns the devices in scope for Golden Config

Once the scope is defined, you will see devices populated through the Golden Config pages.

Performing Config backups

The first and most common use case we'll look at is automated config backups. Backups are required for compliance. After all, we need to compare what is expected to what is actually there (which comes from a network backup). As long as Netmiko (`https://github.com/ktbyers/netmiko`) supports a given device type, it can be backed up by Nautobot.

Nautobot embraces NetDevOps, and in order to do backups in Nautobot, you are required to use a Git repository. This is another example of Nautobot's flexibility to integrate with Git. Backups are not stored locally in Nautobot but rather always pushed and pulled from a Git repository. It is possible to use Git on the local file system, but in most deployments, there is integration to a Git remote.

Golden Config also gives you the flexibility to customize the path where backups are saved. You can define the path using a Jinja template. For example, you may want a basic path that is `{{obj.location.name}}/{{obj.name}}.cfg`, where `obj` is the device, as shown in the following figure:
```

Backup Configuration	
Backup repository	backups ⌄
Backup Path in Jinja Template Form	{{obj.location.name|slugify}}/{{obj.name}}.cfg
	The Jinja path representation of where the backup file will be found. The variable `obj` is available as the device instance object of a given device, as is the case for all Jinja templates. e.g. `{{obj.location.name|slugify}}/{{obj.name}}.cfg`

Figure A3.4 – Backup configuration path template

If you are thinking that you need a compliance solution but already have a backup solution, then you can still use those backups! That is the beauty of Golden Config. You actually don't need to perform backups from Nautobot. As long as the backups are in a Git repo, all you need to do is add that repo to Nautobot, define the path where config files can be found in that repo, and then sync them to Nautobot. Once they are pulled down to Nautobot, they can be used for compliance. This solution allows Nautobot to be used out of band.

As you'd expect, you can also view backups directly from the UI, as shown here using the quick links mentioned:

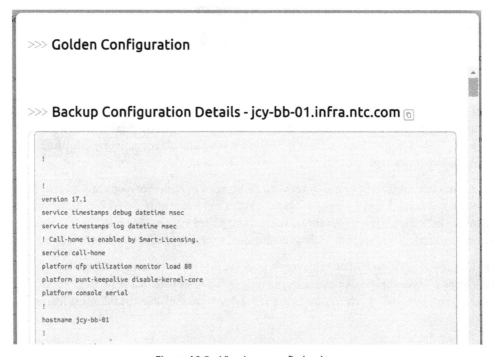

Figure A3.5 – Viewing a config backup

Let's move on to generating configurations with Nautobot.

Generating intended configurations

In order to perform compliance, there needs to be an understanding of what should be configured on the device. It is this configuration that is compared to the backup configuration. This is done in one of two ways: having Nautobot generate configurations or providing existing intended configurations similar to what we just mentioned with regard to backups. Our focus will be on Nautobot generating configurations, as that is where your data is. Similar to backups, if you already have a Git repo that has generated intended configurations, that can be used as a starting point for compliance.

Let's explore Nautobot generating the configurations.

There is great value in allowing Nautobot to generate configurations because when Nautobot generates configurations, it generates them using the data from Nautobot. Think about that for a second and what that means. Instead of deploying a tool that does compliance, another tool that is the Source of Truth, and another tool to generate configs from templates, Nautobot integrates them in a unified fashion. This allows Nautobot to be a full-blown network automation platform as your Source of Truth to generate the intended configurations for your network that is now the source for your compliance. This may even allow you to eliminate standalone configuration compliance tools!

Generating configurations within Nautobot starts with creating a GraphQL query that contains the query needed to collect the data required by your Jinja templates:

Figure A3.6 – Creating a GraphQL query for Golden Config

You can see the association of the GraphQL query to templates in the following figure:

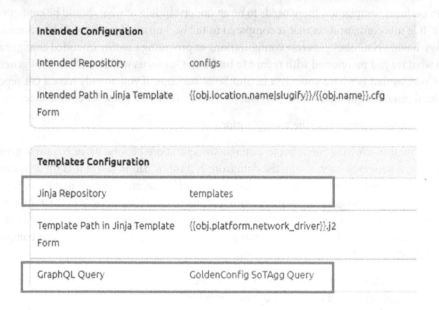

Intended Configuration

Intended Repository	configs
Intended Path in Jinja Template Form	{{obj.location.name|slugify}}/{{obj.name}}.cfg

Templates Configuration

Jinja Repository	templates
Template Path in Jinja Template Form	{{obj.platform.network_driver}}.j2
GraphQL Query	GoldenConfig SoTAgg Query

Figure A3.7 – Associating the GraphQL query to config templates to generate intended configs

You can navigate and check out the GraphQL query in the public demo instance found at `https://demo.nautobot.com`.

It is worth pointing out that generated configurations from templates do not need to be full configurations. They can be partial snippets and stanzas too, which allows a gradual implementation of Nautobot-generated configs.

Similar to backups, you specify Git repositories for both intended configurations and templates. Once configs are created from the data in Nautobot (using the GraphQL query) and the templates, they are pushed to the **Intended Configuration** Git repo, allowing you as a network administrator to view them in Git.

In order to get started doing backups and generating configs in preparation for compliance, you need to do the following:

- Define three Git repositories for backups, intended configurations, and config templates
- Define a dynamic group that is used for the scope of Golden Config operations
- Define a GraphQL query that is used to query Nautobot for the data required by the config templates

Once all set up, you'll have fully populated config settings, as shown here:

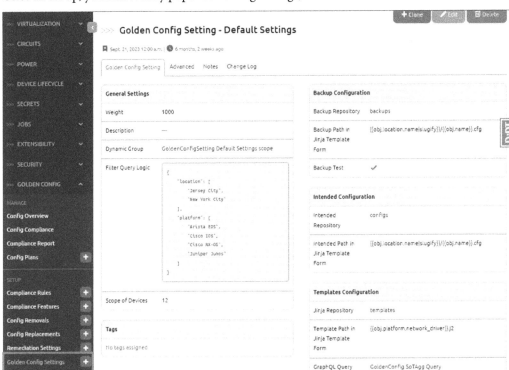

Figure A3.8 – Viewing Golden Config settings

Let's move on to performing compliance.

Performing config compliance

Once Nautobot has backups and intended configurations, it is able to start performing compliance. When all set up and working, you'll see compliance per feature per device. Here is an example of a small network from the **Golden Config** > **Config Compliances** view:

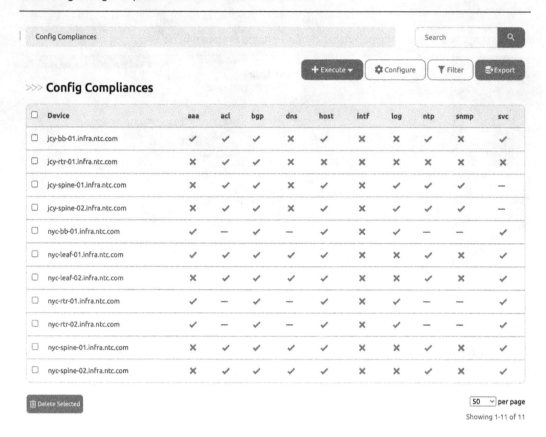

Device	aaa	acl	bgp	dns	host	intf	log	ntp	snmp	svc
jcy-bb-01.infra.ntc.com	✓	✓	✓	✗	✓	✗	✗	✓	✗	✓
jcy-rtr-01.infra.ntc.com	✗	✓	✓	✗	✗	✗	✗	✗	✗	✗
jcy-spine-01.infra.ntc.com	✗	✓	✓	✗	✓	✗	✓	✓	✓	—
jcy-spine-02.infra.ntc.com	✗	✓	✓	✗	✓	✗	✓	✓	✓	—
nyc-bb-01.infra.ntc.com	✓	—	✓	—	✓	✗	✓	—	—	✓
nyc-leaf-01.infra.ntc.com	✓	✓	✓	✓	✓	✗	✗	✓	✗	✓
nyc-leaf-02.infra.ntc.com	✗	✓	✓	✓	✓	✗	✗	✓	✗	✓
nyc-rtr-01.infra.ntc.com	✓	—	✓	—	✓	✗	✓	—	—	✓
nyc-rtr-02.infra.ntc.com	✓	—	✓	—	✓	✗	✓	—	—	✓
nyc-spine-01.infra.ntc.com	✗	✓	✓	✓	✓	✗	✗	✓	✗	✓
nyc-spine-02.infra.ntc.com	✗	✓	✓	✓	✓	✗	✗	✓	✗	✓

Showing 1-11 of 11

Figure A3.9 – Viewing device compliance

The devices shown in the **Config Compliances** table map back to the devices in scope from the initially configured dynamic group. The features (or columns) shown in the table are all user-defined by you, so you can start as small or as big as you'd like. You can literally have one Jinja template for SNMP and just view SNMP compliance to get started. This means your GraphQL query would be a very basic query for gathering just that data! Remember, always start small and basic.

In order to get started with compliance, you need to create **Compliance Rules** and **Compliance Features**, as shown in the navbar:

Figure A3.10 – Compliance rules and features in the navbar

For example, if you want to do compliance for SNMP, you'd first create a compliance feature called SNMP and then a compliance rule for each vendor platform you want to do compliance for. If we were doing SNMP compliance for IOS, NX-OS, Junos, and EOS, this is what the **Compliance Rules** would look like:

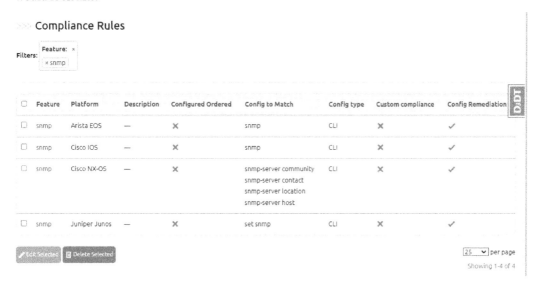

Figure A3.11 – Compliance rules for SNMP

There are a few things to consider by looking at each of the columns in the **Compliance Rules** table. Let's look at each one:

- **Feature**: This is the association back to an abstract user-defined feature that allows us to search and filter based on a given feature.

- **Platform**: Each rule is associated with a Nautobot platform that is shown here.

- **Description**: You can optionally set a description for the rule; this is helpful to document why you have a particular rule.

- **Config Ordered**: This is a Boolean that tells Nautobot whether the ordering of commands matters (e.g., such as for ACLs).

- **Config to Match**: These are the "parents" of the config to match on. For this example, they may seem like partial string matches for SNMP, but in reality, they are the parents. So if commands were nested underneath SNMP commands, that would be in scope for compliance. For example, if you wanted to do a compliance check for BGP configuration, all you would need is `router bgp` as the config to match. Note that the previous screenshot does not use production-grade *configs to match*. It shows different ways to match configs for NX-OS.

- **Config type**: You have options for CLI and JSON. JSON can be used if you are working with API-based systems.

- **Custom compliance**: This is a Boolean to signify if custom compliance is being used. If, for example, you need more advanced logic to know if a device is compliant, you can do that by using custom compliance, e.g., custom Python code and logic can be used. We are not covering custom compliance in this appendix.

- **Config Remediation**: For each compliance rule, you have the option to enable or disable remediation commands from being generated (it does not mean they are automatically sent without human intervention).

Going one step further, Nautobot gives you the flexibility to remove or replace certain configurations before compliance is performed. This is done by navigating to the **Config Removals** and **Config Replacements** menu items:

Figure A3.12 – Removing and replacing configs

Here is an example of each:

>>> **Config Removes**

	Name	Platform	Description	Regex Pattern
☐	ios build config	Cisco IOS	—	Building\s+configuration.*
☐	ios current	Cisco IOS	—	^Current\s+configuration.*
☐	ios_last	Cisco IOS	—	^!\s+Last\s+configuration.*

Figure A3.13 – Removing IOS configuration lines

This example removes particular lines in Cisco configs that always change, have date/time stamps, and display the size of the config. We do not want those in our compliance.

Here is one for config replacements:

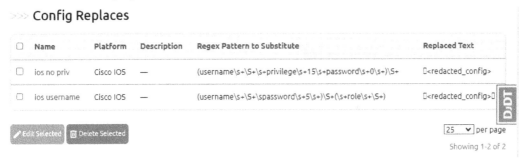

Figure A3.14 – Replacing (and redacting) certain configuration lines

Replacing lines with `redacted` allows us to still perform compliance on those commands being present (or not) without removing them completely. For example, in production, we may want to use Config Replaces for SNMP, so community strings aren't present in compliance reports.

Once everything is set up and compliance is functional, you can view the state of compliance per device by looking at the **Configuration Compliance** tab under the detailed device view:

Figure A3.15 – Viewing config compliance per device

In this tab, you'll have a **Feature Navigation** that will show all compliance rules. Each green and red rectangle is a quick link to the results of that check, and the **Compliant**, **Non-Compliant**, and **Clear** buttons allow you to filter on those:

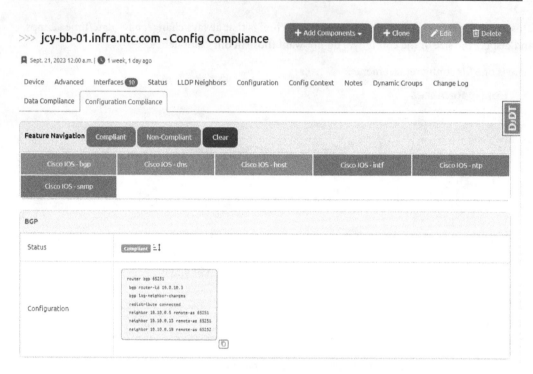

Figure A3.16 – Viewing compliance for a compliant rule and feature

As you can see from the figure, when a device's feature/rule is compliant, you still see the configuration. That configuration is both what is intended and what is actually on the device.

Let's look at an example that is not compliant. There are a few things to call out when viewing the following figure:

- When a feature is not compliant, you see up to five config sections documenting **Intended Configuration**, **Actual Configuration**, **Missing Configuration**, **Extra Configuration**, and **Remediating Configuration** (when enabled).

- **Intended Configuration** is coming from the configuration that is in the Git repo that was mentioned earlier. Assuming the configuration is generated from Nautobot using the GraphQL query, this is very powerful because it is the exact same configuration that can be sent to the device via automation (and Nautobot). You do not need one tool doing compliance and another doing automation. Why not use Nautobot to do both?

- **Actual Configuration** is coming from the configuration backup.

- If there are missing configurations, they are shown in the **Missing Configuration** section.

- If there are extra configurations, they are shown in the **Extra Configuration** section.

- If **Config Remediation** is enabled for a given rule and if remediation is configured for the given platform (Cisco IOS in our example), then you will see the exact commands needed to remediate the device.

Figure A3.17 – Viewing an out-of-compliance feature

In order to start performing compliance, you need to do the following:

- Create at least one compliance feature or as many features as you want to run compliance checks against

- Create one compliance rule per feature and platform you want to run compliance checks against

- Add a Config Removal for each line you want to remove from a given config per platform before compliance is executed

- Add a Config Replacement for each line you want to replace from a given config per platform before compliance is executed

> **Note**
>
> You do not need to start with remediation. You should only start considering remediation once compliance is running as expected.

Now that we've reviewed compliance, let's look at remediation.

Automating config remediation and deployments

In the last section, we looked at SNMP. SNMP was not compliant, and the Config Compliance analysis gave us the **Missing Configuration**, **Extra Configuration**, and **Remediating Configuration**:

Figure A3.18 – Viewing remediation configs

Let's see how these config sections are used when performing remediation.

Config deployments are managed in Nautobot as config plans. You should see **Golden Config** > **Config Plans** in the navbar.

Config plans allow you to have a workflow that best fits your current needs. Not everyone will have the same deployment processes. Consider the following use cases:

- You want to only apply your known good configurations

- You want to only apply your known good configurations for certain parts of your configuration

- You are risk averse and you want to only apply configurations that are missing

- You are risk averse and you want to only apply configurations that are missing for certain parts of your configurations

- You want to not only add missing configurations but also remove extra configurations

- You want to not only add missing configurations but also remove extra configurations for certain parts of your configurations.

- You have one-off configurations you want to deploy to many devices

By separating config plans, each organization is able to use the workflow that best suits its needs.

From a Config Remediation workflow perspective, after seeing there is a non-compliant feature (with remediation enabled), you would want to create a config plan. This is done by navigating to the **Config Plans** page and creating a new config plan. When creating a new plan, you'll see the following form that showcases four different config plan types:

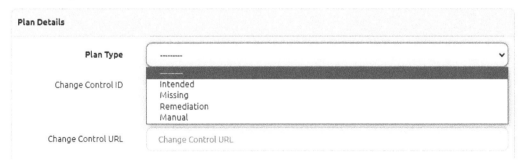

Figure A3.19 – Viewing four different config plan types

Let's review these plan types:

- **Intended**: This will send the commands shown in the **Intended Configuration** section from the compliance view

- **Missing**: This will send the commands shown in the **Missing Configuration** section from the compliance view

- **Remediation**: This will send the commands shown in the **Remediating Configuration** section from the compliance view

- **Manual**: This will send any manual commands you want to send to the device and can be used for ad hoc network automation

With four different plan types, how should you pick one when remediating and configuring a device? It comes down to understanding the feature and how it works on your given platforms.

Let's think about SNMP. You can choose **Intended** or **Missing** as the plan type, but then all we'd ever send is the commands that should be there. What about removing commands that should be there? In order to do that, you need to choose the **Remediation** type.

If you were automating the hostname of a device, it is a single command, and if you send the same command with a different hostname, it is just overridden. Thus, for the hostname, you could choose **Intended**, **Missing**, or **Remediation**:

Figure A3.20 – Reviewing config plan types

There is a plan type that is called **Manual** that lets you send any set of commands to the device(s). When you choose **Manual**, you can also build a Jinja template directly in Nautobot. As an example, consider looping over templates to add a command to each interface.

When creating a plan, you also apply a filter to know the scope of the plan. The form used for the filter is similar to that used when creating config contexts:

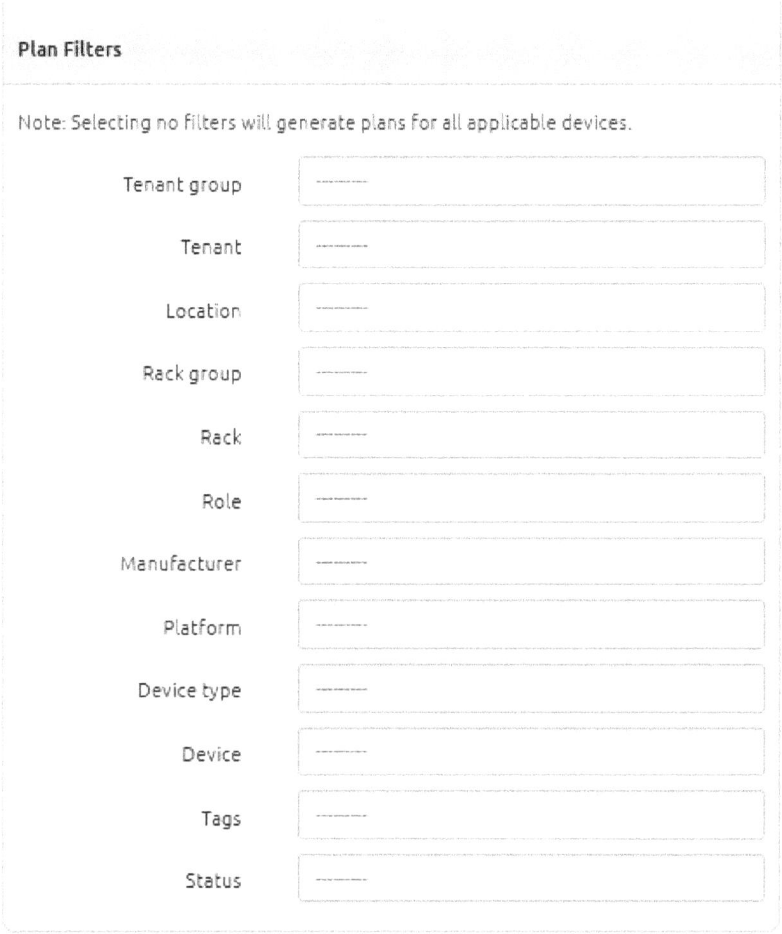

Plan Filters

Note: Selecting no filters will generate plans for all applicable devices.

Tenant group	---------
Tenant	---------
Location	---------
Rack group	---------
Rack	---------
Role	---------
Manufacturer	---------
Platform	---------
Device type	---------
Device	---------
Tags	---------
Status	---------

Figure A3.21 – Choosing a filter for a config plan

This also provides a lot of flexibility to ensure standards are enforced across a location, role, device type, and so on.

Once a plan is created, a pop-up modal shows the links to the job and a new unique plan that was generated for each device in scope within the plan filter:

Generate Config Plans

Job Status:	SUCCESS
Job Results:	↗
Redirect Link:	↗
Details:	Job Completed Successfully. Number of Config Plans generated: 6

Close

Figure A3.22 – Config plan modal

For example, if we need to send commands to all Cisco IOS devices (and we have three of them), three config plan entries would be added to the **Config Plans** table:

>>> **Config Plans**

Filters:
Plan JobResult ID:
× Generate Config Plans started at 2024-04-06 14:12:59.756482+00:00 (SUCCESS)

	Device	Created	Plan Type	Feature	Change Control ID	Change Control URL	Plan Result	Deploy Result	Config Set	Status
☐	jcy-rtr-02.infra.ntc.com	2024-04-06 14:13	Manual	—	—	—	🗒	—	🗋	Not Approved
☐	jcy-rtr-01.infra.ntc.com	2024-04-06 14:13	Manual	—	—	—	🗒	—	🗋	Not Approved
☐	jcy-bb-01.infra.ntc.com	2024-04-06 14:13	Manual	—	—	—	🗒	—	🗋	Not Approved

Deploy Selected   Edit Selected   Delete Selected

25 ⌄ per page
Showing 1-3 of 3

Figure A3.23 – Viewing Config Plans

You can view the commands that will get sent to each device by clicking on the icon within the **Config Set** column.

Take note of the **Status** column—by default, each plan is **Not Approved**. Nautobot requires you to change **Status** to **Approved** before deploying the configurations, ensuring that a human (or machine via API) has approved the configs to be sent.

Once the plans are approved, the proper workflow is to filter for the devices you want to remediate, use the multi-select option, and then click the blue button that says **Deploy Selected**. From there, your devices will be configured and remediated.

> **Pro tip**
> Everything in Golden Config has an API. This means you could use Nautobot as an out-of-band configuration compliance and remediation engine but fetch the commands generated from the API and use other automation tool(s) to do the actual last-mile automation.

Best practices and tips

The following is a high-level list of items to be aware of and best practices when getting started with Nautobot Golden Config:

- After reading this, read the official user docs (`https://docs.nautobot.com/projects/golden-config/en/latest/user/app_overview/`); most questions that new users ask are covered there.

- Determine up front how you want to connect to devices and how credentials will be handled. Will you store them as environment variables or pull them from an enterprise secrets manager platform? Once you determine this, you'll need to update the Nornir settings in the Nautobot config file.

- Define how many devices you want to automate in parallel by configuring it in the Nornir Settings.

- Keep things simple and small as you get started. Don't do remediation on day one. Start with backups, generate configs for one feature, do compliance for one feature, do remediation for one feature, and then expand from there.

- Don't try to get your Jinja templates perfect the first time; let the compliance engine let you know what needs to be adjusted (e.g., the templates, the data, etc.).

- Keep in mind that all configurations can be generated from the data; it may feel like other strategies need to be employed, such as having different rules for different hardware families or roles of configurations, but in reality, that can be handled in Jinja.

- Remember that the philosophy of compliance is to compare the actual config to the intended config.

- Leverage Nautobot's extensibility features of config contexts, custom fields, and (performant) computed fields to ensure you have all the data required from your configuration generation.

- When troubleshooting, remember to use the workflow diagrams in the GitHub Repo wiki that describe each step so you have an understanding of where things went wrong. Here is the backup diagram (`https://github.com/nautobot/nautobot-app-golden-config/wiki/Backup-Job-Workflow-Diagram`). From there, you can navigate to the others.

- Remember there is the ability to have custom compliance and custom remediation when the native capabilities do not fit your business needs.

- Consider how you will manage your secrets in backup and intended configurations and how that affects compliance.

- It is possible to switch between NAPALM and Netmiko dispatchers for retrieving configs and/or sending commands. Based on APIs or SSH being used, you may need to change the dispatcher being used (in **Admin** > **Config**).

- Remediation in Nautobot is done via `hier_config` (`https://github.com/netdevops/hier_config`) or through user-defined custom compliance. If you are using remediation in Nautobot, you should spend time learning about `hier_config`.

- All execution of tasks within Golden Config happens via Nautobot jobs, so you can always navigate to a specific job from the **Job** page and view the results in the **Job Results** page.

This appendix merely scratched the surface with Golden Config. The goal was to showcase its breadth and depth without adding another 100 pages to the book! If you are interested in finding out more, you should definitely install Golden Config and try it out. If you have any questions about anything in the book, be sure to join the NTC Slack at `slack.networktocode.com` and join the `#nautobot` channel!

Index

C

`packtpub.com`

Subscribe to our online digital library for full access to over 7,000 books and videos, as well as industry leading tools to help you plan your personal development and advance your career. For more information, please visit our website.

Why subscribe?

- Spend less time learning and more time coding with practical eBooks and Videos from over 4,000 industry professionals

- Improve your learning with Skill Plans built especially for you

- Get a free eBook or video every month

- Fully searchable for easy access to vital information

- Copy and paste, print, and bookmark content

Did you know that Packt offers eBook versions of every book published, with PDF and ePub files available? You can upgrade to the eBook version at `packtpub.com` and as a print book customer, you are entitled to a discount on the eBook copy. Get in touch with us at `customercare@packtpub.com` for more details.

At `www.packtpub.com`, you can also read a collection of free technical articles, sign up for a range of free newsletters, and receive exclusive discounts and offers on Packt books and eBooks.

Other Books You May Enjoy

If you enjoyed this book, you may be interested in these other books by Packt:

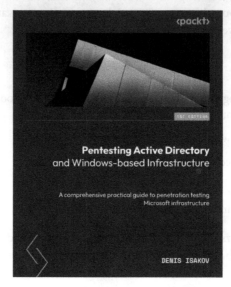

Pentesting Active Directory and Windows-based Infrastructure

Denis Isakov

ISBN: 978-1-80461-136-4

- Understand and adopt the Microsoft infrastructure kill chain methodology

- Attack Windows services, such as Active Directory, Exchange, WSUS, SCCM, AD CS, and SQL Server

- Disappear from the defender's eyesight by tampering with defensive capabilities

- Upskill yourself in offensive OpSec to stay under the radar

- Find out how to detect adversary activities in your Windows environment

- Get to grips with the steps needed to remediate misconfigurations

- Prepare yourself for real-life scenarios by getting hands-on experience with exercises

Security Monitoring with Wazuh

Rajneesh Gupta

ISBN: 978-1-83763-215-2

- Find out how to set up an intrusion detection system with Wazuh

- Get to grips with setting up a file integrity monitoring system

- Deploy Malware Information Sharing Platform (MISP) for threat intelligence automation to detect indicators of compromise (IOCs)

- Explore ways to integrate Shuffle, TheHive, and Cortex to set up security automation

- Apply Wazuh and other open source tools to address your organization's specific needs

- Integrate Osquery with Wazuh to conduct threat hunting

Packt is searching for authors like you

If you're interested in becoming an author for Packt, please visit `authors.packtpub.com` and apply today. We have worked with thousands of developers and tech professionals, just like you, to help them share their insight with the global tech community. You can make a general application, apply for a specific hot topic that we are recruiting an author for, or submit your own idea.

Share your thoughts

Now you've finished *Network Automation with Nautobot*, we'd love to hear your thoughts! Scan the QR code below to go straight to the Amazon review page for this book and share your feedback or leave a review on the site that you purchased it from.

https://packt.link/r/1837637865

Your review is important to us and the tech community and will help us make sure we're delivering excellent quality content.

Download a free PDF copy of this book

Thanks for purchasing this book!

Do you like to read on the go but are unable to carry your print books everywhere?

Is your eBook purchase not compatible with the device of your choice?

Don't worry, now with every Packt book you get a DRM-free PDF version of that book at no cost.

Read anywhere, any place, on any device. Search, copy, and paste code from your favorite technical books directly into your application.

The perks don't stop there, you can get exclusive access to discounts, newsletters, and great free content in your inbox daily

Follow these simple steps to get the benefits:

1. Scan the QR code or visit the link below

https://packt.link/free-ebook/978-1-83763-786-7

2. Submit your proof of purchase
3. That's it! We'll send your free PDF and other benefits to your email directly